LONDON MATHEMATICAL SOCIETY LECTURE NOTE SERIES

Managing Editor: Professor N.J. Hitchin, Mathematical Institute,
University of Oxford, 24–29 St Giles, Oxford OX1 3LB, United Kingdom

The titles below are available from booksellers, or, in case of difficulty, from Cambridge University Press.

46 *p*-adic Analysis: a short course on recent work, N. KOBLITZ
59 Applicable differential geometry, M. CRAMPIN & F.A.E. PIRANI
66 Several complex variables and complex manifolds II, M.J. FIELD
86 Topological topics, I.M. JAMES (ed)
88 FPF ring theory, C. FAITH & S. PAGE
90 Polytopes and symmetry, S.A. ROBERTSON
96 Diophantine equations over function fields, R.C. MASON
97 Varieties of constructive mathematics, D.S. BRIDGES & F. RICHMAN
99 Methods of differential geometry in algebraic topology, M. KAROUBI & C. LERUSTE
100 Stopping time techniques for analysts and probabilists, L. EGGHE
104 Elliptic structures on 3-manifolds, C.B. THOMAS
105 A local spectral theory for closed operators, I. ERDELYI & WANG SHENGWANG
107 Compactification of Siegel moduli schemes, C.-L. CHAI
109 Diophantine analysis, J. LOXTON & A. VAN DER POORTEN (eds)
113 Lectures on the asymptotic theory of ideals, D. REES
116 Representations of algebras, P.J. WEBB (ed)
119 Triangulated categories in the representation theory of finite-dimensional algebras, D. HAPPEL
121 Proceedings of *Groups - St Andrews 1985*, E. ROBERTSON & C. CAMPBELL (eds)
128 Descriptive set theory and the structure of sets of uniqueness, A.S. KECHRIS & A. LOUVEAU
130 Model theory and modules, M. PREST
131 Algebraic, extremal & metric combinatorics, M.-M. DEZA, P. FRANKL & I.G. ROSENBERG (eds)
138 Analysis at Urbana, II, E. BERKSON, T. PECK, & J. UHL (eds)
139 Advances in homotopy theory, S. SALAMON, B. STEER & W. SUTHERLAND (eds)
140 Geometric aspects of Banach spaces, E.M. PEINADOR & A. RODES (eds)
141 Surveys in combinatorics 1989, J. SIEMONS (ed)
144 Introduction to uniform spaces, I.M. JAMES
146 Cohen-Macaulay modules over Cohen-Macaulay rings, Y. YOSHINO
148 Helices and vector bundles, A.N. RUDAKOV *et al*
149 Solitons, nonlinear evolution equations and inverse scattering, M. ABLOWITZ & P. CLARKSON
150 Geometry of low-dimensional manifolds 1, S. DONALDSON & C.B. THOMAS (eds)
151 Geometry of low-dimensional manifolds 2, S. DONALDSON & C.B. THOMAS (eds)
152 Oligomorphic permutation groups, P. CAMERON
153 L-functions and arithmetic, J. COATES & M.J. TAYLOR (eds)
155 Classification theories of polarized varieties, TAKAO FUJITA
158 Geometry of Banach spaces, P.F.X. MÜLLER & W. SCHACHERMAYER (eds)
159 Groups St Andrews 1989 volume 1, C.M. CAMPBELL & E.F. ROBERTSON (eds)
160 Groups St Andrews 1989 volume 2, C.M. CAMPBELL & E.F. ROBERTSON (eds)
161 Lectures on block theory, BURKHARD KÜLSHAMMER
163 Topics in varieties of group representations, S.M. VOVSI
164 Quasi-symmetric designs, M.S. SHRIKANDE & S.S. SANE
166 Surveys in combinatorics, 1991, A.D. KEEDWELL (ed)
168 Representations of algebras, H. TACHIKAWA & S. BRENNER (eds)
169 Boolean function complexity, M.S. PATERSON (ed)
170 Manifolds with singularities and the Adams-Novikov spectral sequence, B. BOTVINNIK
171 Squares, A.R. RAJWADE
172 Algebraic varieties, GEORGE R. KEMPF
173 Discrete groups and geometry, W.J. HARVEY & C. MACLACHLAN (eds)
174 Lectures on mechanics, J.E. MARSDEN
175 Adams memorial symposium on algebraic topology 1, N. RAY & G. WALKER (eds)
176 Adams memorial symposium on algebraic topology 2, N. RAY & G. WALKER (eds)
177 Applications of categories in computer science, M. FOURMAN, P. JOHNSTONE & A. PITTS (eds)
178 Lower K- and L-theory, A. RANICKI
179 Complex projective geometry, G. ELLINGSRUD *et al*
180 Lectures on ergodic theory and Pesin theory on compact manifolds, M. POLLICOTT
181 Geometric group theory I, G.A. NIBLO & M.A. ROLLER (eds)
182 Geometric group theory II, G.A. NIBLO & M.A. ROLLER (eds)
183 Shintani zeta functions, A. YUKIE
184 Arithmetical functions, W. SCHWARZ & J. SPILKER
185 Representations of solvable groups, O. MANZ & T.R. WOLF
186 Complexity: knots, colourings and counting, D.J.A. WELSH
187 Surveys in combinatorics, 1993, K. WALKER (ed)
188 Local analysis for the odd order theorem, H. BENDER & G. GLAUBERMAN
189 Locally presentable and accessible categories, J. ADAMEK & J. ROSICKY
190 Polynomial invariants of finite groups, D.J. BENSON
191 Finite geometry and combinatorics, F. DE CLERCK *et al*
192 Symplectic geometry, D. SALAMON (ed)
194 Independent random variables and rearrangement invariant spaces, M. BRAVERMAN
195 Arithmetic of blowup algebras, WOLMER VASCONCELOS
196 Microlocal analysis for differential operators, A. GRIGIS & J. SJÖSTRAND
197 Two-dimensional homotopy and combinatorial group theory, C. HOG-ANGELONI *et al*
198 The algebraic characterization of geometric 4-manifolds, J.A. HILLMAN
199 Invariant potential theory in the unit ball of $\mathbf{C}^n$, MANFRED STOLL
200 The Grothendieck theory of dessins d'enfant, L. SCHNEPS (ed)
201 Singularities, JEAN-PAUL BRASSELET (ed)

202 The technique of pseudodifferential operators, H.O. CORDES
203 Hochschild cohomology of von Neumann algebras, A. SINCLAIR & R. SMITH
204 Combinatorial and geometric group theory, A.J. DUNCAN, N.D. GILBERT & J. HOWIE (eds)
205 Ergodic theory and its connections with harmonic analysis, K. PETERSEN & I. SALAMA (eds)
207 Groups of Lie type and their geometries, W.M. KANTOR & L. DI MARTINO (eds)
208 Vector bundles in algebraic geometry, N.J. HITCHIN, P. NEWSTEAD & W.M. OXBURY (eds)
209 Arithmetic of diagonal hypersurfaces over finite fields, F.Q. GOUVÊA & N. YUI
210 Hilbert C*-modules, E.C. LANCE
211 Groups 93 Galway / St Andrews I, C.M. CAMPBELL *et al* (eds)
212 Groups 93 Galway / St Andrews II, C.M. CAMPBELL *et al* (eds)
214 Generalised Euler-Jacobi inversion formula and asymptotics beyond all orders, V. KOWALENKO *et al*
215 Number theory 1992–93, S. DAVID (ed)
216 Stochastic partial differential equations, A. ETHERIDGE (ed)
217 Quadratic forms with applications to algebraic geometry and topology, A. PFISTER
218 Surveys in combinatorics, 1995, PETER ROWLINSON (ed)
220 Algebraic set theory, A. JOYAL & I. MOERDIJK
221 Harmonic approximation, S.J. GARDINER
222 Advances in linear logic, J.-Y. GIRARD, Y. LAFONT & L. REGNIER (eds)
223 Analytic semigroups and semilinear initial boundary value problems, KAZUAKI TAIRA
224 Computability, enumerability, unsolvability, S.B. COOPER, T.A. SLAMAN & S.S. WAINER (eds)
225 A mathematical introduction to string theory, S. ALBEVERIO, J. JOST, S. PAYCHA, S. SCARLATTI
226 Novikov conjectures, index theorems and rigidity I, S. FERRY, A. RANICKI & J. ROSENBERG (eds)
227 Novikov conjectures, index theorems and rigidity II, S. FERRY, A. RANICKI & J. ROSENBERG (eds)
228 Ergodic theory of $\mathbf{Z}^d$ actions, M. POLLICOTT & K. SCHMIDT (eds)
229 Ergodicity for infinite dimensional systems, G. DA PRATO & J. ZABCZYK
230 Prolegomena to a middlebrow arithmetic of curves of genus 2, J.W.S. CASSELS & E.V. FLYNN
231 Semigroup theory and its applications, K.H. HOFMANN & M.W. MISLOVE (eds)
232 The descriptive set theory of Polish group actions, H. BECKER & A.S. KECHRIS
233 Finite fields and applications, S. COHEN & H. NIEDERREITER (eds)
234 Introduction to subfactors, V. JONES & V.S. SUNDER
235 Number theory 1993–94, S. DAVID (ed)
236 The James forest, H. FETTER & B. GAMBOA DE BUEN
237 Sieve methods, exponential sums, and their applications in number theory, G.R.H. GREAVES *et al*
238 Representation theory and algebraic geometry, A. MARTSINKOVSKY & G. TODOROV (eds)
239 Clifford algebras and spinors, P. LOUNESTO
240 Stable groups, FRANK O. WAGNER
241 Surveys in combinatorics, 1997, R.A. BAILEY (ed)
242 Geometric Galois actions I, L. SCHNEPS & P. LOCHAK (eds)
243 Geometric Galois actions II, L. SCHNEPS & P. LOCHAK (eds)
244 Model theory of groups and automorphism groups, D. EVANS (ed)
245 Geometry, combinatorial designs and related structures, J.W.P. HIRSCHFELD *et al*
246 *p*-Automorphisms of finite *p*-groups, E.I. KHUKHRO
247 Analytic number theory, Y. MOTOHASHI (ed)
248 Tame topology and o-minimal structures, LOU VAN DEN DRIES
249 The atlas of finite groups: ten years on, ROBERT CURTIS & ROBERT WILSON (eds)
250 Characters and blocks of finite groups, G. NAVARRO
251 Gröbner bases and applications, B. BUCHBERGER & F. WINKLER (eds)
252 Geometry and cohomology in group theory, P. KROPHOLLER, G. NIBLO, R. STÖHR (eds)
253 The *q*-Schur algebra, S. DONKIN
254 Galois representations in arithmetic algebraic geometry, A.J. SCHOLL & R.L. TAYLOR (eds)
255 Symmetries and integrability of difference equations, P.A. CLARKSON & F.W. NIJHOFF (eds)
256 Aspects of Galois theory, HELMUT VÖLKLEIN *et al*
257 An introduction to noncommutative differential geometry and its physical applications 2ed, J. MADORE
258 Sets and proofs, S.B. COOPER & J. TRUSS (eds)
259 Models and computability, S.B. COOPER & J. TRUSS (eds)
260 Groups St Andrews 1997 in Bath, I, C.M. CAMPBELL *et al*
261 Groups St Andrews 1997 in Bath, II, C.M. CAMPBELL *et al*
263 Singularity theory, BILL BRUCE & DAVID MOND (eds)
264 New trends in algebraic geometry, K. HULEK, F. CATANESE, C. PETERS & M. REID (eds)
265 Elliptic curves in cryptography, I. BLAKE, G. SEROUSSI & N. SMART
267 Surveys in combinatorics, 1999, J.D. LAMB & D.A. PREECE (eds)
268 Spectral asymptotics in the semi-classical limit, M. DIMASSI & J. SJÖSTRAND
269 Ergodic theory and topological dynamics, M.B. BEKKA & M. MAYER
270 Analysis on Lie Groups, N.T. VAROPOULOS & S. MUSTAPHA
271 Singular perturbations of differential operators, S. ALBEVERIO & P. KURASOV
272 Character theory for the odd order function, T. PETERFALVI
273 Spectral theory and geometry, E.B. DAVIES & Y. SAFAROV (eds)
274 The Mandlebrot set, theme and variations, TAN LEI (ed)
275 Descriptive set theory and dynamical systems M. FOREMAN *et al*
276 Singularities of plane curves, E. CASAS-ALVERO
277 Computational and geometric aspects of modern algebra, M. D. ATKINSON *et al*
278 Global attractors in abstract parabolic problems, J.W. CHOLEWA & T. DLOTKO
279 Topics in symbolic dynamics and applications, F. BLANCHARD, A. MAASS & A. NOGUEIRA (eds)
280 Characters and automorphism groups of compact Riemann surfaces, THOMAS BREUER
281 Explicit birational geometry of 3-folds, ALESSIO CORTI & MILES REID (eds)
282 Auslander-Buchweitz approximations of equivariant modules, M. HASHIMOTO

London Mathematical Society Lecture Note Series. 286

Clifford Algebras and Spinors

Second Edition

Pertti Lounesto
Helsinki Polytechnic Stadia

CAMBRIDGE
UNIVERSITY PRESS

CAMBRIDGE UNIVERSITY PRESS
Cambridge, New York, Melbourne, Madrid, Cape Town, Singapore, São Paulo

Cambridge University Press
The Edinburgh Building, Cambridge CB2 2RU, UK

Published in the United States of America by Cambridge University Press, New York

www.cambridge.org
Information on this title: www.cambridge.org/9780521005517

First published 1997
Reprinted 1998, 1999
Second edition 2001
Reprinted 2002, 2003

A catalogue record for this publication is available from the British Library

ISBN-13 978-0-521-00551-7 paperback
ISBN-10 0-521-00551-5 paperback

Transferred to digital printing 2006

Contents

Preface

This book is intended for people who are not primarily algebraists, but nonetheless get involved in the subject through other areas – having backgrounds, in say, engineering, physics, geometry or analysis. Readers of this book may have different starting levels, backgrounds and goals.

Chapters 1–2 form a unit for an undergraduate course: Vectors, the scalar product; complex numbers, a geometrical interpretation of the imaginary unit $i = \sqrt{-1}$.

Chapters 3–7 guide the reader through bottlenecks and provide necessary building blocks: Bivectors and the exterior product. Pauli spin matrices and Pauli spinors. Quaternions and the fourth dimension. The cross product is generalized to higher dimensions.

Chapters 8–10 aim to serve readers with different backgrounds: Electromagnetism, special relativity, Dirac theory. The Dirac equation is formulated with complex column spinors, spinors in minimal left ideals and considering spinors as operators.

Chapters 11–13 discuss physical applications of spinors: In the case of an electron, Fierz identities are sufficient to reconstruct spinors from their physical observables, but this is not the case for the neutrino. Boomerangs are introduced to handle neutrinos. A new class of spinors is identified by its bilinear observables: the *flag-dipole spinors* which reside between Weyl, Majorana and Dirac spinors.

Chapters 14–15 are more algebraic than the previous chapters. Clifford algebras are defined for the first time. Finite fields. Isometry classes of quadratic forms and their Witt rings. Tensor products of algebras and Brauer groups are discussed.

Chapters 16–19 view Clifford algebras through matrix algebras: Clifford algebras are given isomorphic images as matrix algebras, Cartan's periodicity of 8, spin groups and their matrix images in lower dimensions, scalar products of spinors with a chessboard of their automorphism groups, Möbius transformations represented by Vahlen matrices.

Chapters 20–23 discuss miscellaneous mathematical topics. A one-variable higher-dimensional generalization of complex analysis: Cauchy's integral formula is generalized to higher dimensions. Multiplication rule of standard basis elements of a Clifford algebra and its relation to Walsh functions. Multivector structure of Clifford algebras and the non-existence of k-vectors, $k \geq 2$, in characteristic 2. In the last chapter, we come into contact with final frontiers science: an exceptional phenomenon in dimension 8, triality, which has no counterpart in any other dimension.

The first parts of initial chapters are accessible without knowledge of other parts of the book – thus a teacher may choose his own path for his lectures on Clifford algebras. The latter parts of the chapters are sometimes more advanced, and can be left as independent study for interested students.

Introduction of the Clifford algebra of multivectors and spinors can be motivated in two different ways, in physics and in geometry:

(i) In physics, the concept of Clifford algebra, as such or in a disguise, is a necessity in the description of electron spin: Spinors cannot be constructed by tensorial methods, in terms of exterior powers of the vector space.

(ii) In geometry, information about orientation of subspaces can be encoded in simple multivectors, which can be added and multiplied. Physicists are familiar with this tool in the special case of one-dimensional subspaces of oriented line-segments, which they manipulate by vectors (not by projection operators, which lose information about orientation).

Acknowledgements

In preparing this book I have enjoyed the help of many friends and colleagues: Jerry Segercrantz from Finland, Peter Morgan, Ian Porteous and Ronald Shaw from England, Rafał Abłamowicz, Michael Becker, Geoffrey Dixon, Tevian Dray, Michael Kinyon and Arvind Raja from the USA, Roland Bacher (Switzerland), Helga Baum (Germany), Bernard Jancewicz (Poland), Alphonse Charlier and Yvon Siret (France), Josep Parra (Spain) and Garret Sobczyk (Mexico).

I thank Ross Jones and Göran Långstedt for revising my English. For their assistance with TEX, I am indebted to Timo Korvola and Martti Nikunen. I appreciate feedback from the readers of the first edition: Leo Dorst, Kynn Jones, Joseph Riel and Perttu Puska.

In particular, I would like to thank Jacques Helmstetter for his comments on my DEA lectures in Grenoble as well as Ismo Lindell at the Electromagnetics Laboratory in Helsinki, and Waldyr Rodrigues at IMECC in Campinas. For financial support, I am indebted to the NOKIA Foundation in Helsinki and CNPq in Brazil.

Helsinki, November 2000 PERTTI LOUNESTO

Mathematical Notation

V	linear space
$\mathbb{F}$	field
$\mathbb{D}$	division ring (typically $\mathbb{R}$, $\mathbb{C}$, $\mathbb{H}$)
$\mathbb{H}$	division ring of quaternions
$\mathbb{O}$	division algebra of octonions
$\mathrm{Mat}(d,\mathbb{F})$	$d \times d$ matrix algebra over $\mathbb{F}$
$\mathrm{Mat}(d,\mathbb{D})$	matrix algebra over $\mathrm{Cen}(\mathbb{D})$ with entries in $\mathbb{D}$
${}^2\mathbb{F}$	double ring $\mathbb{F} \times \mathbb{F}$ of the field $\mathbb{F}$
${}^2\mathrm{Mat}(d,\mathbb{F})$	direct sum $\mathrm{Mat}(d,\mathbb{F}) \oplus \mathrm{Mat}(d,\mathbb{F}) \simeq \mathrm{Mat}(d,{}^2\mathbb{F})$
A	an algebra
$\mathrm{Cen}(A)$	the center of an algebra A
$\mathbb{A}$	$\mathbb{R}$, $\mathbb{C}$, $\mathbb{H}$, ${}^2\mathbb{R}$, or ${}^2\mathbb{H}$
$\mathbb{R}^n$	n-dimensional real linear space
$\mathbb{R}^n = \mathbb{R}^{n,0}$	n-dimensional Euclidean space
$\mathbb{C}^n$	n-dimensional complex linear space
$\mathbb{H}^n$	n-dimensional module over $\mathbb{H}$
$\mathbb{R}^{p,q}$	real quadratic space (p for positive and q for negative)
$\mathcal{C}\ell_{p,q}$	Clifford algebra of $\mathbb{R}^{p,q}$
Q	quadratic form
$\mathcal{C}\ell(Q)$	Clifford algebra of the quadratic form Q; $\mathbf{x}^2 = Q(\mathbf{x})$
$\mathcal{C}\ell_3 \simeq \mathrm{Mat}(2,\mathbb{C})$	Clifford algebra of the Euclidean space $\mathbb{R}^3$
$\mathcal{C}\ell_n = \mathcal{C}\ell_{n,0}$	Clifford algebra of the Euclidean space $\mathbb{R}^n = \mathbb{R}^{n,0}$
$\mathcal{C}\ell_{1,3} \simeq \mathrm{Mat}(2,\mathbb{H})$	Clifford algebra of the Minkowski space $\mathbb{R}^{1,3}$
$\mathcal{C}\ell_{3,1} \simeq \mathrm{Mat}(4,\mathbb{R})$	Clifford algebra of the Minkowski space $\mathbb{R}^{3,1}$
$\hat{u}$	grade involute of $u \in \mathcal{C}\ell(Q) = \mathcal{C}\ell^+(Q) \oplus \mathcal{C}\ell^-(Q)$
$\tilde{u}$	reverse of $u \in \mathcal{C}\ell(Q)$; $\tilde{\mathbf{x}} = \mathbf{x}$ for $\mathbf{x} \in V$
$\bar{u}$	Clifford-conjugate of $u \in \mathcal{C}\ell(Q)$; $\bar{\mathbf{x}} = -\mathbf{x}$ for $\mathbf{x} \in V$
u^*	complex conjugate of u (in chapter 2 also $\bar{u}$)
f	primitive idempotent of $\mathcal{C}\ell_{p,q}$
$S = \mathcal{C}\ell_{p,q} f$	spinor space (minimal left ideal of $\mathcal{C}\ell_{p,q}$)
$\check{S}$, $\check{U}$	left ideal $S \oplus \hat{S}$ or companion of $U \in SO(8)$
$\mathbb{Z}_n$	$\{0, 1, \ldots, n-1\}$ or $\{1, e^{i2\pi/n}, \ldots, e^{i2\pi(n-1)/n}\}$

1

Vectors and Linear Spaces

Vectors provide a mathematical formulation for the notion of direction, thus making direction a part of our mathematical language for describing the physical world. This leads to useful applications in physics and engineering, notably in connection with forces, velocities of motion, and electrical fields. Vectors help us to visualize physical quantities by providing a geometrical interpretation. They also simplify computations by bringing algebra to bear on geometry.

1.1 Scalars and vectors

In geometry and physics and their engineering applications we use two kinds of quantities, scalars and vectors. A **scalar** is a quantity that is determined by its magnitude, measured in units on a suitable scale. [1] For instance, mass, temperature and voltage are scalars.

A **vector** is a quantity that is determined by its direction as well as its magnitude; thus it is a *directed quantity* or a *directed line-segment*. For instance, force, velocity and magnetic intensity are vectors.

We denote vectors by boldface letters $\mathbf{a}, \mathbf{b}, \mathbf{r}$, etc. [or indicate them by arrows, $\vec{a}, \vec{b}, \vec{r}$, etc., especially in dimension 3]. A vector can be depicted by an arrow, a line-segment with a distinguished end point. The two end points are called the initial point (tail) and the terminal point (tip):

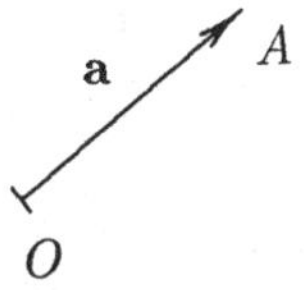

1. length (of the line-segment OA)
2. direction
   - attitude (of the line OA)
   - orientation (from O to A)

The length of a vector $\mathbf{a}$ is denoted by $|\mathbf{a}|$. Two vectors are equal if and only

1 In this chapter scalars are real numbers (elements of $\mathbb{R}$).

if they have the same length and the same direction. Thus,

$$\mathbf{a} = \mathbf{b} \iff |\mathbf{a}| = |\mathbf{b}| \quad \text{and} \quad \mathbf{a} \uparrow\uparrow \mathbf{b}.$$

Two vectors have the same direction, if they are parallel as lines (the same attitude) and similarly aimed (the same orientation). The *zero vector* has length zero, and its direction is unspecified. A *unit vector* $\mathbf{u}$ has length one, $|\mathbf{u}| = 1$. A vector $\mathbf{a}$ and its *opposite* $-\mathbf{a}$ are of equal length and parallel, but have opposite orientations.

1.2 Vector addition and subtraction

Given two vectors $\mathbf{a}$ and $\mathbf{b}$, translate the initial point of $\mathbf{b}$ to the terminal point of $\mathbf{a}$ (without rotating $\mathbf{b}$). Then the sum $\mathbf{a}+\mathbf{b}$ is a vector drawn from the initial point of $\mathbf{a}$ to the terminal point of $\mathbf{b}$. Vector addition can be visualized by the triangle formed by vectors $\mathbf{a}, \mathbf{b}$ and $\mathbf{a}+\mathbf{b}$.

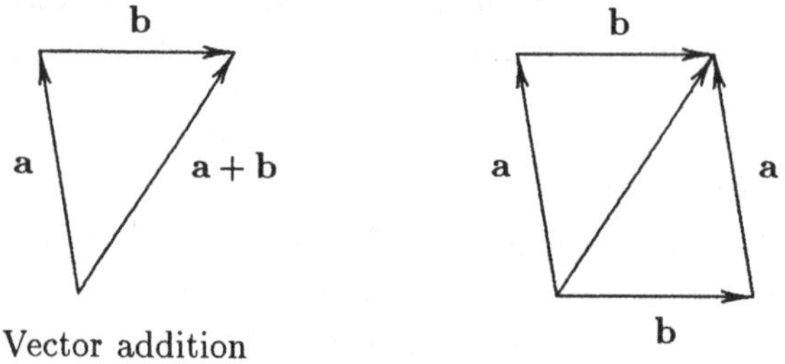

Vector addition

Vector addition is commutative, $\mathbf{a}+\mathbf{b} = \mathbf{b}+\mathbf{a}$, as can be seen by inspection of the parallelogram with $\mathbf{a}$ and $\mathbf{b}$ as sides. It is also associative, $(\mathbf{a}+\mathbf{b})+\mathbf{c} = \mathbf{a}+(\mathbf{b}+\mathbf{c})$, and such that two opposite vectors cancel each other, $\mathbf{a}+(-\mathbf{a}) = 0$.

Instead of $\mathbf{a}+(-\mathbf{b})$ we simply write the difference as $\mathbf{a}-\mathbf{b}$. Note the order in $\overrightarrow{BA} = \overrightarrow{OA} - \overrightarrow{OB}$ when $\mathbf{a} = \overrightarrow{OA}$ and $\mathbf{b} = \overrightarrow{OB}$.

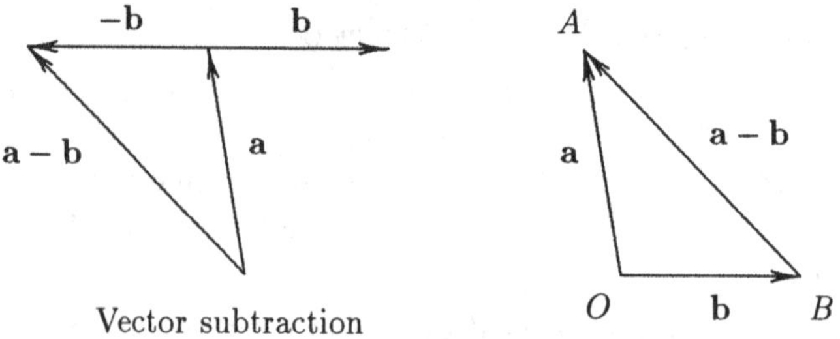

Vector subtraction

Remark. To qualify as vectors, quantities must have more than just direction and magnitude – they must also satisfy certain rules of combination. For instance, a rotation can be characterized by a direction $\mathbf{a}$, the axis of rotation, and a magnitude $\alpha = |\mathbf{a}|$, the angle of rotation, but rotations are not vectors because their composition fails to satisfy the commutative rule of vector addition, $\mathbf{a}+\mathbf{b} = \mathbf{b}+\mathbf{a}$. The lack of commutativity of the composition of rotations can be verified by turning a box around two of its horizontal axes by 90°:

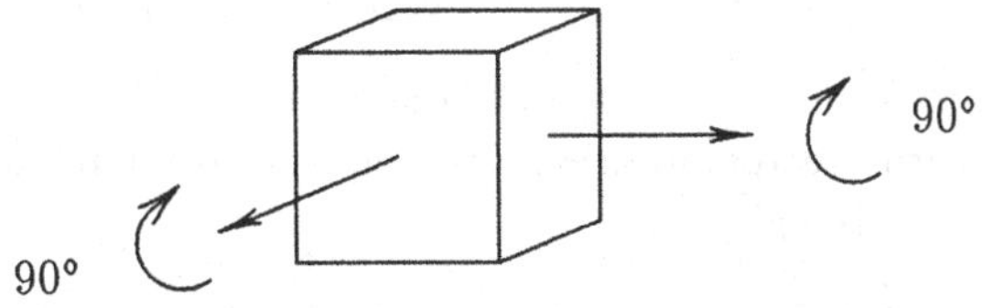

The terminal attitude of the box depends on the order of operations. The axis of the composite rotation is not even horizontal, so that neither $\mathbf{a}+\mathbf{b}$ nor $\mathbf{b}+\mathbf{a}$ can represent the composite rotation. We conclude that rotation angles are not vectors – they are a different kind of directed quantities. ∎

1.3 Multiplication by numbers (scalars)

Instead of $\mathbf{a}+\mathbf{a}$ we write $2\mathbf{a}$, etc., and agree that $(-1)\mathbf{a} = -\mathbf{a}$, the opposite of $\mathbf{a}$. This suggests the following definition for multiplication of vectors $\mathbf{a}$ by real numbers $\lambda \in \mathbb{R}$: the vector $\lambda\mathbf{a}$ has length $|\lambda\mathbf{a}| = |\lambda||\mathbf{a}|$ and direction given by (for $\mathbf{a} \neq 0$)

$$\lambda\mathbf{a} \uparrow\uparrow \mathbf{a} \quad \text{if} \quad \lambda > 0,$$
$$\lambda\mathbf{a} \uparrow\downarrow \mathbf{a} \quad \text{if} \quad \lambda < 0.$$

Numbers multiplying vectors are called *scalars*. Multiplication by scalars, or *scalar multiplication*, satisfies distributivity, $\lambda(\mathbf{a}+\mathbf{b}) = \lambda\mathbf{a}+\lambda\mathbf{b}$, $(\lambda+\mu)\mathbf{a} = \lambda\mathbf{a}+\mu\mathbf{a}$, associativity, $(\lambda\mu)\mathbf{a} = \lambda(\mu\mathbf{a})$, and the unit property, $1\mathbf{a} = \mathbf{a}$, for all real numbers λ, μ and vectors $\mathbf{a}, \mathbf{b}$.

1.4 Bases and coordinates

In the plane any two non-parallel vectors $\mathbf{e}_1, \mathbf{e}_2$ form a *basis* so that an arbitrary vector in the plane can be uniquely expressed as a linear combination $\mathbf{a} = a_1\mathbf{e}_1 + a_2\mathbf{e}_2$. The numbers a_1, a_2 are called *coordinates* or *components* of the vector $\mathbf{a}$ with respect to the basis $\{\mathbf{e}_1, \mathbf{e}_2\}$.

When a basis has been chosen, vectors can be expressed in terms of the coordinates alone, for instance,

$$\mathbf{e}_1 = (1,0), \quad \mathbf{e}_2 = (0,1), \quad \mathbf{a} = (a_1, a_2).$$

If we single out a distinguished point, the origin O, we can use vectors to label the points A by $\mathbf{a} = \overrightarrow{OA}$. In the *coordinate system* fixed by O and $\{\mathbf{e}_1, \mathbf{e}_2\}$ we can denote points and vectors in a similar manner,

$$\text{point} \quad A = (a_1, a_2), \qquad \text{vector} \quad \mathbf{a} = (a_1, a_2),$$

since all the vectors have a common initial point O.

In coordinate form vector addition and multiplication by scalars are just coordinate-wise operations:

$$(a_1, a_2) + (b_1, b_2) = (a_1 + b_1, a_2 + b_2),$$
$$\lambda(a_1, a_2) = (\lambda a_1, \lambda a_2).$$

Conversely, we may start from the set $\mathbb{R} \times \mathbb{R} = \{(x, y) \mid x, y \in \mathbb{R}\}$, and equip it with component-wise addition and multiplication by scalars. This construction introduces a real *linear structure* on the set $\mathbb{R}^2 = \mathbb{R} \times \mathbb{R}$ making it a 2-dimensional real *linear space* $\mathbb{R}^2$. The real linear structure allows us to view the set $\mathbb{R}^2$ intuitively as a plane, the *vector plane* $\mathbb{R}^2$. The two unit points on the axes give the *standard basis*

$$\mathbf{e}_1 = (1,0), \quad \mathbf{e}_2 = (0,1)$$

of the 2-dimensional linear space $\mathbb{R}^2$.

In our ordinary space a basis is formed by three non-zero vectors $\mathbf{e}_1, \mathbf{e}_2, \mathbf{e}_3$ which are not in the same plane. An arbitrary vector $\mathbf{a}$ can be uniquely represented as a linear combination of the basis vectors:

$$\mathbf{a} = a_1\mathbf{e}_1 + a_2\mathbf{e}_2 + a_3\mathbf{e}_3.$$

The numbers a_1, a_2, a_3 are coordinates [2] in the basis $\{\mathbf{e}_1, \mathbf{e}_2, \mathbf{e}_3\}$. Conversely, coordinate-wise addition and scalar multiplication make the set

$$\mathbb{R} \times \mathbb{R} \times \mathbb{R} = \{(x, y, z) \mid x, y, z \in \mathbb{R}\}$$

a 3-dimensional real *linear space* or *vector space* $\mathbb{R}^3$. In a coordinate system fixed by the origin O and a standard basis $\{\mathbf{e}_1, \mathbf{e}_2, \mathbf{e}_3\}$ a point $P = (x, y, z)$ and its *position vector*

$$\overrightarrow{OP} = x\mathbf{e}_1 + y\mathbf{e}_2 + z\mathbf{e}_3$$

have the same coordinates. [3]

2 Some authors speak about components of vectors and coordinates of points.

3 Since a vector beginning at the origin is completely determined by its endpoints, we will sometimes refer to the *point* $\mathbf{r}$ rather than to the *endpoint of the vector* $\mathbf{r}$.

1.5 Linear spaces and linear functions

Above we introduced vectors by visualizing them without specifying the grounds of our study. In an axiomatic approach, one starts with a set whose elements satisfy certain characteristic rules. Vectors then become elements of a mathematical object called a linear space or a vector space V. In a linear space vectors can be added to each other but not multiplied by each other. Instead, vectors are multiplied by numbers, in this context called scalars. [4]

Formally, we begin with a set V and the field of real numbers $\mathbb{R}$. We associate with each pair of elements $\mathbf{a}, \mathbf{b} \in V$ a unique element in V, called the *sum* and denoted by $\mathbf{a}+\mathbf{b}$, and to each $\mathbf{a} \in V$ and each real number $\lambda \in \mathbb{R}$ we associate a unique element in V, called the *scalar multiple* and denoted by $\lambda\mathbf{a}$. The set V is called a **linear space** V over $\mathbb{R}$ if the usual rules of addition are satisfied for all $\mathbf{a}, \mathbf{b}, \mathbf{c} \in V$

$$\begin{aligned} &\mathbf{a}+\mathbf{b} = \mathbf{b}+\mathbf{a} && \text{commutativity} \\ &(\mathbf{a}+\mathbf{b})+\mathbf{c} = \mathbf{a}+(\mathbf{b}+\mathbf{c}) && \text{associativity} \\ &\mathbf{a}+0 = \mathbf{a} && \text{zero-vector } 0 \\ &\mathbf{a}+(-\mathbf{a}) = 0 && \text{opposite vector } -\mathbf{a} \end{aligned}$$

and if the scalar multiplication satisfies

$$\begin{aligned} &\left.\begin{aligned} \lambda(\mathbf{a}+\mathbf{b}) &= \lambda\mathbf{a}+\lambda\mathbf{b} \\ (\lambda+\mu)\mathbf{a} &= \lambda\mathbf{a}+\mu\mathbf{a} \end{aligned}\right\} && \text{distributivity} \\ &(\lambda\mu)\mathbf{a} = \lambda(\mu\mathbf{a}) && \text{associativity} \\ &1\mathbf{a} = \mathbf{a} && \text{unit property} \end{aligned}$$

for all $\lambda, \mu \in \mathbb{R}$ and $\mathbf{a}, \mathbf{b} \in V$. The elements of V are called *vectors*, and the linear space V is also called a vector space. The above axioms of a linear space set up a real *linear structure* on V.

A subset U of a linear space V is called a linear *subspace* of V if it is closed under the operations of a linear space:

$$\begin{aligned} &\mathbf{a}+\mathbf{b} \in U && \text{for } \mathbf{a}, \mathbf{b} \in U, \\ &\lambda\mathbf{a} \in U && \text{for } \lambda \in \mathbb{R},\ \mathbf{a} \in U. \end{aligned}$$

For instance, $\mathbb{R}^2$ is a subspace of $\mathbb{R}^3$.

A function $L : U \to V$ between two linear spaces U and V is said to be *linear* if for any $\mathbf{a}, \mathbf{b} \in U$ and $\lambda \in \mathbb{R}$,

$$\begin{aligned} &L(\mathbf{a}+\mathbf{b}) = L(\mathbf{a}) + L(\mathbf{b}) \quad \text{and} \\ &L(\lambda\mathbf{a}) = \lambda L(\mathbf{a}). \end{aligned}$$

4 Vectors are not scalars, and scalars are not vectors. Vectors belong to a linear space V, and scalars belong to a field $\mathbb{F}$. In this chapter $\mathbb{F} = \mathbb{R}$.

Linear functions preserve the linear structure. A linear function $V \to V$ is called a linear transformation or an *endomorphism*. An invertible linear function $U \to V$ is a *linear isomorphism*, denoted by $U \simeq V$. [5]

The set of linear functions $U \to V$ is itself a linear space. A composition of linear functions is also a linear function. The set of linear transformations $V \to V$ is a ring denoted by $\mathrm{End}(V)$. Since the endomorphism ring $\mathrm{End}(V)$ is also a linear space over $\mathbb{R}$, it is an associative algebra over $\mathbb{R}$, denoted by $\mathrm{End}_{\mathbb{R}}(V)$. [6]

1.6 Linear independence; dimension

A vector $\mathbf{b} \in V$ is said to be a *linear combination* of vectors $\mathbf{a}_1, \mathbf{a}_2, \ldots, \mathbf{a}_k$ if it can be written as a sum of multiples of the vectors $\mathbf{a}_1, \mathbf{a}_2, \ldots, \mathbf{a}_k$, that is,

$$\mathbf{b} = \lambda_1 \mathbf{a}_1 + \lambda_2 \mathbf{a}_2 + \cdots + \lambda_k \mathbf{a}_k \quad \text{where} \quad \lambda_1, \lambda_2, \ldots, \lambda_k \in \mathbb{R}.$$

A set of vectors $\{\mathbf{a}_1, \mathbf{a}_2, \ldots, \mathbf{a}_k\}$ is said to be *linearly independent* if none of the vectors can be written as a linear combination of the other vectors. In other words, a set of vectors $\{\mathbf{a}_1, \mathbf{a}_2, \ldots, \mathbf{a}_k\}$ is linearly independent if $\lambda_1 = \lambda_2 = \ldots = \lambda_k = 0$ is the only set of real numbers satisfying

$$\lambda_1 \mathbf{a}_1 + \lambda_2 \mathbf{a}_2 + \cdots + \lambda_k \mathbf{a}_k = 0.$$

In a linear combination

$$\mathbf{b} = \lambda_1 \mathbf{a}_1 + \lambda_2 \mathbf{a}_2 + \cdots + \lambda_k \mathbf{a}_k$$

of linearly independent vectors $\mathbf{a}_1, \mathbf{a}_2, \ldots, \mathbf{a}_k$ the numbers $\lambda_1, \lambda_2, \ldots, \lambda_k$ are unique; we call them the *coordinates* of $\mathbf{b}$.

Linear combinations of $\{\mathbf{a}_1, \mathbf{a}_2, \ldots, \mathbf{a}_k\} \subset V$ form a subspace of V; we say that this subspace is *spanned* by $\{\mathbf{a}_1, \mathbf{a}_2, \ldots, \mathbf{a}_k\}$. A linearly independent set $\{\mathbf{a}_1, \mathbf{a}_2, \ldots, \mathbf{a}_k\} \subset V$ which spans V is said to be a *basis* of V. All the bases for V have the same number of elements called the *dimension* of V.

Quadratic Structures

Concepts such as *distance* or *angle* are *not* inherent in the concept of a linear structure alone. For instance, it is meaningless to say that two lines in the linear space $\mathbb{R}^2$ meet each other at right angles, or that there is a basis of

5 Finite-dimensional real linear spaces are isomorphic if they are of the same dimension.

6 A ring R is a set with the usual addition and an associative multiplication $R \times R \to R$ which is distributive with respect to the addition. An algebra A is a linear space with a bilinear product $A \times A \to A$.

equally long vectors $\mathbf{e}_1, \mathbf{e}_2$ in $\mathbb{R}^2$. The linear structure allows comparison of lengths of parallel vectors, but it does not enable comparison of lengths of non-parallel vectors. For this, an extra structure is needed, namely the metric or quadratic structure.

The quadratic structure on a linear space $\mathbb{R}^n$ brings along an algebra which makes it possible to calculate with geometric objects. In the rest of this chapter we shall study such a geometric algebra associated with the Euclidean plane $\mathbb{R}^2$.

1.7 Scalar product

We will associate with two vectors a real number, the *scalar product* $\mathbf{a} \cdot \mathbf{b} \in \mathbb{R}$ of $\mathbf{a}, \mathbf{b} \in \mathbb{R}^2$. This scalar valued product of $\mathbf{a} = a_1\mathbf{e}_1 + a_2\mathbf{e}_2$ and $\mathbf{b} = b_1\mathbf{e}_1 + b_2\mathbf{e}_2$ is defined as

in coordinates	$\mathbf{a} \cdot \mathbf{b} = a_1 b_1 + a_2 b_2$
geometrically	$\mathbf{a} \cdot \mathbf{b} = \lvert\mathbf{a}\rvert\lvert\mathbf{b}\rvert \cos\varphi$

where φ $[0 \leq \varphi \leq 180°]$ is the angle between $\mathbf{a}$ and $\mathbf{b}$. The geometrical construction depends on the prior introduction of lengths and angles. Instead, the coordinate approach can be used to define the length

$$|\mathbf{a}| = \sqrt{\mathbf{a} \cdot \mathbf{a}},$$

which equals $|\mathbf{a}| = \sqrt{a_1^2 + a_2^2}$, and the angle given by

$$\cos\varphi = \frac{\mathbf{a} \cdot \mathbf{b}}{|\mathbf{a}||\mathbf{b}|}.$$

Two vectors $\mathbf{a}$ and $\mathbf{b}$ are said to be *orthogonal*, if $\mathbf{a} \cdot \mathbf{b} = 0$. A vector of length one, $|\mathbf{a}| = 1$, is called a *unit vector*. For instance, the standard basis vectors $\mathbf{e}_1 = (1,0)$, $\mathbf{e}_2 = (0,1)$ are orthogonal unit vectors, and so form an *orthonormal basis* for $\mathbb{R}^2$.

The scalar product can be characterized by its properties:

$$\left.\begin{aligned} &(\mathbf{a}+\mathbf{b}) \cdot \mathbf{c} = \mathbf{a} \cdot \mathbf{c} + \mathbf{b} \cdot \mathbf{c} \\ &(\lambda\mathbf{a}) \cdot \mathbf{b} = \lambda(\mathbf{a} \cdot \mathbf{b}) \end{aligned}\right\} \quad \text{linear in the first factor}$$

$$\mathbf{a} \cdot \mathbf{b} = \mathbf{b} \cdot \mathbf{a} \qquad \text{symmetric}$$

$$\mathbf{a} \cdot \mathbf{a} > 0 \quad \text{for} \quad \mathbf{a} \neq 0 \qquad \text{positive definite.}$$

Symmetry and linearity with respect to the first factor together imply bilinearity, that is, linearity with respect to both factors. The real linear space $\mathbb{R}^2$ endowed with a bilinear, symmetric and positive definite product is called a *Euclidean plane* $\mathbb{R}^2$.

All Euclidean planes are isometric [7] to the one with the metric/norm

$$\mathbf{r} = x\mathbf{e}_1 + y\mathbf{e}_2 \to |\mathbf{r}| = \sqrt{x^2 + y^2}.$$

In the rest of this chapter we assume this metric structure on our vector plane $\mathbb{R}^2$.

Remark. The quadratic form $\mathbf{r} = x\mathbf{e}_1 + y\mathbf{e}_2 \to |\mathbf{r}|^2 = x^2 + y^2$ enables us to compare lengths of non-parallel line-segments. The linear structure by itself allows only comparison of parallel line-segments. ∎

1.8 The Clifford product of vectors; the bivector

It would be useful to have a multiplication of vectors satisfying the same axioms as the multiplication of real numbers – distributivity, associativity and commutativity – and require that the norm is preserved in multiplication, $|\mathbf{ab}| = |\mathbf{a}||\mathbf{b}|$. Since this is impossible in dimensions $n \geq 3$, we will settle for distributivity and associativity, but drop commutativity. However, we will attach a geometrical meaning to the lack of commutativity.

Take two orthogonal unit vectors $\mathbf{e}_1$ and $\mathbf{e}_2$ in the vector plane $\mathbb{R}^2$. The length of the vector $\mathbf{r} = x\mathbf{e}_1 + y\mathbf{e}_2$ is $|\mathbf{r}| = \sqrt{x^2 + y^2}$. If the vector $\mathbf{r}$ is multiplied by itself, $\mathbf{rr} = \mathbf{r}^2$, [8] a natural choice is to require that the product equals the square of the length of $\mathbf{r}$,

$$\mathbf{r}^2 = |\mathbf{r}|^2.$$

In coordinate form, we introduce a product for vectors in such a way that

$$(x\mathbf{e}_1 + y\mathbf{e}_2)^2 = x^2 + y^2.$$

Use the distributive rule without assuming commutativity to obtain

$$x^2\mathbf{e}_1^2 + y^2\mathbf{e}_2^2 + xy(\mathbf{e}_1\mathbf{e}_2 + \mathbf{e}_2\mathbf{e}_1) = x^2 + y^2.$$

This is satisfied if the orthogonal unit vectors $\mathbf{e}_1, \mathbf{e}_2$ obey the multiplication rules

$\mathbf{e}_1^2 = \mathbf{e}_2^2 = 1$ $\mathbf{e}_1\mathbf{e}_2 = -\mathbf{e}_2\mathbf{e}_1$	which correspond to	$\lvert\mathbf{e}_1\rvert = \lvert\mathbf{e}_2\rvert = 1$ $\mathbf{e}_1 \perp \mathbf{e}_2$

Use associativity to calculate the square $(\mathbf{e}_1\mathbf{e}_2)^2 = -\mathbf{e}_1^2\mathbf{e}_2^2 = -1$. Since the square of the product $\mathbf{e}_1\mathbf{e}_2$ is negative, it follows that $\mathbf{e}_1\mathbf{e}_2$ is neither a scalar

7 An isometry of quadratic forms is a linear function $f : V \to V'$ such that $Q'(f(\mathbf{a})) = Q(\mathbf{a})$ for all $\mathbf{a} \in V$.

8 The scalar product $\mathbf{a} \cdot \mathbf{b}$ is not the same as the Clifford product $\mathbf{ab}$. Instead, the two products are related by $\mathbf{a} \cdot \mathbf{b} = \frac{1}{2}(\mathbf{ab} + \mathbf{ba})$.

nor a vector. The product is a new kind of unit, called a **bivector**, representing the oriented plane area of the square with sides $\mathbf{e}_1$ and $\mathbf{e}_2$. Write for short $\mathbf{e}_{12} = \mathbf{e}_1\mathbf{e}_2$.

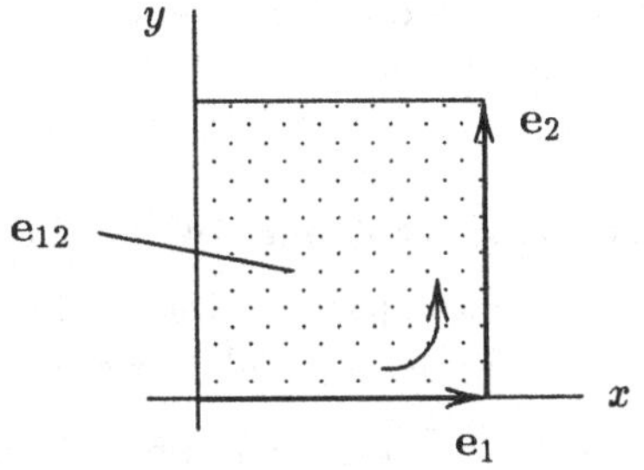

We define the *Clifford product* of two vectors $\mathbf{a} = a_1\mathbf{e}_1 + a_2\mathbf{e}_2$ and $\mathbf{b} = b_1\mathbf{e}_1 + b_2\mathbf{e}_2$ to be $\mathbf{ab} = a_1b_1 + a_2b_2 + (a_1b_2 - a_2b_1)\mathbf{e}_{12}$, a sum of a scalar and a bivector.

1.9 The Clifford algebra $C\ell_2$

The four elements

1	scalar
$\mathbf{e}_1$, $\mathbf{e}_2$	vectors
$\mathbf{e}_{12}$	bivector

form a basis for the **Clifford algebra $C\ell_2$** [9] of the vector plane $\mathbb{R}^2$, that is, an arbitrary element

$$u = u_0 + u_1\mathbf{e}_1 + u_2\mathbf{e}_2 + u_{12}\mathbf{e}_{12} \quad \text{in} \quad C\ell_2$$

is a linear combination of a scalar u_0, a vector $u_1\mathbf{e}_1 + u_2\mathbf{e}_2$ and a bivector $u_{12}\mathbf{e}_{12}$. [10]

Example. Compute $\mathbf{e}_1\mathbf{e}_{12} = \mathbf{e}_1\mathbf{e}_1\mathbf{e}_2 = \mathbf{e}_2$, $\mathbf{e}_{12}\mathbf{e}_1 = \mathbf{e}_1\mathbf{e}_2\mathbf{e}_1 = -\mathbf{e}_1^2\mathbf{e}_2 = -\mathbf{e}_2$, $\mathbf{e}_2\mathbf{e}_{12} = \mathbf{e}_2\mathbf{e}_1\mathbf{e}_2 = -\mathbf{e}_1\mathbf{e}_2^2 = -\mathbf{e}_1$ and $\mathbf{e}_{12}\mathbf{e}_2 = \mathbf{e}_1\mathbf{e}_2^2 = \mathbf{e}_1$. Note in particular that $\mathbf{e}_{12}$ anticommutes with both $\mathbf{e}_1$ and $\mathbf{e}_2$. ∎

The Clifford algebra $C\ell_2$ is a 4-dimensional real linear space with basis elements

9 These algebras were invented by William Kingdon Clifford (1845-1879). The first announcement of the result was issued in a talk in 1876, which was published posthumously in 1882. The first publication of the invention came out in another paper in 1878.

10 The Clifford algebra $C\ell_n$ of $\mathbb{R}^n$ contains 0-vectors (or scalars), 1-vectors (or just vectors), 2-vectors, ..., n-vectors. The aggregates of k-vectors give the linear space $C\ell_n$ a multivector structure $C\ell_n = \mathbb{R} \oplus \mathbb{R}^n \oplus \bigwedge^2 \mathbb{R}^n \oplus \ldots \oplus \bigwedge^n \mathbb{R}^n$.

$1, \mathbf{e}_1, \mathbf{e}_2, \mathbf{e}_{12}$ which have the multiplication table

	$\mathbf{e}_1$	$\mathbf{e}_2$	$\mathbf{e}_{12}$
$\mathbf{e}_1$	1	$\mathbf{e}_{12}$	$\mathbf{e}_2$
$\mathbf{e}_2$	$-\mathbf{e}_{12}$	1	$-\mathbf{e}_1$
$\mathbf{e}_{12}$	$-\mathbf{e}_2$	$\mathbf{e}_1$	-1

1.10 Exterior product = bivector part of the Clifford product

Extracting the scalar and bivector parts of the Clifford product we have as products of two vectors $\mathbf{a} = a_1\mathbf{e}_1 + a_2\mathbf{e}_2$ and $\mathbf{b} = b_1\mathbf{e}_1 + b_2\mathbf{e}_2$

$$\mathbf{a} \cdot \mathbf{b} = a_1 b_1 + a_2 b_2, \qquad \text{the scalar product 'a dot b'},$$
$$\mathbf{a} \wedge \mathbf{b} = (a_1 b_2 - a_2 b_1)\mathbf{e}_{12}, \qquad \text{the exterior product 'a wedge b'}.$$

The bivector $\mathbf{a} \wedge \mathbf{b}$ represents the oriented plane segment of the parallelogram with sides $\mathbf{a}$ and $\mathbf{b}$. The area of this parallelogram is $|a_1 b_2 - a_2 b_1|$, and we will take the *magnitude* of the bivector $\mathbf{a} \wedge \mathbf{b}$ to be this area $|\mathbf{a} \wedge \mathbf{b}| = |a_1 b_2 - a_2 b_1|$.

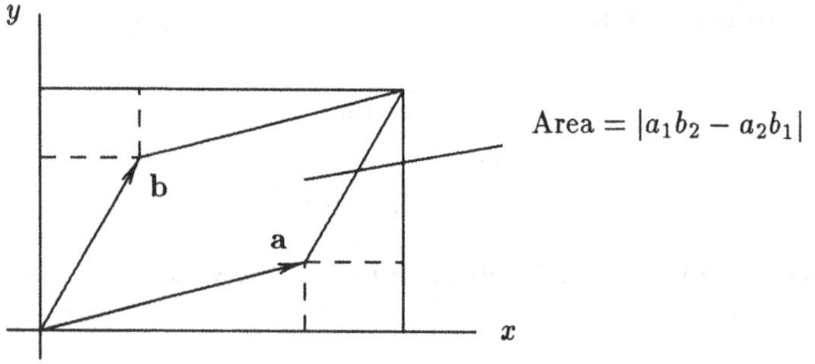

The parallelogram can be regarded as a kind of geometrical product of its sides:

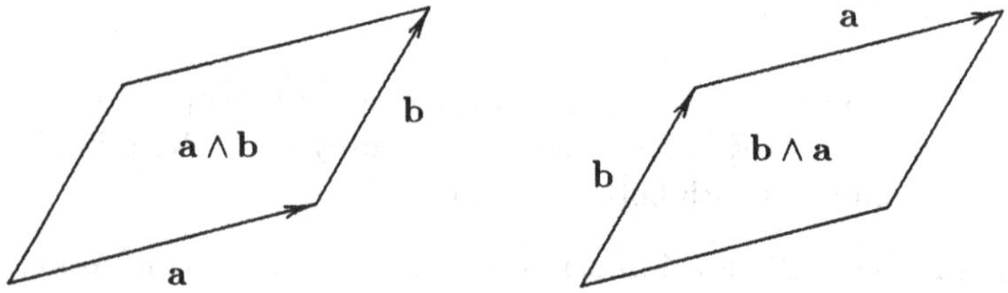

The bivectors $\mathbf{a} \wedge \mathbf{b}$ and $\mathbf{b} \wedge \mathbf{a}$ have the same magnitude but opposite senses of rotation. This can be expressed simply by writing

$$\mathbf{a} \wedge \mathbf{b} = -\mathbf{b} \wedge \mathbf{a}.$$

Using the multiplication table of the Clifford algebra $\mathcal{C}\ell_2$ we notice that the Clifford product

$$(a_1\mathbf{e}_1 + a_2\mathbf{e}_2)(b_1\mathbf{e}_1 + b_2\mathbf{e}_2) = a_1b_1 + a_2b_2 + (a_1b_2 - a_2b_1)\mathbf{e}_{12}$$

of two vectors $\mathbf{a} = a_1\mathbf{e}_1 + a_2\mathbf{e}_2$ and $\mathbf{b} = b_1\mathbf{e}_1 + b_2\mathbf{e}_2$ is a sum of a scalar $\mathbf{a} \cdot \mathbf{b} = a_1b_1 + a_2b_2$ and a bivector $\mathbf{a} \wedge \mathbf{b} = (a_1b_2 - a_2b_1)\mathbf{e}_{12}$. [11] In an equation,

$$\mathbf{ab} = \mathbf{a} \cdot \mathbf{b} + \mathbf{a} \wedge \mathbf{b}. \tag{a}$$

The commutative rule $\mathbf{a} \cdot \mathbf{b} = \mathbf{b} \cdot \mathbf{a}$ together with the anticommutative rule $\mathbf{a} \wedge \mathbf{b} = -\mathbf{b} \wedge \mathbf{a}$ implies a relation between $\mathbf{ab}$ and $\mathbf{ba}$. Thus,

$$\mathbf{ba} = \mathbf{a} \cdot \mathbf{b} - \mathbf{a} \wedge \mathbf{b}. \tag{b}$$

Adding and subtracting equations (a) and (b), we find

$$\mathbf{a} \cdot \mathbf{b} = \frac{1}{2}(\mathbf{ab} + \mathbf{ba}) \quad \text{and} \quad \mathbf{a} \wedge \mathbf{b} = \frac{1}{2}(\mathbf{ab} - \mathbf{ba}).$$

Two vectors $\mathbf{a}$ and $\mathbf{b}$ are parallel, $\mathbf{a} \parallel \mathbf{b}$, when they commute, $\mathbf{ab} = \mathbf{ba}$, that is, $\mathbf{a} \wedge \mathbf{b} = 0$ or $a_1b_2 = a_2b_1$, and orthogonal, $\mathbf{a} \perp \mathbf{b}$, when they anticommute, $\mathbf{ab} = -\mathbf{ba}$, that is, $\mathbf{a} \cdot \mathbf{b} = 0$. Thus,

$$\begin{aligned} \mathbf{ab} = \mathbf{ba} \quad &\Longleftrightarrow \quad \mathbf{a} \parallel \mathbf{b} \quad \Longleftrightarrow \quad \mathbf{a} \wedge \mathbf{b} = 0 \quad \Longleftrightarrow \quad \mathbf{ab} = \mathbf{a} \cdot \mathbf{b}, \\ \mathbf{ab} = -\mathbf{ba} \quad &\Longleftrightarrow \quad \mathbf{a} \perp \mathbf{b} \quad \Longleftrightarrow \quad \mathbf{a} \cdot \mathbf{b} = 0 \quad \Longleftrightarrow \quad \mathbf{ab} = \mathbf{a} \wedge \mathbf{b}. \end{aligned}$$

1.11 Components of a vector in given directions

Consider decomposing a vector $\mathbf{r}$ into two components, one parallel to $\mathbf{a}$ and the other parallel to $\mathbf{b}$, where $\mathbf{a} \nparallel \mathbf{b}$. This means determining the coefficients α and β in the decomposition $\mathbf{r} = \alpha\mathbf{a} + \beta\mathbf{b}$. The coefficient α may be obtained by forming the exterior product $\mathbf{r} \wedge \mathbf{b} = (\alpha\mathbf{a} + \beta\mathbf{b}) \wedge \mathbf{b}$ and using $\mathbf{b} \wedge \mathbf{b} = 0$; this results in $\mathbf{r} \wedge \mathbf{b} = \alpha(\mathbf{a} \wedge \mathbf{b})$. Similarly, $\mathbf{a} \wedge \mathbf{r} = \beta(\mathbf{a} \wedge \mathbf{b})$. In the last two equations both sides are multiples of $\mathbf{e}_{12}$ and we may write, symbolically, [12]

$$\alpha = \frac{\mathbf{r} \wedge \mathbf{b}}{\mathbf{a} \wedge \mathbf{b}}, \qquad \beta = \frac{\mathbf{a} \wedge \mathbf{r}}{\mathbf{a} \wedge \mathbf{b}}.$$

---

11 The bivector valued exterior product $\mathbf{a} \wedge \mathbf{b} = (a_1b_2 - a_2b_1)\mathbf{e}_{12}$, which represents a plane area, should not be confused with the vector valued cross product $\mathbf{a} \times \mathbf{b} = (a_1b_2 - a_2b_1)\mathbf{e}_3$, which represents a line segment.

12 As an element of the exterior algebra $\bigwedge\mathbb{R}^2$ the bivector $\mathbf{a} \wedge \mathbf{b}$ is not invertible. As an element of the Clifford algebra $\mathcal{C}\ell_2$ a non-zero bivector $\mathbf{a} \wedge \mathbf{b}$ is invertible, but since the multiplication in $\mathcal{C}\ell_2$ is non-commutative, it is more appropriate to write

$$\alpha = (\mathbf{r} \wedge \mathbf{b})(\mathbf{a} \wedge \mathbf{b})^{-1} \quad \text{and} \quad \beta = (\mathbf{a} \wedge \mathbf{r})(\mathbf{a} \wedge \mathbf{b})^{-1}.$$

However, since $\mathbf{r} \wedge \mathbf{b}$, $\mathbf{a} \wedge \mathbf{r}$ and $\mathbf{a} \wedge \mathbf{b}$ commute, our notation is also acceptable.

The coefficients α and β could be obtained visually by comparing the oriented areas (instead of lengths) in the following figure:

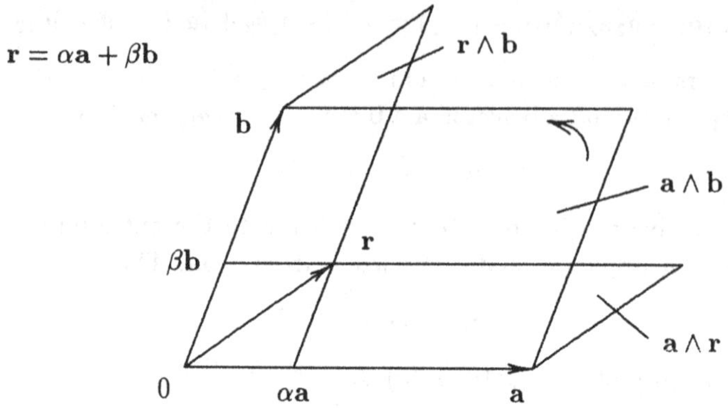

Exercise 5

1.12 Perpendicular projections and reflections

Let us calculate the component of $\mathbf{a}$ in the direction of $\mathbf{b}$ when the two vectors diverge by an angle φ, $0 < \varphi < 180°$. The parallel component $\mathbf{a}_{\|}$ is a scalar multiple of the unit vector $\mathbf{b}/|\mathbf{b}|$:

$$\mathbf{a}_{\|} = |\mathbf{a}|\cos\varphi\frac{\mathbf{b}}{|\mathbf{b}|} = |\mathbf{a}||\mathbf{b}|\cos\varphi\frac{\mathbf{b}}{|\mathbf{b}|^2}.$$

In other words, the parallel component $\mathbf{a}_{\|}$ is the scalar product $\mathbf{a}\cdot\mathbf{b} = |\mathbf{a}||\mathbf{b}|\cos\varphi$ multiplied by the vector $\mathbf{b}^{-1} = \mathbf{b}/|\mathbf{b}|^2$, called the inverse [13] of the vector $\mathbf{b}$. Thus,

$$\mathbf{a}_{\|} = (\mathbf{a}\cdot\mathbf{b})\frac{\mathbf{b}}{|\mathbf{b}|^2} = (\mathbf{a}\cdot\mathbf{b})\mathbf{b}^{-1}.$$

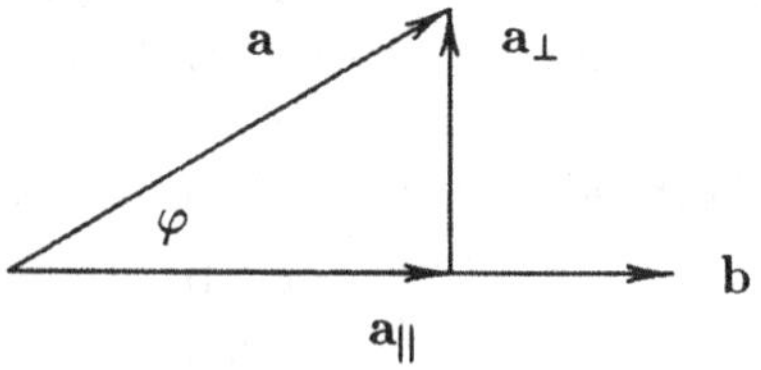

The last formula tells us that the length of $\mathbf{b}$ is irrelevant when projecting into the direction of $\mathbf{b}$.

The perpendicular component $\mathbf{a}_{\perp}$ is given by the difference

$$\mathbf{a}_{\perp} = \mathbf{a} - \mathbf{a}_{\|} = \mathbf{a} - (\mathbf{a}\cdot\mathbf{b})\mathbf{b}^{-1} = (\mathbf{ab} - \mathbf{a}\cdot\mathbf{b})\mathbf{b}^{-1} = (\mathbf{a}\wedge\mathbf{b})\mathbf{b}^{-1}.$$

13 The inverse $\mathbf{b}^{-1}$ of a non-zero vector $\mathbf{b} \in \mathbb{R}^2 \subset C\ell_2$ satisfies $\mathbf{b}^{-1}\mathbf{b} = \mathbf{bb}^{-1} = 1$ in the Clifford algebra $C\ell_2$. A vector and its inverse are parallel vectors.

Note that the bivector $\mathbf{e}_{12}$ anticommutes with all the vectors in the $\mathbf{e}_1\mathbf{e}_2$-plane, therefore

$$(\mathbf{a} \wedge \mathbf{b})\mathbf{b}^{-1} = -\mathbf{b}^{-1}(\mathbf{a} \wedge \mathbf{b}) = \mathbf{b}^{-1}(\mathbf{b} \wedge \mathbf{a}) = -(\mathbf{b} \wedge \mathbf{a})\mathbf{b}^{-1}.$$

The area of the parallelogram with sides $\mathbf{a}, \mathbf{b}$ is seen to be

$$|\mathbf{a}_\perp \mathbf{b}| = |\mathbf{a} \wedge \mathbf{b}| = |\mathbf{a}||\mathbf{b}| \sin \varphi$$

where $0 < \varphi < 180°$.

The reflection of $\mathbf{r}$ across the line $\mathbf{a}$ is obtained by sending $\mathbf{r} = \mathbf{r}_\| + \mathbf{r}_\perp$ to $\mathbf{r}' = \mathbf{r}_\| - \mathbf{r}_\perp$, where $\mathbf{r}_\| = (\mathbf{r} \cdot \mathbf{a})\mathbf{a}^{-1}$. The mirror image $\mathbf{r}'$ of $\mathbf{r}$ with respect to $\mathbf{a}$ is then

$$\begin{aligned}\mathbf{r}' &= (\mathbf{r} \cdot \mathbf{a})\mathbf{a}^{-1} - (\mathbf{r} \wedge \mathbf{a})\mathbf{a}^{-1} \\ &= (\mathbf{r} \cdot \mathbf{a} - \mathbf{r} \wedge \mathbf{a})\mathbf{a}^{-1} \\ &= (\mathbf{a} \cdot \mathbf{r} + \mathbf{a} \wedge \mathbf{r})\mathbf{a}^{-1} \\ &= \mathbf{a}\mathbf{r}\mathbf{a}^{-1}\end{aligned}$$

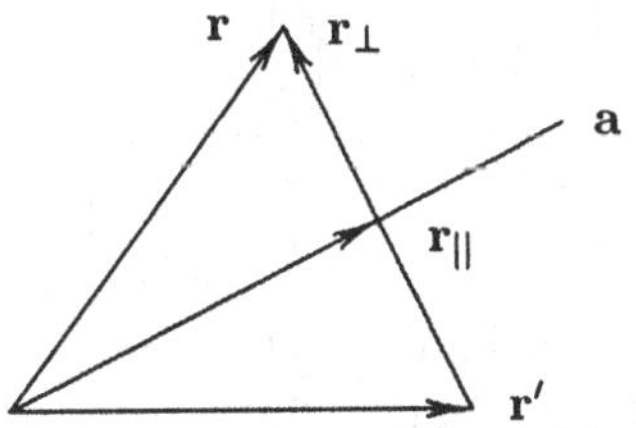

and further

$$\begin{aligned}\mathbf{r}' &= (2\mathbf{a} \cdot \mathbf{r} - \mathbf{r}\mathbf{a})\mathbf{a}^{-1} \\ &= 2\frac{\mathbf{a} \cdot \mathbf{r}}{\mathbf{a}^2}\mathbf{a} - \mathbf{r}.\end{aligned}$$

The formula $\mathbf{r}' = \mathbf{a}\mathbf{r}\mathbf{a}^{-1}$ can be obtained directly using only commutation properties of the Clifford product: decompose $\mathbf{r} = \mathbf{r}_\| + \mathbf{r}_\perp$, where $\mathbf{a}\mathbf{r}_\|\mathbf{a}^{-1} = \mathbf{r}_\|\mathbf{a}\mathbf{a}^{-1} = \mathbf{r}_\|$, while $\mathbf{a}\mathbf{r}_\perp\mathbf{a}^{-1} = -\mathbf{r}_\perp\mathbf{a}\mathbf{a}^{-1} = -\mathbf{r}_\perp$.

The composition of two reflections, first across $\mathbf{a}$ and then across $\mathbf{b}$, is given by

$$\mathbf{r} \to \mathbf{r}' = \mathbf{a}\mathbf{r}\mathbf{a}^{-1} \to \mathbf{r}'' = \mathbf{b}\mathbf{r}'\mathbf{b}^{-1} = \mathbf{b}(\mathbf{a}\mathbf{r}\mathbf{a}^{-1})\mathbf{b}^{-1} = (\mathbf{b}\mathbf{a})\mathbf{r}(\mathbf{b}\mathbf{a})^{-1}.$$

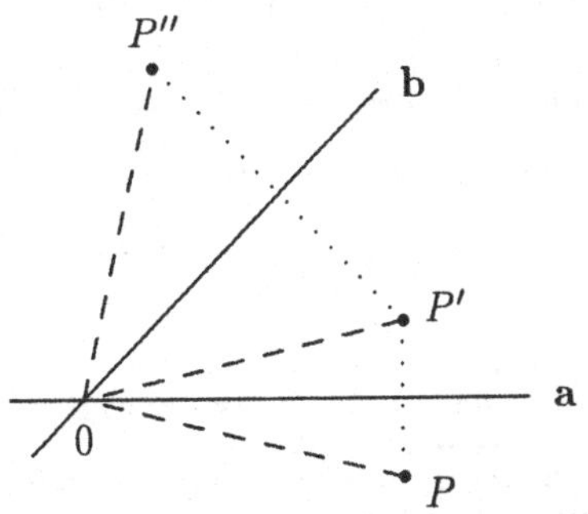

The composite of these two reflections is a rotation by twice the angle between $\mathbf{a}$ and $\mathbf{b}$. As a consequence, if a triangle ABC with angles α, β, γ is turned about its vertices A, B, C by the angles $2\alpha, 2\beta, 2\gamma$ in the same direction, the result is an identity rotation.

Exercises 6,7

1.13 Matrix representation of $C\ell_2$

In this chapter we have introduced the Clifford algebra $C\ell_2$ of the Euclidean plane $\mathbb{R}^2$. The Clifford algebra $C\ell_2$ is a 4-dimensional algebra over the reals $\mathbb{R}$. It is isomorphic, as an associative algebra, to the matrix algebra of real 2×2-matrices $\mathrm{Mat}(2, \mathbb{R})$, as can be seen by the correspondences

$$1 \simeq \begin{pmatrix} 1 & 0 \\ 0 & 1 \end{pmatrix},$$

$$\mathbf{e}_1 \simeq \begin{pmatrix} 1 & 0 \\ 0 & -1 \end{pmatrix}, \qquad \mathbf{e}_2 \simeq \begin{pmatrix} 0 & 1 \\ 1 & 0 \end{pmatrix},$$

$$\mathbf{e}_{12} \simeq \begin{pmatrix} 0 & 1 \\ -1 & 0 \end{pmatrix}.$$

However, in the Clifford algebra $C\ell_2$ there is more structure than in the matrix algebra $\mathrm{Mat}(2, \mathbb{R})$. In the Clifford algebra $C\ell_2$ we have singled out by definition a privileged subspace, namely the subspace of vectors or 1-vectors $\mathbb{R}^2 \subset C\ell_2$. No similar privileged subspace is incorporated in the definition of the matrix algebra $\mathrm{Mat}(2, \mathbb{R})$. [14]

For arbitrary elements the above correspondences mean that

$$u_0 + u_1\mathbf{e}_1 + u_2\mathbf{e}_2 + u_{12}\mathbf{e}_{12} \simeq \begin{pmatrix} u_0 + u_1 & u_2 + u_{12} \\ u_2 - u_{12} & u_0 - u_1 \end{pmatrix}$$

and

$$\frac{1}{2}[(a+d) + (a-d)\mathbf{e}_1 + (b+c)\mathbf{e}_2 + (b-c)\mathbf{e}_{12}] \simeq \begin{pmatrix} a & b \\ c & d \end{pmatrix}.$$

In this representation the transpose of a matrix,

$$\begin{pmatrix} a & b \\ c & d \end{pmatrix}^{\mathsf{T}} = \begin{pmatrix} a & c \\ b & d \end{pmatrix},$$

corresponds to the *reverse*

$$\tilde{u} = u_0 + u_1\mathbf{e}_1 + u_2\mathbf{e}_2 - u_{12}\mathbf{e}_{12}$$

14 For instance, we might choose $\mathbf{u}_1 = \sqrt{2}\mathbf{e}_1 + \mathbf{e}_{12}$, $\mathbf{u}_2 = \mathbf{e}_2$. This also results in the commutation relations $\mathbf{u}_1^2 = 1$, $\mathbf{u}_2^2 = 1$, $\mathbf{u}_1\mathbf{u}_2 + \mathbf{u}_2\mathbf{u}_1 = 0$, which define a different representation of $C\ell_2$ as $\mathrm{Mat}(2, \mathbb{R})$.

of $u = u_0 + u_1\mathbf{e}_1 + u_2\mathbf{e}_2 + u_{12}\mathbf{e}_{12}$ in $C\ell_2$. The complementary (or adjoint) matrix

$$\begin{pmatrix} d & -b \\ -c & a \end{pmatrix} \left[= (ad-bc)\begin{pmatrix} a & b \\ c & d \end{pmatrix}^{-1} \quad \text{for} \quad ad-bc \neq 0 \right]$$

corresponds to the *Clifford-conjugate* [15]

$$\bar{u} = u_0 - u_1\mathbf{e}_1 - u_2\mathbf{e}_2 - u_{12}\mathbf{e}_{12}.$$

The reversion and Clifford-conjugation are anti-involutions, that is, involutory anti-automorphisms,

$$\tilde{\tilde{u}} = u, \quad \widetilde{uv} = \tilde{v}\tilde{u},$$
$$\bar{\bar{u}} = u, \quad \overline{uv} = \bar{v}\bar{u}.$$

We still need the *grade involute*

$$\hat{u} = u_0 - u_1\mathbf{e}_1 - u_2\mathbf{e}_2 + u_{12}\mathbf{e}_{12}$$

for which $\hat{u} = \tilde{u}^{-} = \bar{u}^{\sim}$.

Exercises

1. Let $a = \mathbf{e}_2 - \mathbf{e}_{12}$, $b = \mathbf{e}_1 + \mathbf{e}_2$, $c = 1 + \mathbf{e}_2$. Compute ab, ac. What did you learn by completing this computation?
2. Let $a = \mathbf{e}_2 + \mathbf{e}_{12}$, $b = \frac{1}{2}(1 + \mathbf{e}_1)$. Compute ab, ba. What did you learn?
3. Let $a = 1 + \mathbf{e}_1$, $b = -1 + \mathbf{e}_1$, $c = \mathbf{e}_1 + \mathbf{e}_2$. Compute ab, ba, ac, ca, bc and cb. What did you learn?
4. Let $a = \frac{1}{2}(1 + \mathbf{e}_1)$, $b = \mathbf{e}_1 + \mathbf{e}_{12}$. Compute a^2, b^2.
5. Let $\mathbf{a} = \mathbf{e}_1 - 2\mathbf{e}_2$, $\mathbf{b} = \mathbf{e}_1 + \mathbf{e}_2$, $\mathbf{r} = 5\mathbf{e}_1 - \mathbf{e}_2$. Compute α, β in the decomposition $\mathbf{r} = \alpha\mathbf{a} + \beta\mathbf{b}$.
6. Let $\mathbf{a} = 8\mathbf{e}_1 - \mathbf{e}_2$, $\mathbf{b} = 2\mathbf{e}_1 + \mathbf{e}_2$. Compute $\mathbf{a}_{\|}$, $\mathbf{a}_{\perp}$.
7. Let $\mathbf{r} = 4\mathbf{e}_1 - 3\mathbf{e}_2$, $\mathbf{a} = 3\mathbf{e}_1 - \mathbf{e}_2$, $\mathbf{b} = 2\mathbf{e}_1 + \mathbf{e}_2$. Reflect first $\mathbf{r}$ across $\mathbf{a}$ and then the result across $\mathbf{b}$.
8. Show that for any $u \in C\ell_2$, $u\bar{u} = \bar{u}u \in \mathbb{R}$, and that u is invertible, if $u\bar{u} \neq 0$, with inverse
$$u^{-1} = \frac{\bar{u}}{u\bar{u}}.$$
9. Let $u = 1 + \mathbf{e}_1 + \mathbf{e}_{12}$. Compute u^{-1}. Show that
$u^{-1} = \hat{u}(u\hat{u})^{-1} \neq (u\hat{u})^{-1}\hat{u}$, $u^{-1} = (\hat{u}u)^{-1}\hat{u} \neq \hat{u}(\hat{u}u)^{-1}$ and
$u^{-1} = \tilde{u}(u\tilde{u})^{-1} \neq (u\tilde{u})^{-1}\tilde{u}$, $u^{-1} = (\tilde{u}u)^{-1}\tilde{u} \neq \tilde{u}(\tilde{u}u)^{-1}$.

15 In some countries a vector $\mathbf{u} = u_1\mathbf{e}_1 + u_2\mathbf{e}_2 \in \mathbb{R}^2$ is denoted by $\bar{u}$ in handwriting, but this practice clashes with our notation for the Clifford-conjugate.

10. Consider the four anti-involutions of $\mathrm{Mat}(2,\mathbb{R})$ sending

$$\begin{pmatrix} a & b \\ c & d \end{pmatrix} \text{ to } \begin{pmatrix} a & c \\ b & d \end{pmatrix}, \begin{pmatrix} a & -c \\ -b & d \end{pmatrix}, \begin{pmatrix} d & b \\ c & a \end{pmatrix}, \begin{pmatrix} d & -b \\ -c & a \end{pmatrix}.$$

Define two anti-automorphisms α, β to be similar, if there is an intertwining automorphism γ such that $\alpha\gamma = \gamma\beta$. Determine which ones of these four anti-involutions are similar or dissimilar to each other. Hint: keep track of what happens to the matrices

$$\begin{pmatrix} 1 & 0 \\ 0 & -1 \end{pmatrix}, \begin{pmatrix} 0 & 1 \\ 1 & 0 \end{pmatrix}, \begin{pmatrix} 0 & -1 \\ 1 & 0 \end{pmatrix}$$

with squares I, I, and $-I$.

Remark. In completing the exercises, note that an arbitrary element of $\mathcal{C}\ell_2$ is most easily perceived when written in the order of increasing indices as $u_0 + u_1\mathbf{e}_1 + u_2\mathbf{e}_2 + u_{12}\mathbf{e}_{12}$. ∎

Solutions

1. $ab = ac = 1 - \mathbf{e}_1 + \mathbf{e}_2 - \mathbf{e}_{12}$; one can learn that $ab = ac \not\Rightarrow b = c$.
2. $ab = 0$, $ba = \mathbf{e}_2 + \mathbf{e}_{12}$; one can learn that $ab = 0 \not\Rightarrow ba = 0$ (and also that $ba = a \not\Rightarrow b = 1$).
3. $ab = ba = 0$, $ac = 1 + \mathbf{e}_1 + \mathbf{e}_2 + \mathbf{e}_{12}$, $ca = 1 + \mathbf{e}_1 + \mathbf{e}_2 - \mathbf{e}_{12}$, $bc = 1 - \mathbf{e}_1 - \mathbf{e}_2 + \mathbf{e}_{12}$, $cb = 1 - \mathbf{e}_1 - \mathbf{e}_2 - \mathbf{e}_{12}$; one can learn that $ab = ba = 0 \not\Rightarrow ac = 0$ or $ca = 0$.
4. $a^2 = a$, $b^2 = 0$.
5. $\mathbf{r} = 2\mathbf{a} + 3\mathbf{b}$.
6. $\mathbf{a}_{\|} = 6\mathbf{e}_1 + 3\mathbf{e}_2$, $\mathbf{a}_{\perp} = 2\mathbf{e}_1 - 4\mathbf{e}_2$.
7. $\mathbf{r}' = \mathbf{ara}^{-1} = 5\mathbf{e}_1$, $\mathbf{r}'' = \mathbf{br}'\mathbf{b}^{-1} = 3\mathbf{e}_1 + 4\mathbf{e}_2$.
8. $u\bar{u} = \bar{u}u = u_0^2 - u_1^2 - u_2^2 + u_{12}^2 \in \mathbb{R}$.
9. $u^{-1} = 1 - \mathbf{e}_1 - \mathbf{e}_{12}$ and $(u\hat{u})^{-1}\hat{u} = \tilde{u}(\tilde{u}u)^{-1} = 1 + 3\mathbf{e}_1 - 4\mathbf{e}_2 - 5\mathbf{e}_{12}$ and $\hat{u}(\hat{u}u)^{-1} = (u\tilde{u})^{-1}\tilde{u} = 1 + 3\mathbf{e}_1 + 4\mathbf{e}_2 - 5\mathbf{e}_{12}$.
10. Only two of the anti-involutions are similar,

$$\alpha\begin{pmatrix} a & b \\ c & d \end{pmatrix} = \begin{pmatrix} a & -c \\ -b & d \end{pmatrix}, \quad \beta\begin{pmatrix} a & b \\ c & d \end{pmatrix} = \begin{pmatrix} d & b \\ c & a \end{pmatrix},$$

as can be seen by choosing the intertwining automorphism

$$\gamma\begin{pmatrix} a & b \\ c & d \end{pmatrix} = \frac{1}{\sqrt{2}}\begin{pmatrix} 1 & -1 \\ 1 & 1 \end{pmatrix}\begin{pmatrix} a & b \\ c & d \end{pmatrix}\frac{1}{\sqrt{2}}\begin{pmatrix} 1 & 1 \\ -1 & 1 \end{pmatrix}$$

for which $\alpha\gamma = \gamma\beta$.

Bibliography

W.K. Clifford: Applications of Grassmann's extensive algebra. *Amer. J. Math.* **1** (1878), 350-358.

W.K. Clifford: On the classification of geometric algebras; pp. 397-401 in R. Tucker (ed.): *Mathematical Papers by William Kingdon Clifford*, Macmillan, London, 1882. (Reprinted by Chelsea, New York, 1968.) Title of talk announced already in *Proc. London Math. Soc.* **7** (1876), p. 135.

M.J. Crowe: *A History of Vector Analysis*. University of Notre Dame Press, 1967. Reprinted by Dover, New York, 1985.

D.C. Lay: *Linear Algebra and its Applications. Instructor's Edition*. Addison-Wesley, Reading, MA, 1994.

M. Riesz: *Clifford Numbers and Spinors*. The Institute for Fluid Dynamics and Applied Mathematics, Lecture Series No. **38**, University of Maryland, 1958. Reprinted as facsimile (eds.: E.F. Bolinder, P. Lounesto) by Kluwer, Dordrecht, The Netherlands, 1993.

G. Strang: *Introduction to Linear Algebra*. Wellesley-Cambridge Press, Cambridge, MA, 1993.

2

Complex Numbers

The feature distinguishing the complex numbers from the real numbers is that the complex numbers contain a square root of -1 called the *imaginary unit* $i = \sqrt{-1}$. [1] Complex numbers are of the form

$$z = x + iy$$

where $x, y \in \mathbb{R}$ and i satisfies $i^2 = -1$. The real numbers x, y are called the *real part* $x = \text{Re}(z)$ and the *imaginary part* $y = \text{Im}(z)$. To each ordered pair of real numbers x, y there corresponds a unique complex number $x + iy$.

A complex number $x + iy$ can be represented graphically as a point with rectangular coordinates (x, y). The xy-plane, where the complex numbers are represented, is called the *complex plane* $\mathbb{C}$. Its x-axis is the *real axis* and y-axis the *imaginary axis*.

A complex number $z = x + iy$ has an opposite $-z = -x - iy$ and a *complex conjugate* $\bar{z} = x - iy$, [2] obtained by changing the sign of the imaginary part.

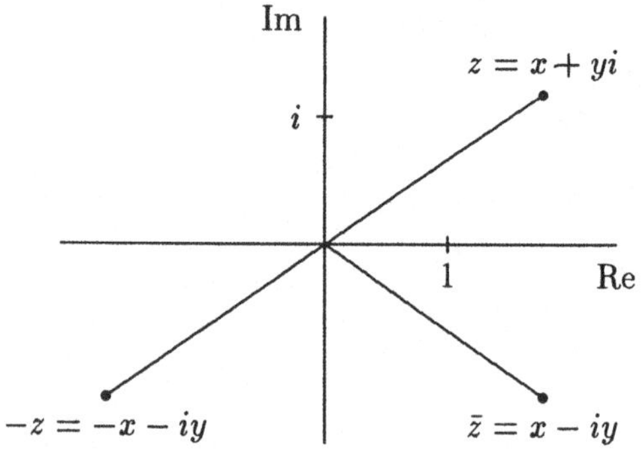

1 Electrical engineers denote the square root of -1 by $j = \sqrt{-1}$.
2 In quantum mechanics the complex conjugate is denoted by $z^* = x - iy$.

The sum of two complex numbers is computed by adding separately the real parts and the imaginary parts:

$$(x_1 + iy_1) + (x_2 + iy_2) = (x_1 + x_2) + i(y_1 + y_2).$$

Addition of complex numbers can be illustrated by the parallelogram law of vector addition.

The product of two complex numbers is usually defined to be

$$(x_1 + iy_1)(x_2 + iy_2) = x_1x_2 - y_1y_2 + i(x_1y_2 + y_1x_2),$$

although this result is also a consequence of distributivity, associativity and the replacement $i^2 = -1$.

Examples. 1. $i^3 = -i$, $i^4 = 1$, $i^5 = i$. 2. $(1+i)^2 = 2i$. ▮

The product of a complex number $z = x + iy$ and its complex conjugate $\bar{z} = x - iy$ is a real number $z\bar{z} = x^2 + y^2$. Since this real number is non-zero for $z \neq 0$, we may introduce the inverse

$$z^{-1} = \frac{\bar{z}}{z\bar{z}}$$

or in coordinate form

$$\frac{1}{x+iy} = \frac{x-iy}{x^2+y^2}.$$

Division is carried out as multiplication by the inverse: $z_1/z_2 = z_1z_2^{-1}$.

If we introduce polar coordinates r, φ in the complex plane by setting $x = r\cos\varphi$ and $y = r\sin\varphi$, then the complex number $z = x + iy$ can be written as

$$z = r(\cos\varphi + i\sin\varphi).$$

This is the *polar form* of z. [3] The distance $r = \sqrt{x^2+y^2}$ from z to 0 is denoted by $|z|$ and called the *norm* of z. Thus [4]

$$|z| = \sqrt{z\bar{z}}.$$

The real number φ is called the *phase-angle* or *argument* of z [sometimes all the real numbers $\varphi + 2\pi k$, $k \in \mathbb{Z}$, are assigned to the same phase-angle].

The familiar addition rules for the sine and cosine result in the polar form of multiplication,

$$z_1z_2 = r_1r_2[\cos(\varphi_1 + \varphi_2) + i\sin(\varphi_1 + \varphi_2)],$$

---

3 Electrical engineers denote the polar form by $r\underline{/\varphi}$.

4 The scalar product $\mathrm{Re}(z_1\bar{z}_2)$ is compatible with the norm $|z|$. Incidentally, $\mathrm{Im}(z_1\bar{z}_2)$ measures the signed area of the parallelogram determined by z_1 and z_2.

of complex numbers

$$z_1 = r_1(\cos\varphi_1 + i\sin\varphi_1) \quad \text{and} \quad z_2 = r_2(\cos\varphi_2 + i\sin\varphi_2).$$

Thus, the norm of a product is the product of the norms,

$$|z_1 z_2| = |z_1||z_2|,$$

and the phase-angle of a product is the sum of the phase-angles (mod 2π).

The exponential function can be defined everywhere in the complex plane by

$$\exp(z) = 1 + z + \frac{z^2}{2} + \frac{z^3}{6} + \ldots + \frac{z^k}{k!} + \ldots$$

We write $e^z = \exp(z)$. The series expansions of trigonometric functions result in *Euler's formula*

$$e^{i\varphi} = \cos\varphi + i\sin\varphi$$

which allows us to abbreviate $z = r(\cos\varphi + i\sin\varphi)$ as $z = re^{i\varphi}$.

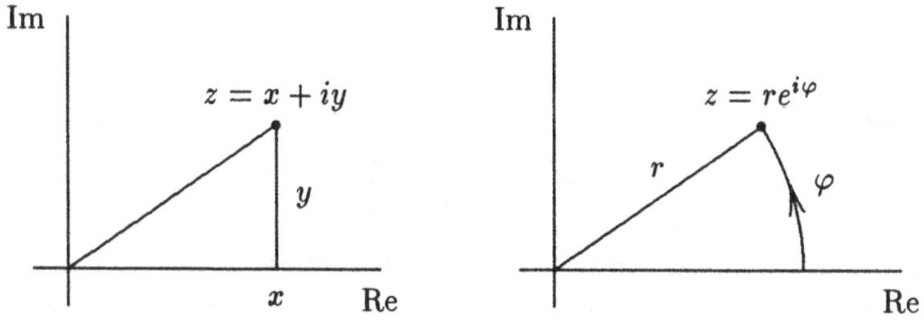

The exponential form of multiplication seems natural:

$$(r_1 e^{i\varphi_1})(r_2 e^{i\varphi_2}) = (r_1 r_2) e^{i(\varphi_1+\varphi_2)}.$$

Powers and roots are computed as

$$(re^{i\varphi})^n = r^n e^{in\varphi} \quad \text{and} \quad \sqrt[n]{re^{i\varphi}} = \sqrt[n]{r} e^{i\varphi/n + i2\pi k/n}, \; k \in \mathbb{Z}_n.$$

Examples. $(1+i)^{-1} = \frac{1}{2}(1-i)$, $\sqrt{i} = \pm\frac{1}{\sqrt{2}}(1+i)$, $e^{i\pi/2} = i$. ∎

2.1 The field $\mathbb{C}$ versus the real algebra $\mathbb{C}$

Numbers are elements of a mathematical object called a field. In a field numbers can be both added and multiplied. The usual rules of addition

$a + b = b + a$	commutativity
$(a + b) + c = a + (b + c)$	associativity
$a + 0 = a$	zero 0
$a + (-a) = 0$	opposite $-a$ of a

are satisfied for all numbers a, b, c in a field $\mathbb{F}$. The multiplication satisfies

$$\left.\begin{array}{l}(a+b)c = ac+bc\\ a(b+c) = ab+ac\end{array}\right\} \quad \text{distributivity}$$

$$\begin{array}{ll}(ab)c = a(bc) & \text{associativity}\\ 1a = a & \text{unity } 1\\ aa^{-1} = 1 & \text{inverse } a^{-1} \text{ of } a \neq 0\\ ab = ba & \text{commutativity}\end{array}$$

for all numbers a, b, c in a field $\mathbb{F}$. The above rules of addition and multiplication make up the *axioms* of a **field** $\mathbb{F}$.

Examples of fields are the fields of real numbers $\mathbb{R}$, complex numbers $\mathbb{C}$, rationals $\mathbb{Q}$, and the finite fields $\mathbb{F}_q$ where $q = p^m$ with a prime p. [5]

It is tempting to regard $\mathbb{R}$ as a unique subfield in $\mathbb{C}$. However, $\mathbb{C}$ contains several, infinitely many, subfields isomorphic to $\mathbb{R}$; choosing one means introducing a real linear structure on $\mathbb{C}$, obtained by restricting a in the product $\mathbb{C} \times \mathbb{C} \to \mathbb{C}$, $(a, b) \to ab$ to be real, $a \in \mathbb{R}$. Such extra structure turns the field $\mathbb{C}$ into a real algebra $\mathbb{C}$.

Definition. An *algebra* over a field $\mathbb{F}$ is a linear space A over $\mathbb{F}$ together with a bilinear [6] function $A \times A \to A$, $(a, b) \to ab$. [7] ∎

To distinguish the field $\mathbb{C}$ from a real algebra $\mathbb{C}$ let us construct $\mathbb{C}$ as the set $\mathbb{R} \times \mathbb{R}$ of all ordered pairs of real numbers $z = (x, y)$ with addition and multiplication defined as

$$(x_1, y_1) + (x_2, y_2) = (x_1 + x_2, y_1 + y_2) \quad \text{and}$$
$$(x_1, y_1)(x_2, y_2) = (x_1x_2 - y_1y_2, x_1y_2 + x_2y_1).$$

The set $\mathbb{R} \times \mathbb{R}$ together with the above addition and multiplication rules makes up the *field* $\mathbb{C}$. The imaginary unit $(0, 1)$ satisfies $(0, 1)^2 = (-1, 0)$.

Since $(x_1, 0) + (x_2, 0) = (x_1 + x_2, 0)$ and $(x_1, 0)(x_2, 0) = (x_1x_2, 0)$, the real field $\mathbb{R}$ is contained in $\mathbb{C}$ as a subfield by $\mathbb{R} \to \mathbb{C}$, $x \to (x, 0)$. If we restrict multiplication so that one factor is in this distinguished copy of $\mathbb{R}$,

$$(\lambda, 0)(x, y) = (\lambda x, \lambda y),$$

then we actually introduce a *real linear structure* on the set $\mathbb{R}^2 = \mathbb{R} \times \mathbb{R}$. This

---

5 The finite fields $\mathbb{F}_q$, where $q = p^m$ with a prime p, are called Galois fields $GF(p^m)$.

6 Bilinear means linear with respect to both arguments. This implies distributivity. In other words, distributivity has no independent meaning for an algebra.

7 Note that associativity is not assumed.

real linear structure allows us to view the field of complex numbers intuitively as the complex plane $\mathbb{C}$. [8]

The above construction of $\mathbb{C}$ as the real linear space $\mathbb{R}^2$ brings in more structure than just the field structure: it makes $\mathbb{C}$ an *algebra* over $\mathbb{R}$. [9] We often identify $\mathbb{R}$ with the subfield $\{(x,0) \mid x \in \mathbb{R}\}$ of $\mathbb{C}$, and denote the standard basis of $\mathbb{R}^2$ by $1 = (1,0)$, $i = (0,1)$ in $\mathbb{C}$.

A function $\alpha : \mathbb{C} \to \mathbb{C}$ is an *automorphism of the field* $\mathbb{C}$ if it preserves addition and multiplication,

$$\alpha(z_1 + z_2) = \alpha(z_1) + \alpha(z_2),$$
$$\alpha(z_1 z_2) = \alpha(z_1)\alpha(z_2),$$

as well as the unity, $\alpha(1) = 1$. A function $\alpha : \mathbb{C} \to \mathbb{C}$ is an *automorphism of the real algebra* $\mathbb{C}$ if it preserves the real linear structure and multiplication (of complex numbers),

$$\alpha(z_1 + z_2) = \alpha(z_1) + \alpha(z_2), \quad \alpha(\lambda z) = \lambda\alpha(z), \quad \lambda \in \mathbb{R},$$
$$\alpha(z_1 z_2) = \alpha(z_1)\alpha(z_2),$$

as well as the unity, $\alpha(1) = 1$.

The field $\mathbb{C}$ has an infinity of automorphisms. In contrast, the only automorphisms of the real algebra $\mathbb{C}$ are the identity automorphism and complex conjugation.

Theorem. Complex conjugation is the only field automorphism of $\mathbb{C}$ which is different from the identity but preserves a fixed subfield $\mathbb{R}$.

Proof. First, note that $\alpha(i) = \pm i$ for any field automorphism α of $\mathbb{C}$, since $\alpha(i)^2 = \alpha(i^2) = \alpha(-1) = -1$. If $\alpha : \mathbb{C} \to \mathbb{C}$ is a field automorphism such that $\alpha(\mathbb{R}) \subset \mathbb{R}$, then $\alpha(x) = x$ for all $x \in \mathbb{R}$, because the only automorphism of the real field is the identity. It then follows that, for all $x + iy$ with $x, y \in \mathbb{R}$,

$$\alpha(x + iy) = \alpha(x) + \alpha(i)\alpha(y) = x + \alpha(i)y$$

where $\alpha(i) = \pm i$. The case $\alpha(i) = i$ gives the identity automorphism, and the case $\alpha(i) = -i$ gives complex conjugation. ∎

The other automorphisms of the field $\mathbb{C}$ send a real subfield $\mathbb{R}$ onto an isomorphic copy of $\mathbb{R}$, which is necessarily different from the original subfield $\mathbb{R}$. However, any field automorphism of $\mathbb{C}$ fixes point-wise the rational subfield $\mathbb{Q}$.

---

8 The geometric view of complex numbers is connected with the structure of $\mathbb{C}$ as a real algebra, and not so much as a field.

9 In the above construction we introduced a field structure into the real linear space $\mathbb{R}^2$ and arrived at an algebra $\mathbb{C}$ over $\mathbb{R}$, or equivalently at a field $\mathbb{C}$ with a distinguished subfield $\mathbb{R}$.

Example. It is known that there is a field automorphism of $\mathbb{C}$ sending $\sqrt{2}$ to $-\sqrt{2}$ and $\sqrt[4]{2}$ to $i\sqrt[4]{2}$, but no one has been able to construct such an automorphism explicitly since its existence proof calls for the axiom of choice. ∎

If a field automorphism of $\mathbb{C}$ is neither the identity nor a complex conjugation, then it sends some irrational numbers outside $\mathbb{R}$, and permutes an infinity of subfields all isomorphic with $\mathbb{R}$. Related to each real subfield there is a unique complex conjugation across that subfield, and all such automorphisms of finite order are complex conjugations for some real subfield. The image $\alpha(\mathbb{R})$ under such an automorphism α of a distinguished real subfield $\mathbb{R}$ is dense in $\mathbb{C}$ [in the topology of the metric $|z| = \sqrt{z\bar{z}}$ given by the complex conjugation across $\mathbb{R}$]. This can be seen as follows: An automorphism α must satisfy $\alpha(rx) = r\alpha(x)$ when $r \in \mathbb{Q}$. So if there is an irrational $x \in \mathbb{R}$ with $t = \alpha(x) \notin \mathbb{R}$, and necessarily $t \notin \mathbb{Q} + i\mathbb{Q}$, the image $\alpha(\mathbb{R})$ of $\mathbb{R}$ contains all numbers of the form $\alpha(r + sx) = r + st$ with $r, s \in \mathbb{Q}$. This is a dense set in $\mathbb{C}$.

The above discussion indicates that **there is no unique complex conjugation in the field of complex numbers**, and that the field structure of $\mathbb{C}$ does not fix by itself the subfield $\mathbb{R}$ of $\mathbb{C}$. The field injection $\mathbb{R} \to \mathbb{C}$ is an extra piece of structure added on top of the field $\mathbb{C}$. If a privileged real subfield $\mathbb{R}$ is singled out in $\mathbb{C}$, it brings along a real linear structure on $\mathbb{C}$, and a unique complex conjugation across $\mathbb{R}$, which then naturally imports a metric structure to $\mathbb{C}$.

Our main interest in complex numbers in this book is $\mathbb{C}$ as a real algebra, not so much as a field.

2.2 The double-ring $^2\mathbb{R}$ of $\mathbb{R}$

There is more than one interesting bilinear product (or algebra structure) on the linear space $\mathbb{R}^2$. For instance, component-wise multiplication

$$(x_1, y_1)(x_2, y_2) = (x_1x_2, y_1y_2)$$

results in the double-ring $^2\mathbb{R}$ of $\mathbb{R}$. The only automorphisms of the real algebra $^2\mathbb{R}$ are the identity and the *swap*

$$^2\mathbb{R} \to {}^2\mathbb{R}, \quad (\lambda, \mu) \to \operatorname{swap}(\lambda, \mu) = (\mu, \lambda).$$

The swap acts like the complex conjugation of $\mathbb{C}$, since

$$\operatorname{swap}[a(1,1) + b(1,-1)] = a(1,1) - b(1,-1).$$

The multiplicative unity $1 = (1,1)$ and the reflected element $j = (1,-1)$ are now related by $j^2 = 1$.

Alternatively and equivalently we may consider pairs of real numbers $(a, b) \in \mathbb{R}^2$ as Study numbers

$$a + jb, \quad j^2 = 1, \; j \neq 1.$$

Study numbers have Study conjugate $(a + jb)^- = a - jb$, Lorentz squared norm $(a + jb)(a - jb) = a^2 - b^2$, and the hyperbolic polar form $a + jb = \rho(\cosh \chi + j \sinh \chi)$ for $a^2 - b^2 \geq 0$. [10] In products Lorentz squared norms are preserved and hyperbolic angles added. Study numbers have the matrix representation

$$a + jb \simeq \begin{pmatrix} a & b \\ b & a \end{pmatrix}.$$

Exercise 4

2.3 Representation by means of real 2×2-matrices

Complex numbers were constructed as ordered pairs of real numbers. Thus we can replace

$$z = x + iy \quad \text{in} \quad \mathbb{C} \quad \text{by} \quad \begin{pmatrix} x \\ y \end{pmatrix} \quad \text{in} \quad \mathbb{R}^2,$$

making explicit the real linear structure on $\mathbb{C}$. The product of two complex numbers $c = a + ib$ and z,

$$cz = ax - by + i(bx + ay),$$

can be replaced by / factored as

$$\begin{pmatrix} ax - by \\ bx + ay \end{pmatrix} = \begin{pmatrix} a & -b \\ b & a \end{pmatrix} \begin{pmatrix} x \\ y \end{pmatrix} = \begin{pmatrix} a & -b \\ b & a \end{pmatrix} \begin{pmatrix} x & -y \\ y & x \end{pmatrix} \begin{pmatrix} 1 \\ 0 \end{pmatrix}.$$

One is thus led to consider representing complex numbers by certain real 2×2-matrices in $\mathrm{Mat}(2, \mathbb{R})$: [11]

$$\mathbb{C} \to \mathrm{Mat}(2, \mathbb{R}), \quad a + ib \to \begin{pmatrix} a & -b \\ b & a \end{pmatrix}.$$

10 The linear space $\mathbb{R}^2$ endowed with an indefinite quadratic form $(a, b) \to a^2 - b^2$ is the hyperbolic quadratic space $\mathbb{R}^{1,1}$. The Clifford algebra of $\mathbb{R}^{1,1}$ is $\mathcal{C}\ell_{1,1}$ which has Study numbers as the even subalgebra $\mathcal{C}\ell^+_{1,1}$.

11 In this matrix representation, the complex conjugate of a complex number becomes the transpose of the matrix and the (squared) norm becomes the determinant. The norm is preserved under similarity transformations, but 'transposition = complex conjugation' is only preserved under similarities by orthogonal matrices.

The multiplicative unity 1 and the imaginary unit i in $\mathbb{C}$ are represented by the matrices

$$I = \begin{pmatrix} 1 & 0 \\ 0 & 1 \end{pmatrix} \quad \text{and} \quad J = \begin{pmatrix} 0 & -1 \\ 1 & 0 \end{pmatrix}.$$

However, this is not the only linear representation of $\mathbb{C}$ in $\mathrm{Mat}(2, \mathbb{R})$. A similarity transformation by an invertible matrix U, $\det U \neq 0$, sends the representative of the imaginary unit J to another 'imaginary unit' $J' = UJU^{-1}$ in $\mathrm{Mat}(2, \mathbb{R})$.

Example. Choosing $U = \left(\begin{smallmatrix} 1 & 1 \\ 0 & 1 \end{smallmatrix}\right)$, we find $J' = \left(\begin{smallmatrix} 1 & -2 \\ 1 & -1 \end{smallmatrix}\right)$, and the matrix representation $x + iy \to \left(\begin{smallmatrix} x+y & -2y \\ y & x-y \end{smallmatrix}\right)$. ∎

Geometric Interpretation of $i = \sqrt{-1}$

In the rest of this chapter we shall study introduction of complex numbers by means of the Clifford algebra $C\ell_2$ of the Euclidean plane $\mathbb{R}^2$. This approach gives the imaginary unit $i = \sqrt{-1}$ various geometrical meanings. We will see that i represents

(i) an oriented plane area in $\mathbb{R}^2$,
(ii) a quarter turn of $\mathbb{R}^2$.

The Euclidean plane $\mathbb{R}^2$ has a quadratic form

$$\mathbf{r} = x\mathbf{e}_1 + y\mathbf{e}_2 \to |\mathbf{r}|^2 = x^2 + y^2.$$

We introduce an associative product of vectors such that

$$\mathbf{r}^2 = |\mathbf{r}|^2 \quad \text{or} \quad (x\mathbf{e}_1 + y\mathbf{e}_2)^2 = x^2 + y^2.$$

Using distributivity this results in the multiplication rules

$$\mathbf{e}_1^2 = \mathbf{e}_2^2 = 1, \quad \mathbf{e}_1\mathbf{e}_2 = -\mathbf{e}_2\mathbf{e}_1.$$

The element $\mathbf{e}_1\mathbf{e}_2$ satisfies

$$\boxed{(\mathbf{e}_1\mathbf{e}_2)^2 = -1}$$

and therefore cannot be a scalar or a vector. It is an example of a bivector, the *unit bivector*. Denote it for short by $\mathbf{e}_{12} = \mathbf{e}_1\mathbf{e}_2$.

2.4 $\mathbb{C}$ as the even Clifford algebra Cl_2^+

The Clifford algebra Cl_2 is a 4-dimensional real algebra with a basis $\{1, \mathbf{e}_1, \mathbf{e}_2, \mathbf{e}_{12}\}$. The basis elements obey the multiplication table

	$\mathbf{e}_1$	$\mathbf{e}_2$	$\mathbf{e}_{12}$
$\mathbf{e}_1$	1	$\mathbf{e}_{12}$	$\mathbf{e}_2$
$\mathbf{e}_2$	$-\mathbf{e}_{12}$	1	$-\mathbf{e}_1$
$\mathbf{e}_{12}$	$-\mathbf{e}_2$	$\mathbf{e}_1$	-1

The basis elements span the subspaces consisting of [12]

1	$\mathbb{R}$	scalars
$\mathbf{e}_1, \mathbf{e}_2$	$\mathbb{R}^2$	vectors
$\mathbf{e}_{12}$	$\bigwedge^2 \mathbb{R}^2$	bivectors.

Thus, the Clifford algebra Cl_2 contains copies of $\mathbb{R}$ and $\mathbb{R}^2$, and it is a direct sum of its subspaces of elements of degrees 0,1,2:

$$Cl_2 = \mathbb{R} \oplus \mathbb{R}^2 \oplus \bigwedge^2 \mathbb{R}^2.$$

The Clifford algebra is also a direct sum $Cl_2 = Cl_2^+ \oplus Cl_2^-$ of its

$$\text{even part} \quad Cl_2^+ = \mathbb{R} \oplus \bigwedge^2 \mathbb{R}^2,$$
$$\text{odd part} \quad Cl_2^- = \mathbb{R}^2.$$

The even part is not only a subspace but also a subalgebra. It consists of elements of the form $x + y\mathbf{e}_{12}$ where $x, y \in \mathbb{R}$ and $\mathbf{e}_{12}^2 = -1$. Thus, the even subalgebra $Cl_2^+ = \mathbb{R} \oplus \bigwedge^2 \mathbb{R}^2$ of Cl_2 is isomorphic to $\mathbb{C}$. The unit bivector $\mathbf{e}_{12}$ shares the basic property of the square root i of -1, that is $i^2 = -1$, and we could write $i = \mathbf{e}_{12}$. It should be noted, however, that our imaginary unit $\mathbf{e}_{12}$ anticommutes with $\mathbf{e}_1$ and $\mathbf{e}_2$ and thus $\mathbf{e}_{12}$ *anticommutes with every vector in the* $\mathbf{e}_1\mathbf{e}_2$*-plane*: [13]

$$\mathbf{r}\mathbf{e}_{12} = -\mathbf{e}_{12}\mathbf{r} \quad \text{for} \quad \mathbf{r} = x\mathbf{e}_1 + y\mathbf{e}_2 \quad \text{and} \quad \mathbf{e}_{12} = \mathbf{e}_1\mathbf{e}_2.$$

---

12 In higher dimensions the Clifford algebra Cl_n of $\mathbb{R}^n$ is a sum of its subspaces of k-vectors: $Cl_n = \mathbb{R} \oplus \mathbb{R}^n \oplus \bigwedge^2 \mathbb{R}^n \oplus \ldots \oplus \bigwedge^n \mathbb{R}^n$.

13 In a complex linear space, or complex algebra, where scalars are complex numbers, the imaginary unit commutes with all the vectors, $i\mathbf{r} = \mathbf{r}i$.

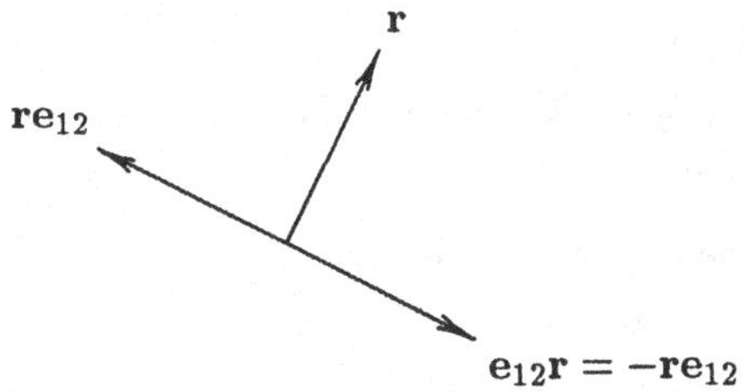

2.5 Imaginary unit = the unit bivector

Multiplying the vector $\mathbf{r} = x\mathbf{e}_1 + y\mathbf{e}_2$ by the unit bivector $\mathbf{e}_{12}$ gives another vector $\mathbf{r}\mathbf{e}_{12} = x\mathbf{e}_2 - y\mathbf{e}_1$ which is perpendicular to $\mathbf{r}$. The function $\mathbf{r} \to \mathbf{r}\mathbf{e}_{12}$ is a left turn, and the effect of two left turns $[\mathbf{e}_{12} \cdot \mathbf{e}_{12}]$ is to reverse direction $[-1]$; or, in a more picturesque way, is a *U*-turn. The statement '$\mathbf{e}_{12}^2 = -1$' is just an arithmetic version of the obvious geometric fact that the sum of two right angles, $90° + 90°$, is a straight angle, $180°$. In the vector plane $\mathbb{R}^2$ the sense of rotation depends on what side the vector $\mathbf{r} = x\mathbf{e}_1 + y\mathbf{e}_2$ is multiplied by $\mathbf{e}_{12}$ so that the rotation $\mathbf{r} \to \mathbf{e}_{12}\mathbf{r} = y\mathbf{e}_1 - x\mathbf{e}_2$ is clockwise and $\mathbf{r} \to \mathbf{r}\mathbf{e}_{12} = -y\mathbf{e}_1 + x\mathbf{e}_2$ is counter-clockwise.

In the complex plane $\mathbb{C} = \mathbb{R} \oplus \bigwedge^2 \mathbb{R}^2$ both the rotations sending $z = x + y\mathbf{e}_{12}$ to $\mathbf{e}_{12}z$ and $z\mathbf{e}_{12}$ are counter-clockwise. Multiplying a complex number $z = x + y\mathbf{e}_{12}$ by the unit bivector $\mathbf{e}_{12}$ results in a left turn, $z\mathbf{e}_{12} = -y + x\mathbf{e}_{12}$, and the effect of two left turns $[\mathbf{e}_{12} \cdot \mathbf{e}_{12}]$ is direction reversal $[-1]$; that is a half-turn in the complex plane $\mathbb{C}$:

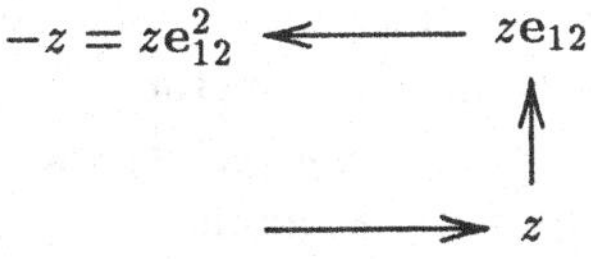

The square root of -1 has two distinct geometric roles in $\mathbb{R}^2$: it is the generator of rotations, $i = \mathbf{e}_1\mathbf{e}_2 \in \mathcal{C}\ell_2^+$, and it represents a unit oriented plane area $\mathbf{e}_1 \wedge \mathbf{e}_2 \in \bigwedge^2 \mathbb{R}^2$. [14]

A complex number $z = x + y\mathbf{e}_{12} \in \mathbb{R} \oplus \bigwedge^2 \mathbb{R}^2$ is a sum of

- a real number $x = \mathrm{Re}(z)$ and
- a bivector $y\mathbf{e}_{12} = \mathbf{e}_{12}\,\mathrm{Im}(z)$.

14 In an n-dimensional vector space $\mathbb{R}^n$ rotations can be represented by multiplications in Clifford algebras $\mathcal{C}\ell_n$, while certain simple elements of the exterior algebra $\bigwedge \mathbb{R}^n = \mathbb{R} \oplus \mathbb{R}^n \oplus \bigwedge^2 \mathbb{R}^n \oplus \cdots \oplus \bigwedge^n \mathbb{R}^n$ represent oriented subspaces of dimensions $0, 1, 2, \ldots, n$.

2.6 Even and odd parts

The Clifford algebra $\mathcal{C}\ell_2$ of $\mathbb{R}^2$ contains both the complex plane $\mathbb{C}$ and the vector plane $\mathbb{R}^2$ so that

$$\begin{array}{ll} \mathbb{R}^2 & \text{is spanned by } \mathbf{e}_1 \text{ and } \mathbf{e}_2, \\ \mathbb{C} & \text{is spanned by } 1 \text{ and } \mathbf{e}_{12}. \end{array}$$

The only common point of the two planes is the zero 0. The two planes are both parts of the same algebra $\mathcal{C}\ell_2$. The vector plane $\mathbb{R}^2$ and the complex field $\mathbb{C}$ are incorporated as separate substructures in the Clifford algebra $\mathcal{C}\ell_2 = \mathcal{C}\ell_2^+ \oplus \mathcal{C}\ell_2^-$ so that the complex plane $\mathbb{C}$ is the *even part* $\mathcal{C}\ell_2^+$ and the vector plane $\mathbb{R}^2$ is the *odd part* $\mathcal{C}\ell_2^-$.

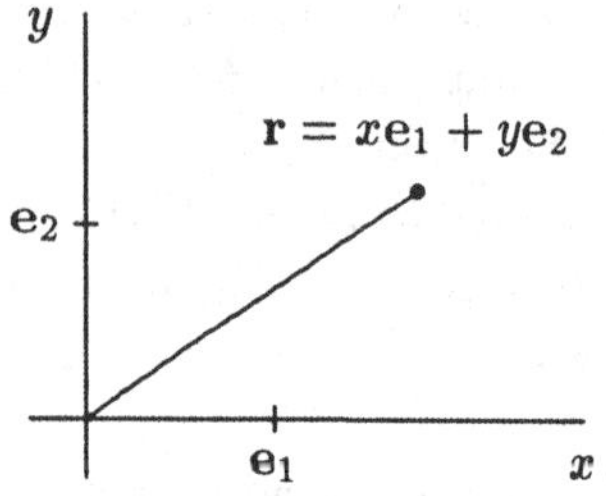

Vector plane $\mathbb{R}^2 = \mathcal{C}\ell_2^-$

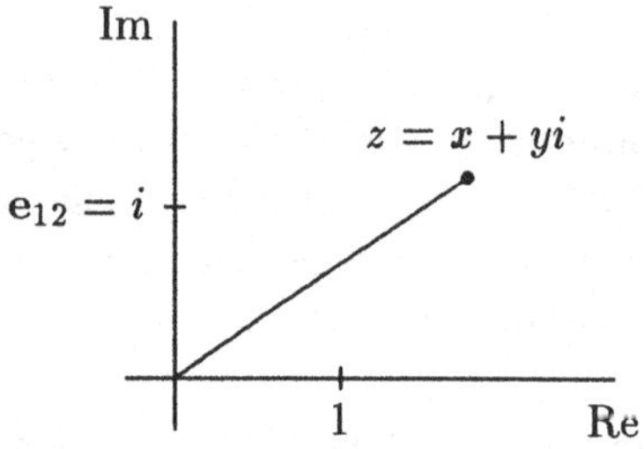

Complex plane $\mathbb{C} = \mathcal{C}\ell_2^+$

The names even and odd mean that the elements are products of an even or odd number of vectors. Parity considerations show that

- complex number times complex number is a complex number,
- vector times complex number is a vector,
- complex number times vector is a vector, and
- vector times vector is a complex number.

The above observations can be expressed by the inclusions

$$\begin{aligned} \mathcal{C}\ell_2^+ \mathcal{C}\ell_2^+ &\subset \mathcal{C}\ell_2^+, \\ \mathcal{C}\ell_2^- \mathcal{C}\ell_2^+ &\subset \mathcal{C}\ell_2^-, \\ \mathcal{C}\ell_2^+ \mathcal{C}\ell_2^- &\subset \mathcal{C}\ell_2^-, \\ \mathcal{C}\ell_2^- \mathcal{C}\ell_2^- &\subset \mathcal{C}\ell_2^+. \end{aligned}$$

By writing $(\mathcal{C}\ell_2)_0 = \mathcal{C}\ell_2^+$ and $(\mathcal{C}\ell_2)_1 = \mathcal{C}\ell_2^-$, this can be further condensed to $(\mathcal{C}\ell_2)_j(\mathcal{C}\ell_2)_k \subset (\mathcal{C}\ell_2)_{j+k}$, where j, k are added modulo 2. These observations are expressed by saying that the Clifford algebra $\mathcal{C}\ell_2$ has an *even-odd grading* or that it is graded over $\mathbb{Z}_2 = \{0, 1\}$. [15]

---

15 We have already met a $\mathbb{Z}_2$-graded algebra, namely the real algebra $\mathbb{C} = \mathbb{R} \oplus i\mathbb{R}$ with even part $\mathbb{R} = \mathrm{Re}(\mathbb{C})$ and odd part $i\mathbb{R} = i\,\mathrm{Im}(\mathbb{C})$.

2.7 Involutions and the norm

The Clifford algebra $\mathcal{C}\ell_2$ has three involutions similar to complex conjugation in $\mathbb{C}$. For an element $u = \langle u\rangle_0 + \langle u\rangle_1 + \langle u\rangle_2 \in \mathcal{C}\ell_2$, $\langle u\rangle_k \in \bigwedge^k \mathbb{R}^2$, we define

$$\begin{array}{ll} \text{grade involution} & \hat{u} = \langle u\rangle_0 - \langle u\rangle_1 + \langle u\rangle_2, \\ \text{reversion} & \tilde{u} = \langle u\rangle_0 + \langle u\rangle_1 - \langle u\rangle_2, \\ \text{Clifford-conjugation} & \bar{u} = \langle u\rangle_0 - \langle u\rangle_1 - \langle u\rangle_2. \end{array}$$

The grade involution is an automorphism, $\widehat{uv} = \hat{u}\hat{v}$, while the reversion and the Clifford-conjugation are anti-automorphisms, $\widetilde{uv} = \tilde{v}\tilde{u}$, $\overline{uv} = \bar{v}\bar{u}$.

For a complex number $z = x + y\mathbf{e}_{12}$ the complex conjugation $z \to \bar{z} = x - y\mathbf{e}_{12}$ is a restriction of the Clifford-conjugation $u \to \bar{u}$ in $\mathcal{C}\ell_2$ and also of the reversion $u \to \tilde{u}$ in $\mathcal{C}\ell_2$. Likewise, the norm $|z| = \sqrt{x^2 + y^2}$ in $\mathbb{C}$, obtained as the square root of $z\bar{z} = x^2 + y^2$, is a restriction of the norm $|u| = \sqrt{\langle u\tilde{u}\rangle_0}$ in $\mathcal{C}\ell_2$.

A complex number is a product of its norm $r = |z|$ and its phase-factor $\cos\varphi + \mathbf{e}_{12}\sin\varphi$, where $x = r\cos\varphi$ and $y = r\sin\varphi$. The expression $z = r(\cos\varphi + \mathbf{e}_{12}\sin\varphi)$ can be abbreviated as $z = r\exp(\mathbf{e}_{12}\varphi)$, and read as '$r$ in phase φ.'

2.8 Vectors multiplied by complex numbers

The product of a vector $\mathbf{r} = x\mathbf{e}_1 + y\mathbf{e}_2$ and a unit complex number $e^{\mathbf{i}\varphi} = \cos\varphi + \mathbf{i}\sin\varphi$, where for short $\mathbf{i} = \mathbf{e}_{12}$, is another vector in the $\mathbf{e}_1\mathbf{e}_2$-plane:

$$\mathbf{r}\cos\varphi + \mathbf{r}\mathbf{i}\sin\varphi = \mathbf{r}e^{\mathbf{i}\varphi}.$$

The vector $\mathbf{r}\mathbf{i} = x\mathbf{e}_2 - y\mathbf{e}_1$ is perpendicular to $\mathbf{r}$ so that a rotation to the left by $\pi/2$ carries $\mathbf{r}$ to $\mathbf{r}\mathbf{i}$.

Since the unit bivector $\mathbf{i}$ anticommutes with every vector $\mathbf{r}$ in the $\mathbf{e}_1\mathbf{e}_2$-plane, the rotated vector could also be expressed as

$$\mathbf{r}\cos\varphi + \mathbf{r}\mathbf{i}\sin\varphi = \mathbf{r}\cos\varphi - \mathbf{i}\mathbf{r}\sin\varphi = e^{-\mathbf{i}\varphi}\mathbf{r}.$$

Furthermore, we have $\cos\varphi + \mathbf{i}\sin\varphi = (\cos\frac{\varphi}{2} + \mathbf{i}\sin\frac{\varphi}{2})^2$ and thus the rotated vector also has the form $s^{-1}\mathbf{r}s$ where $s = e^{\mathbf{i}\varphi/2}$ and $s^{-1} = e^{-\mathbf{i}\varphi/2}$. The rotation of $\mathbf{r}$ to the left by the angle φ will then result in $\mathbf{r}z = z^{-1}\mathbf{r} = s^{-1}\mathbf{r}s$ where $z = e^{\mathbf{i}\varphi}$, $z^{-1} = e^{-\mathbf{i}\varphi}$ and $s^2 = z$. There are two complex numbers s and $-s$ which result in the same rotation $s^{-1}\mathbf{r}s = (-s)^{-1}\mathbf{r}(-s)$. In other words, there are two complex numbers which produce the same final result but via different actions.

$s = e^{\mathbf{i}\varphi/2}$

$-s = e^{-\mathbf{i}(2\pi-\varphi)/2} = e^{\mathbf{i}\varphi/2}e^{-\mathbf{i}\pi}$

$\boxed{e^{\mathbf{i}\pi} = -1}$

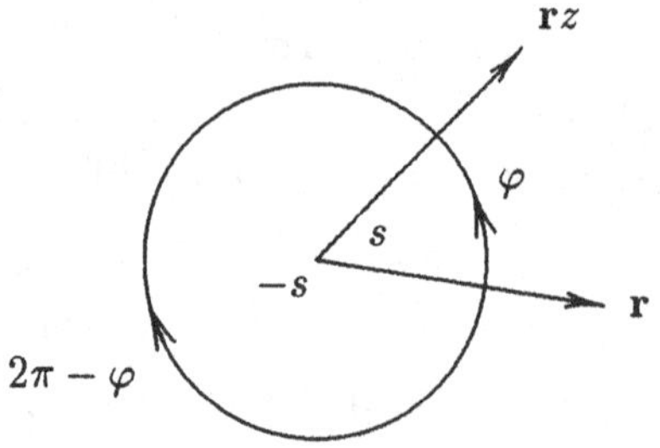

2.9 The group Spin(2)

The unit complex numbers $z \in \mathbb{C}$, $|z| = 1$, form the *unit circle* $S^1 = \{z \in \mathbb{C} \mid |z| = 1\}$, which with multiplication of complex numbers as the product becomes the *unitary group* $U(1) = \{z \in \mathbb{C} \mid z\bar{z} = 1\}$. A counter-clockwise rotation of the complex plane $\mathbb{C}$ by an angle φ can be represented by complex number multiplication:

$$x + iy \to (\cos\varphi + i\sin\varphi)(x + iy), \quad \cos\varphi + i\sin\varphi \in U(1).$$

A counter-clockwise rotation of the vector plane $\mathbb{R}^2$ by an angle φ can be represented by a matrix multiplication:

$$\begin{pmatrix} x \\ y \end{pmatrix} \to \begin{pmatrix} \cos\varphi & -\sin\varphi \\ \sin\varphi & \cos\varphi \end{pmatrix} \begin{pmatrix} x \\ y \end{pmatrix}, \quad \begin{pmatrix} \cos\varphi & -\sin\varphi \\ \sin\varphi & \cos\varphi \end{pmatrix} \in SO(2)$$

where $SO(2) = \{R \in \mathrm{Mat}(2,\mathbb{R}) \mid R^{\mathsf{T}}R = I,\ \det R = 1\}$, the rotation group. The rotation group $SO(2)$ is isomorphic to the unitary group $U(1)$.

Rotations of $\mathbb{R}^2$ can also be represented by Clifford multiplication: [16]

$$x\mathbf{e}_1 + y\mathbf{e}_2 \to (\cos\frac{\varphi}{2} + \mathbf{e}_{12}\sin\frac{\varphi}{2})^{-1}(x\mathbf{e}_1 + y\mathbf{e}_2)(\cos\frac{\varphi}{2} + \mathbf{e}_{12}\sin\frac{\varphi}{2})$$

where $\cos\frac{\varphi}{2} + \mathbf{e}_{12}\sin\frac{\varphi}{2} \in \mathbf{Spin}(2) = \{s \in C\ell_2^+ \mid s\bar{s} = 1\}$, the spin group. The fact that two opposite elements of the spin group $\mathbf{Spin}(2)$ represent the same rotation in $SO(2)$ is expressed by saying that $\mathbf{Spin}(2)$ is a *two-fold* [17] cover of $SO(2)$, and written as $\mathbf{Spin}(2)/\{\pm 1\} \simeq SO(2)$. Although $SO(2)$ and $\mathbf{Spin}(2)$ act differently on $\mathbb{R}^2$, they are isomorphic as abstract groups, that is,

16 We use this particular form to represent the rotation because the expression $x\mathbf{e}_1 + y\mathbf{e}_2 \to (\cos\frac{\varphi}{2}+\mathbf{e}_{12}\sin\frac{\varphi}{2})^{-1}(x\mathbf{e}_1+y\mathbf{e}_2)(\cos\frac{\varphi}{2}+\mathbf{e}_{12}\sin\frac{\varphi}{2})$ can be generalized to higher dimensions. The expression $x\mathbf{e}_1 + y\mathbf{e}_2 \to (x\mathbf{e}_1 + y\mathbf{e}_2)(\cos\varphi + \mathbf{e}_{12}\sin\varphi)$ is not generalizable to higher-dimensional rotations.

17 You are already familiar with two-fold covers: 1. A position of the hands of your watch corresponds to two positions of the Sun. 2. A rotating mirror turns half the angle of the image. 3. Circulating a coin one full turn around another makes the coin turn twice around its center.

$\mathbf{Spin}(2) \simeq SO(2)$. [18]

Exercise 6

History

Imaginary numbers first appeared around 1540, when Tartaglia and Cardano expressed real roots of a cubic equation in terms of conjugate complex numbers. The first one to represent complex numbers by points on a plane was a Norwegian surveyor, Caspar Wessel, in 1798. He posited an imaginary axis perpendicular to the axis of real numbers. This configuration came to be known as the Argand diagram, although Argand's contribution was an interpretation of $i = \sqrt{-1}$ as a rotation by a right angle in the plane. Complex numbers got their name from Gauss, and their formal definition as pairs of real numbers is due to Hamilton in 1833 (first published 1837).

Exercises

1. $(3+4i)^{-1}$, $\sqrt{3+4i}$, $\sqrt[4]{-4}$, $\sqrt[3]{-i}$, $\log(-1+i)$.
2. Let $z_k = e^{i\,2\pi k/n}$, $k = 1, 2, \ldots, n-1$. Compute $(1-z_1)(1-z_2)\cdots(1-z_{n-1})$.
3. An *ordering* of a field $\mathbb{F}$ is an assignment of a subset $P \subset \mathbb{F}$ such that

   (i) $0 \notin P$,
   (ii) for all non-zero $a \in \mathbb{F}$ either $a \in P$ or $-a \in P$, but not both,
   (iii) $a + b \in P$ and $ab \in P$ for all $a, b \in P$.

   It is customary to call P the set of *positive* numbers, and the set $-P = \{-a \mid a \in P\}$ the set of *negative* numbers. The statement $a - b \in P$ is also written $a > b$ (and $a - b \in P \cup \{0\}$ is written $a \geq b$). Show that the field $\mathbb{C}$ cannot be ordered.
4. Two automorphisms α, β of an algebra are similar if there exists an intertwining automorphism γ such that $\alpha\gamma = \gamma\beta$. The identity automorphism is similar only to itself.

a) Show that the two involutions of the real algebra $\mathbb{C}$ are dissimilar, and that the two involutions of the real algebra ${}^2\mathbb{R}$ are dissimilar.

b) Show that the two involutions $\alpha(\lambda, \mu) = (\mu, \lambda)$ and $\beta(\lambda, \mu) = (\bar{\mu}, \bar{\lambda})$ are similar involutions of the real or complex algebra ${}^2\mathbb{C}$ [that is, find an intertwining automorphism γ of ${}^2\mathbb{C}$ such that $\alpha\gamma = \gamma\beta$].

5. A rotation is called rational if it sends a vector with rational coordinates to

18 Both $SO(2)$ and $\mathbf{Spin}(2)$ are homeomorphic to S^1.

another vector with rational coordinates. Determine all the rational rotations of $\mathbb{R}^2$. Hint: $R \in SO(2) \setminus \{-I\}$ can be written in the form $R = (I + A)(I - A)^{-1}$ where $A^\top = -A$.

6. Write $\tilde{u} = \langle u\rangle_0 + \langle u\rangle_1 - \langle u\rangle_2$ for $u = \langle u\rangle_0 + \langle u\rangle_1 + \langle u\rangle_2 \in C\ell_2$, $\langle u\rangle_k \in \bigwedge^k \mathbb{R}^2$. Let $\mathbf{Pin}(2) = \{u \in C\ell_2 \mid \tilde{u}u = 1\}$, $\mathbb{R}^2 \to \mathbb{R}^2$, $\mathbf{x} \to R(\mathbf{x}) = u\mathbf{x}u^{-1}$, and $O(2) = \{R \in \mathrm{Mat}(2,\mathbb{R}) \mid R^\top R = I\}$. Show that $\mathbf{Pin}(2)/\{\pm 1\} \simeq O(2)$ and $\mathbf{Pin}(2) \simeq O(2)$.
7. Show that a 2-dimensional real algebra with unity 1 is both commutative and associative. Hint: First show that there is a basis $\{1, a\}$ such that $a^2 = \alpha 1$, $\alpha \in \mathbb{R}$.
8. Show that a 2-dimensional real algebra with unity 1 and no zero-divisors [$ab = 0$ implies $a = 0$ or $b = 0$] is isomorphic to $\mathbb{C}$.

Solutions

1. $\frac{1}{5}(3 - 4i)$, $\pm(2 + i)$, $\pm 1 \pm i$, $\sqrt[3]{-i} = \{i, \pm\frac{\sqrt{3}}{2} - i\frac{1}{2}\}$, $\log(-1 + i) = \frac{1}{2}\log 2 + i\frac{3\pi}{4} + i\,2\pi k$.
2. Note that the roots of $x^n - 1 = 0$ are $z_k = e^{i\,2\pi k/n}$, $k = 0, 1, \ldots, n - 1$. Therefore $(x - z_0)(x - z_1)(x - z_2)\cdots(x - z_{n-1}) = x^n - 1$. Define $f(x) = (x - z_1)(x - z_2)\cdots(x - z_{n-1})$ which equals
$$f(x) = \frac{x^n - 1}{x - 1} \quad \text{for} \quad x \neq 1$$
and $f(x) = x^{n-1} + \ldots + x + 1$ in general. Compute $f(1) = n$.
3. In an ordered field non-zero numbers have positive squares, and the sum of such squares is positive, and therefore non-zero. The equality $i^2 + 1 = 0$ in $\mathbb{C}$ can also be written as $i^2 + 1^2 = 0$, which excludes the inequality $i^2 + 1^2 > 0$. Consequently, it is impossible to order the field $\mathbb{C}$.
4. b) Choose $\gamma(\lambda, \mu) = (\bar{\lambda}, \mu)$ or $\gamma(\lambda, \mu) = (\lambda, \bar{\mu})$ to find $\alpha\gamma = \gamma\beta$.

Bibliography

L.V. Ahlfors: *Complex Analysis*. McGraw-Hill, New York, 1953.

E. Cartan (exposé d'après l'article allemand de E. Study): Nombres complexes; pp. 329-468 in J. Molk (red.): *Encyclopédie des sciences mathématiques*, Tome **I**, vol. **1**, Fasc. **4**, art. **I5**, 1908. Reprinted in E. Cartan: *Œuvres complètes*, Partie II. Gauthier-Villars, Paris, 1953, pp. 107-246.

R.V. Churchill, J.W. Brown, R.F. Verhey: *Complex Variables and Applications*. McGraw-Hill, New York, 1984.

H.-D. Ebbinghaus et al. (eds.): *Numbers*. Springer, New York, 1991.

B. L. van der Waerden: *A History of Algebra*. Springer, Berlin, 1985.

3

Bivectors and the Exterior Algebra

There are other kinds of directed quantities besides vectors, most notably bivectors. For instance, a moment of a force, angular velocity of a rotating body, and magnetic induction can be described with bivectors. In three dimensions bivectors are dual to vectors, and their use can be circumvented. Scalars, vectors, bivectors and the volume element span the exterior algebra $\bigwedge\mathbb{R}^3$, which provides a multivector structure for the Clifford algebra $\mathcal{C}\ell_3$ of the Euclidean space $\mathbb{R}^3$.

3.1 Bivectors as directed plane segments

In three dimensions bivectors are oriented plane segments, which have a direction and a magnitude, the area of the plane segment. Two bivectors have the same direction if they are on parallel planes (the same attitude) and are similarly oriented (the same sense of rotation).

Vector (directed line segment)

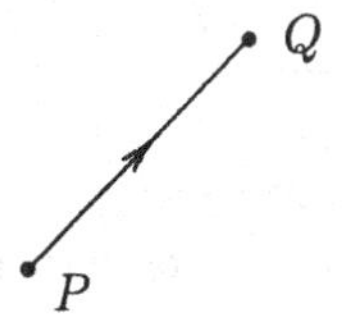

1. magnitude (length of PQ)
2. direction
   - attitude (line PQ)
   - orientation (toward the point Q)

Bivector (directed plane segment)

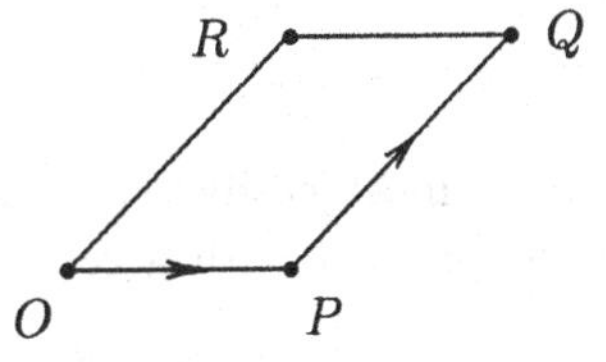

1. magnitude (area of $OPQR$)
2. direction
   - attitude (plane OPQ)
   - orientation (sense of rotation)

Bivectors are denoted by boldface capital letters $\mathbf{A}, \mathbf{B}$, etc. [1] The area or norm of a bivector $\mathbf{A}$ is denoted by $|\mathbf{A}|$. Two bivectors $\mathbf{A}$ and $\mathbf{B}$ in parallel planes have the same attitude, and we write $\mathbf{A} \parallel \mathbf{B}$. Parallel bivectors $\mathbf{A}$ and $\mathbf{B}$ can be regarded as directed angles turning either the same way, $\mathbf{A} \uparrow\uparrow \mathbf{B}$, or the opposite way, $\mathbf{A} \uparrow\downarrow \mathbf{B}$. If two plane segments have the same area and the same direction (= parallel planes with the same sense of rotation), then the bivectors are equal:

$$\mathbf{A} = \mathbf{B} \quad \Longleftrightarrow \quad |\mathbf{A}| = |\mathbf{B}| \quad \text{and} \quad \mathbf{A} \uparrow\uparrow \mathbf{B}$$

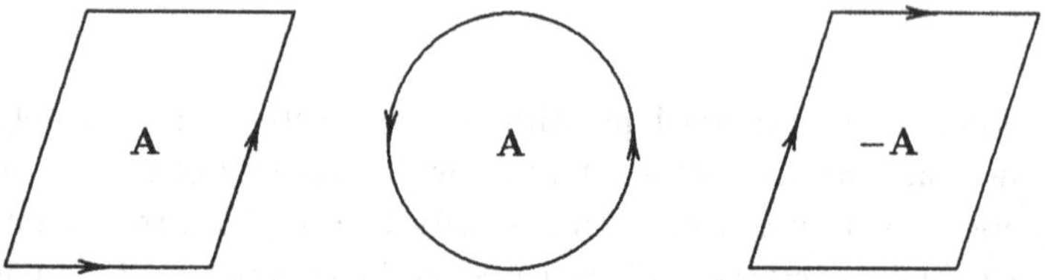

A bivector $\mathbf{A}$ and its *opposite* $-\mathbf{A}$ are of equal area and parallel, but have opposite orientations. A *unit bivector* $\mathbf{A}$ has area one, $|\mathbf{A}| = 1$.

The shape of the area is irrelevant.

Representing a bivector as an oriented parallelogram suggests that a bivector can be thought of as a geometrical product of vectors along its sides. With this in mind we introduce the *exterior product* $\mathbf{a} \wedge \mathbf{b}$ of two vectors $\mathbf{a}$ and $\mathbf{b}$ as the bivector obtained by sweeping $\mathbf{b}$ along $\mathbf{a}$.

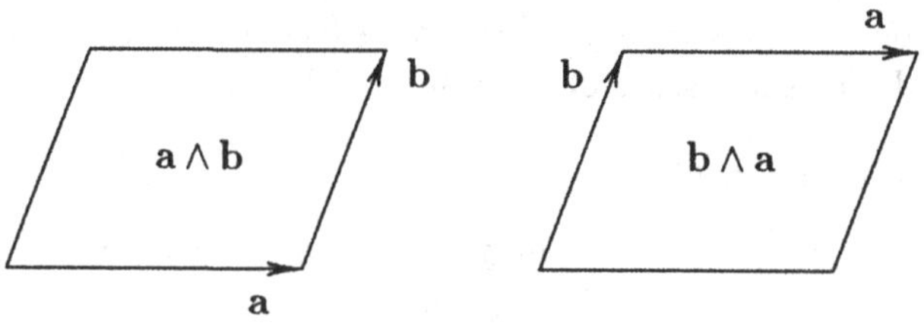

The bivectors $\mathbf{a} \wedge \mathbf{b}$ and $\mathbf{b} \wedge \mathbf{a}$ have the same area and the same attitude but opposite senses of rotations. This can be simply expressed by writing

$$\mathbf{a} \wedge \mathbf{b} = -\mathbf{b} \wedge \mathbf{a}.$$

3.2 Addition of bivectors

The geometric interpretation of bivector addition is most easily seen when the bivectors are expressed in terms of the exterior product with a common

[1] In handwriting, bivectors can be distinguished by an angle on top of the letter, $\overset{\angle}{A}$, $\overset{\angle}{B}$.

vector factor. In three dimensions this is always possible because any two planes will either be parallel or intersect along a common line. [2] Thus let $\mathbf{A} = \mathbf{a} \wedge \mathbf{c}$ and $\mathbf{B} = \mathbf{b} \wedge \mathbf{c}$; then the bivector $\mathbf{A} + \mathbf{B}$ is defined so that $\mathbf{A}+\mathbf{B} = \mathbf{a}\wedge\mathbf{c}+\mathbf{b}\wedge\mathbf{c} = (\mathbf{a}+\mathbf{b})\wedge\mathbf{c}$. The geometric significance of this can be depicted as follows:

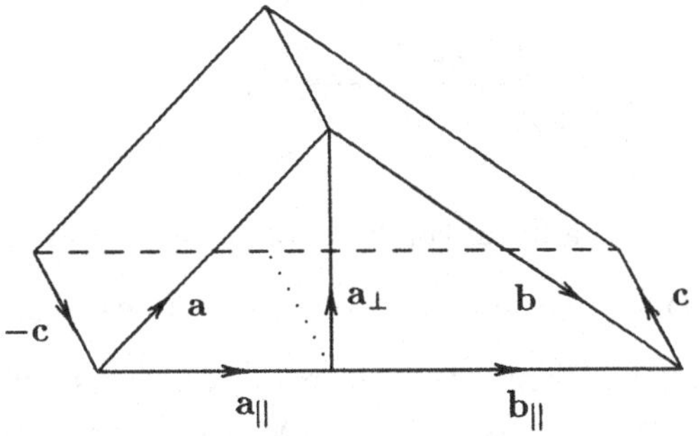

By decomposing the vectors $\mathbf{a}$ and $\mathbf{b}$ into components parallel and perpendicular to $\mathbf{a}+\mathbf{b}$, [3] so that

$$\mathbf{a} = \mathbf{a}_{\|} + \mathbf{a}_{\perp} \quad \text{and} \quad \mathbf{b} = \mathbf{b}_{\|} + \mathbf{b}_{\perp}$$

where $\mathbf{b}_{\perp} = -\mathbf{a}_{\perp}$, we are able to reduce the general addition of bivectors in three dimensions to the addition of coplanar bivectors. This is evident in the equality

$$\mathbf{a}\wedge\mathbf{c}+\mathbf{b}\wedge\mathbf{c} = (\mathbf{a}+\mathbf{b})\wedge\mathbf{c} = (\mathbf{a}_{\|}+\mathbf{b}_{\|})\wedge\mathbf{c} = \mathbf{a}_{\|}\wedge\mathbf{c}+\mathbf{b}_{\|}\wedge\mathbf{c}.$$

3.3 Basis of the linear space of bivectors

Bivectors can be added and multiplied by scalars. This way the set of bivectors becomes a linear space, denoted by $\bigwedge^2 \mathbb{R}^3$. A basis for the linear space $\bigwedge^2 \mathbb{R}^3$ can be constructed by means of a basis $\{\mathbf{e}_1, \mathbf{e}_2, \mathbf{e}_3\}$ of the linear space $\mathbb{R}^3$. The oriented plane segments of the coordinate planes, obtained by taking the exterior products

$$\mathbf{e}_1\wedge\mathbf{e}_2, \ \mathbf{e}_1\wedge\mathbf{e}_3, \ \mathbf{e}_2\wedge\mathbf{e}_3,$$

2 The two bivectors are first translated in the affine space $\mathbb{R}^3$ so that they induce opposite orientations to their common edge, that is, the terminal side of $\mathbf{A} = \mathbf{a}\wedge\mathbf{c}$ is opposite to the initial side of $\mathbf{B} = (-\mathbf{c})\wedge\mathbf{b}$.

3 A depiction of addition of bivectors does not require a metric, or perpendicular components. It is sufficient that one component of both $\mathbf{a}$ and $\mathbf{b}$ is parallel to $\mathbf{a}+\mathbf{b}$, so that the two components sum up to $\mathbf{a}+\mathbf{b}$, while the other component can be any non-parallel component.

form a basis for the linear space of bivectors $\bigwedge^2 \mathbb{R}^3$.

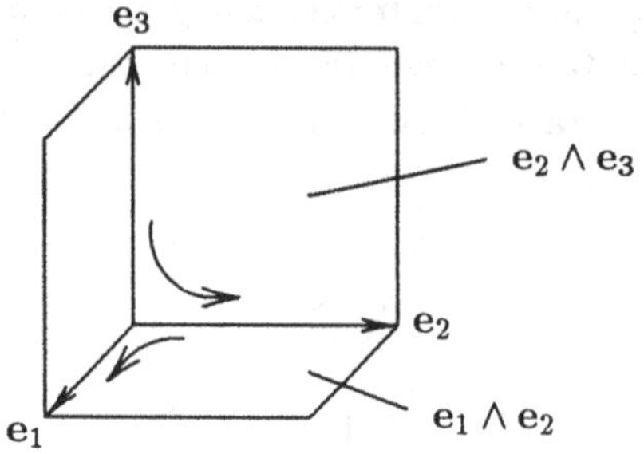

An arbitrary bivector is a linear combination of the basis elements,

$$\mathbf{B} = B_{12}\mathbf{e}_1 \wedge \mathbf{e}_2 + B_{13}\mathbf{e}_1 \wedge \mathbf{e}_3 + B_{23}\mathbf{e}_2 \wedge \mathbf{e}_3,$$

and such linear combinations form the space of bivectors $\bigwedge^2 \mathbb{R}^3$. [4] The construction of bivectors calls only for a linear structure, and no metric is needed.

The scalar product on a Euclidean space $\mathbb{R}^3$ extends to a symmetric bilinear product on the space of bivectors $\bigwedge^2 \mathbb{R}^3$,

$$\langle \mathbf{x}_1 \wedge \mathbf{x}_2, \mathbf{y}_1 \wedge \mathbf{y}_2 \rangle = \begin{vmatrix} \mathbf{x}_1 \cdot \mathbf{y}_1 & \mathbf{x}_1 \cdot \mathbf{y}_2 \\ \mathbf{x}_2 \cdot \mathbf{y}_1 & \mathbf{x}_2 \cdot \mathbf{y}_2 \end{vmatrix}.$$

In particular, $\langle \mathbf{a} \wedge \mathbf{b}, \mathbf{a} \wedge \mathbf{b} \rangle = |\mathbf{a}|^2|\mathbf{b}|^2 - (\mathbf{a} \cdot \mathbf{b})^2$. The norm or area of $\mathbf{B} = B_{12}\mathbf{e}_1 \wedge \mathbf{e}_2 + B_{13}\mathbf{e}_1 \wedge \mathbf{e}_3 + B_{23}\mathbf{e}_2 \wedge \mathbf{e}_3$ is seen to be

$$|\mathbf{B}| = \sqrt{\langle \mathbf{B}, \mathbf{B} \rangle} = \sqrt{B_{12}^2 + B_{13}^2 + B_{23}^2}.$$

3.4 The oriented volume element

The exterior product $\mathbf{a} \wedge \mathbf{b} \wedge \mathbf{c}$ of three vectors $\mathbf{a} = a_1\mathbf{e}_1 + a_2\mathbf{e}_2 + a_3\mathbf{e}_3$, $\mathbf{b} = b_1\mathbf{e}_1 + b_2\mathbf{e}_2 + b_3\mathbf{e}_3$ and $\mathbf{c} = c_1\mathbf{e}_1 + c_2\mathbf{e}_2 + c_3\mathbf{e}_3$ represents the oriented volume of the parallelepiped with edges $\mathbf{a}, \mathbf{b}, \mathbf{c}$:

$$\mathbf{a} \wedge \mathbf{b} \wedge \mathbf{c} = \begin{vmatrix} a_1 & a_2 & a_3 \\ b_1 & b_2 & b_3 \\ c_1 & c_2 & c_3 \end{vmatrix} \mathbf{e}_1 \wedge \mathbf{e}_2 \wedge \mathbf{e}_3.$$

It is an element of the 1-dimensional linear space of 3-vectors $\bigwedge^3 \mathbb{R}^3$ with basis $\mathbf{e}_1 \wedge \mathbf{e}_2 \wedge \mathbf{e}_3$. The exterior product is associative,

$$(\mathbf{a} \wedge \mathbf{b}) \wedge \mathbf{c} = \mathbf{a} \wedge (\mathbf{b} \wedge \mathbf{c}),$$

4 In three dimensions all bivectors are simple, that is, they are exterior products of two vectors, $\mathbf{B} = \mathbf{x} \wedge \mathbf{y}$ for some $\mathbf{x}, \mathbf{y} \in \mathbb{R}^3$. This is no longer true in four dimensions; for instance $\mathbf{e}_1 \wedge \mathbf{e}_2 + \mathbf{e}_3 \wedge \mathbf{e}_4$ is not simple.

and antisymmetric,

$$\begin{aligned}\mathbf{a}\wedge\mathbf{b}\wedge\mathbf{c} &= \mathbf{b}\wedge\mathbf{c}\wedge\mathbf{a} = \mathbf{c}\wedge\mathbf{a}\wedge\mathbf{b}\\ &= -\mathbf{c}\wedge\mathbf{b}\wedge\mathbf{a} = -\mathbf{a}\wedge\mathbf{c}\wedge\mathbf{b} = -\mathbf{b}\wedge\mathbf{a}\wedge\mathbf{c}\end{aligned}$$

for $\mathbf{a},\mathbf{b},\mathbf{c}\in\mathbb{R}^3$.

The exterior product of the orthogonal unit vectors $\mathbf{e}_1,\mathbf{e}_2,\mathbf{e}_3\in\mathbb{R}^3$ is the unit oriented volume element $\mathbf{e}_1\wedge\mathbf{e}_2\wedge\mathbf{e}_3\in\bigwedge^3\mathbb{R}^3$. The norm or volume $|\mathbf{V}|$ of a 3-vector [5]

$$\mathbf{V} = V\mathbf{e}_1\wedge\mathbf{e}_2\wedge\mathbf{e}_3$$

is $|\mathbf{V}| = |V|$, that is, $|V\mathbf{e}_1\wedge\mathbf{e}_2\wedge\mathbf{e}_3| = V$ for $V\geq 0$ and $|V\mathbf{e}_1\wedge\mathbf{e}_2\wedge\mathbf{e}_3| = -V$ for $V<0$.

More formally, the scalar product on $\mathbb{R}^3$ extends to a symmetric bilinear product on $\bigwedge^3\mathbb{R}^3$ by

$$<\mathbf{x}_1\wedge\mathbf{x}_2\wedge\mathbf{x}_3,\mathbf{y}_1\wedge\mathbf{y}_2\wedge\mathbf{y}_3> = \begin{vmatrix}\mathbf{x}_1\cdot\mathbf{y}_1 & \mathbf{x}_1\cdot\mathbf{y}_2 & \mathbf{x}_1\cdot\mathbf{y}_3\\ \mathbf{x}_2\cdot\mathbf{y}_1 & \mathbf{x}_2\cdot\mathbf{y}_2 & \mathbf{x}_2\cdot\mathbf{y}_3\\ \mathbf{x}_3\cdot\mathbf{y}_1 & \mathbf{x}_3\cdot\mathbf{y}_2 & \mathbf{x}_3\cdot\mathbf{y}_3\end{vmatrix}$$

giving the norm as $|\mathbf{V}| = \sqrt{<\mathbf{V},\mathbf{V}>}$.

3.5 The cross product

Let $\mathbf{a} = a_1\mathbf{e}_1 + a_2\mathbf{e}_2 + a_3\mathbf{e}_3$ and $\mathbf{b} = b_1\mathbf{e}_1 + b_2\mathbf{e}_2 + b_3\mathbf{e}_3$. The bivector

$$\mathbf{a}\wedge\mathbf{b} = (a_2b_3 - a_3b_2)\mathbf{e}_2\wedge\mathbf{e}_3 + (a_3b_1 - a_1b_3)\mathbf{e}_3\wedge\mathbf{e}_1 + (a_1b_2 - a_2b_1)\mathbf{e}_1\wedge\mathbf{e}_2$$

can be expressed as a 'determinant'

$$\mathbf{a}\wedge\mathbf{b} = \begin{vmatrix}\mathbf{e}_2\wedge\mathbf{e}_3 & \mathbf{e}_3\wedge\mathbf{e}_1 & \mathbf{e}_1\wedge\mathbf{e}_2\\ a_1 & a_2 & a_3\\ b_1 & b_2 & b_3\end{vmatrix}.$$

It is customary to introduce a vector with the same coordinates. Thus, we define the *cross product*

$$\mathbf{a}\times\mathbf{b} = (a_2b_3 - a_3b_2)\mathbf{e}_1 + (a_3b_1 - a_1b_3)\mathbf{e}_2 + (a_1b_2 - a_2b_1)\mathbf{e}_3$$

of $\mathbf{a}$ and $\mathbf{b}$. The cross product can also be represented by a 'determinant'

$$\mathbf{a}\times\mathbf{b} = \begin{vmatrix}\mathbf{e}_1 & \mathbf{e}_2 & \mathbf{e}_3\\ a_1 & a_2 & a_3\\ b_1 & b_2 & b_3\end{vmatrix}.$$

5 V is a real number, positive or negative, while $\mathbf{V}$ is a 3-vector. The usual volume is $|V|$.

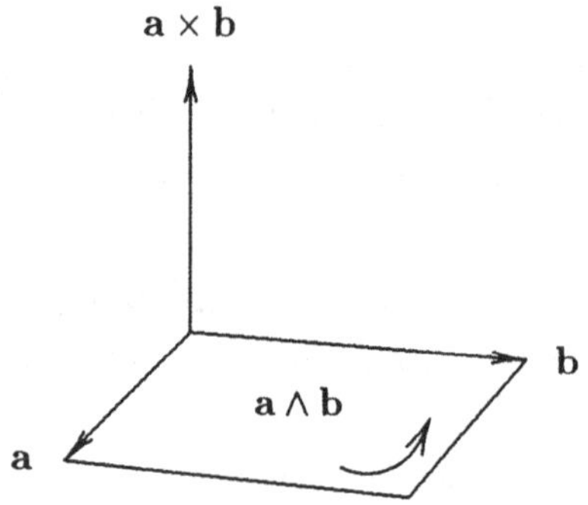

The direction of $\mathbf{a}\times\mathbf{b}$ is perpendicular to the plane of $\mathbf{a}\wedge\mathbf{b}$ and the length/norm of $\mathbf{a}\times\mathbf{b}$ equals the area/norm of $\mathbf{a}\wedge\mathbf{b}$,

$$|\mathbf{a}\times\mathbf{b}| = |\mathbf{a}\wedge\mathbf{b}| = |\mathbf{a}||\mathbf{b}|\sin\varphi$$

where φ, $0 \leq \varphi \leq 180°$, is the angle between $\mathbf{a}$ and $\mathbf{b}$.

In spite of the resemblance between the determinant expressions for the exterior product $\mathbf{a}\wedge\mathbf{b}$ and the cross product $\mathbf{a}\times\mathbf{b}$ there is a difference: the exterior product does not require a metric while the cross product requires or induces a metric. The metric gets involved in positioning the vector $\mathbf{a}\times\mathbf{b}$ perpendicular to the bivector $\mathbf{a}\wedge\mathbf{b}$.

3.6 The Hodge dual

Since the vector space $\mathbb{R}^3$ and the bivector space $\bigwedge^2\mathbb{R}^3$ are both of dimension 3, they are linearly isomorphic. We can use the metric on the vector space $\mathbb{R}^3$ to set up a standard isomorphism between the two linear spaces, the Hodge dual sending a vector $\mathbf{a}\in\mathbb{R}^3$ to a bivector $\star\mathbf{a}\in\bigwedge^2\mathbb{R}^3$, defined by

$$\mathbf{b}\wedge\star\mathbf{a} = (\mathbf{b}\cdot\mathbf{a})\mathbf{e}_1\wedge\mathbf{e}_2\wedge\mathbf{e}_3 \quad \text{for all} \quad \mathbf{b}\in\mathbb{R}^3.$$

The Hodge dual depends not only on the metric but also on the choice of orientation – it is customary to use a right-handed and orthonormal basis $\{\mathbf{e}_1,\mathbf{e}_2,\mathbf{e}_3\}$.

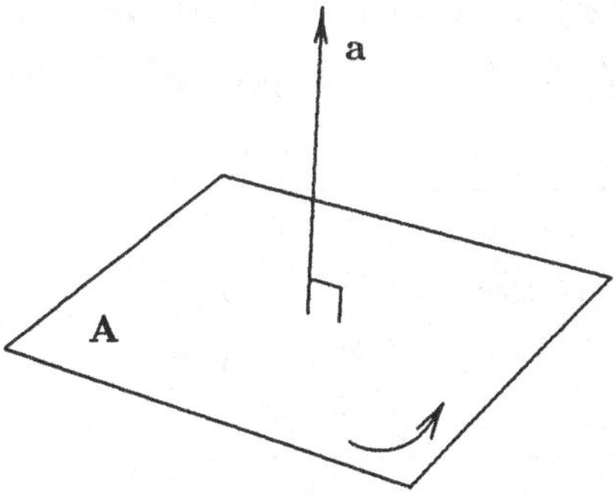

Vector $\mathbf{a}$ and its dual bivector $\mathbf{A} = \mathbf{a}\mathbf{e}_{123}$

Thus, we have assigned to each vector

$$\mathbf{a} = a_1\mathbf{e}_1 + a_2\mathbf{e}_2 + a_3\mathbf{e}_3 \in \mathbb{R}^3$$

a bivector

$$\mathbf{A} = \star\mathbf{a} = a_1\mathbf{e}_2 \wedge \mathbf{e}_3 + a_2\mathbf{e}_3 \wedge \mathbf{e}_1 + a_3\mathbf{e}_1 \wedge \mathbf{e}_2 \in \bigwedge^2 \mathbb{R}^3.$$

Using the induced metric on the bivector space $\bigwedge^2 \mathbb{R}^3$ we can extend the Hodge dual to a mapping sending a bivector $\mathbf{A} \in \bigwedge^2 \mathbb{R}^3$ to a vector $\star\mathbf{A} \in \mathbb{R}^3$, defined by

$$\mathbf{B} \wedge \star\mathbf{A} = \langle \mathbf{B}, \mathbf{A} \rangle \mathbf{e}_1 \wedge \mathbf{e}_2 \wedge \mathbf{e}_3 \quad \text{for all} \quad \mathbf{B} \in \bigwedge^2 \mathbb{R}^3.$$

Using duality, the relation between the cross product and the exterior product can be written as [6]

$$\mathbf{a} \wedge \mathbf{b} = \star(\mathbf{a} \times \mathbf{b}),$$
$$\mathbf{a} \times \mathbf{b} = \star(\mathbf{a} \wedge \mathbf{b}).$$

6 In terms of the Clifford algebra $C\ell_3$ the relation between the exterior product and the cross product can be written as

$$\mathbf{a} \wedge \mathbf{b} = (\mathbf{a} \times \mathbf{b})\mathbf{e}_{123},$$
$$\mathbf{a} \times \mathbf{b} = -(\mathbf{a} \wedge \mathbf{b})\mathbf{e}_{123}.$$

The metric gets involved in multiplying by $\mathbf{e}_{123} = \mathbf{e}_1\mathbf{e}_2\mathbf{e}_3$. Using the Clifford algebra $C\ell_3$ the Hodge dual can be computed as $\star u = \tilde{u}\mathbf{e}_{123}$. This gives rise to the *Clifford dual* defined as $u\mathbf{e}_{123}$ for $u \in C\ell_3$. Later we will see that in actual computations the Clifford dual is more convenient than the Hodge dual (although in three dimensions the Hodge dual happens to be symmetric/involutory).

3.7 The exterior algebra and the Clifford algebra

The exterior algebra $\bigwedge \mathbb{R}^3$ of the linear space $\mathbb{R}^3$ is a direct sum of the

subspaces of		with basis
scalars	$\mathbb{R}$	1
vectors	$\mathbb{R}^3$	$\mathbf{e}_1, \mathbf{e}_2, \mathbf{e}_3$
bivectors	$\bigwedge^2 \mathbb{R}^3$	$\mathbf{e}_1 \wedge \mathbf{e}_2, \mathbf{e}_1 \wedge \mathbf{e}_3, \mathbf{e}_2 \wedge \mathbf{e}_3$
volume elements	$\bigwedge^3 \mathbb{R}^3$	$\mathbf{e}_1 \wedge \mathbf{e}_2 \wedge \mathbf{e}_3$

We also write $\mathbb{R} = \bigwedge^0 \mathbb{R}^3$ and $\mathbb{R}^3 = \bigwedge^1 \mathbb{R}^3$. Thus, $\bigwedge \mathbb{R}^3$ is a direct sum of its subspaces of homogeneous degrees $0, 1, 2, 3$:

$$\bigwedge \mathbb{R}^3 = \mathbb{R} \oplus \mathbb{R}^3 \oplus \bigwedge^2 \mathbb{R}^3 \oplus \bigwedge^3 \mathbb{R}^3.$$

The dimensions of $\mathbb{R}$, $\mathbb{R}^3$, $\bigwedge^2 \mathbb{R}^3$, $\bigwedge^3 \mathbb{R}^3$ and $\bigwedge \mathbb{R}^3$ are $1, 3, 3, 1$ and $2^3 = 8$, respectively.

The exterior algebra $\bigwedge \mathbb{R}^3$ is an associative algebra with unity 1 satisfying

$$\boxed{\begin{array}{l} \mathbf{e}_i \wedge \mathbf{e}_j = -\mathbf{e}_j \wedge \mathbf{e}_i \quad \text{for} \quad i \neq j \\ \mathbf{e}_i \wedge \mathbf{e}_i = 0 \end{array}}$$

for a basis $\{\mathbf{e}_1, \mathbf{e}_2, \mathbf{e}_3\}$ of the linear space $\mathbb{R}^3$. The exterior product of two homogeneous elements satisfies

$$\mathbf{a} \wedge \mathbf{b} \in \bigwedge^{i+j} \mathbb{R}^3 \quad \text{for} \quad \mathbf{a} \in \bigwedge^{i} \mathbb{R}^3, \ \mathbf{b} \in \bigwedge^{j} \mathbb{R}^3.$$

The product of two elements u and v in the Clifford algebra $\mathcal{C}\ell_3$ of the Euclidean space $\mathbb{R}^3$ is denoted by juxtaposition, uv, to distinguish it from the exterior product $u \wedge v$. An orthonormal basis $\{\mathbf{e}_1, \mathbf{e}_2, \mathbf{e}_3\}$ of the Euclidean space $\mathbb{R}^3 \subset \mathcal{C}\ell_3$ satisfies [7]

$$\boxed{\begin{array}{l} \mathbf{e}_i \mathbf{e}_j = -\mathbf{e}_j \mathbf{e}_i \quad \text{for} \quad i \neq j \\ \mathbf{e}_i \mathbf{e}_i = 1 \end{array}}$$

7 These rules were invented by W.K. Clifford in 1882. In an earlier paper Clifford 1878 had considered an associative algebra of dimension 8 with the rules $\mathbf{e}_i \mathbf{e}_i = -1$ for $i = 1, 2, 3$.

and generates a basis of $\mathcal{C}\ell_3$, corresponding to a basis of $\bigwedge \mathbb{R}^3$,

$\mathcal{C}\ell_3$	$\bigwedge \mathbb{R}^3$
1	1
$\mathbf{e}_1,\ \mathbf{e}_2,\ \mathbf{e}_3$	$\mathbf{e}_1,\ \mathbf{e}_2,\ \mathbf{e}_3$
$\mathbf{e}_1\mathbf{e}_2,\ \mathbf{e}_1\mathbf{e}_3,\ \mathbf{e}_2\mathbf{e}_3$	$\mathbf{e}_1 \wedge \mathbf{e}_2,\ \mathbf{e}_1 \wedge \mathbf{e}_3,\ \mathbf{e}_2 \wedge \mathbf{e}_3$
$\mathbf{e}_1\mathbf{e}_2\mathbf{e}_3$	$\mathbf{e}_1 \wedge \mathbf{e}_2 \wedge \mathbf{e}_3$

The above correspondences induce an identification of the linear spaces $\mathcal{C}\ell_3$ and $\bigwedge \mathbb{R}^3$, and we shall write

$$\mathcal{C}\ell_3 = \mathbb{R} \oplus \mathbb{R}^3 \oplus \bigwedge^2 \mathbb{R}^3 \oplus \bigwedge^3 \mathbb{R}^3.$$

This decomposition introduces a *multivector structure* into the Clifford algebra $\mathcal{C}\ell_3$. The multivector structure is unique, that is, an arbitrary element $u \in \mathcal{C}\ell_3$ can be uniquely decomposed into a sum of k-vectors, the k-vector parts $\langle u \rangle_k$ of u,

$$u = \langle u \rangle_0 + \langle u \rangle_1 + \langle u \rangle_2 + \langle u \rangle_3 \quad \text{where} \quad \langle u \rangle_k \in \bigwedge^k \mathbb{R}^3.$$

3.8 The Clifford product of two vectors

A new kind of product called the *Clifford product* of vectors $\mathbf{a}$ and $\mathbf{b}$ is obtained by adding the scalar $\mathbf{a} \cdot \mathbf{b}$ and the bivector $\mathbf{a} \wedge \mathbf{b}$:

$$\mathbf{ab} = \mathbf{a} \cdot \mathbf{b} + \mathbf{a} \wedge \mathbf{b}.$$

The commutative rule $\mathbf{a} \cdot \mathbf{b} = \mathbf{b} \cdot \mathbf{a}$ together with the anticommutative rule $\mathbf{a} \wedge \mathbf{b} = -\mathbf{b} \wedge \mathbf{a}$ implies a relation between $\mathbf{ab}$ and $\mathbf{ba}$. Thus,

$$\mathbf{ba} = \mathbf{a} \cdot \mathbf{b} - \mathbf{a} \wedge \mathbf{b}.$$

Two vectors $\mathbf{a}$ and $\mathbf{b}$ are parallel, $\mathbf{a} \parallel \mathbf{b}$, when their product is commutative, $\mathbf{ab} = \mathbf{ba}$, and perpendicular, $\mathbf{a} \perp \mathbf{b}$, when their product is anticommutative, $\mathbf{ab} = -\mathbf{ba}$.

Note that if $\mathbf{a}$ is decomposed into components parallel, $\mathbf{a}_{\parallel}$, and perpendicular, $\mathbf{a}_{\perp}$, to $\mathbf{b}$, then $\mathbf{ab} = \mathbf{a}_{\parallel}\mathbf{b} + \mathbf{a}_{\perp}\mathbf{b} = \mathbf{a} \cdot \mathbf{b} + \mathbf{a} \wedge \mathbf{b}$.

$$\mathbf{a} \cdot \mathbf{b} = \tfrac{1}{2}(\mathbf{ab} + \mathbf{ba})$$
$$\mathbf{a} \wedge \mathbf{b} = \tfrac{1}{2}(\mathbf{ab} - \mathbf{ba})$$
$$\mathbf{a} \cdot \mathbf{b} = |\mathbf{a}||\mathbf{b}| \cos\varphi$$
$$|\mathbf{a} \wedge \mathbf{b}| = |\mathbf{a}||\mathbf{b}| \sin\varphi$$

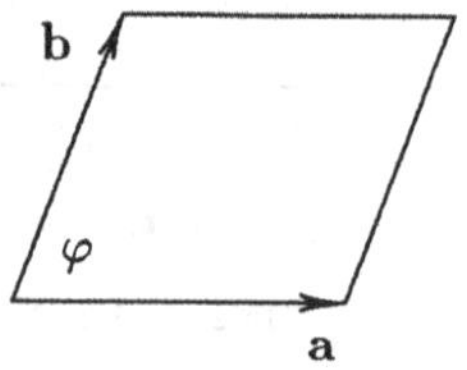

Compute the product $\mathbf{abba}$ to get $\mathbf{a}^2\mathbf{b}^2 = (\mathbf{a} \cdot \mathbf{b})^2 - (\mathbf{a} \wedge \mathbf{b})^2$ and use $(\mathbf{a} \wedge \mathbf{b})^2 = -|\mathbf{a} \wedge \mathbf{b}|^2$ to obtain the identity

$$\mathbf{a}^2\mathbf{b}^2 = (\mathbf{a} \cdot \mathbf{b})^2 + |\mathbf{a} \wedge \mathbf{b}|^2.$$

3.9 Even and odd parts

The Clifford algebra is, like the exterior algebra, a direct sum of two of its subspaces,

$$\text{the even part} \quad \mathbb{R} \oplus \textstyle\bigwedge^2 \mathbb{R}^3,$$
$$\text{the odd part} \quad \mathbb{R}^3 \oplus \textstyle\bigwedge^3 \mathbb{R}^3.$$

For both algebras the even part is also a subalgebra. The even subalgebra $(\bigwedge \mathbb{R}^3)^+ = \mathbb{R} \oplus \bigwedge^2 \mathbb{R}^3$ of $\bigwedge \mathbb{R}^3$ is commutative, but the even subalgebra $\mathcal{C}\ell_3^+ = \mathbb{R} \oplus \bigwedge^2 \mathbb{R}^3$ of $\mathcal{C}\ell_3$ is not commutative; instead it is isomorphic to the quaternion algebra: $\mathbb{H} \simeq \mathcal{C}\ell_3^+$. The odd parts are denoted by $\mathcal{C}\ell_3^-$ and $(\bigwedge \mathbb{R}^3)^-$.

3.10 The center

The center of an algebra consists of those elements which commute with all the elements of the algebra. The center $\mathrm{Cen}(\mathcal{C}\ell_3) = \mathbb{R} \oplus \bigwedge^3 \mathbb{R}^3$ of $\mathcal{C}\ell_3$ is isomorphic to $\mathbb{C}$, and the center of $\bigwedge \mathbb{R}^3$ is $\mathrm{Cen}(\bigwedge \mathbb{R}^3) = \mathbb{R} \oplus \bigwedge^2 \mathbb{R}^3 \oplus \bigwedge^3 \mathbb{R}^3$.

3.11 Gradings and the multivector structure

The exterior products of homogeneous elements satisfy the relations

$$\mathbf{a} \wedge \mathbf{b} \in \bigwedge^{i+j} \mathbb{R}^3 \quad \text{for} \quad \mathbf{a} \in \bigwedge^{i} \mathbb{R}^3 \quad \text{and} \quad \mathbf{b} \in \bigwedge^{j} \mathbb{R}^3.$$

Such a property of an algebra is usually referred to by saying that the algebra is graded over the index group $\mathbb{Z}$. We shall refer to this property of the exterior algebra $\bigwedge \mathbb{R}^3$ as the *dimension grading*, because simple homogeneous elements represent subspaces of specified dimension. The homogeneous elements in $\bigwedge \mathbb{R}^3$ satisfy

$$\mathbf{a} \wedge \mathbf{b} = (-1)^{ij} \mathbf{b} \wedge \mathbf{a} \quad \text{for} \quad \mathbf{a} \in \bigwedge^i \mathbb{R}^3, \ \mathbf{b} \in \bigwedge^j \mathbb{R}^3,$$

that is, the exterior algebra $\bigwedge \mathbb{R}^3$ is *graded commutative.* [8]

The Clifford products of even and odd subspaces satisfy the inclusion relations

$$\begin{aligned} &\mathcal{C}\ell_3^+ \mathcal{C}\ell_3^+ \subset \mathcal{C}\ell_3^+, \quad &\mathcal{C}\ell_3^+ \mathcal{C}\ell_3^- \subset \mathcal{C}\ell_3^-, \\ &\mathcal{C}\ell_3^- \mathcal{C}\ell_3^+ \subset \mathcal{C}\ell_3^-, \quad &\mathcal{C}\ell_3^- \mathcal{C}\ell_3^- \subset \mathcal{C}\ell_3^+. \end{aligned}$$

These relations can be summarized by saying that the Clifford algebra $\mathcal{C}\ell_3$ has an *even-odd grading,* or that it is graded over the index group $\mathbb{Z}_2 = \{0, 1\}$.

The exterior algebra $\bigwedge \mathbb{R}^3$ is also even-odd graded.

The Clifford algebra $\mathcal{C}\ell_3$ is not graded over $\mathbb{Z}$. However, we can reconstruct the exterior product from the Clifford product in a unique manner. We shall refer to the dimension grading of the associated exterior algebra by saying that the Clifford algebra has a *multivector structure.* Recall that $\mathbb{R}$ and $\mathbb{R}^3$ have, by definition, unique copies in $\mathcal{C}\ell_3$. The exterior product of two vectors equals the antisymmetric part of their Clifford product,

$$\mathbf{x} \wedge \mathbf{y} = \frac{1}{2}(\mathbf{xy} - \mathbf{yx}) \in \bigwedge^2 \mathbb{R}^3 \quad \text{for} \quad \mathbf{x}, \mathbf{y} \in \mathbb{R}^3,$$

whence the space of bivectors $\bigwedge^2 \mathbb{R}^3$ has a unique copy in $\mathcal{C}\ell_3$. The subspace of 3-vectors $\bigwedge^3 \mathbb{R}^3$ can be uniquely reconstructed within $\mathcal{C}\ell_3$ by a completely antisymmetrized Clifford product

$$\mathbf{x} \wedge \mathbf{y} \wedge \mathbf{z} = \frac{1}{6}(\mathbf{xyz} + \mathbf{yzx} + \mathbf{zxy} - \mathbf{zyx} - \mathbf{xzy} - \mathbf{yxz}) \in \bigwedge^3 \mathbb{R}^3$$

of three vectors $\mathbf{x}, \mathbf{y}, \mathbf{z} \in \mathbb{R}^3$.

Thus, we have established a linear isomorphism sending $\bigwedge \mathbb{R}^3$ to $\mathcal{C}\ell_3$ defined

8 The graded opposite algebra of $\bigwedge \mathbb{R}^3$ is the linear space $\bigwedge \mathbb{R}^3$ with a new product $u \circ v$ defined by

$$(u_0 + u_1) \circ (v_0 + v_1) = v_0 \wedge u_0 + v_0 \wedge u_1 + v_1 \wedge u_0 - v_1 \wedge u_1$$

for $u_0, v_0 \in (\bigwedge \mathbb{R}^3)^+$ and $u_1, v_1 \in (\bigwedge \mathbb{R}^3)^-$. Since $\bigwedge \mathbb{R}^3$ is graded commutative, that is $u \circ v = u \wedge v$, the graded opposite of $\bigwedge \mathbb{R}^3$ is just $\bigwedge \mathbb{R}^3$.

for k-vectors:

$\bigwedge\mathbb{R}^3$		$C\ell_3$
α	$=$	$\alpha \in \mathbb{R}$
$\mathbf{x}$	$=$	$\mathbf{x} \in \mathbb{R}^3$
$\mathbf{x} \wedge \mathbf{y}$	$=$	$\frac{1}{2}(\mathbf{xy} - \mathbf{yx}) \in \bigwedge^2 \mathbb{R}^3$
$\mathbf{x} \wedge \mathbf{y} \wedge \mathbf{z}$	$=$	$\frac{1}{6}(\mathbf{xyz} + \mathbf{yzx} + \mathbf{zxy} - \mathbf{zyx} - \mathbf{xzy} - \mathbf{yxz}) \in \bigwedge^3 \mathbb{R}^3$

There is another construction of the subspace of 3-vectors $\bigwedge^3 \mathbb{R}^3$, obtained by using the reversion, $\mathbf{x} \wedge \mathbf{y} \wedge \mathbf{z} = \frac{1}{2}(\mathbf{xyz} - \mathbf{zyx}) \in \bigwedge^3 \mathbb{R}^3$ for $\mathbf{x}, \mathbf{y}, \mathbf{z} \in \mathbb{R}^3$, related to the following recursive construction, via an intermediate step in $\bigwedge^2 \mathbb{R}^3$:

$$\mathbf{x} \wedge \mathbf{B} = \frac{1}{2}(\mathbf{xB} + \mathbf{Bx}) \in \bigwedge^3 \mathbb{R}^3 \quad \text{for} \quad \mathbf{x} \in \mathbb{R}^3,\ \mathbf{B} \in \bigwedge^2 \mathbb{R}^3.$$

3.12 Products of vectors and bivectors, visualization

A vector $\mathbf{a} \in \mathbb{R}^3$ and a bivector $\mathbf{B} \in \bigwedge \mathbb{R}^3$ can be multiplied to give a 3-vector $\mathbf{a} \wedge \mathbf{B} = \mathbf{B} \wedge \mathbf{a} \in \bigwedge^3 \mathbb{R}^3$. The exterior product of a vector and a bivector can be depicted as an oriented volume:

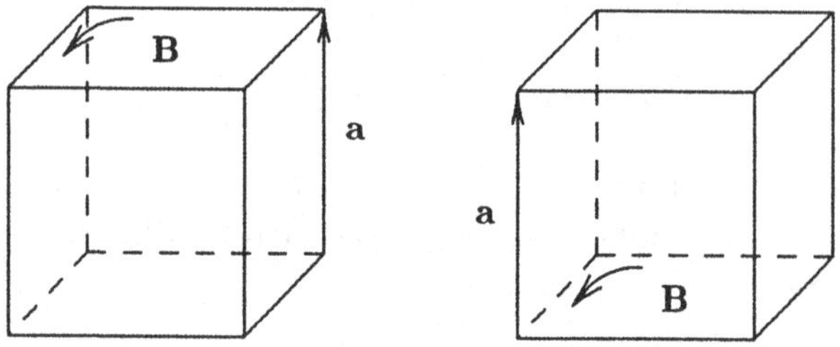

$\mathbf{a} \wedge \mathbf{B}$ $\qquad$ $\mathbf{B} \wedge \mathbf{a}$

The orientation is obtained by putting the arrows in succession. The commutativity of the exterior product $\mathbf{a} \wedge \mathbf{B} = \mathbf{B} \wedge \mathbf{a}$ means that the screws of $\mathbf{a} \wedge \mathbf{B}$ and $\mathbf{B} \wedge \mathbf{a}$ can be rotated onto each other (without reflection).

A vector $\mathbf{x} \in \mathbb{R}^3$ and a bivector $\mathbf{B} \in \bigwedge^2 \mathbb{R}^3$ can also be multiplied so that the result is a vector $\mathbf{B} \llcorner \mathbf{x} \in \mathbb{R}^3$. Consider a vector $\mathbf{x}$ tilted by an angle φ out of the plane of a bivector $\mathbf{B}$. Let $\mathbf{a}$ be the orthogonal projection of $\mathbf{x}$ in the plane of $\mathbf{B}$. Then $|\mathbf{a}| = |\mathbf{x}| \cos\varphi$. The right *contraction* of the bivector $\mathbf{B}$ by the vector $\mathbf{x}$ is a vector $\mathbf{y} = \mathbf{B} \llcorner \mathbf{x}$ in the plane of $\mathbf{B}$ such that

(i) $|\mathbf{y}| = |\mathbf{B}||\mathbf{a}|$,

(ii) $\mathbf{y} \perp \mathbf{a}$ and $\mathbf{a} \wedge \mathbf{y} \uparrow\uparrow \mathbf{B}$.

By convention, we agree that

$$\mathbf{x} \rfloor \mathbf{B} = -\mathbf{B} \lfloor \mathbf{x},$$

that is, the left and right contractions have opposite signs.

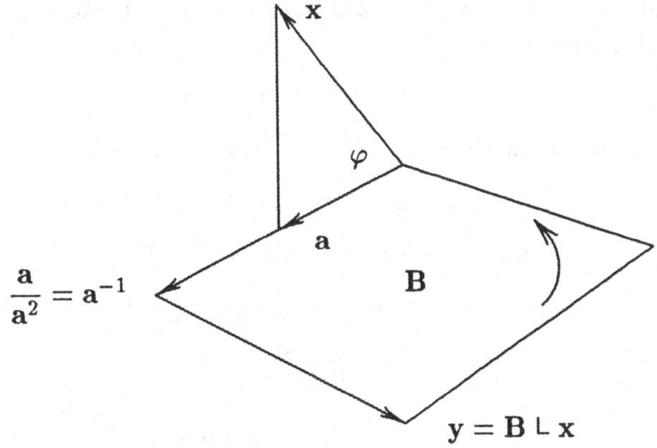

[The inverse vector $\mathbf{a}^{-1}$ of $\mathbf{a}$ has a geometrical meaning in this figure: it gives the area of the rectangle, $|\mathbf{B}| = |\mathbf{a}^{-1}||\mathbf{y}|$.]

Write $\mathbf{x}_{\|} = \mathbf{a}$ and $\mathbf{x}_{\perp} = \mathbf{x} - \mathbf{x}_{\|}$. Then $\mathbf{x} \rfloor \mathbf{B} = \mathbf{x}_{\|}\mathbf{B}$ and $\mathbf{x} \wedge \mathbf{B} = \mathbf{x}_{\perp}\mathbf{B}$ so that

$\mathbf{x}_{\|} = (\mathbf{x} \rfloor \mathbf{B})\mathbf{B}^{-1}$	parallel component
$\mathbf{x}_{\perp} = (\mathbf{x} \wedge \mathbf{B})\mathbf{B}^{-1}$	perpendicular component

where $\mathbf{B}^{-1} = \mathbf{B}/\mathbf{B}^2$, $\mathbf{B}^2 = -|\mathbf{B}|^2$.

3.13 Contractions and the derivation

The Clifford product of two vectors $\mathbf{a}$ and $\mathbf{b}$ is a sum of a scalar $\mathbf{a} \cdot \mathbf{b}$ and a bivector $\mathbf{a} \wedge \mathbf{b}$,

$$\mathbf{ab} = \mathbf{a} \cdot \mathbf{b} + \mathbf{a} \wedge \mathbf{b},$$

so that the terms on the right hand side can be recaptured from the Clifford product:

$$\mathbf{a} \cdot \mathbf{b} = \frac{1}{2}(\mathbf{ab} + \mathbf{ba}), \qquad \mathbf{a} \wedge \mathbf{b} = \frac{1}{2}(\mathbf{ab} - \mathbf{ba}).$$

The product of a vector $\mathbf{a}$ and a bivector $\mathbf{B}$ is a sum of a vector and a 3-vector:

$$\mathbf{aB} = \mathbf{a} \rfloor \mathbf{B} + \mathbf{a} \wedge \mathbf{B}$$

where

$$\mathbf{a} \rfloor \mathbf{B} = \frac{1}{2}(\mathbf{aB} - \mathbf{Ba}), \qquad \mathbf{a} \wedge \mathbf{B} = \frac{1}{2}(\mathbf{aB} + \mathbf{Ba}).$$

In general, the Clifford product of a vector $\mathbf{x} \in \mathbb{R}^3$ and an arbitrary element $u \in \mathcal{C}\ell_3$ can be decomposed into a sum of the left contraction and the exterior product as follows: [9]

$$\mathbf{x}u = \mathbf{x} \rfloor u + \mathbf{x} \wedge u$$

where we can write, in the case where u is a k-vector in $\bigwedge^k \mathbb{R}^3$,

$$\mathbf{x} \rfloor u = \tfrac{1}{2}(\mathbf{x}u - (-1)^k u\mathbf{x}) \in \textstyle\bigwedge^{k-1} \mathbb{R}^3,$$
$$\mathbf{x} \wedge u = \tfrac{1}{2}(\mathbf{x}u + (-1)^k u\mathbf{x}) \in \textstyle\bigwedge^{k+1} \mathbb{R}^3.$$

The exterior product and the left contraction by a homogeneous element, respectively, raise or lower the degree, that is,

$$\mathbf{a} \wedge \mathbf{b} \in \bigwedge^{i+j} \mathbb{R}^3, \qquad \mathbf{a} \rfloor \mathbf{b} \in \bigwedge^{j-i} \mathbb{R}^3$$

for $\mathbf{a} \in \bigwedge^i \mathbb{R}^3$ and $\mathbf{b} \in \bigwedge^j \mathbb{R}^3$.

The left contraction can be obtained from the exterior product and the Clifford product as follows:

$$u \rfloor v = [u \wedge (v\mathbf{e}_{123})]\mathbf{e}_{123}^{-1}.$$

This means that the left contraction is dual to the exterior product. The *left contraction* can be directly defined by its characteristic properties

1) $\mathbf{x} \rfloor \mathbf{y} = \mathbf{x} \cdot \mathbf{y}$,
2) $\mathbf{x} \rfloor (u \wedge v) = (\mathbf{x} \rfloor u) \wedge v + \hat{u} \wedge (\mathbf{x} \rfloor v)$,
3) $(u \wedge v) \rfloor w = u \rfloor (v \rfloor w)$,

where $\mathbf{x}, \mathbf{y} \in \mathbb{R}^3$ and $u, v, w \in \bigwedge \mathbb{R}^3$. Recalling that $\hat{u} = (-1)^k u$ for $u \in \bigwedge^k \mathbb{R}^3$, the second rule can also be written as

$$\mathbf{x} \rfloor (u \wedge v) = (\mathbf{x} \rfloor u) \wedge v + (-1)^k u \wedge (\mathbf{x} \rfloor v),$$

when $u \in \bigwedge^k \mathbb{R}^3$. The second rule means that the left contraction by a vector is a *derivation* of the exterior algebra $\bigwedge \mathbb{R}^3$. It happens that the left contraction by a vector is also a derivation of the Clifford algebra, that is,

$$\mathbf{x} \rfloor (uv) = (\mathbf{x} \rfloor u)v + \hat{u}(\mathbf{x} \rfloor v) \quad \text{for} \quad \mathbf{x} \in \mathbb{R}^3,\ u, v \in \mathcal{C}\ell_3.$$

9 A scalar product on $\mathbb{R}^3 \subset \bigwedge \mathbb{R}^3$ induces a contraction on $\bigwedge \mathbb{R}^3$ which can be used to introduce a new product $\mathbf{x}u = \mathbf{x} \rfloor u + \mathbf{x} \wedge u$ for $\mathbf{x} \in \mathbb{R}^3$ and $u \in \bigwedge \mathbb{R}^3$, which extends by linearity and associativity to all of $\bigwedge \mathbb{R}^3$. The linear space $\bigwedge \mathbb{R}^3$ provided with this new product is the Clifford algebra $\mathcal{C}\ell_3$.

3.14 The Clifford algebra versus the exterior algebra

Both the Clifford algebra $\mathcal{C}\ell_3$ and the exterior algebra $\bigwedge \mathbb{R}^3$ contain a copy of $\mathbb{R}^3$, which enables application of calculations to the geometry of $\mathbb{R}^3$. The feature distinguishing $\mathcal{C}\ell_3$ from $\bigwedge \mathbb{R}^3$ is that the Clifford multiplication of vectors preserves the norm, $|\mathbf{ab}| = |\mathbf{a}||\mathbf{b}|$ for all $\mathbf{a}, \mathbf{b} \in \mathbb{R}^3$, whereas $|\mathbf{a} \wedge \mathbf{b}| \leq |\mathbf{a}||\mathbf{b}|$. The equality $|\mathbf{ab}| = |\mathbf{a}||\mathbf{b}|$ enables more calculations to be carried out in $\mathbb{R}^3$, most notably rotations become represented as operations within one algebra, the Clifford algebra $\mathcal{C}\ell_3$.

Historical survey

The exterior algebra $\bigwedge \mathbb{R}^3$ of the linear space $\mathbb{R}^3$ was constructed by Grassmann in 1844. Grassmann's exterior algebra $\bigwedge \mathbb{R}^3$ has a basis

$$\begin{aligned}&1\\&\mathbf{e}_1, \mathbf{e}_2, \mathbf{e}_3\\&\mathbf{e}_1 \wedge \mathbf{e}_2, \mathbf{e}_1 \wedge \mathbf{e}_3, \mathbf{e}_2 \wedge \mathbf{e}_3\\&\mathbf{e}_1 \wedge \mathbf{e}_2 \wedge \mathbf{e}_3\end{aligned}$$

satisfying the multiplication rules

$$\begin{aligned}&\mathbf{e}_i \wedge \mathbf{e}_j = -\mathbf{e}_j \wedge \mathbf{e}_i \quad \text{for} \quad i \neq j,\\&\mathbf{e}_i \wedge \mathbf{e}_i = 0.\end{aligned}$$

Clifford introduced a new product into the exterior algebra; he kept the first rule

$$\mathbf{e}_i\mathbf{e}_j = -\mathbf{e}_j\mathbf{e}_i \quad \text{for} \quad i \neq j,$$

that is $\mathbf{e}_i\mathbf{e}_j = \mathbf{e}_i \wedge \mathbf{e}_j$, but replaced the second rule by

$$\begin{aligned}\mathbf{e}_i\mathbf{e}_i &= 1 \quad \text{in 1882,} \quad \text{and}\\ \mathbf{e}_i\mathbf{e}_i &= -1 \quad \text{in 1878.}\end{aligned}$$

These two algebras generated are Clifford's geometric algebras

$$\mathcal{C}\ell_3 = \mathcal{C}\ell_{3,0} \simeq \mathrm{Mat}(2, \mathbb{C}) \quad \text{and} \quad \mathcal{C}\ell_{0,3} \simeq \mathbb{H} \oplus \mathbb{H}$$

of the positive definite and negative definite quadratic spaces $\mathbb{R}^3 = \mathbb{R}^{3,0}$ and $\mathbb{R}^{0,3}$, respectively.

Exercises

1. Find the area of the triangle with vertices $(1, -4, -6)$, $(5, -4, -2)$ and $(0, 0, 0)$.

2. Find the volume of the parallelepiped with edges $\mathbf{a} = 2\mathbf{e}_1 - 3\mathbf{e}_2 + 4\mathbf{e}_3$, $\mathbf{b} = \mathbf{e}_1 + 2\mathbf{e}_2 - \mathbf{e}_3$, $\mathbf{c} = 3\mathbf{e}_1 - \mathbf{e}_2 + 2\mathbf{e}_3$.
3. Compute the square of the volume element $\mathbf{e}_{123} = \mathbf{e}_1\mathbf{e}_2\mathbf{e}_3$ (square with respect to the Clifford product).
4. Show that $\mathbf{e}_{123}$ commutes with $\mathbf{e}_1, \mathbf{e}_2, \mathbf{e}_3$.
5. Find the inverse of the bivector $\mathbf{B} = 3\mathbf{e}_{12} + \mathbf{e}_{23}$ (inverse with respect to the Clifford product).
6. Let $\mathbf{a} = 2\mathbf{e}_1 + 3\mathbf{e}_2 + 7\mathbf{e}_3$ and $\mathbf{B} = 4\mathbf{e}_{12} + 5\mathbf{e}_{13} - \mathbf{e}_{23}$. Compute $\mathbf{a} \wedge \mathbf{B}$ and $\mathbf{a} \,\lrcorner\, \mathbf{B}$.
7. Let $\mathbf{a} = 3\mathbf{e}_1 + 4\mathbf{e}_2 + 7\mathbf{e}_3$ and $\mathbf{B} = 7\mathbf{e}_{12} + \mathbf{e}_{13}$. Compute the perpendicular and parallel components of $\mathbf{a}$ in the plane of $\mathbf{B}$.
8. Show that the Clifford product of a bivector $\mathbf{B} \in \bigwedge^2 \mathbb{R}^3$ and an arbitrary element $u \in \mathcal{C}\ell_3$ can be decomposed as

$$\mathbf{B}u = \mathbf{B} \,\lrcorner\, u + \frac{1}{2}(\mathbf{B}u - u\mathbf{B}) + \mathbf{B} \wedge u.$$

9. Reconstruct the dot product $\mathbf{a} \cdot \mathbf{b}$ with the help of the cross product $\mathbf{a} \times \mathbf{b}$ and the exterior product $\mathbf{a} \wedge \mathbf{b}$. Hint: $\mathbf{a} \times (\mathbf{a} \times \mathbf{b}) = (\mathbf{a} \cdot \mathbf{b})\mathbf{a} - \mathbf{a}^2\mathbf{b}$.

Define the right contraction by $u \,\llcorner\, v = \mathbf{e}_{123}^{-1}[(\mathbf{e}_{123}u) \wedge v]$ for $u, v \in \mathcal{C}\ell_3$.

10. Show that the following properties – the characteristic properties – of the right contraction hold:

$$\begin{aligned} &1) \quad \mathbf{x} \,\llcorner\, \mathbf{y} = \mathbf{x} \cdot \mathbf{y}, \\ &2) \quad (u \wedge v) \,\llcorner\, \mathbf{x} = u \wedge (v \,\llcorner\, \mathbf{x}) + (u \,\llcorner\, \mathbf{x}) \wedge \hat{v}, \\ &3) \quad u \,\llcorner\, (v \wedge w) = (u \,\llcorner\, v) \,\llcorner\, w, \end{aligned}$$

   for $\mathbf{x}, \mathbf{y} \in \mathbb{R}^3$ and $u, v, w \in \bigwedge \mathbb{R}^3$.
11. Show that $\mathbf{a} \,\llcorner\, \mathbf{b} \in \bigwedge^{i-j} \mathbb{R}^3$ for $\mathbf{a} \in \bigwedge^i \mathbb{R}^3$ and $\mathbf{b} \in \bigwedge^j \mathbb{R}^3$.
12. Show that $(u \,\lrcorner\, v) \,\llcorner\, w = u \,\lrcorner\, (v \,\llcorner\, w)$.
13. Show that $u \,\lrcorner\, v = \star(\star^{-1}(v) \wedge \tilde{u})$ and $u \,\llcorner\, v = \star^{-1}(\tilde{v} \wedge \star(u))$.
14. Show that

$$u\mathbf{x} = u \,\llcorner\, \mathbf{x} + u \wedge \mathbf{x}$$

   where, for a k-vector $u \in \bigwedge^k \mathbb{R}^3$,

$$\begin{aligned} u \,\llcorner\, \mathbf{x} &= \tfrac{1}{2}(u\mathbf{x} - (-1)^k \mathbf{x}u) \in \textstyle\bigwedge^{k-1} \mathbb{R}^3, \\ u \wedge \mathbf{x} &= \tfrac{1}{2}(u\mathbf{x} + (-1)^k \mathbf{x}u) \in \textstyle\bigwedge^{k+1} \mathbb{R}^3. \end{aligned}$$

15. Show that $u \wedge v - v \wedge u \in \bigwedge^2 \mathbb{R}^3$ and $uv - vu \in \mathbb{R}^3 \oplus \bigwedge^2 \mathbb{R}^3$.

Let $\mathbf{a} \in \mathbb{R}^3$, $\mathbf{B} \in \bigwedge^2 \mathbb{R}^3$, $u = 1 + \mathbf{a} + \mathbf{B}$.

16. The exterior inverse of u is $u^{\wedge(-1)} = 1 - \mathbf{a} - \mathbf{B} + \alpha\mathbf{a}\wedge\mathbf{B}$ with some $\alpha \in \mathbb{R}$. Determine α. Hint: use power series or $u \wedge u^{\wedge(-1)} = 1$.
17. The exterior square root of u is $u^{\wedge(1/2)} = 1 + \frac{1}{2}\mathbf{a} + \frac{1}{2}\mathbf{B} + \beta\mathbf{a}\wedge\mathbf{B}$ with some $\beta \in \mathbb{R}$. Determine β. Hint: $u^{\wedge(1/2)} \wedge u^{\wedge(1/2)} = u$.
18. Show that $1 \rfloor u = u$ for all $u \in \bigwedge\mathbb{R}^3$.

Solutions

1. $\mathbf{a} = \mathbf{e}_1 - 4\mathbf{e}_2 - 6\mathbf{e}_3$, $\mathbf{b} = 5\mathbf{e}_1 - 4\mathbf{e}_2 - 2\mathbf{e}_3$, $\mathbf{a}\wedge\mathbf{b} = 16\mathbf{e}_{12} + 28\mathbf{e}_{13} - 16\mathbf{e}_{23}$, $\frac{1}{2}|\mathbf{a}\wedge\mathbf{b}| = \frac{1}{2}\sqrt{16^2+28^2+16^2} = 18$.
2. $\mathbf{a}\wedge\mathbf{b}\wedge\mathbf{c} = -7\mathbf{e}_{123}$, $|\mathbf{a}\wedge\mathbf{b}\wedge\mathbf{c}| = 7$.
3. $\mathbf{e}_{123}^2 = -1$.
5. $\mathbf{B}^2 = -10$, $|\mathbf{B}| = \sqrt{10}$, $\mathbf{B}^{-1} = -\frac{1}{10}(3\mathbf{e}_{12} + \mathbf{e}_{23})$.
6. $\mathbf{a}\wedge\mathbf{B} = 11\mathbf{e}_{123}$, $\mathbf{a}\rfloor\mathbf{B} = -47\mathbf{e}_1 + 15\mathbf{e}_2 + 7\mathbf{e}_3$.
7. $\mathbf{a}_\perp = -0.9\mathbf{e}_2 + 6.3\mathbf{e}_3$, $\mathbf{a}_\| = 3\mathbf{e}_1 + 4.9\mathbf{e}_2 + 0.7\mathbf{e}_3$.
9. Take a wedge product with $\mathbf{b}$ to obtain $(\mathbf{a}\times(\mathbf{a}\times\mathbf{b}))\wedge\mathbf{b} = (\mathbf{a}\cdot\mathbf{b})(\mathbf{a}\wedge\mathbf{b})$, and

$$\mathbf{a}\cdot\mathbf{b} = \frac{(\mathbf{a}\times(\mathbf{a}\times\mathbf{b}))\wedge\mathbf{b}}{\mathbf{a}\wedge\mathbf{b}} \quad \text{for} \quad \mathbf{a}\nparallel\mathbf{b}$$

(the division is carried out in the Clifford algebra $\mathcal{C}\ell_3$, or it is just a ratio of two parallel bivectors).
16. $\alpha = 2$.
17. $\beta = -\frac{1}{4}$.
18. $1\rfloor u = (1\wedge 1)\rfloor u = 1\rfloor(1\rfloor u)$ and so the contraction by 1 is a projection with eigenvalues 0 and 1. The only idempotents of $\bigwedge\mathbb{R}^3$ are 0 and 1, and so $1\rfloor u = 0$ or $1\rfloor u = u$, identically. The latter must be chosen, since $1\rfloor(\mathbf{x}\cdot\mathbf{y}) = 1\rfloor(\mathbf{x}\rfloor\mathbf{y}) = (1\wedge\mathbf{x})\rfloor\mathbf{y} = \mathbf{x}\rfloor\mathbf{y} = \mathbf{x}\cdot\mathbf{y} \neq 0$ for some $\mathbf{x},\mathbf{y}\in\mathbb{R}^3$.

Bibliography

R. Deheuvels: *Formes quadratiques et groupes classiques.* Presses Universitaires de France, Paris, 1981.

J. Dieudonné: The tragedy of Grassmann. *Linear and Multilinear Algebra* **8** (1979), 1-14.

W. Greub: *Multilinear Algebra*, 2nd edn. Springer, Berlin, 1978.

J. Helmstetter: Algèbres de Clifford et algèbres de Weyl. *Cahiers Math.* **25**, Montpellier, 1982.

G. Sobczyk: *Vector Calculus with Complex Variables.* Spring Hill College, Mobile, AL, 1982.

I. Stewart: Hermann Grassmann was right. *Nature* **321**, 1 May (1986), 17.

D. Sturmfels: *Algorithms of Invariant Theory.* Springer, Wien, 1993.

4

Pauli Spin Matrices and Spinors

In classical mechanics kinetic energy $\frac{1}{2}mv^2 = \frac{p^2}{2m}$, $\vec{p} = m\vec{v}$, and potential energy $W = W(\vec{r})$ sum up to the total energy [1]

$$E = \frac{p^2}{2m} + W.$$

Inserting differential operators for total energy and momentum,

$$E = i\hbar\frac{\partial}{\partial t} \quad \text{and} \quad \vec{p} = -i\hbar\nabla,$$

into the above equation results in the *Schrödinger equation* [2]

$$i\hbar\frac{\partial\psi}{\partial t} = -\frac{\hbar^2}{2m}\nabla^2\psi + W\psi,$$

a quantum mechanical description of the electron. The Schrödinger equation explains all atomic phenomena except those involving magnetism and relativity.

The wave function ψ is complex valued, $\psi(\vec{r}, t) \in \mathbb{C}$. The square norm $|\psi|^2$ integrated over a region in space gives the probability of finding the electron in that region. [3]

The Stern & Gerlach experiment, in 1922, showed that a beam of silver atoms splits in two in a magnetic field [there were two distinct spots on the screen, instead of a smear of silver along a line]. Uhlenbeck & Goudsmit in 1925 proposed that silver atoms and the electron have an intrinsic angular momentum, the *spin*. The spin interacts with the magnetic field, and the electron goes up or down according as the spin is parallel or opposite to the vertical magnetic field.

1 This holds in a conservative system.

2 The Schrödinger equation arose out of the hypothesis that if light has both wave and particle properties, then perhaps particles might have wave properties such as interference and diffraction.

3 This is the Born interpretation.

In an electromagnetic field $\vec{E}, \vec{B}$ with potentials $V, \vec{A}$ the Schrödinger equation becomes [4]

$$i\hbar\frac{\partial\psi}{\partial t} = \frac{1}{2m}[(-i\hbar\nabla - e\vec{A})^2]\psi - eV\psi, \tag{1}$$

or after 'squaring'

$$i\hbar\frac{\partial\psi}{\partial t} = \frac{1}{2m}[-\hbar^2\nabla^2 + e^2A^2 + i\hbar e(\nabla\cdot\vec{A} + \vec{A}\cdot\nabla)]\psi - eV\psi.$$

This equation does not yet involve the spin of the electron. The differential operator, known as the generalized momentum,

$$\vec{\pi} = \vec{p} - e\vec{A} \quad \text{where} \quad \vec{p} = -i\hbar\nabla$$

is such that its components $\pi_k = p_k - eA_k$ satisfy the commutation relations

$$\pi_1\pi_2 - \pi_2\pi_1 = i\hbar eB_3 \quad \text{(permute 1, 2, 3 cyclically)}.$$

Pauli 1927 introduced the spin into quantum mechanics by adding a new term into the Schrödinger equation. The *Pauli spin matrices*

$$\sigma_1 = \begin{pmatrix} 0 & 1 \\ 1 & 0 \end{pmatrix}, \quad \sigma_2 = \begin{pmatrix} 0 & -i \\ i & 0 \end{pmatrix}, \quad \sigma_3 = \begin{pmatrix} 1 & 0 \\ 0 & -1 \end{pmatrix}$$

satisfy

$$\sigma_1\sigma_2 = i\sigma_3 \quad \text{(permute 1, 2, 3 cyclically)}$$

and the anticommutation relations

$$\sigma_j\sigma_k + \sigma_k\sigma_j = 2\delta_{jk}I.$$

Applying the above commutation and anticommutation relations, and temporarily using the old-fashioned notation

$$\vec{\sigma}\cdot\vec{\pi} = \sigma_1\pi_1 + \sigma_2\pi_2 + \sigma_3\pi_3,$$

we may see that

$$(\vec{\sigma}\cdot\vec{\pi})^2 = \pi^2 - \hbar e(\vec{\sigma}\cdot\vec{B})$$

where

$$\pi^2 = p^2 + e^2A^2 - e(\vec{p}\cdot\vec{A} + \vec{A}\cdot\vec{p}).$$

Pauli replaced π^2 by $(\vec{\sigma}\cdot\vec{\pi})^2$ in equation (1):

$$i\hbar\frac{\partial\psi}{\partial t} = \frac{1}{2m}[\pi^2 - \hbar e(\vec{\sigma}\cdot\vec{B})]\psi - eV\psi.$$

4 A Schrödinger equation with $W = 0$ is brought into this form by a gauge transformation $\psi(\vec{r}, t) \to \varphi(\vec{r}, t)e^{i\alpha(\vec{r}, t)}$, when $eV = \hbar\frac{\partial\alpha}{\partial t}$ and $e\vec{A} = \hbar\nabla\alpha$.

This Schrödinger-Pauli equation describes the spin by virtue of the term

$$\frac{\hbar e}{2m}(\vec{\sigma} \cdot \vec{B}).$$

The matrix $\vec{\sigma} \cdot \vec{B}$ operates on two-component column matrices with entries in $\mathbb{C}$. The wave function sends space-time points to *Pauli spinors*

$$\psi(\vec{r},t) = \begin{pmatrix} \psi_1 \\ \psi_2 \end{pmatrix}, \quad \psi_1, \psi_2 \in \mathbb{C},$$

that is, it has values in the complex linear space $\mathbb{C}^2$.

The Schrödinger-Pauli equation in the Clifford algebra $\mathcal{C}\ell_3$. The multiplication rules of the Pauli spin matrices $\sigma_1, \sigma_2, \sigma_3 \in \text{Mat}(2, \mathbb{C})$ imply the matrix identity

$$(\vec{\sigma} \cdot \vec{B})^2 = (B_1^2 + B_2^2 + B_3^2)I.$$

Thus, we may regard the set of traceless Hermitian matrices as a Euclidean space $\mathbb{R}^3$ with an orthonormal basis $\{\sigma_1, \sigma_2, \sigma_3\}$.

The length (of the representative) of a vector $\vec{B}$ is preserved under a similarity transformation $U(\vec{\sigma} \cdot \vec{B})U^{-1}$ by a special unitary matrix $U \in SU(2)$,

$$SU(2) = \{U \in \text{Mat}(2, \mathbb{C}) \mid U^\dagger U = I, \ \det U = 1\}.$$

In this way, not only vectors but also rotations become represented within the matrix algebra $\text{Mat}(2, \mathbb{C})$. In fact, each rotation $R \in SO(3)$ becomes represented by two matrices $\pm U \in SU(2)$, and we say that $SU(2)$ is a two-fold covering of $SO(3)$:

$$SO(3) \simeq \frac{SU(2)}{\{\pm I\}}.$$

Pauli spinors could also be replaced by square matrices with only the first column being non-zero,

$$\psi = \begin{pmatrix} \psi_1 & 0 \\ \psi_2 & 0 \end{pmatrix}, \quad \psi_1, \psi_2 \in \mathbb{C}.$$

Such square matrix spinors form a *left ideal* S of the matrix algebra $\text{Mat}(2, \mathbb{C})$, that is, for $U \in \text{Mat}(2, \mathbb{C})$ and $\psi \in S$ we also have $U\psi \in S$. [5]

The matrix algebra $\text{Mat}(2, \mathbb{C})$ is an isomorphic image of the Clifford algebra $\mathcal{C}\ell_3$ of the Euclidean space $\mathbb{R}^3$. Thus, not only vectors in $\mathbb{R}$ and rotations in

5 The left ideal can be written as $S = \text{Mat}(2, \mathbb{C})f$, where $f = \frac{1}{2}(I + \sigma_3)$ is an idempotent satisfying $f^2 = f$. The idempotent is primitive and the left ideal is minimal.

$SO(3)$ have representatives in $\mathcal{C}\ell_3$, but also spinor spaces or spinor representations of the rotation group $SO(3)$ [6] can be constructed within the Clifford algebra $\mathcal{C}\ell_3$. [7]

In the notation of the Clifford algebra $\mathcal{C}\ell_3$ we could describe Pauli's achievement by saying that he replaced $\pi^2 = \vec{\pi} \cdot \vec{\pi}$ by $\vec{\pi}^2 = \vec{\pi} \cdot \vec{\pi} + \vec{\pi} \wedge \vec{\pi} = \pi^2 - \hbar e \vec{B}$ and came across the equation

$$i\hbar \frac{\partial \psi}{\partial t} = \frac{1}{2m}[\pi^2 - \hbar e \vec{B}]\psi - eV\psi$$

where $\vec{B} \in \mathbb{R}^3 \subset \mathcal{C}\ell_3$ and $\psi(\vec{r}, t) \in S = \mathcal{C}\ell_3 f$, $f = \frac{1}{2}(1 + \mathbf{e}_3)$. All the arguments and functions now have values in one algebra, which will facilitate numerical computations.

In this chapter we shall study more closely the Clifford algebra $\mathcal{C}\ell_3$ and the spin group $\mathbf{Spin}(3)$, and reformulate once more the Schrödinger-Pauli equation in terms of $\mathcal{C}\ell_3$.

4.1 Orthogonal unit vectors, orthonormal basis

The 3-dimensional Euclidean space $\mathbb{R}^3$ has a basis consisting of three orthogonal unit vectors $\mathbf{e}_1, \mathbf{e}_2, \mathbf{e}_3$. The *Clifford algebra* $\mathcal{C}\ell_3$ of $\mathbb{R}^3$ is the real associative algebra generated by the set $\{\mathbf{e}_1, \mathbf{e}_2, \mathbf{e}_3\}$ satisfying the relations

$$\mathbf{e}_1^2 = 1, \quad \mathbf{e}_2^2 = 1, \quad \mathbf{e}_3^2 = 1,$$
$$\mathbf{e}_1\mathbf{e}_2 = -\mathbf{e}_2\mathbf{e}_1, \quad \mathbf{e}_1\mathbf{e}_3 = -\mathbf{e}_3\mathbf{e}_1, \quad \mathbf{e}_2\mathbf{e}_3 = -\mathbf{e}_3\mathbf{e}_2.$$

The Clifford algebra $\mathcal{C}\ell_3$ is 8-dimensional with the following basis:

1	the scalar
$\mathbf{e}_1, \mathbf{e}_2, \mathbf{e}_3$	vectors
$\mathbf{e}_1\mathbf{e}_2, \mathbf{e}_1\mathbf{e}_3, \mathbf{e}_2\mathbf{e}_3$	bivectors
$\mathbf{e}_1\mathbf{e}_2\mathbf{e}_3$	a volume element.

We abbreviate the unit bivectors as $\mathbf{e}_{ij} = \mathbf{e}_i\mathbf{e}_j$, when $i \neq j$, and the unit oriented volume element as $\mathbf{e}_{123} = \mathbf{e}_1\mathbf{e}_2\mathbf{e}_3$. An arbitrary element in $\mathcal{C}\ell_3$ is a sum of a scalar, a vector, a bivector and a volume element, and can be written as $\alpha + \mathbf{a} + \mathbf{b}\mathbf{e}_{123} + \beta\mathbf{e}_{123}$, where $\alpha, \beta \in \mathbb{R}$ and $\mathbf{a}, \mathbf{b} \in \mathbb{R}^3$.

Example. Compute the product $\mathbf{e}_{12}\mathbf{e}_{13}$. By definition $\mathbf{e}_{12}\mathbf{e}_{13} = (\mathbf{e}_1\mathbf{e}_2)(\mathbf{e}_1\mathbf{e}_3)$

6 Actually, spinor representations are representations of the universal covering group $SU(2) \simeq \mathbf{Spin}(3)$ of $SO(3)$. The spinor representations cannot be reached by tensor methods, as irreducible components of tensor products of antisymmetric powers of $\mathbb{R}^3$.

7 The orthogonal group $O(3)$ also has a non-trivial covering group $\mathbf{Pin}(3)$ residing within $\mathcal{C}\ell_3$.

and by associativity $(\mathbf{e}_1\mathbf{e}_2)(\mathbf{e}_1\mathbf{e}_3) = \mathbf{e}_1\mathbf{e}_2\mathbf{e}_1\mathbf{e}_3$. Use anticommutativity, $\mathbf{e}_1\mathbf{e}_2 = -\mathbf{e}_2\mathbf{e}_1$, and substitute $\mathbf{e}_1^2 = 1$ to get $\mathbf{e}_1\mathbf{e}_2\mathbf{e}_1\mathbf{e}_3 = -\mathbf{e}_1^2\mathbf{e}_2\mathbf{e}_3 = -\mathbf{e}_{23}$. ∎

Imaginary units. The three unit bivectors $\mathbf{e}_1\mathbf{e}_2$, $\mathbf{e}_1\mathbf{e}_3$, $\mathbf{e}_2\mathbf{e}_3$ represent unit oriented plane segments as well as generators of rotations in the coordinate planes, and share the basic property of the imaginary unit, $(\mathbf{e}_i\mathbf{e}_j)^2 = -1$ for $i \neq j$. The oriented volume element $\mathbf{e}_1\mathbf{e}_2\mathbf{e}_3$ also shares the basic property of the imaginary unit, $(\mathbf{e}_1\mathbf{e}_2\mathbf{e}_3)^2 = -1$, and furthermore it commutes with all the elements in $\mathcal{C}\ell_3$. The unit oriented volume element $\mathbf{e}_1\mathbf{e}_2\mathbf{e}_3$ represents the duality operator, which swaps plane segments and line segments orthogonal to the plane segments. ∎

4.2 Matrix representation of $\mathcal{C}\ell_3$

The set of 2×2-matrices with complex numbers as entries is denoted by $\mathrm{Mat}(2,\mathbb{C})$. Mostly we shall regard this set as a *real* algebra with scalar multiplication taken over the real numbers in $\mathbb{R}$ although the matrix entries are in the complex field $\mathbb{C}$. The Pauli spin matrices

$$\sigma_1 = \begin{pmatrix} 0 & 1 \\ 1 & 0 \end{pmatrix}, \quad \sigma_2 = \begin{pmatrix} 0 & -i \\ i & 0 \end{pmatrix}, \quad \sigma_3 = \begin{pmatrix} 1 & 0 \\ 0 & -1 \end{pmatrix}$$

satisfy the multiplication rules

$$\begin{aligned} &\sigma_1^2 = \sigma_2^2 = \sigma_3^2 = I \quad \text{and} \\ &\sigma_1\sigma_2 = i\sigma_3 = -\sigma_2\sigma_1, \\ &\sigma_3\sigma_1 = i\sigma_2 = -\sigma_1\sigma_3, \\ &\sigma_2\sigma_3 = i\sigma_1 = -\sigma_3\sigma_2. \end{aligned}$$

They also generate the real algebra $\mathrm{Mat}(2,\mathbb{C})$. The correspondences $\mathbf{e}_1 \simeq \sigma_1$, $\mathbf{e}_2 \simeq \sigma_2$, $\mathbf{e}_3 \simeq \sigma_3$ establish an isomorphism between the real algebras, $\mathcal{C}\ell_3 \simeq \mathrm{Mat}(2,\mathbb{C})$, with the following correspondences of the basis elements:

$\mathrm{Mat}(2,\mathbb{C})$	$\mathcal{C}\ell_3$
I	1
$\sigma_1,\ \sigma_2,\ \sigma_3$	$\mathbf{e}_1,\ \mathbf{e}_2,\ \mathbf{e}_3$
$\sigma_1\sigma_2,\ \sigma_1\sigma_3,\ \sigma_2\sigma_3$	$\mathbf{e}_{12},\ \mathbf{e}_{13},\ \mathbf{e}_{23}$
$\sigma_1\sigma_2\sigma_3$	$\mathbf{e}_{123}$

Note that $\mathbf{e}_{ij} = -\mathbf{e}_{ji}$ for $i \neq j$. The essential difference between the Clifford algebra $\mathcal{C}\ell_3$ and its matrix image $\mathrm{Mat}(2,\mathbb{C})$ is that in the Clifford algebra $\mathcal{C}\ell_3$ we will, in its definition, distinguish a particular subspace, the vector space $\mathbb{R}^3$, in which the square of a vector equals its length squared, that is, $\mathbf{r}^2 = |\mathbf{r}|^2$. No such distinguished subspace has been singled out in the definition of the matrix algebra $\mathrm{Mat}(2,\mathbb{C})$. Instead, we have chosen the traceless Hermitian matrices to represent $\mathbb{R}^3$, and thereby added extra structure to $\mathrm{Mat}(2,\mathbb{C})$. [8]

4.3 The center of $C\ell_3$

The element $\mathbf{e}_{123}$ commutes with all the vectors $\mathbf{e}_1, \mathbf{e}_2, \mathbf{e}_3$ and therefore with every element of $C\ell_3$. In other words, elements of the form

$$x + y\mathbf{e}_{123} \simeq \begin{pmatrix} x+iy & 0 \\ 0 & x+iy \end{pmatrix}$$

commute with all the elements in $C\ell_3$. The subalgebra of scalars and 3-vectors

$$\mathbb{R} \oplus \bigwedge^3 \mathbb{R}^3 = \{x + y\mathbf{e}_{123} \mid x,\ y \in \mathbb{R}\}$$

is the *center* $\mathrm{Cen}(C\ell_3)$ of $C\ell_3$, that is, it consists of those elements of $C\ell_3$ which commute with every element of $C\ell_3$. Note that $\sigma_1\sigma_2\sigma_3 = iI$. Since $\mathbf{e}_{123}^2 = -1$, the center of $C\ell_3$ is isomorphic to the complex field $\mathbb{C}$, that is,

$$\mathrm{Cen}(C\ell_3) = \mathbb{R} \oplus \bigwedge^3 \mathbb{R}^3 \simeq \mathbb{C}.$$

4.4 The even subalgebra $C\ell_3^+$

The elements 1 and $\mathbf{e}_{12} = \mathbf{e}_1\mathbf{e}_2$, $\mathbf{e}_{13} = \mathbf{e}_1\mathbf{e}_3$, $\mathbf{e}_{23} = \mathbf{e}_2\mathbf{e}_3$ are called *even*, because they are products of an even number of vectors. The even elements are represented by the following matrices:

$$w + x\mathbf{e}_{23} + y\mathbf{e}_{31} + z\mathbf{e}_{12} \simeq \begin{pmatrix} w+iz & ix+y \\ ix-y & w-iz \end{pmatrix}.$$

The even elements form a real subspace

$$\begin{aligned} \mathbb{R} \oplus \bigwedge^2 \mathbb{R}^3 &= \{w + x\mathbf{e}_{23} + y\mathbf{e}_{31} + z\mathbf{e}_{12} \mid w,x,y,z \in \mathbb{R}\} \\ &\simeq \{wI + xi\sigma_1 + yi\sigma_2 + zi\sigma_3 \mid w,x,y,z \in \mathbb{R}\} \end{aligned}$$

8 We could also have chosen, for the representatives of the anticommuting (and therefore orthogonal) unit vectors in $\mathbb{R}^3$, the following matrices:

$$u_1 = \frac{1}{4}\begin{pmatrix} 3i & 5 \\ 5 & -3i \end{pmatrix}, \quad u_2 = \begin{pmatrix} 0 & -i \\ i & 0 \end{pmatrix}, \quad u_3 = \frac{1}{4}\begin{pmatrix} 5 & -3i \\ -3i & -5 \end{pmatrix},$$

that is, $u_1 = \frac{1}{4}(5\sigma_1 + 3\sigma_1\sigma_2)$, $u_2 = \sigma_2$, $u_3 = \frac{1}{4}(5\sigma_3 - 3\sigma_2\sigma_3)$. These matrices are non-Hermitian and satisfy $u_ju_k + u_ku_j = 2\delta_{jk}I$.

which is closed under multiplication. Thus, the subspace $\mathbb{R} \oplus \bigwedge^2 \mathbb{R}^3$ is a subalgebra, called the *even subalgebra* of $\mathcal{C}\ell_3$. We will denote the even subalgebra by even$(\mathcal{C}\ell_3)$ or for short by $\mathcal{C}\ell_3^+$. The even subalgebra is isomorphic to the division ring of quaternions $\mathbb{H}$, as can be seen by the following correspondences:

$\mathbb{H}$	$\mathcal{C}\ell_3^+$
i	$-\mathbf{e}_{23}$
j	$-\mathbf{e}_{31}$
k	$-\mathbf{e}_{12}$

Remark. The Clifford algebra $\mathcal{C}\ell_3$ contains two subalgebras, isomorphic to $\mathbb{C}$ [the center] and $\mathbb{H}$ [the even subalgebra], in such a way that [temporarily we denote these subalgebras by their isomorphic images]

1. $ab = ba$ for $a \in \mathbb{C}$ and $b \in \mathbb{H}$,
2. $\mathcal{C}\ell_3$ is generated as a real algebra by $\mathbb{C}$ and $\mathbb{H}$,
3. $(\dim \mathbb{C})(\dim \mathbb{H}) = \dim \mathcal{C}\ell_3$.

These three observations can be expressed as

$$\mathbb{C} \otimes \mathbb{H} \simeq \mathcal{C}\ell_3. \qquad \blacksquare$$

4.5 Involutions of $\mathcal{C}\ell_3$

The Clifford algebra $\mathcal{C}\ell_3$ has three involutions similar to complex conjugation. Take an arbitrary element

$$u = \langle u \rangle_0 + \langle u \rangle_1 + \langle u \rangle_2 + \langle u \rangle_3 \quad \text{in} \quad \mathcal{C}\ell_3,$$

written as a sum of a scalar $\langle u \rangle_0$, a vector $\langle u \rangle_1$, a bivector $\langle u \rangle_2$ and a volume element $\langle u \rangle_3$. We introduce the following involutions:

$$\begin{aligned} \hat{u} &= \langle u \rangle_0 - \langle u \rangle_1 + \langle u \rangle_2 - \langle u \rangle_3, && \text{grade involution,} \\ \tilde{u} &= \langle u \rangle_0 + \langle u \rangle_1 - \langle u \rangle_2 - \langle u \rangle_3, && \text{reversion,} \\ \bar{u} &= \langle u \rangle_0 - \langle u \rangle_1 - \langle u \rangle_2 + \langle u \rangle_3, && \text{Clifford-conjugation.} \end{aligned}$$

Clifford-conjugation is a composition of the two other involutions: $\bar{u} = \hat{u}^{\tilde{}} = \tilde{u}^{\hat{}}$.

The correspondences $\sigma_1 \simeq \mathbf{e}_1$, $\sigma_2 \simeq \mathbf{e}_2$, $\sigma_3 \simeq \mathbf{e}_3$ fix the following representations for the involutions:

$$u \simeq \begin{pmatrix} a & b \\ c & d \end{pmatrix}, \qquad a, b, c, d \in \mathbb{C},$$

$$\hat{u} \simeq \begin{pmatrix} d^* & -c^* \\ -b^* & a^* \end{pmatrix}, \qquad \tilde{u} \simeq \begin{pmatrix} a^* & c^* \\ b^* & d^* \end{pmatrix}, \qquad \bar{u} \simeq \begin{pmatrix} d & -b \\ -c & a \end{pmatrix},$$

where the asterisk denotes complex conjugation. We recognize that the reverse $\tilde{u}$ is represented by the Hermitian conjugate $u^\dagger$ and the Clifford-conjugate $\bar{u}$ by the matrix $u^{-1}\det u \in \mathrm{Mat}(2,\mathbb{R})$ [for an invertible u].

The grade involution is an automorphism, that is,

$$\widehat{uv} = \hat{u}\hat{v},$$

while the reversion and the conjugation are anti-automorphisms, that is,

$$\widetilde{uv} = \tilde{v}\tilde{u} \quad \text{and} \quad \overline{uv} = \bar{v}\bar{u}.$$

The grade involution induces the even-odd grading of $\mathcal{C}\ell_3 = \mathcal{C}\ell_3^+ \oplus \mathcal{C}\ell_3^-$.

The reversion can be used to extend the norm from $\mathbb{R}^3$ to all of $\mathcal{C}\ell_3$ by setting

$$|u|^2 = \langle u\tilde{u}\rangle_0.$$

The norm of

$$u = u_0 + u_1\mathbf{e}_1 + u_2\mathbf{e}_2 + u_3\mathbf{e}_3 + u_{12}\mathbf{e}_{12} + u_{13}\mathbf{e}_{13} + u_{23}\mathbf{e}_{23} + u_{123}\mathbf{e}_{123}$$

can be obtained from

$$|u|^2 = |u_0|^2 + |u_1|^2 + |u_2|^2 + |u_3|^2 + |u_{12}|^2 + |u_{13}|^2 + |u_{23}|^2 + |u_{123}|^2.$$

The norm satisfies the inequality

$$|uv| \le \sqrt{2}|u||v| \quad \text{for} \quad u, v \in \mathcal{C}\ell_3.$$

The conjugation can be used to determine the inverse

$$u^{-1} = \frac{\bar{u}}{u\bar{u}}$$

of $u \in \mathcal{C}\ell_3$, $u\bar{u} \neq 0$. The element $u\bar{u} = \bar{u}u$ is in the center $\mathbb{R} \oplus \bigwedge^3 \mathbb{R}^3$ of $\mathcal{C}\ell_3$, so that division by it is unambiguous.

4.6 Reflections and rotations

In the Euclidean space $\mathbb{R}^3$ the vectors $\mathbf{r}$ and $\mathbf{ara}^{-1} = 2(\mathbf{a}\cdot\mathbf{r})\mathbf{a}^{-1} - \mathbf{r}$ are symmetric with respect to the axis $\mathbf{a}$ [use the definition of the Clifford product, $\mathbf{ar} + \mathbf{ra} = 2\mathbf{a}\cdot\mathbf{r}$]. The opposite of $\mathbf{ara}^{-1}$, the vector

$$-\mathbf{ara}^{-1} = \mathbf{r} - 2\frac{\mathbf{a}\cdot\mathbf{r}}{\mathbf{a}^2}\mathbf{a},$$

is obtained by reflecting $\mathbf{r}$ across the mirror perpendicular to $\mathbf{a}$ [reflection across the plane $\mathbf{ae}_{123}$].

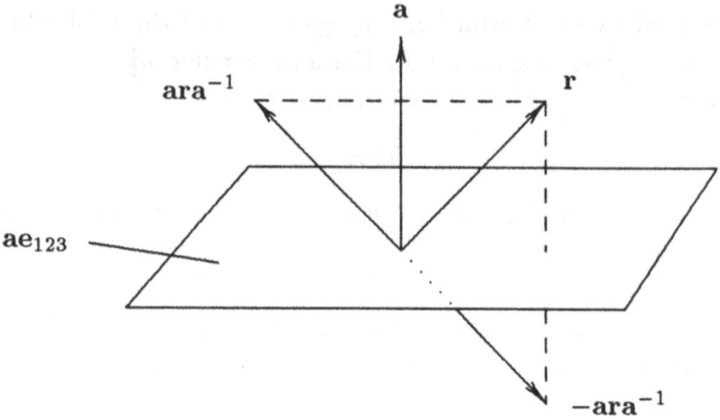

Two successive reflections in planes perpendicular to $\mathbf{a}$ and $\mathbf{b}$ result in a rotation $\mathbf{r} \to \mathbf{bara}^{-1}\mathbf{b}^{-1}$ around the axis which is perpendicular to both $\mathbf{a}$ and $\mathbf{b}$. Indeed, $\mathbf{r}$ can be decomposed as $\mathbf{r} = \mathbf{r}_{\|} + \mathbf{r}_{\perp}$ where $\mathbf{r}_{\|}$ and $\mathbf{r}_{\perp}$ are parallel and perpendicular, respectively, to the plane of $\mathbf{a}$ and $\mathbf{b}$. The perpendicular component $\mathbf{r}_{\perp}$ remains invariant under both the reflections while the two successive reflections together rotate the parallel component $\mathbf{r}_{\|}$ in the plane of $\mathbf{a}$ and $\mathbf{b}$ by twice the angle between $\mathbf{a}$ and $\mathbf{b}$.

Consider a vector $\mathbf{a} = a_1\mathbf{e}_1 + a_2\mathbf{e}_2 + a_3\mathbf{e}_3$ and the bivector $\mathbf{ae}_{123} = a_1\mathbf{e}_{23} + a_2\mathbf{e}_{31} + a_3\mathbf{e}_{12}$ dual to $\mathbf{a}$. The vector $\mathbf{a}$ has positive square

$$\mathbf{a}^2 = |\mathbf{a}|^2, \quad \text{where} \quad |\mathbf{a}| = \sqrt{a_1^2 + a_2^2 + a_3^2},$$

but the bivector $\mathbf{ae}_{123}$ has negative square

$$(\mathbf{ae}_{123})^2 = -|\mathbf{a}|^2.$$

It follows that

$$\exp(\mathbf{ae}_{123}) = \cos\alpha + \mathbf{e}_{123}\frac{\mathbf{a}}{\alpha}\sin\alpha$$

where $\alpha = |\mathbf{a}|$. A spatial rotation of the vector $\mathbf{r} = x\mathbf{e}_1 + y\mathbf{e}_2 + z\mathbf{e}_3$ around the axis $\mathbf{a}$ by the angle α is given by

$$\mathbf{r} \to a\mathbf{r}a^{-1}, \qquad a = \exp(\tfrac{1}{2}\mathbf{ae}_{123}).$$

The sense of the rotation is clockwise when regarded from the arrow-head of $\mathbf{a}$. The axis of two consecutive rotations around the axes $\mathbf{a}$ and $\mathbf{b}$ is given by the *Rodrigues formula*

$$\mathbf{c}' = \frac{\mathbf{a}' + \mathbf{b}' + \mathbf{a}' \times \mathbf{b}'}{1 - \mathbf{a}' \cdot \mathbf{b}'} \quad \text{where} \quad \mathbf{a}' = \frac{\mathbf{a}}{\alpha}\tan\frac{\alpha}{2}.$$

This result is obtained by dividing both sides of the formula

$$\exp(\tfrac{1}{2}\mathbf{c}\mathbf{e}_{123}) = \exp(\tfrac{1}{2}\mathbf{b}\mathbf{e}_{123})\exp(\tfrac{1}{2}\mathbf{a}\mathbf{e}_{123})$$

by their scalar parts and then by inspecting the bivector parts.

4.7 The group Spin(3)

The Clifford algebra $\mathcal{C}\ell_3$ of $\mathbb{R}^3$ can be employed to construct the universal covering group for the rotation group $SO(3)$ of $\mathbb{R}^3$. A vector $\mathbf{x} \in \mathbb{R}^3$ can be rotated by the formula

$$\mathbb{R}^3 \to \mathbb{R}^3, \ \mathbf{x} \to \rho(s)\mathbf{x} = s\mathbf{x}s^{-1}$$

where s is an element of the group

$$\mathbf{Spin}(3) = \{s \in \mathcal{C}\ell_3 \mid \tilde{s}s = 1, \ \bar{s}s = 1\}.$$

The group **Spin**(3), called the *spin group*, is a two-fold covering group of the rotation group $SO(3)$.

In the matrix formulation provided by the Pauli spin matrices, the spin group **Spin**(3) has an isomorphic image, the special unitary group

$$SU(2) = \{s \in \mathrm{Mat}(2,\mathbb{C}) \mid s^\dagger s = I, \ \det s = 1\}.$$

For an element $s \in SU(2)$ the function $\mathbf{x} \to \rho(s)\mathbf{x} = s\mathbf{x}s^\dagger$ is a rotation of the Euclidean space of traceless Hermitian matrices,

$$\{\mathbf{x} \in \mathrm{Mat}(2,\mathbb{C}) \mid \mathrm{trace}(\mathbf{x}) = 0, \ \mathbf{x}^\dagger = \mathbf{x}\} \simeq \mathbb{R}^3.$$

Every element in $SO(3)$ can be represented by a matrix in $SU(2)$. There are two matrices s and $-s$ in $SU(2)$ representing the same rotation $R = \rho(\pm s) \in SO(3)$. In other words, the group homomorphism $\rho : \mathbf{Spin}(3) \to SO(3)$ is surjective with kernel $\{\pm 1\}$. This can be depicted by a sequence

$$1 \longrightarrow \{\pm 1\} \longrightarrow \mathbf{Spin}(3) \xrightarrow{\rho} SO(3) \longrightarrow 1$$

which is exact, that is, the image of a homomorphism coincides with the kernel of the successive homomorphism.

The spin group **Spin**(3) is a universal cover of the rotation group $SO(3)$, that is, the Lie group **Spin**(3) is simply connected. [9] The group $SO(3)$ is doubly connected. [10]

9 A Lie group is simply connected if it is connected and every loop in the group can be shrunk to a point.

10 Rotations in $SO(3)$ can be represented by vectors $\mathbf{a} \in \mathbb{R}^3$, $|\mathbf{a}| \leq \pi$. Each rotation, $|\mathbf{a}| < \pi$, has a unique representative, and each half-turn, $|\mathbf{a}| = \pi$, is represented twice, $\pm\mathbf{a}$. A loop connecting the identity and a half-turn does not shrink to a point.

4.8 Pauli spinors

In the non-relativistic theory of the spinning electron one considers column matrices, the *Pauli spinors*

$$\psi = \begin{pmatrix} \psi_1 \\ \psi_2 \end{pmatrix} \in \mathbb{C}^2 \quad \text{where} \quad \psi_1, \psi_2 \in \mathbb{C}.$$

An isomorphic complex linear space is obtained if one replaces Pauli spinors by the *square matrix spinors*

$$\psi = \begin{pmatrix} \psi_1 & 0 \\ \psi_2 & 0 \end{pmatrix}$$

where only the first column is non-zero. The fact that only the first column is non-zero can be expressed as

$$\psi \in \mathrm{Mat}(2, \mathbb{C})f \quad \text{where} \quad f = \begin{pmatrix} 1 & 0 \\ 0 & 0 \end{pmatrix}.$$

We shall regard the correspondences $\mathbf{e}_1 \simeq \sigma_1$, $\mathbf{e}_2 \simeq \sigma_2$, $\mathbf{e}_3 \simeq \sigma_3$ as an identification between $\mathcal{C}\ell_3$ and $\mathrm{Mat}(2, \mathbb{C})$. If we multiply $\psi \in \mathrm{Mat}(2, \mathbb{C})f$ on the left by an arbitrary element $u \in \mathcal{C}\ell_3 = \mathrm{Mat}(2, \mathbb{C})$, then the result is also of the same type:

$$\begin{pmatrix} u_{11} & u_{12} \\ u_{21} & u_{22} \end{pmatrix} \begin{pmatrix} \psi_1 & 0 \\ \psi_2 & 0 \end{pmatrix} = \begin{pmatrix} \varphi_1 & 0 \\ \varphi_2 & 0 \end{pmatrix}.$$

Such matrices, with only the first column being non-zero, form a *left ideal* S of $\mathcal{C}\ell_3$, that is,

$$u\psi \in S \quad \text{for all} \quad u \in \mathcal{C}\ell_3 \quad \text{and} \quad \psi \in S \subset \mathcal{C}\ell_3.$$

This left ideal S of $\mathcal{C}\ell_3$ contains no left ideal other than S itself and the zero ideal $\{0\}$. Such a left ideal is called *minimal* in $\mathcal{C}\ell_3$.

As a real linear space, S has a basis $\{f_0, f_1, f_2, f_3\}$ where

$$\begin{aligned}
f_0 &= \tfrac{1}{2}(1 + \mathbf{e}_3) &\simeq& \begin{pmatrix} 1 & 0 \\ 0 & 0 \end{pmatrix}, \\
f_1 &= \tfrac{1}{2}(\mathbf{e}_{23} + \mathbf{e}_2) &\simeq& \begin{pmatrix} 0 & 0 \\ i & 0 \end{pmatrix}, \\
f_2 &= \tfrac{1}{2}(\mathbf{e}_{31} - \mathbf{e}_1) &\simeq& \begin{pmatrix} 0 & 0 \\ -1 & 0 \end{pmatrix}, \\
f_3 &= \tfrac{1}{2}(\mathbf{e}_{12} + \mathbf{e}_{123}) &\simeq& \begin{pmatrix} i & 0 \\ 0 & 0 \end{pmatrix}.
\end{aligned}$$

The element $f = f_0$ is an *idempotent*, that is, $f^2 = f$.

The subset

$$\mathbb{F} = fC\ell_3 f \simeq \left\{ \begin{pmatrix} c & 0 \\ 0 & 0 \end{pmatrix} \middle| \, c \in \mathbb{C} \right\}$$

of $C\ell_3$ is a subring with unity f, that is, $af = fa$ for $a \in \mathbb{F}$. None of the elements of $\mathbb{F}$ is invertible as an element of $C\ell_3$, but for each non-zero $a \in \mathbb{F}$ there is a unique $b \in \mathbb{F}$ such that $ab = f$. Thus, $\mathbb{F}$ is a *division ring* with unity f [this follows from the idempotent f being *primitive* in $C\ell_3$]. As a 2-dimensional real division algebra $\mathbb{F}$ must be isomorphic to $\mathbb{C}$. The isomorphism $\mathbb{F} \simeq \mathbb{C}$ is seen by the equation $f_3^2 = -f_0$ relating the basis elements $\{f_0, f_3\}$ of the real algebra $\mathbb{F}$.

Comment. The multiplication of an element ψ of the real linear space S on the left by an arbitrary even element $u \in C\ell_3^+$, expressed in coordinate form in the basis $\{f_0, f_1, f_2, f_3\}$,

$$u\psi = (u_0 + u_1\mathbf{e}_{23} + u_2\mathbf{e}_{31} + u_3\mathbf{e}_{23})(\psi_0 f_0 + \psi_1 f_1 + \psi_2 f_2 + \psi_3 f_3),$$

corresponds to the matrix multiplication

$$u\psi \simeq \begin{pmatrix} u_0 & -u_1 & -u_2 & -u_3 \\ u_1 & u_0 & u_3 & -u_2 \\ u_2 & -u_3 & u_0 & u_1 \\ u_3 & u_2 & -u_1 & u_0 \end{pmatrix} \begin{pmatrix} \psi_0 \\ \psi_1 \\ \psi_2 \\ \psi_3 \end{pmatrix}.$$

The square matrices corresponding to the left multiplication by even elements constitute a subring of $\mathrm{Mat}(4, \mathbb{R})$; this subring is an isomorphic image of the quaternion ring $\mathbb{H}$. ∎

The minimal left ideal

$$S = C\ell_3 f \simeq \left\{ \begin{pmatrix} \psi_1 & 0 \\ \psi_2 & 0 \end{pmatrix} \middle| \, \psi_1, \psi_2 \in \mathbb{C} \right\}$$

has a natural right $\mathbb{F}$-linear structure defined by

$$S \times \mathbb{F} \to S, \ (\psi, \lambda) \to \psi\lambda.$$

We shall provide the minimal left ideal S with this right $\mathbb{F}$-linear structure, and call it a *spinor space*. [11]

The map $C\ell_3 \to \mathrm{End}_{\mathbb{F}} S$, $u \to \tau(u)$, where $\tau(u)$ is defined by the relation $\tau(u)\psi = u\psi$, is a real algebra isomorphism. Employing the basis $\{f_0, -f_2\}$ for the $\mathbb{F}$-linear space S, the elements $\tau(\mathbf{e}_1), \tau(\mathbf{e}_2), \tau(\mathbf{e}_3)$ will be represented by the matrices $\sigma_1, \sigma_2, \sigma_3$. In this way the Pauli matrices are reproduced.

11 Note that multiplying a matrix ψ in S, a left ideal, on the left by $\lambda \in \mathbb{F}$ does not result in a left $\mathbb{F}$-linear structure.

There is a natural way to introduce scalar products on the spinor space $S \subset C\ell_3$. First, note that for all $\psi, \varphi \in S$ the product

$$\tilde{\psi}\varphi \simeq \begin{pmatrix} \psi_1^* & \psi_2^* \\ 0 & 0 \end{pmatrix} \begin{pmatrix} \varphi_1 & 0 \\ \varphi_2 & 0 \end{pmatrix} = \begin{pmatrix} \psi_1^*\varphi_1 + \psi_2^*\varphi_2 & 0 \\ 0 & 0 \end{pmatrix}$$

falls in the division ring $\mathbb{F}$ ($z \to z^*$ means complex conjugation). To show that the map

$$S \times S \to \mathbb{F}, \ (\psi, \varphi) \to \tilde{\psi}\varphi$$

defines a scalar product we only have to verify that the reversion $\psi \to \tilde{\psi}$ is a right-to-left $\mathbb{F}$-semilinear map. For all $\psi \in S$, $\lambda \in \mathbb{F}$ we have $(\psi\lambda)^{\sim} = \tilde{\lambda}\tilde{\psi}$ where the map $\lambda \to \tilde{\lambda}$ is an anti-involution of the division algebra $\mathbb{F}$ (actually complex conjugation).

Multiplying a spinor $\psi \in S \subset C\ell_3$ by an element $s \in C\ell_3$ is a right $\mathbb{F}$-linear transformation $S \to S$, $\psi \to s\psi$. The automorphism group of the scalar product is formed by those right $\mathbb{F}$-linear transformations which preserve the scalar product, that is,

$$(s\psi)^{\sim}(s\varphi) = \tilde{\psi}\varphi \quad \text{for all} \quad \psi, \varphi \in S.$$

The automorphism group of the scalar product $\tilde{\psi}\varphi$ is seen to be the group $\{s \in C\ell_3 \mid \tilde{s}s = 1\}$ which is isomorphic to the group of unitary 2×2-matrices,

$$U(2) = \{s \in \mathrm{Mat}(2, \mathbb{C}) \mid s^\dagger s = I\}.$$

We can also use the Clifford conjugate $u \to \bar{u}$ of $C\ell_3$ to introduce a scalar product for spinors. In this case, the element

$$\bar{\psi}\varphi \simeq \begin{pmatrix} 0 & 0 \\ -\psi_2 & \psi_1 \end{pmatrix} \begin{pmatrix} \varphi_1 & 0 \\ \varphi_2 & 0 \end{pmatrix} = \begin{pmatrix} 0 & 0 \\ \psi_1\varphi_2 - \psi_2\varphi_1 & 0 \end{pmatrix}$$

does not appear in the division ring $\mathbb{F} = fC\ell_3 f$. However, we can find an invertible element $a \in C\ell_3$ so that $a\bar{\psi}\varphi \in \mathbb{F}$, e.g. $a = \mathbf{e}_1$ or $a = \mathbf{e}_{31}$. The map

$$S \times S \to \mathbb{F}, \ (\psi, \varphi) \to a\bar{\psi}\varphi$$

defines a scalar product. Writing

$$J = \begin{pmatrix} 0 & 1 \\ -1 & 0 \end{pmatrix}$$

we find that $a\bar{\psi}\varphi \simeq \tau(\psi)^{\top} J \tau(\varphi)$. Hence, the automorphism group $\{s \in C\ell_3 \mid \bar{s}s = 1\}$ of the scalar product $a\bar{\psi}\varphi$ is the group of symplectic 2×2-matrices,

$$Sp(2, \mathbb{C}) = \{s \in \mathrm{Mat}(2, \mathbb{C}) \mid s^{\top} J s = J\}.$$

4.9 Spinor operators

Up till now spinors have been objects which have been operated upon. Next we will replace such passive spinors by active spinor operators. Instead of spinors

$$\psi = \begin{pmatrix} \psi_1 & 0 \\ \psi_2 & 0 \end{pmatrix} \in \mathcal{C}\ell_3 f$$

in minimal left ideals we will consider the following even elements:

$$\Psi = 2\,\mathrm{even}(\psi) = \begin{pmatrix} \psi_1 & -\psi_2^* \\ \psi_2 & \psi_1^* \end{pmatrix} \in \mathcal{C}\ell_3^+,$$

also computed as $\Psi = \psi + \hat{\psi}$ for $\psi \in \mathcal{C}\ell_3 f$. Classically, the expectation values of the components of the spin have been determined in terms of the column spinor $\psi \in \mathbb{C}^2$ by computing the following three real numbers:

$$s_1 = \psi^\dagger \sigma_1 \psi, \quad s_2 = \psi^\dagger \sigma_2 \psi, \quad s_3 = \psi^\dagger \sigma_3 \psi.$$

In terms of $\psi \in \mathcal{C}\ell_3 f$ this computation could be repeated as

$$s_1 = 2\langle \psi \mathbf{e}_1 \tilde{\psi} \rangle_0, \quad s_2 = 2\langle \psi \mathbf{e}_2 \tilde{\psi} \rangle_0, \quad s_3 = 2\langle \psi \mathbf{e}_3 \tilde{\psi} \rangle_0.$$

However, in terms of $\Psi \in \mathcal{C}\ell_3^+$ we may compute $\mathbf{s} = s_1\mathbf{e}_1 + s_2\mathbf{e}_2 + s_3\mathbf{e}_3$ directly as

$$\mathbf{s} = \Psi \mathbf{e}_3 \tilde{\Psi}.$$

Since Ψ acts here like an operator, we call it a *spinor operator*. It should be emphasized that not only did we get all the components of the spin vector $\mathbf{s}$ at one stroke, but we also got the entity $\mathbf{s}$ as a whole.

Remark. The mapping $\mathcal{C}\ell_3^+ \to \mathbb{R}^3$, $\Psi \to \Psi\sigma_3\Psi^\dagger = \Psi\mathbf{e}_3\tilde{\Psi}$ is the *KS*-transformation (introduced by Kustaanheimo & Stiefel 1965) for spinor regularization of Kepler motion, and its restriction to norm-one spinor operators Ψ satisfying $\Psi\tilde{\Psi} = 1$ (or equivalently $\Psi\Psi^\dagger = I$) results in a Hopf fibration $S^3 \to S^2$ (the matrix $\Psi\sigma_3\Psi^\dagger$ is both unitary and involutory and represents a reflection of the spinor space with axis ψ).

The above mapping should not be confused with the 'Cartan map', see Cartan 1966 p. 41 and Keller & Rodríguez-Romo 1991 p. 1591. A 'Cartan map' $\mathbb{C}^2 \times \mathbb{C}^2 \to \mathcal{C}\ell_3$, $(\psi, \varphi) \to 2\psi\mathbf{e}_1\bar{\varphi}$, where $\mathbb{C}^2 = \mathcal{C}\ell_3 f$, sends a pair of square matrix spinors to a complex 4-vector $x_0 + \mathbf{x}$,

$$x_0 = -(\psi_1\varphi_2 - \psi_2\varphi_1), \quad \mathbf{x} = \begin{pmatrix} \psi_1\varphi_1 - \psi_2\varphi_2 \\ i(\psi_1\varphi_1 + \psi_2\varphi_2) \\ -(\psi_1\varphi_2 + \psi_2\varphi_1) \end{pmatrix}.$$

When $\psi = \varphi$, $\mathbf{x}^2 = 0$. ∎

Note also that $\mathrm{trace}(\psi\psi^\dagger) = 2\langle \psi\tilde{\psi} \rangle_0 = \Psi\tilde{\Psi}$ which equals $\Psi\bar{\Psi} = \det(\Psi)$.

In operator form the Schrödinger-Pauli equation

$$i\hbar\frac{\partial\Psi}{\partial t} = \frac{1}{2m}\pi^2\Psi - \frac{\hbar e}{2m}\vec{B}\Psi\mathbf{e}_3 - eV\Psi$$

shows explicitly the quantization direction $\mathbf{e}_3$ of the spin. The explicit occurrence of $\mathbf{e}_3$ is due to the injection $\mathbb{C}^2 \to C\ell_3 f$, $\psi \to \Psi$; technically $2\,\mathrm{even}(\vec{B}\psi) = \vec{B}\Psi\mathbf{e}_3$. If we rotate the system $90°$ around the y-axis, counter-clockwise as seen from the positive y-axis, then vectors and spinors transform to

$$\vec{B}' = u\vec{B}u^{-1} \quad \text{and} \quad \Psi' = u\Psi \quad \text{where} \quad u = \exp(\frac{\pi}{4}\mathbf{e}_{13}),$$

and the Pauli equation transforms to

$$i\hbar\frac{\partial\Psi'}{\partial t} = \frac{1}{2m}\pi'^2\Psi' - \frac{\hbar e}{2m}\vec{B}'\Psi'\mathbf{e}_3 - eV\Psi'.$$

If this equation is multiplied on the right by u^{-1}, then $\mathbf{e}_3$ goes to $\mathbf{e}_1 = u\mathbf{e}_3u^{-1}$, and the equation looks like

$$i\hbar\frac{\partial\Psi''}{\partial t} = \frac{1}{2m}\pi'^2\Psi'' - \frac{\hbar e}{2m}\vec{B}'\Psi''\mathbf{e}_1 - eV\Psi'',$$

where $\Psi'' = u\Psi u^{-1}$. Both the transformation laws give the same values for observables, that is, $\Psi'\mathbf{e}_3\tilde{\Psi}' = \Psi''\mathbf{e}_1\tilde{\Psi}''$.

Exercises

1. Compute the square of $\mathbf{a} + \mathbf{b}\mathbf{e}_{123}$ where $\mathbf{a}, \mathbf{b} \in \mathbb{R}^3$.
2. Compute p^2, q^2 and pq for $p = \frac{1}{2}(1 + \mathbf{e}_3)$ and $q = \frac{1}{2}(1 - \mathbf{e}_3)$.
3. Compute the squares of $\frac{1}{2}(1 + \mathbf{e}_3) \pm \frac{1}{2}(1 - \mathbf{e}_3)\mathbf{e}_{12}$.
4. Find all the four square roots of $\cos\varphi + \mathbf{e}_{12}\sin\varphi$. Hint: $\mathbf{e}_{12}\mathbf{e}_3 = \mathbf{e}_3\mathbf{e}_{12}$.
5. Find the exponentials of $\pm\frac{\pi}{2}(1 - \mathbf{e}_3)\mathbf{e}_{12}$. Hint: $\mathbf{e}_{12}$ and $\mathbf{e}_{123}$ commute [or $q = \frac{1}{2}(1 - \mathbf{e}_3)$ is an idempotent satisfying $q^2 = q$].
6. Let $u = \alpha + \mathbf{a} + \mathbf{b}\mathbf{e}_{123} + \beta\mathbf{e}_{123}$ [$\alpha, \beta \in \mathbb{R}$ and $\mathbf{a}, \mathbf{b} \in \mathbb{R}^3$]. Compute $u\bar{u}$.
7. Find the inverse of $u = \alpha + \mathbf{a} + \mathbf{b}\mathbf{e}_{123} + \beta\mathbf{e}_{123}$. Hint: $u\bar{u}$ is of the form $x + y\mathbf{e}_{123}$, $x, y \in \mathbb{R}$.
8. Find the exponential of $u = \alpha + \mathbf{a} + \mathbf{b}\mathbf{e}_{123} + \beta\mathbf{e}_{123}$. Hint: compute $(\mathbf{a} + \mathbf{b}\mathbf{e}_{123})^2$.
9. Show that each non-zero even element in $C\ell_3^+$ is invertible.
10. Show that $u\tilde{u} \in \mathbb{R} \oplus \mathbb{R}^3$ for all $u \in C\ell_3$.
11. Show that $|ua\tilde{u}| = |u|^2|\mathbf{a}|$ for $\mathbf{a} \in \mathbb{R}^3$, $u \in \mathbb{R} \oplus \bigwedge^2\mathbb{R}^3$.
12. Show that the norm on $C\ell_3$, defined by $|u|^2 = \langle u\tilde{u}\rangle_0$, agrees with the

norm given by $|u|^2 = <u,u>$ where the symmetric bilinear product is determined by

$$<\alpha,\beta> = \alpha\beta \quad \text{for} \quad \alpha,\beta \in \mathbb{R},$$
$$<\mathbf{a},\mathbf{b}> = \mathbf{a}\cdot\mathbf{b} \quad \text{for} \quad \mathbf{a},\mathbf{b} \in \mathbb{R}^3$$

and by

$$<\mathbf{x}_1 \wedge \ldots \wedge \mathbf{x}_k, \mathbf{y}_1 \wedge \ldots \wedge \mathbf{y}_k> = \begin{vmatrix} \mathbf{x}_1\cdot\mathbf{y}_1 & \ldots & \mathbf{x}_1\cdot\mathbf{y}_k \\ \vdots & \ddots & \vdots \\ \mathbf{x}_k\cdot\mathbf{y}_1 & \ldots & \mathbf{x}_k\cdot\mathbf{y}_k \end{vmatrix}$$

in $\bigwedge^k \mathbb{R}^3$, $k \geq 2$. [One also needs to assume orthogonality of the components in $\mathcal{C}\ell_3 = \mathbb{R} \oplus \mathbb{R}^3 \oplus \bigwedge^2 \mathbb{R}^3 \oplus \bigwedge^3 \mathbb{R}^3$.]

13. Show that the reflection across the plane of the bivector $\mathbf{A}$ is obtained by $\mathbf{r} \to \mathbf{r}' = -\mathbf{A}\mathbf{r}\mathbf{A}^{-1}$.
14. Let $\mathbf{x},\mathbf{y},\mathbf{z} \in \mathbb{R}^3$. Compute $\langle\mathbf{xyz}\rangle_1$ and $\langle\mathbf{xyz}\rangle_3$. Hint: use reversion.

Solutions

1. $(\mathbf{a} + \mathbf{b}\mathbf{e}_{123})^2 = \mathbf{a}\cdot\mathbf{a} - \mathbf{b}\cdot\mathbf{b} + 2(\mathbf{a}\cdot\mathbf{b})\mathbf{e}_{123}$.
2. $p^2 = p$ and $q^2 = q$, that is, p and q are idempotents; and $pq = 0$ [and so there are zero-divisors in the Clifford algebra $\mathcal{C}\ell_3$].
3. $\mathbf{e}_3$ [this shows that vectors can have square roots].
4. $\pm(\cos\frac{\varphi}{2} + \mathbf{e}_{12}\sin\frac{\varphi}{2})$, $\pm\mathbf{e}_3(\cos\frac{\varphi}{2} + \mathbf{e}_{12}\sin\frac{\varphi}{2})$.
5. $\mathbf{e}_3$ [this shows that vectors also have logarithms].
6. $\alpha^2 - \beta^2 - \mathbf{a}\cdot\mathbf{a} + \mathbf{b}\cdot\mathbf{b} + 2(\alpha\beta - \mathbf{a}\cdot\mathbf{b})\mathbf{e}_{123}$.
8. Denote $r = \sqrt{(\mathbf{a}+\mathbf{b}\mathbf{e}_{123})^2} \in \mathbb{R} \oplus \bigwedge^3 \mathbb{R}^3$, $v = (\mathbf{a}+\mathbf{b}\mathbf{e}_{123})/r$, $v^2 = 1$. Then $\exp(u) = \exp(\alpha + \beta\mathbf{e}_{123})[\frac{1}{2}(1+v)\exp(r) + \frac{1}{2}(1-v)\exp(-r)]$ when $r \neq 0$. When $r = 0$: $\exp(u) = \exp(\alpha+\beta\mathbf{e}_{123})(1 + \mathbf{a} + \mathbf{b}\mathbf{e}_{123})$.
10. $u = \alpha + \mathbf{a} + \mathbf{b}\mathbf{e}_{123} + \beta\mathbf{e}_{123}$, $u\tilde{u} = \alpha^2 + \beta^2 + \mathbf{a}^2 + \mathbf{b}^2 + 2(\alpha\mathbf{a} + \beta\mathbf{b} + \mathbf{a}\times\mathbf{b})$ which is in $\mathbb{R} \oplus \mathbb{R}^3$. Direct proof:

$$(u\tilde{u})\tilde{} = \tilde{\tilde{u}}\tilde{u} = u\tilde{u}$$

which implies $u\tilde{u} \in \mathbb{R} \oplus \mathbb{R}^3$, since the reversion sends bivectors and 3-vectors to their opposites.

13. Decompose $\mathbf{r}$ into components parallel, $\mathbf{r}_{||}$, and perpendicular, $\mathbf{r}_\perp$, to $\mathbf{A}$, and note that $\mathbf{A}$ anticommutes with vectors in its plane,
$\mathbf{A}(\mathbf{r}_{||} + \mathbf{r}_\perp) = (-\mathbf{r}_{||} + \mathbf{r}_\perp)\mathbf{A}$. Then
$\mathbf{A}(\mathbf{r}_{||} + \mathbf{r}_\perp)\mathbf{A}^{-1} = (-\mathbf{r}_{||} + \mathbf{r}_\perp)\mathbf{A}\mathbf{A}^{-1} = -\mathbf{r}'$.
14. First, $(\mathbf{xyz})\tilde{} = \mathbf{zyx}$ and $(\mathbf{xyz})\tilde{} = \langle\mathbf{xyz}\rangle_1 - \langle\mathbf{xyz}\rangle_3$. Therefore,

$\langle\mathbf{xyz}\rangle_1 = \frac{1}{2}(\mathbf{xyz} + \mathbf{zyx})$ and $\langle\mathbf{xyz}\rangle_3 = \frac{1}{2}(\mathbf{xyz} - \mathbf{zyx})$, and also
$\langle\mathbf{xyz}\rangle_1 = (\mathbf{y}\cdot\mathbf{z})\mathbf{x} - (\mathbf{z}\cdot\mathbf{x})\mathbf{y} + (\mathbf{x}\cdot\mathbf{y})\mathbf{z}$ and $\langle\mathbf{xyz}\rangle_3 = \mathbf{x}\wedge\mathbf{y}\wedge\mathbf{z}$.

Bibliography

W. Baylis: *Theoretical Methods in the Physical Sciences: an Introduction to Problem Solving with MAPLE V*. Birkhäuser, Boston, MA, 1994.

H.A Bethe, E.E. Salpeter: *Quantum Mechanics of One- and Two-Electron Atoms*. Springer, Berlin, 1957.

E. Cartan: *The Theory of Spinors*. The M.I.T. Press, Cambridge, MA, 1966.

A. Charlier, A. Bérard, M.-F. Charlier, D. Fristot: *Tensors and the Clifford Algebra, Applications to the Physics of Bosons and Fermions*. Marcel Dekker, New York, 1992.

R.C. Feynman, R.B. Leighton, M. Sands: *The Feynman Lectures on Physics, Vol. III, Quantum Mechanics*. Addison-Wesley, Reading, MA, 1965.

D. Hestenes: *Space-Time Algebra*. Gordon and Breach, New York, 1966, 1987, 1992.

J. Keller, S. Rodríguez-Romo: Multivectorial generalization of the Cartan map. *J. Math. Phys.* **32** (1991), 1591-1598.

E. Merzbacher: *Quantum Mechanics*. Wiley, New York, 1970.

W. Pauli: Zur Quantenmechanik des magnetischen Elektrons. *Z. Physik* **43** (1927), 601-623.

M. Riesz: *Clifford Numbers and Spinors*. The Institute for Fluid Dynamics and Applied Mathematics, Lecture Series No. **38**, University of Maryland, 1958. Reprinted as facsimile (eds.: E.F. Bolinder, P. Lounesto) by Kluwer, Dordrecht, The Netherlands, 1993.

5

Quaternions

We saw in the chapter on *Complex Numbers* that it is convenient to use the real algebra of complex numbers $\mathbb{C}$ to represent the rotation group $SO(2)$ of the plane $\mathbb{R}^2$. In this chapter we shall study rotations of the 3-dimensional space $\mathbb{R}^3$. The composition of spatial rotations is no longer commutative, and we need a non-commutative multiplication to represent the rotation group $SO(3)$. This can be done within the real algebra of 3×3-matrices $\mathrm{Mat}(3, \mathbb{R})$, or by the real algebra of **quaternions**, $\mathbb{H}$, invented by Hamilton.

The complex plane $\mathbb{C}$ is a real linear space $\mathbb{R}^2$, and multiplication by a complex number $c = a + ib$, that is, the map $\mathbb{C} \to \mathbb{C}$, $z \to cz$, may be regarded as a real linear map with matrix $\begin{pmatrix} a & -b \\ b & a \end{pmatrix}$ operating on $\begin{pmatrix} x \\ y \end{pmatrix}$ in $\mathbb{R}^2$. The complex plane is also a real quadratic space $\mathbb{R}^{2,0}$, in short $\mathbb{R}^2$, with a quadratic form

$$\mathbb{C} \to \mathbb{R},\ z = x + iy \to z\bar{z} = x^2 + y^2,$$

and norm $|z| = \sqrt{z\bar{z}}$. Multiplication of complex numbers preserves the norm, that is, $|cz| = |c||z|$ for all c, $z \in \mathbb{C}$, and so multiplication by c is a rotation of $\mathbb{R}^2$ if, and only if, $|c| = 1$. Conversely, any rotation of $\mathbb{R}^2$ can be represented by a unit complex number c, $|c| = 1$, in $\mathbb{C}$. The unit complex numbers form a group

$$U(1) = \{c \in \mathbb{C} \mid c\bar{c} = 1\},$$

called the *unitary group*, which is isomorphic to the rotation group $SO(2) = \{U \in \mathrm{Mat}(2, \mathbb{R}) \mid U^\top U = I,\ \det U = 1\}$, that is, $U(1) \simeq SO(2)$. The unitary group $U(1)$ can be visualized as the *unit circle*

$$S^1 = \{x + iy \in \mathbb{C} \mid x^2 + y^2 = 1\}$$

of the complex plane $\mathbb{C}$.

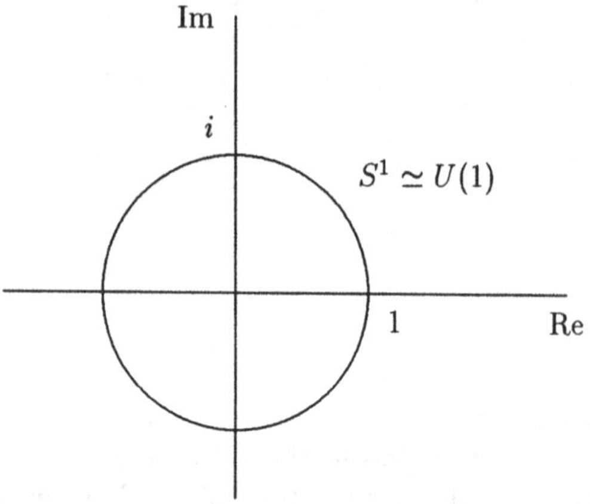

Similarly, the algebra of quaternions $\mathbb{H}$ may be used to represent rotations of the 3-dimensional space $\mathbb{R}^3$. It will turn out that quaternions are also convenient to represent the rotations of the 4-dimensional space $\mathbb{R}^4$.

Quaternions as hypercomplex numbers

Quaternions are generalized complex numbers of the form $q = w + ix + jy + kz$ where w, x, y, z are real numbers and the generalized imaginary units i, j, k satisfy the following multiplication rules:

$$i^2 = j^2 = k^2 = -1,$$
$$ij = k = -ji, \quad jk = i = -kj, \quad ki = j = -ik.$$

Note that the multiplication is by definition **non-commutative**. One can show that quaternion multiplication is *associative*. The above multiplication rules can be condensed into the following form:

$$i^2 = j^2 = k^2 = ijk = -1$$

where in the last identity we have omitted parentheses and thereby tacitly assumed associativity.

The generalized imaginary units will be denoted either by i, j, k or by $\mathbf{i}, \mathbf{j}, \mathbf{k}$. They have two different roles: they act as generators of

rotations, that is, they are bivectors, or
translations, that is, they are vectors.

This distinction is not clear-cut since bivectors are dual to vectors in $\mathbb{R}^3$.

5.1 Pure part and cross product

A quaternion $q = w + \mathbf{i}x + \mathbf{j}y + \mathbf{k}z$ is a sum of a scalar and a vector, called the *real part*, $\text{Re}(q) = w \in \mathbb{R}$, and the *pure part*, $\text{Pu}(q) = \mathbf{i}x + \mathbf{j}y + \mathbf{k}z \in \mathbb{R}^3$. The quaternions form a 4-dimensional real linear space $\mathbb{H}$ which contains the real axis $\mathbb{R}$ and a 3-dimensional real linear space $\mathbb{R}^3$ so that $\mathbb{H} = \mathbb{R} \oplus \mathbb{R}^3$. We denote the pure part also by a boldface letter so that $q = q_0 + \mathbf{q}$ where $q_0 \in \mathbb{R}$ and $\mathbf{q} = \mathbf{i}q_1 + \mathbf{j}q_2 + \mathbf{k}q_3 \in \mathbb{R}^3$. The real linear space $\mathbb{R} \oplus \mathbb{R}^3$ with the quaternion product is an associative algebra over $\mathbb{R}$ called the *quaternion algebra* $\mathbb{H}$. The product of two quaternions $a = a_0 + \mathbf{a}$ and $b = b_0 + \mathbf{b}$ can be written as

$$ab = a_0 b_0 - \mathbf{a} \cdot \mathbf{b} + a_0 \mathbf{b} + \mathbf{a} b_0 + \mathbf{a} \times \mathbf{b}.$$

A quaternion $q = q_0 + \mathbf{q}$ is *pure* if its real part vanishes, $q_0 = 0$, so that $q = \mathbf{q} \in \mathbb{R}^3$. A product of two pure quaternions $\mathbf{a} = \mathbf{i}a_1 + \mathbf{j}a_2 + \mathbf{k}a_3$ and $\mathbf{b} = \mathbf{i}b_1 + \mathbf{j}b_2 + \mathbf{k}b_3$ is a sum of a real number and a pure quaternion:

$$\mathbf{ab} = -\mathbf{a} \cdot \mathbf{b} + \mathbf{a} \times \mathbf{b}$$

where we recognize the scalar product $\mathbf{a} \cdot \mathbf{b} = a_1 b_1 + a_2 b_2 + a_3 b_3$ and the cross product $\mathbf{a} \times \mathbf{b} = \mathbf{i}(a_2 b_3 - a_3 b_2) + \mathbf{j}(a_3 b_1 - a_1 b_3) + \mathbf{k}(a_1 b_2 - a_2 b_1)$.

The vector space $\mathbb{R}^3$ with the cross product $\mathbf{a} \times \mathbf{b}$ is a real algebra, that is, it is a real linear space with a bilinear map

$$\mathbb{R}^3 \times \mathbb{R}^3 \to \mathbb{R}^3, \quad (\mathbf{a}, \mathbf{b}) \to \mathbf{a} \times \mathbf{b}.$$

The cross product satisfies two rules

$$\begin{aligned} &\mathbf{a} \times \mathbf{b} = -\mathbf{b} \times \mathbf{a}, \\ &\mathbf{a} \times (\mathbf{b} \times \mathbf{c}) + \mathbf{b} \times (\mathbf{c} \times \mathbf{a}) + \mathbf{c} \times (\mathbf{a} \times \mathbf{b}) = 0, \end{aligned}$$

the latter being called the Jacobi identity; this makes $\mathbb{R}^3$ with the cross product a *Lie algebra*. In particular, the cross product is not associative, $\mathbf{a} \times (\mathbf{b} \times \mathbf{c}) \neq (\mathbf{a} \times \mathbf{b}) \times \mathbf{c}$.

We can reobtain the cross product of two pure quaternions $\mathbf{a}, \mathbf{b} \in \mathbb{R}^3$ as the pure part of their quaternion product: $\mathbf{a} \times \mathbf{b} = \text{Pu}(\mathbf{ab})$.

5.2 Quaternion conjugate, norm and inverse

The conjugate $\bar{q}$ of a quaternion $q = w + \mathbf{i}x + \mathbf{j}y + \mathbf{k}z$ is obtained by changing the sign of the pure part:

$$\bar{q} = w - \mathbf{i}x - \mathbf{j}y - \mathbf{k}z.$$

We shall also refer to $\bar{q}$ as the *quaternion conjugate* of q. The conjugation is an anti-automorphism of $\mathbb{H}$; $\overline{ab} = \bar{b}\bar{a}$ for $a, b \in \mathbb{H}$.

A quaternion q multiplied by its conjugate $\bar{q}$ results in a real number $q\bar{q} = w^2 + x^2 + y^2 + z^2$ called the square norm of $q = w + \mathbf{i}x + \mathbf{j}y + \mathbf{k}z$. The *norm* $|q|$ of q is given by $|q| = \sqrt{q\bar{q}}$ so that

$$|w + \mathbf{i}x + \mathbf{j}y + \mathbf{k}z| = \sqrt{w^2 + x^2 + y^2 + z^2}.$$

The norm of a product of two quaternions a and b is the product of their norms – as an equation, $|ab| = |a||b|$ for $a, b \in \mathbb{H}$ – which turns $\mathbb{H}$ into a normed algebra.

The *inverse* q^{-1} of a non-zero quaternion q is obtained by $q^{-1} = \bar{q}/|q|^2$ or more explicitly by

$$\frac{1}{w + \mathbf{i}x + \mathbf{j}y + \mathbf{k}z} = \frac{w - \mathbf{i}x - \mathbf{j}y - \mathbf{k}z}{w^2 + x^2 + y^2 + z^2}.$$

In particular, $ab = 0$ implies $a = 0$ or $b = 0$, which means that the quaternion algebra is a division algebra (or that the ring of quaternions is a division ring).

5.3 The center of $\mathbb{H}$

The set of those elements in $\mathbb{H}$ which commute with every element of $\mathbb{H}$ forms the *center* of $\mathbb{H}$,

$$\mathrm{Cen}(\mathbb{H}) = \{w \in \mathbb{H} \mid wq = qw \text{ for all } q \in \mathbb{H}\}.$$

The center is of course closed under multiplication. The center of the division ring $\mathbb{H}$ is isomorphic to the field of real numbers $\mathbb{R}$. In contrast to the case of the complex field $\mathbb{C}$, the real axis in $\mathbb{H}$ is the unique subfield which is the center of $\mathbb{H}$.

5.4 Rotations in three dimensions

Take a pure quaternion or a vector

$$\mathbf{r} = \mathbf{i}x + \mathbf{j}y + \mathbf{k}z \in \mathbb{R}^3, \quad \text{where} \quad \mathbb{H} = \mathbb{R} \oplus \mathbb{R}^3,$$

of length $|\mathbf{r}| = \sqrt{x^2 + y^2 + z^2}$. For a non-zero quaternion $a \in \mathbb{H}$, the expression $a\mathbf{r}a^{-1}$ is again a pure quaternion with the same length, that is,

$$a\mathbf{r}a^{-1} \in \mathbb{R}^3 \quad \text{and} \quad |a\mathbf{r}a^{-1}| = |\mathbf{r}|.$$

In other words, the mapping

$$\mathbb{R}^3 \to \mathbb{R}^3, \quad \mathbf{r} \to a\mathbf{r}a^{-1}$$

is a rotation of the quadratic space of pure quaternions $\mathbb{R}^3$. Each rotation in $SO(3) = \{U \in \mathrm{Mat}(3, \mathbb{R}) \mid U^\top U = I, \ \det U = 1\}$ can be so represented, and there are two unit quaternions a and $-a$ representing the same rotation, $a\mathbf{r}a^{-1} = (-a)\mathbf{r}(-a)^{-1}$. In other words, the sphere of unit quaternions,

$$S^3 = \{q \in \mathbb{H} \mid |q| = 1\},$$

is a two-fold covering group of $SO(3)$, that is, $SO(3) \simeq S^3/\{\pm 1\}$.

A rotation has three parameters in dimension 3. In other words, $SO(3)$ and S^3 are 3-dimensional manifolds. The three parameters are the angle of rotation and the two direction cosines of the axis of rotation.

To find the axis of this rotation we take a unit quaternion a, $|a| = 1$, and write it in the form $a = e^{\mathbf{a}/2}$ where $\mathbf{a} \in \mathbb{R}^3$. Note that

$$e^{\mathbf{a}/2} = \cos\frac{\alpha}{2} + \frac{\mathbf{a}}{\alpha}\sin\frac{\alpha}{2}$$

where $\alpha = |\mathbf{a}|$. The rotation $\mathbf{r} \to a\mathbf{r}a^{-1}$ turns $\mathbf{r}$ about the axis $\mathbf{a}$ by the angle α. The sense of the rotation is counter-clockwise when regarded from the arrow-head of $\mathbf{a}$.

The composite of two consecutive rotations, first around $\mathbf{a}$ by the angle $\alpha = |\mathbf{a}|$ and then around $\mathbf{b}$ by the angle $\beta = |\mathbf{b}|$, is again a rotation around some axis, say $\mathbf{c}$. The axis of the composite rotation can be found by inspection of the real and pure parts of the formula $e^{\mathbf{c}/2} = e^{\mathbf{b}/2}e^{\mathbf{a}/2}$. Divide both sides by their real parts and substitute

$$\mathbf{c}' = \frac{\mathbf{c}}{\gamma}\tan\frac{\gamma}{2}, \quad \text{where} \quad \gamma = |\mathbf{c}|,$$

to obtain the *Rodrigues formula*

$$\mathbf{c}' = \frac{\mathbf{a}' + \mathbf{b}' - \mathbf{a}' \times \mathbf{b}'}{1 - \mathbf{a}' \cdot \mathbf{b}'}.$$

5.5 Rotations in four dimensions

The mapping $\mathbb{H} \to \mathbb{H}$, $q \to aqb^{-1}$, where $a, b \in \mathbb{H}$ are unit quaternions $|a| = |b| = 1$, is a rotation of the 4-dimensional space $\mathbb{R}^4 = \mathbb{H}$. In other words, the real linear mapping

$$\mathbb{H} \to \mathbb{H}, \quad q \to aqb^{-1}, \quad \text{where} \quad a, b \in \mathbb{H} \quad \text{and} \quad |a| = |b| = 1,$$

is a rotation of $\mathbb{R}^4$. Each rotation in $SO(4)$ can be so represented, and there are two elements (a, b) and $(-a, -b)$ in $S^3 \times S^3$ representing the same rotation, that is, $aqb^{-1} = (-a)q(-b)^{-1}$. In other words, the group $S^3 \times S^3$ is a two-fold covering group of $SO(4)$, that is,

$$SO(4) \simeq \frac{S^3 \times S^3}{\{(1,1), (-1,-1)\}}.$$

A rotation in dimension 4 can be represented by a pair of unit quaternions, and so it has six parameters, in other words, $\dim SO(4) = \dim(S^3 \times S^3) = 6$. A rotation has two completely orthogonal invariant planes; both the invariant planes can turn arbitrarily; this takes two parameters. Fixing a plane in $\mathbb{R}^4$ takes the remaining four parameters: three parameters for a unit vector in S^3, plus two parameters for another orthogonal unit vector in S^2, minus one parameter for rotating the pairs of such vectors in the plane.

5.6 Matrix representation of quaternion multiplication

The product of two quaternions $q = w+\mathbf{i}x+\mathbf{j}y+\mathbf{k}z$ and $u = u_0+\mathbf{i}u_1+\mathbf{j}u_2+\mathbf{k}u_3$ can be represented by matrix multiplication:

$$\begin{pmatrix} w & -x & -y & -z \\ x & w & -z & y \\ y & z & w & -x \\ z & -y & x & w \end{pmatrix} \begin{pmatrix} u_0 \\ u_1 \\ u_2 \\ u_3 \end{pmatrix} = \begin{pmatrix} v_0 \\ v_1 \\ v_2 \\ v_3 \end{pmatrix}$$

where $qu = v$. Swapping the multiplication to the right, that is, $uq = v'$, gives a partially transformed matrix:

$$\begin{pmatrix} w & -x & -y & -z \\ x & w & z & -y \\ y & -z & w & x \\ z & y & -x & w \end{pmatrix} \begin{pmatrix} u_0 \\ u_1 \\ u_2 \\ u_3 \end{pmatrix} = \begin{pmatrix} v_0' \\ v_1' \\ v_2' \\ v_3' \end{pmatrix}.$$

Let us denote the above matrices respectively by L_q and R_q, that is,

$$L_q(u) = qu \ (= v) \quad \text{and} \quad R_q(u) = uq \ (= v').$$

We find that [1]

$$L_\mathbf{i} L_\mathbf{j} L_\mathbf{k} = -I \quad \text{and} \quad R_\mathbf{i} R_\mathbf{j} R_\mathbf{k} = I.$$

The sets $\{L_q \in \mathrm{Mat}(4,\mathbb{R}) \mid q \in \mathbb{H}\}$ and $\{R_q \in \mathrm{Mat}(4,\mathbb{R}) \mid q \in \mathbb{H}\}$ form two subalgebras of $\mathrm{Mat}(4,\mathbb{R})$, both isomorphic to $\mathbb{H}$. For two arbitrary quaternions $a, b \in \mathbb{H}$ these two matrix representatives commute, that is, $L_a R_b = R_b L_a$. Any real 4×4-matrix is a linear combination of matrices of the form $L_a R_b$. The above observations together with $(\dim \mathbb{H})^2 = \dim \mathrm{Mat}(4,\mathbb{R})$ imply that

$$\mathrm{Mat}(4,\mathbb{R}) \simeq \mathbb{H} \otimes \mathbb{H},$$

or more informatively $\mathrm{Mat}(4,\mathbb{R}) = \mathbb{H} \otimes \mathbb{H}^*$. [2]

1 Note that $R_\mathbf{i}^\top R_\mathbf{j}^\top R_\mathbf{k}^\top = -I$.

2 For unit quaternions $a, b \in \mathbb{H}$ such that $|a| = |b| = 1$ we may choose $L_a \in Q$ and $R_b \in Q^*$ or alternatively $L_a \in Q^*$ and $R_b \in Q$. For a discussion about the meaning of Q and Q^*, see the chapter on *The Fourth Dimension*.

Take a matrix of the form $U = L_a R_b$ in $\text{Mat}(4, \mathbb{R})$. Then $U^\top U = |a|^2|b|^2 I$, but in general $U + U^\top \neq \alpha I$. Take a matrix of the form $V = L_a + R_b$ in $\text{Mat}(4, \mathbb{R})$. Then $V + V^\top = 2(\text{Re}(a) + \text{Re}(b))I$, but in general $V^\top V \neq \beta I$. Conversely, if $U \in \text{Mat}(4, \mathbb{R})$ is such that $U + U^\top = \alpha I$ and $U^\top U = \beta I$ then the matrix U belongs either to $\mathbb{H}$ or to $\mathbb{H}^*$.

Besides real 4×4-matrices, quaternions can also be represented by complex 2×2-matrices:

$$w + \mathbf{i}x + \mathbf{j}y + \mathbf{k}z \simeq \begin{pmatrix} w + iz & ix + y \\ ix - y & w - iz \end{pmatrix}.$$

The orthogonal unit vectors $\mathbf{i}, \mathbf{j}, \mathbf{k}$ are represented by matrices obtained by multiplying each of the Pauli matrices $\sigma_1, \sigma_2, \sigma_3$ by $i = \sqrt{-1}$:

$$\mathbf{i} \simeq \begin{pmatrix} 0 & i \\ i & 0 \end{pmatrix}, \quad \mathbf{j} \simeq \begin{pmatrix} 0 & 1 \\ -1 & 0 \end{pmatrix}, \quad \mathbf{k} \simeq \begin{pmatrix} i & 0 \\ 0 & -i \end{pmatrix}.$$

5.7 Linear spaces over $\mathbb{H}$

Much of the theory of linear spaces over commutative fields extends to $\mathbb{H}$. Because of the non-commutativity of $\mathbb{H}$ it is, however, necessary to distinguish between two types of linear spaces over $\mathbb{H}$, namely *right* linear spaces and *left* linear spaces.

A *right* linear space over $\mathbb{H}$ consists of an additive group V and a map

$$V \times \mathbb{H} \to V, \quad (\mathbf{x}, \lambda) \to \mathbf{x}\lambda$$

such that the usual distributivity and unity axioms hold and such that, for all $\lambda, \mu \in \mathbb{H}$ and $\mathbf{x} \in V$,

$$(\mathbf{x}\lambda)\mu = \mathbf{x}(\lambda\mu).$$

A *left* linear space over $\mathbb{H}$ consists of an additive group V and a map

$$\mathbb{H} \times V \to V, \ (\lambda, \mathbf{x}) \to \lambda\mathbf{x}$$

such that the usual distributivity and unity axioms hold and such that, for all $\lambda, \mu \in \mathbb{H}$ and $\mathbf{x} \in V$,

$$\lambda(\mu\mathbf{x}) = (\lambda\mu)\mathbf{x}.$$

A mapping $L : V \to U$ between two right linear spaces V and U is a *right linear map* if it respects addition and, for all $\mathbf{x} \in V$, $\lambda \in \mathbb{H}$, $L(\mathbf{x}\lambda) = (L(\mathbf{x}))\lambda$.

Comment. In the matrix form the above definition means that

$$\begin{pmatrix} a & b \\ c & d \end{pmatrix} \left[\begin{pmatrix} x_1 \\ x_2 \end{pmatrix} \lambda \right] = \begin{pmatrix} a & b \\ c & d \end{pmatrix} \begin{pmatrix} x_1\lambda \\ x_2\lambda \end{pmatrix} = \left[\begin{pmatrix} a & b \\ c & d \end{pmatrix} \begin{pmatrix} x_1 \\ x_2 \end{pmatrix} \right] \lambda. \quad \blacksquare$$

Remark. Although there are linear spaces over $\mathbb{H}$, there are no algebras over $\mathbb{H}$, since non-commutativity of $\mathbb{H}$ precludes bilinearity over $\mathbb{H}$: $\lambda(xy) = (\lambda x)y \neq (x\lambda)y = x(\lambda y) \neq x(y\lambda) = (xy)\lambda$. ∎

5.8 Function theory of quaternion variables

The richness of complex analysis suggests that there might be a function theory of quaternion variables. There are several different ways to generalize the theory of complex variables to the theory of quaternion functions of quaternion variables, $f : \mathbb{H} \to \mathbb{H}$. However, many generalizations are uninteresting, the classes of functions are too small or too large. In the following we will first eliminate the uninteresting generalizations.

First, consider quaternion differentiable functions such that

$$f'(q) = \lim_{h\to 0}[f(q+h) - f(q)]h^{-1}, \quad \text{where} \quad q, h \in \mathbb{H},$$

exists. The derivative $f'(q)$ is a real linear function

$$\mathbb{R}^4 \to \mathbb{R}^4 : h \to f'(q)h$$

corresponding to multiplication by a quaternion $a \in \mathbb{H}$ on the left, $f'(q)h = ah$ for $h \in \mathbb{H} = \mathbb{R}^4$. However, since $ah \neq ha$ the only quaternion differentiable functions are the affine right $\mathbb{H}$-linear functions

$$f(q) = aq + b \quad \text{where} \quad a, b \in \mathbb{H}.$$

We conclude that the set of quaternion differentiable functions reduces to a small and uninteresting set.

Second, if we consider power series in a quaternion variable $q = w + \mathbf{i}x + \mathbf{j}y + \mathbf{k}z$, then we get the set of all power series in the four real variables w, x, y, z. For instance, the coordinates are first-order functions

$$\begin{aligned} w &= \tfrac{1}{4}(q - \mathbf{i}q\mathbf{i} - \mathbf{j}q\mathbf{j} - \mathbf{k}q\mathbf{k}), \\ x &= \tfrac{1}{4}(q - \mathbf{i}q\mathbf{i} + \mathbf{j}q\mathbf{j} + \mathbf{k}q\mathbf{k})\mathbf{i}^{-1}, \\ y &= \tfrac{1}{4}(q + \mathbf{i}q\mathbf{i} - \mathbf{j}q\mathbf{j} + \mathbf{k}q\mathbf{k})\mathbf{j}^{-1}, \\ z &= \tfrac{1}{4}(q + \mathbf{i}q\mathbf{i} + \mathbf{j}q\mathbf{j} - \mathbf{k}q\mathbf{k})\mathbf{k}^{-1}, \end{aligned}$$

and so the set of power series in q, with left and right quaternion coefficients, is the set of all power series in the real variables w, x, y, z. This set is too big to be interesting.

Third, we could consider power series in q with real coefficients, that is, functions of type $f(q) = a_0 + a_1 q + a_2 q^2 + \ldots$ where $a_0, a_1, a_2, \ldots$ are real.

Restrict such a function to the complex subfield $\mathbb{C} \subset \mathbb{H}$, and send $z = x + iy$ to $f(z) = u+iv$, where $u = u(x, y)$ and $v = v(x, y)$. Decompose the quaternion q into real and vector parts, $q = q_0 + \mathbf{q}$, and note that $\mathbf{q}/|\mathbf{q}|$ is a generalized imaginary unit, $(\mathbf{q}/|\mathbf{q}|)^2 = -1$. Then

$$f(q_0 + \mathbf{q}) = u(q_0, |\mathbf{q}|) + \frac{\mathbf{q}}{|\mathbf{q}|} v(q_0, |\mathbf{q}|).$$

So this generalization just rotates the graph of $\mathbb{C} \to \mathbb{C}$, $z \to f(z)$, or rather makes $i = \mathbf{i}$ sweep all of $S^2 = \{\mathbf{r} \in \mathbb{R}^3 \mid |\mathbf{r}| = 1\}$, and thus gives only (a subclass of) axially symmetric functions.

Fourth, we could consider functions which are conformal almost everywhere in $\mathbb{R}^4$. This leads to Möbius transformations of $\mathbb{R}^4$, or its one-point compactification $\mathbb{R}^4 \cup \{\infty\}$. The Möbius transformations are compositions of the four mappings sending q to

aqb^{-1}	$a, b \in S^3$	rotations
$q + b$	$b \in \mathbb{H}$	translations
$q\lambda$	$\lambda > 0$	dilations
$(q^{-1} + c)^{-1}$	$c \in \mathbb{H}$	transversions.

A nice thing about quaternions is that all Möbius transformations of $\mathbb{R}^4$ can be written in the form $(aq + b)(cq + d)^{-1}$, where $a, b, c, d \in \mathbb{H}$.

Fifth, we could focus our attention on a generalization of the Cauchy-Riemann equations,

$$\frac{\partial f}{\partial w} + \mathbf{i}\frac{\partial f}{\partial x} + \mathbf{j}\frac{\partial f}{\partial y} + \mathbf{k}\frac{\partial f}{\partial z} = 0 \quad \text{where} \quad f : \mathbb{H} \to \mathbb{H}.$$

Using the differential operator

$$\nabla = \mathbf{i}\frac{\partial}{\partial x} + \mathbf{j}\frac{\partial}{\partial y} + \mathbf{k}\frac{\partial}{\partial z}$$

the above equation can be put into the form

$$\frac{\partial f_0}{\partial w} + \frac{\partial \mathbf{f}}{\partial w} + \nabla f_0 - \nabla \cdot \mathbf{f} + \nabla \times \mathbf{f} = 0$$

where $f = f_0 + \mathbf{f}$ with $f_0 : \mathbb{H} \to \mathbb{R}$ and $\mathbf{f} : \mathbb{H} \to \mathbb{R}^3$. This decomposes into scalar and vector parts

$$\frac{\partial f_0}{\partial w} - \nabla \cdot \mathbf{f} = 0 \quad \text{and} \quad \frac{\partial \mathbf{f}}{\partial w} + \nabla f_0 + \nabla \times \mathbf{f} = 0.$$

There are three linearly independent first-order solutions to these equations

$$q_x = x - \mathbf{i}w, \; q_y = y - \mathbf{j}w, \; q_z = z - \mathbf{k}w.$$

Higher-order homogeneous solutions are linear combinations of symmetrized products of q_x, q_y, q_z. For instance, the symmetrized product of degrees $2, 1, 0$ with respect to q_x, q_y, q_z is seen to be

$$q_x^2 q_y + q_x q_y q_x + q_y q_x^2 = 3(x^2 - w^2)y - 6wxy\mathbf{i} + (w^3 - 3wx^2)\mathbf{j}.$$

This already shows that the last alternative results in an interesting class of new functions, to some extent analogous to the class of holomorphic functions of a complex variable.

Historical survey

Hamilton invented his quaternions in 1843 when he tried to introduce a product for vectors in $\mathbb{R}^3$ similar to the product of complex numbers in $\mathbb{C}$. The present-day formalism of vector algebra was extracted out of the quaternion product of two vectors, $\mathbf{ab} = -\mathbf{a} \cdot \mathbf{b} + \mathbf{a} \times \mathbf{b}$, by Gibbs in 1901.

Hamilton tried to find an algebraic system which would do for the space $\mathbb{R}^3$ the same thing as complex numbers do for the plane $\mathbb{R}^2$. In particular, Hamilton wanted to find a multiplication rule for triplets $\mathbf{a} = a_1\mathbf{i} + a_2\mathbf{j} + a_3\mathbf{k}$ and $\mathbf{b} = b_1\mathbf{i} + b_2\mathbf{j} + b_3\mathbf{k}$ so that $|\mathbf{ab}| = |\mathbf{a}||\mathbf{b}|$, that is, a multiplicative product of vectors $\mathbf{a}, \mathbf{b} \in \mathbb{R}^3$. However, no such bilinear products exist (at least not over the rationals), since $3 \times 21 = 63 \neq n_1^2 + n_2^2 + n_3^2$ for any integers n_1, n_2, n_3 though $3 = 1^2 + 1^2 + 1^2$ and $21 = 1^2 + 2^2 + 4^2$ (no integer of the form $4^a(8b+7)$, with $a \geq 0$, $b \geq 0$, is a sum of three squares, a result of Legendre in 1830).

Hamilton also tried to find a generalized complex number system in three dimensions. However, no such associative hypercomplex numbers exist in three dimensions. This can be seen by considering generalized imaginary units $\mathbf{i}$ and $\mathbf{j}$ such that $\mathbf{i}^2 = \mathbf{j}^2 = -1$, and such that $1, \mathbf{i}, \mathbf{j}$ span $\mathbb{R}^3$. [3] The product must be of the form $\mathbf{ij} = \alpha + \mathbf{i}\beta + \mathbf{j}\gamma$ for some real α, β, γ. Then

$$\begin{aligned}\mathbf{i}(\mathbf{ij}) &= \mathbf{i}\alpha - \beta + (\mathbf{ij})\gamma = \mathbf{i}\alpha - \beta + (\alpha + \mathbf{i}\beta + \mathbf{j}\gamma)\gamma \\ &= -\beta + \alpha\gamma + \mathbf{i}(\alpha + \beta\gamma) + \mathbf{j}\gamma^2,\end{aligned}$$

whereas by associativity $\mathbf{i}(\mathbf{ij}) = \mathbf{i}^2\mathbf{j} = -\mathbf{j}$ which leads to a contradiction since $\gamma^2 \geq 0$ for all real γ.

Hamilton's great idea was to go to four dimensions and consider elements of the form $q = w + \mathbf{i}x + \mathbf{j}y + \mathbf{k}z$ where the hypercomplex units $\mathbf{i}, \mathbf{j}, \mathbf{k}$ satisfy the following **non-commutative** multiplication rules

$$\begin{aligned}&\mathbf{i}^2 = \mathbf{j}^2 = \mathbf{k}^2 = -1, \\ &\mathbf{ij} = \mathbf{k} = -\mathbf{ji},\ \mathbf{jk} = \mathbf{i} = -\mathbf{kj},\ \mathbf{ki} = \mathbf{j} = -\mathbf{ik}.\end{aligned}$$

3 Actually, it is not necessary to assume that $\mathbf{j}^2 = -1$. The computation shows that there is no embedding $\mathbb{C} \subset \mathbb{R}^3$, where $\mathbb{R}^3$ is an associative algebra.

Hamilton named his four-component elements **quaternions**. Quaternions form a division ring which we have denoted by $\mathbb{H}$ in honor of Hamilton.

Cayley in 1845 was the first one to publish the quaternionic representation of rotations of $\mathbb{R}^3 \to \mathbb{R}^3$, $\mathbf{r} \to a\mathbf{r}a^{-1}$, but he mentioned that the result was known to Hamilton. Cayley, in 1855, also discovered the quaternionic representation of 4-dimensional rotations:

$$\mathbb{R}^4 \to \mathbb{R}^4, \ q \to aqb^{-1},$$

where we have identified $\mathbb{R}^4 = \mathbb{H}$.

The differential operator $\nabla = \mathbf{i}\frac{\partial}{\partial x} + \mathbf{j}\frac{\partial}{\partial y} + \mathbf{k}\frac{\partial}{\partial z}$ is due to Hamilton, although his symbol for nabla was turned 30°. The first one to study solutions of

$$\frac{\partial f}{\partial w} + \mathbf{i}\frac{\partial f}{\partial x} + \mathbf{j}\frac{\partial f}{\partial y} + \mathbf{k}\frac{\partial f}{\partial z} = 0, \quad \text{where} \quad f : \mathbb{H} \to \mathbb{H},$$

was Fueter 1935.

Comment

The quaternion formalism might seem awkward to a physicist or an engineer, for two reasons: first, the squares of $\mathbf{i}, \mathbf{j}, \mathbf{k}$ are negative, $\mathbf{i}^2 = \mathbf{j}^2 = \mathbf{k}^2 = -1$, and second, one invokes a 4-dimensional space which is beyond our ability of visualization.

Exercises

1. Let $\mathbf{u}$ be a unit vector in $\mathbb{R}^3$, $|\mathbf{u}| = 1$. Show that $\mathbb{R}^3 \to \mathbb{R}^3$, $\mathbf{x} \to \mathbf{uxu}$ is a reflection across the plane $\mathbf{u}^\perp$.
2. Determine square roots of the quaternion $q = q_0 + \mathbf{q}$.
3. Hurwitz integral quaternions $q = w + ix + jy + kz$ are $\mathbb{Z}$-linear combinations of i, j, k and $\frac{1}{2}(1 + i + j + k)$, that is, either all w, x, y, z are integers or of the form $n + \frac{1}{2}$. Show that $|q|^2$ is an integer, and that the set

$$\{w + ix + jy + kx \mid w, x, y, z \in \mathbb{Z} \quad \text{or} \quad w, x, y, z \in \mathbb{Z} + \tfrac{1}{2}\}$$

is closed under multiplication.
4. Clearly, $ab = ba$ implies $e^a e^b = e^{a+b}$, but does $e^a e^b = e^{a+b}$ imply $ab = ba$?
5. Denote

$$\begin{pmatrix} a & b \\ c & d \end{pmatrix}^- = \begin{pmatrix} \bar{a} & \bar{b} \\ \bar{c} & \bar{d} \end{pmatrix}.$$

Show that

$$\begin{pmatrix} a & b \\ c & d \end{pmatrix}^{-1} = \frac{1}{\Delta}\begin{pmatrix} a\bar{d}d - b\bar{d}c & c\bar{b}b - d\bar{b}a \\ b\bar{c}c - a\bar{c}d & d\bar{a}a - c\bar{a}b \end{pmatrix}^{-}$$

for a non-zero $\Delta = |a|^2|d|^2 + |b|^2|c|^2 - 2\,\mathrm{Re}(a\,\bar{c}\,d\,\bar{b})$.

6. Verify that only one of the matrices

$$a = \begin{pmatrix} 1 & i \\ j & k \end{pmatrix} \quad \text{and} \quad b = \begin{pmatrix} 1 & j \\ i & k \end{pmatrix}$$

is invertible.

7. Does an involutory automorphism of the real algebra $\mathrm{Mat}(2, \mathbb{H})$ necessarily send a diagonal matrix of the form

$$\begin{pmatrix} a & 0 \\ 0 & a \end{pmatrix} \quad \text{where} \quad a \in \mathbb{H}$$

to a diagonal matrix?

8. Suppose A $(\neq \mathbb{R})$ is a simple real associative algebra of dimension ≤ 4 with center $\mathbb{R}$. Show that A is $\mathbb{H}$ or $\mathrm{Mat}(2, \mathbb{R})$.
9. Suppose A $(\neq \mathbb{R})$ is a simple real associative algebra with center $\mathbb{R}$ and an anti-automorphism $x \to \alpha(x)$ such that $x + \alpha(x) \in \mathbb{R}$ and $x\alpha(x) \in \mathbb{R}$. Show that A is $\mathbb{H}$ or $\mathrm{Mat}(2, \mathbb{R})$.
10. Show that all the subgroups of $Q_8 = \{\pm 1, \pm i, \pm j, \pm k\}$ are normal, that is, for a subgroup $H \subset Q_8$ and elements $g \in Q_8$, $h \in H$, $ghg^{-1} \in H$.
11. Take two vectors $\mathbf{a}, \mathbf{b}$ in $\mathbb{R}^3$, such that $|\mathbf{a}| = |\mathbf{b}|$, and $a = e^{\mathbf{a}}$, $b = e^{\mathbf{b}}$ in S^3. Determine the point-wise invariant plane of the simple rotation $q \to aqb^{-1}$ of $\mathbb{R}^4$.

Solutions

2. If $q = 0$, then there is only one square root, 0. If $\mathbf{q} = 0$, $q_0 > 0$, then there are two square roots, $\pm\sqrt{q_0}$. If $\mathbf{q} = 0$, $q_0 < 0$, then there is an infinity of square roots, $\sqrt{-q_0}\,\mathbf{u}$, where $\mathbf{u}$ is a unit pure quaternion $\mathbf{u} \in \mathbb{R}^3 \subset \mathbb{H}$, $|\mathbf{u}| = 1$. If $\mathbf{q} \neq 0$, then there are two square roots,

$$\sqrt{\tfrac{1}{2}(|q| + q_0)} + \frac{\mathbf{q}}{|\mathbf{q}|}\sqrt{\tfrac{1}{2}(|q| - q_0)}$$

and its opposite.

4. Hint: consider the quaternions $a = 3\pi i$ and $b = 4\pi j$, or the matrices

$$a = 3\pi\begin{pmatrix} 0 & -1 \\ 1 & 0 \end{pmatrix} \quad \text{and} \quad b = 4\pi\begin{pmatrix} 0 & i \\ i & 0 \end{pmatrix}.$$

6. a is invertible, but b is not.
11. If $a = b^{-1}$, the point-wise invariant plane is $\mathbf{a}^{\perp}$ in $\mathbb{R}^3$. Otherwise the point-wise invariant plane is spanned by $\mathbf{a} + \mathbf{b}$ and $|\mathbf{a}||\mathbf{b}| - \mathbf{ab} = |\mathbf{a}||\mathbf{b}| + \mathbf{a} \cdot \mathbf{b} - \mathbf{a} \times \mathbf{b}$.

Bibliography

S.L. Altmann: *Rotations, Quaternions, and Double Groups*. Oxford University Press, Oxford, 1986.

G.M. Dixon: *Division Algebras: Octonions, Quaternions, Complex Numbers and the Algebraic Design of Physics*. Kluwer, Dordrecht, The Netherlands, 1994.

P. du Val: *Homographies, Quaternions and Rotations*. Oxford University Press, Oxford, 1964.

H.-D. Ebbinghaus et al. (eds.): *Numbers*. Springer, New York, 1991.

R. Fueter: Über die analytische Darstellung der regulären Funktionen einer Quaternionenvariablen. *Comment. Math. Helv.* **8** (1935), 371-378.

K. Gürlebeck, W. Sprössig: *Quaternionic Analysis and Elliptic Boundary Value Problems*. Akademie-Verlag, Berlin, 1989. Birkhäuser, Basel, 1990.

W.R. Hamilton: *Elements of Quaternions*. Longmans Green, London, 1866. Chelsea, New York, 1969.

T.L. Hankins: *Sir William Rowan Hamilton*. Johns Hopkins University Press, Baltimore, MD, 1980.

K. Imaeda: *Quaternionic Formulation of Classical Electrodynamics and the Theory of Functions of a Biquaternion Variable*. Department of Electronic Science, Okayama University of Science, 1983.

I.R. Porteous: *Topological Geometry*. Van Nostrand Reinhold, London, 1969. Cambridge University Press, Cambridge, 1981.

M. Riesz: *Clifford Numbers and Spinors*. The Institute for Fluid Dynamics and Applied Mathematics, Lecture Series No. **38**, University of Maryland, 1958. Reprinted as facsimile (eds.: E.F. Bolinder, P. Lounesto) by Kluwer, Dordrecht, The Netherlands, 1993.

A. Sudbery: Quaternionic Analysis. *Math. Proc. Cambridge Philos. Soc.* **85** (1979), 199-225.

6

The Fourth Dimension

In this chapter we study the geometry of the Euclidean space $\mathbb{R}^4$. The purpose is to help readers to get a solid view, or as solid a view as possible, of the first dimension beyond our ability to visualize. This is an important intermediate step in scrutinizing higher dimensions. We start by reviewing regular figures in lower dimensions.

6.1 Regular polygons in $\mathbb{R}^2$

The equilateral triangle, the square, the regular pentagon, ..., are regular polygons. We shall also call them a 3-cell, 4-cell, 5-cell, ..., denoted by $\{3\}$, $\{4\}$, $\{5\}$, ..., respectively. Therefore, we call a regular p-gon a p-cell, denoted by $\{p\}$. As p grows toward infinity, we get in the limit an ∞-cell, where the line is divided into line segments of equal length. As a degenerate case we get a 2-cell, which is bounded by 2 line segments in the same place. The interior angle of a regular p-gon at a vertex is $(1 - 2/p)\pi$.

6.2 Regular polyhedra in $\mathbb{R}^3$

A regular polyhedron is a convex polyhedron bounded by congruent regular polygons, for instance, by p-gons. The number of regular p-gons meeting at a vertex is the same, say q; it satisfies

$$q\left(1 - \frac{2}{p}\right)\pi < 2\pi,$$

because the sum of angles of faces meeting at a vertex cannot exceed 2π. The above inequality can also be written in the form

$$\frac{1}{p} + \frac{1}{q} > \frac{1}{2}.$$

The same result is obtained by inspection of the topological properties of a regular polyhedron: the numbers N_0, N_1, N_2 of vertices, edges and faces satisfy the Euler formula:

$$N_0 - N_1 + N_2 = 2.$$

On the other hand, each edge of a regular polyhedron is a boundary of two faces, each with p sides, so that $2N_1 = pN_2$; and a vertex is a meeting point of q edges, each with 2 end points, so that $qN_0 = 2N_1$. The above inequality is a consequence of the Euler formula and the equation

$$qN_0 = 2N_1 = pN_2.$$

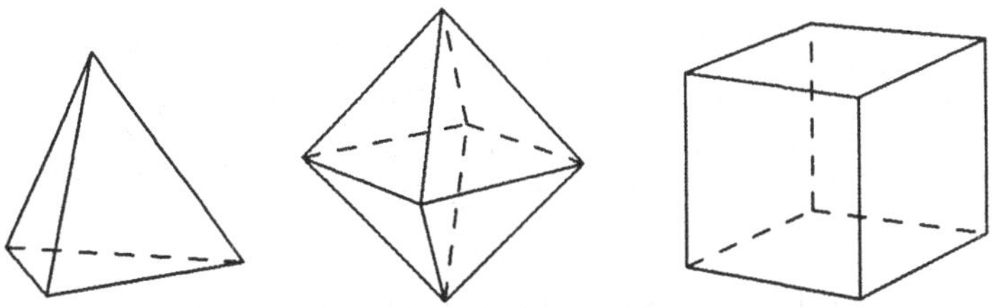

3,3 3,4 4,3

A regular polyhedron ($p, q \geq 3$) must satisfy the foregoing inequality, and so only a few pairs p, q are possible. These regular polyhedra are called Platonic solids, or p, q-cells with *Schläfli* symbols $\{p, q\}$. There are five Platonic solids.

Name	$\{p, q\}$	N_0	N_1	N_2
Tetrahedron	$\{3, 3\}$	4	6	4
Octahedron	$\{3, 4\}$	6	12	8
Cube	$\{4, 3\}$	8	12	6
Icosahedron	$\{3, 5\}$	12	30	20
Dodecahedron	$\{5, 3\}$	20	30	12

When $q = 2$ in the above inequality we get a dihedron with Schläfli symbol $\{p, 2\}$. A dihedron is bounded by two regular polygons positioned in the same place.

When a plane is covered by regular polygons so that at each vertex there meet q regular p-gons, we are solving the equation

$$\frac{1}{p} + \frac{1}{q} = \frac{1}{2}.$$

There are three solutions to the above equation; they have Schläfli symbols

$\{4,4\}$, $\{3,6\}$ and $\{6,3\}$ corresponding to tilings of the plane by squares, equilateral triangles and regular hexagons. These regular tilings are called *tessellations*.

6.3 Regular polytopes in $\mathbb{R}^4$

A polyhedron is regular if its faces and vertices (= parts of the polyhedron near a vertex point) are regular. A regular polyhedron with Schläfli symbol $\{p, q\}$ has p-cells as faces and q-cells as vertices. A vertex is regular, if a plane cuts off a regular polygon whose central normal passes through the vertex.

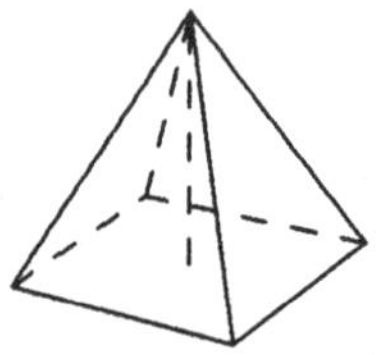

A regular vertex

A polytope is a higher-dimensional analog of a polyhedron. A polytope is regular if its faces and vertices are regular. A 4-dimensional regular polytope with p, q-cells as faces must have q, r-cells as vertices. This drops the number of 4-dimensional regular polytopes from $5^2 = 25$ to 11. The sum of the solid angles of the faces meeting at a vertex cannot exceed 4π. As a consequence, there remain six possible combinations of p, q and q, r. A closer inspection shows that all these six combinations are in fact 4-dimensional regular polytopes; we shall call them p, q, r-cells with Schläfli symbols $\{p, q, r\}$.

$\{p,q,r\}$	N_0	N_1	N_2	N_3	Face	Vertex
$\{3,3,3\}$	5	10	10	5	Tetrahedron	Tetrahedron
$\{3,3,4\}$	8	24	32	16	Tetrahedron	Octahedron
$\{4,3,3\}$	16	32	24	8	Cube	Tetrahedron
$\{3,4,3\}$	24	96	96	24	Octahedron	Cube
$\{3,3,5\}$	120	720	1200	600	Tetrahedron	Icosahedron
$\{5,3,3\}$	600	1200	720	120	Dodecahedron	Tetrahedron

There are the regular simplex $\{3,3,3\}$ and the hypercube $\{4,3,3\}$, also called a tesseract. There is the octahedron analog $\{3,3,4\}$, a dipyramid with octahedron as a basis. There are the analogs of the icosahedron and the dodecahedron, $\{3,3,5\}$ and $\{5,3,3\}$; and there is an extra regular polytope $\{3,4,3\}$.

The 3-dimensional space can be filled with cubes, a configuration with

Schläfli symbol $\{4,3,4\}$. The 4-dimensional space can be filled with hypercubes, dipyramids and the extra regular polytope, configurations with Schläfli symbols $\{4,3,3,4\}$, $\{3,3,4,3\}$ and $\{3,4,3,3\}$.

In a higher-dimensional space, $n > 4$, there are only the regular simplex, dipyramid and hypercube, and it can only be filled with hypercubes.

6.4 The spheres

A circle with radius r in $\mathbb{R}^2$ has circumference $2\pi r$ and area πr^2. A sphere with radius r in $\mathbb{R}^3$ has surface $4\pi r^2$ and volume $\frac{4}{3}\pi r^3$. A hypersphere with radius r in $\mathbb{R}^4$ has 3-dimensional surface $2\pi^2 r^3$ and 4-dimensional hypervolume $\frac{1}{2}\pi^2 r^4$. For lower-dimensional spheres we have the following table:

n	surface	volume
1	2	$2r$
2	$2\pi r$	πr^2
3	$4\pi r^2$	$\frac{4}{3}\pi r^3$
4	$2\pi^2 r^3$	$\frac{1}{2}\pi^2 r^4$
5	$\frac{8}{3}\pi^2 r^4$	$\frac{8}{15}\pi^2 r^5$

If the volume of the sphere in $\mathbb{R}^n$ is denoted by $\omega_n r^n$ then its surface is $n\omega_n r^{n-1}$. Observe a rule $m\omega_m r^{m-1} = 2\pi r \cdot \omega_n r^n$ between the surface in dimension $m = n+2$ and the volume in dimension n. This leads to the recursion

$$\omega_{n+2} = \frac{2\pi\omega_n}{n+2}$$

and the formula

$$\omega_n = \frac{\pi^{n/2}}{(n/2)!}$$

which can be computed for odd n by recalling that $(1/2)! = \sqrt{\pi}/2$.

6.5 Rotations in four dimensions

Let A be an antisymmetric 4×4-matrix, that is, $A \in \mathrm{Mat}(4,\mathbb{R})$, $A^\top = -A$. Then the matrix e^A represents a rotation of the 4-dimensional Euclidean space $\mathbb{R}^4$. In general, a rotation of $\mathbb{R}^4$ has two invariant planes which are completely orthogonal; in particular they have only one point in common. The antisymmetric matrix A has imaginary eigenvalues, say $\pm i\alpha$ and $\pm i\beta$, the eigenvalues of the rotation matrix e^A are unit complex numbers $e^{\pm i\alpha}$ and $e^{\pm i\beta}$, and the invariant planes turn by angles α and β under e^A. First, assume that $\alpha > \beta \geq 0$ (and $\alpha < \pi$). Each vector is turned through at least an angle β and at most an angle α. In the case $\beta = 0$ we have a simple rotation leaving one plane point-wise fixed. If β/α is rational, then $e^{tA} = I$ for some $t > 0$. If β/α is irrational, then $e^{tA} \neq I$ for any $t > 0$.

By the Cayley-Hamilton theorem e^A is a linear combination of the matrices I, A, A^2 and A^3 so that

$$e^A = h_0 I + h_1 A + h_2 A^2 + h_3 A^3$$

and direct computation shows that

$$\begin{aligned}
h_0 &= \frac{1}{\alpha^2 - \beta^2}(\alpha^2 \cos\beta - \beta^2 \cos\alpha), \\
h_1 &= \frac{1}{\alpha^2 - \beta^2}\left(\frac{\alpha^2}{\beta}\sin\beta - \frac{\beta^2}{\alpha}\sin\alpha\right), \\
h_2 &= \frac{1}{\alpha^2 - \beta^2}(\cos\beta - \cos\alpha), \\
h_3 &= \frac{1}{\alpha^2 - \beta^2}\left(\frac{1}{\beta}\sin\beta - \frac{1}{\alpha}\sin\alpha\right).
\end{aligned}$$

Letting α now approach β and computing the coefficients in the limit give

$$\begin{aligned}
\lim_{\alpha\to\beta} e^A = {} & I(\cos\alpha + \frac{\alpha}{2}\sin\alpha) \\
& + \frac{A}{\alpha}\left(\frac{3}{2}\sin\alpha - \frac{\alpha}{2}\cos\alpha\right) \\
& + \frac{A^2}{\alpha^2}\left(\frac{\alpha}{2}\sin\alpha\right) \\
& + \frac{A^3}{\alpha^3}\left(\frac{1}{2}\sin\alpha - \frac{\alpha}{2}\cos\alpha\right).
\end{aligned}$$

Observe that in the limit $A^2 = -\alpha^2 I$, which cancels some terms and results in

$$\lim_{\alpha\to\beta} e^A = I\cos\alpha + \frac{A}{\alpha}\sin\alpha.$$

These rotations with only one rotation angle α have a whole bundle of invariant rotation planes. In fact, every point of $\mathbb{R}^4$ stays in some invariant plane, but not every plane of $\mathbb{R}^4$ is an invariant plane of e^A.

If a rotation U of $\mathbb{R}^4$ has rotation angles α and β we shall denote it by $U(\alpha, \beta)$. Consider the set $\mathcal{J} = \{U(\alpha,\beta) \in SO(4) \mid \alpha = \beta\}$ and the relation '$\sim$' in the set $\mathcal{J}' = \mathcal{J} \setminus \{I, -I\}$,

$$U \sim V \Longleftrightarrow UV \in \mathcal{J},$$

which can be seen to be an equivalence relation. The equivalence class of a matrix $U \in \mathcal{J}'$ is the set $\{X \in \mathcal{J}' \mid X \sim U\}$. This equivalence class together with the center $\{I, -I\}$ of the rotation group $SO(4)$ forms a subgroup of $SO(4)$, denoted in the sequel by the letter Q. Also $(\mathcal{J} \setminus Q) \cup \{I, -I\}$ is a subgroup of $SO(4)$; denote it by Q^*. Observe that $UV = VU$ for $U \in Q$ and $V \in Q^*$. It can be shown that Q and Q^* are isomorphic to the group of unit quaternions $S^3 = \{q \in \mathbb{H} \mid |q| = 1\}$.

Each rotation $L \in SO(4)$ of $\mathbb{R}^4$ can be written in the form $L = UV$, where $U \in Q$, $V \in Q^*$. The rotation angles of L are $\alpha \pm \beta$ when the rotation angles of U and V are α and β. A pair of completely orthogonal planes, both with a fixed sense of rotation, induces a pair of senses of rotations for all pairs of completely orthogonal planes. There are two classes of such pairs of oriented planes: those of the type Q and those of type Q^*.

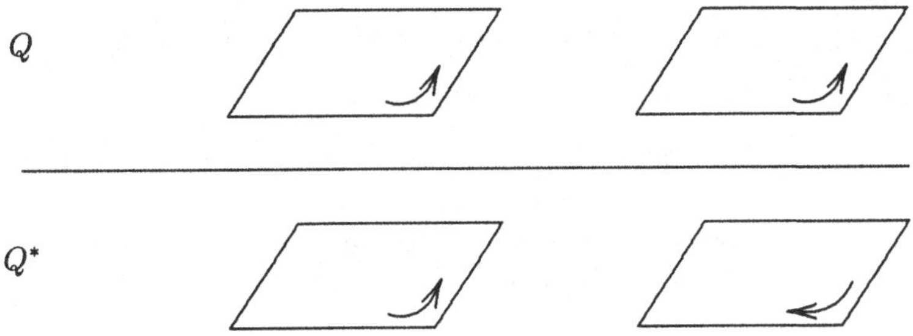

Furthermore, we have an isomorphism of algebras,

$$\mathbb{H} \simeq \{\lambda q \mid \lambda > 0,\ q \in Q\} \cup \{0\},$$

which we shall regard as an identification. Introduce the algebra

$$\mathbb{H}^* = \{\lambda q \mid \lambda > 0,\ q \in Q^*\} \cup \{0\}.$$

and observe an isomorphism of algebras, $\mathbb{H} \simeq \mathbb{H}^*$.

6.6 Rotating ball in $\mathbb{R}^4$

A rotating ball in $\mathbb{R}^3$ has an axis of rotation, like the axis going through the North and South Poles, and a plane of rotation, like the plane of the equator. A rotating ball in $\mathbb{R}^4$ has two planes of rotation, which are completely orthogonal to each other in the sense that they have only one point in common. Let the angular velocities in these planes be bivectors ω_1 and ω_2. The total angular velocity is a bivector $\omega = \omega_1 + \omega_2$. The velocity $\mathbf{v}$ of a point $\mathbf{x}$ on the surface of the ball is

$$\mathbf{v} = \mathbf{x} \lrcorner \omega_1 + \mathbf{x} \lrcorner \omega_2.$$

Assume that φ is the angle between the direction $\mathbf{x}$ and the plane of ω_1. Then

$$|\mathbf{v}| = |\mathbf{x}|\sqrt{|\omega_1|^2(\cos\varphi)^2 + |\omega_2|^2(\sin\varphi)^2}.$$

Therefore, the local angular velocity $|\mathbf{v}|/|\mathbf{x}|$ is always between $|\omega_1|$ and $|\omega_2|$.

If $|\omega_1| = |\omega_2|$, then every point on the sphere is rotating at the same velocity and furthermore every point is travelling along some great circle, that is, everybody is living on an equator!

6.7 The Clifford algebra $C\ell_4$

The Clifford algebra $C\ell_4$ of $\mathbb{R}^4$ with an orthonormal basis $\{\mathbf{e}_1, \mathbf{e}_2, \mathbf{e}_3, \mathbf{e}_4\}$ is generated by the relations

$$\mathbf{e}_1^2 = \mathbf{e}_2^2 = \mathbf{e}_3^2 = \mathbf{e}_4^2 = 1 \quad \text{and} \quad \mathbf{e}_i\mathbf{e}_j = -\mathbf{e}_j\mathbf{e}_i \quad \text{for} \quad i \neq j.$$

It is a 16-dimensional algebra with basis consisting of

scalar	1
vectors	$\mathbf{e}_1, \mathbf{e}_2, \mathbf{e}_3, \mathbf{e}_4$
bivectors	$\mathbf{e}_{12}, \mathbf{e}_{13}, \mathbf{e}_{14}, \mathbf{e}_{23}, \mathbf{e}_{24}, \mathbf{e}_{34}$
3-vectors	$\mathbf{e}_{123}, \mathbf{e}_{124}, \mathbf{e}_{134}, \mathbf{e}_{234}$
volume element	$\mathbf{e}_{1234}$

where $\mathbf{e}_{ij} = \mathbf{e}_i\mathbf{e}_j$ for $i \neq j$ and $\mathbf{e}_{1234} = \mathbf{e}_1\mathbf{e}_2\mathbf{e}_3\mathbf{e}_4$.

An arbitrary element $u \in C\ell_4$ is a sum of its k-vector parts:

$$u = \langle u\rangle_0 + \langle u\rangle_1 + \langle u\rangle_2 + \langle u\rangle_3 + \langle u\rangle_4 \quad \text{where} \quad \langle u\rangle_k \in \bigwedge^k \mathbb{R}^4.$$

There are three important involutions of $C\ell_4$:

$$\begin{aligned}
\hat{u} &= \langle u\rangle_0 - \langle u\rangle_1 + \langle u\rangle_2 - \langle u\rangle_3 + \langle u\rangle_4 \quad &&\text{grade involution}\\
\tilde{u} &= \langle u\rangle_0 + \langle u\rangle_1 - \langle u\rangle_2 - \langle u\rangle_3 + \langle u\rangle_4 \quad &&\text{reversion}\\
\bar{u} &= \langle u\rangle_0 - \langle u\rangle_1 - \langle u\rangle_2 + \langle u\rangle_3 + \langle u\rangle_4 \quad &&\text{Clifford-conjugation.}
\end{aligned}$$

The Clifford algebra $C\ell_4$ is isomorphic to the real algebra of 2×2-matrices $\text{Mat}(2, \mathbb{H})$ with quaternions as entries,

$$\mathbf{e}_1 = \begin{pmatrix} 0 & -i \\ i & 0 \end{pmatrix}, \quad \mathbf{e}_2 = \begin{pmatrix} 0 & -j \\ j & 0 \end{pmatrix}, \quad \mathbf{e}_3 = \begin{pmatrix} 0 & -k \\ k & 0 \end{pmatrix}, \quad \mathbf{e}_4 = \begin{pmatrix} 0 & 1 \\ 1 & 0 \end{pmatrix}.$$

6.8 Bivectors in $\bigwedge^2 \mathbb{R}^4 \subset C\ell_4$

The essential difference between 3-dimensional and 4-dimensional spaces is that bivectors are no longer products of two vectors. Instead, bivectors are sums of products of two vectors in $\mathbb{R}^4$. In the 3-dimensional space $\mathbb{R}^3$ there are only *simple* bivectors, that is, all the bivectors represent a plane. In the 4-dimensional space $\mathbb{R}^4$ this is not the case any more.

Example. The bivector $\mathbf{B} = \mathbf{e}_{12} + \mathbf{e}_{34} \in \bigwedge^2 \mathbb{R}^4$ is not simple. For all simple elements the square is real, but $\mathbf{B}^2 = -2 + 2\mathbf{e}_{1234} \notin \mathbb{R}$. ∎

If the square of a bivector is real, then it is simple. [1]

Usually a bivector in $\bigwedge^2 \mathbb{R}^4$ can be uniquely written as a sum of two simple bivectors, which represent completely orthogonal planes. There is an exception to this uniqueness, crucial to the study of four dimensions: If the simple components of a bivector have equal squares, that is equal norms, then the decomposition to a sum of simple components is not unique.

Example. The bivector $\mathbf{e}_1\mathbf{e}_2 + \mathbf{e}_3\mathbf{e}_4$ can also be decomposed into a sum of two completely orthogonal bivectors as follows:

$$\mathbf{e}_1\mathbf{e}_2 + \mathbf{e}_3\mathbf{e}_4 = \frac{1}{2}(\mathbf{e}_1 + \mathbf{e}_3)(\mathbf{e}_2 + \mathbf{e}_4) + \frac{1}{2}(\mathbf{e}_1 - \mathbf{e}_3)(\mathbf{e}_2 - \mathbf{e}_4).$$ ∎

6.9 The group Spin(4) and its Lie algebra

The group $\mathbf{Spin}(4) = \{s \in \mathcal{C}\ell_4^+ \mid s\tilde{s} = 1\}$ is a two-fold covering group of the rotation group $SO(4)$ so that the map

$$\mathbb{R}^4 \to \mathbb{R}^4, \ \mathbf{x} \to s\mathbf{x}s^{-1}, \quad \text{where} \quad s \in \mathbf{Spin}(4),$$

is a rotation, and each rotation can be so represented, the same rotation being obtained by s and $-s$. The Lie algebra of $\mathbf{Spin}(4)$ is the subspace of bivectors $\bigwedge^2 \mathbb{R}^4$ with commutator product as the product. The two sets of basis bivectors

$$\begin{matrix} \frac{1}{4}(\mathbf{e}_{23} + \mathbf{e}_{14}) & & \frac{1}{4}(\mathbf{e}_{23} - \mathbf{e}_{14}) \\ \frac{1}{4}(\mathbf{e}_{31} + \mathbf{e}_{24}) & \text{and} & \frac{1}{4}(\mathbf{e}_{31} - \mathbf{e}_{24}) \\ \frac{1}{4}(\mathbf{e}_{12} + \mathbf{e}_{34}) & & \frac{1}{4}(\mathbf{e}_{12} - \mathbf{e}_{34}) \end{matrix}$$

in $\bigwedge^2 \mathbb{R}^4 \subset \mathcal{C}\ell_4$ both span a Lie algebra isomorphic to the subspace $\bigwedge^2 \mathbb{R}^3 \subset \mathcal{C}\ell_3$ with basis $\{\frac{1}{2}\mathbf{e}_{23}, \frac{1}{2}\mathbf{e}_{31}, \frac{1}{2}\mathbf{e}_{12}\}$, that is, they satisfy the same commutation relations. In other words, the Lie algebras

$$\frac{1}{2}(1 - \mathbf{e}_{1234}) \bigwedge^2 \mathbb{R}^4 \quad \text{and} \quad \frac{1}{2}(1 + \mathbf{e}_{1234}) \bigwedge^2 \mathbb{R}^4$$

---

1 Although the square of a 3-vector is real, it need not be simple. For instance, $\mathbf{V} = \mathbf{e}_{123} + \mathbf{e}_{456} \in \bigwedge^3 \mathbb{R}^6$ is not simple [this can be seen by computing $\mathbf{V}\mathbf{e}_i\mathbf{V}^{-1}$, $i = 1, 2, \ldots, 6$, and observing that they are not all vectors].

are both isomorphic to $\bigwedge^2 \mathbb{R}^3$. The two subspaces $\frac{1}{2}(1 \pm \mathbf{e}_{1234}) \bigwedge^2 \mathbb{R}^4$ of $\mathcal{C}\ell_4$ annihilate each other, and consequently,

$$\bigwedge^2 \mathbb{R}^4 \simeq \bigwedge^2 \mathbb{R}^3 \oplus \bigwedge^2 \mathbb{R}^3.$$

At the group level this means the isomorphism

$$\mathbf{Spin}(4) \simeq \mathbf{Spin}(3) \times \mathbf{Spin}(3)$$

where $\mathbf{Spin}(3) \simeq S^3 \simeq SU(2)$.

6.10 The mapping $\mathbf{F} \to (1+\mathbf{F})(1-\mathbf{F})^{-1}$ for $\mathbf{F} \in \bigwedge^2 \mathbb{R}^4$

The exponential $e^{\mathbf{F}/2} \in \mathbf{Spin}(4)$ of a bivector $\mathbf{F} \in \bigwedge^2 \mathbb{R}^4$ corresponds to the rotation $e^A \in SO(4)$, where $A(\mathbf{x}) = \mathbf{F} \llcorner \mathbf{x}$, for $\mathbf{x} \in \mathbb{R}^4$. Every rotation of $\mathbb{R}^4$ can be so represented, and the two elements $\pm e^{\mathbf{F}/2}$ represent the same rotation.

The exterior exponential $e^{\mathbf{F}}_{\wedge} = 1 + \mathbf{F} + \frac{1}{2}\mathbf{F} \wedge \mathbf{F}$ of a bivector $\mathbf{F} \in \bigwedge^2 \mathbb{R}^4$ is a multiple of an element in $\mathbf{Spin}(4)$, that is,

$$\frac{e^{\mathbf{F}}_{\wedge}}{|e^{\mathbf{F}}_{\wedge}|} \in \mathbf{Spin}(4).$$

Up to a sign, every element in $\mathbf{Spin}(4)$ can be so represented, except $\pm \mathbf{e}_{1234}$. The exterior exponential $e^{\mathbf{F}}_{\wedge}$ of the bivector $\mathbf{F}$ corresponds to the rotation $(I+A)(I-A)^{-1}$; every rotation of $\mathbb{R}^4$ can be so represented, except $-I$.

The above observations raise the question: What is the rotation corresponding to $(1+\mathbf{F})(1-\mathbf{F})^{-1} \in \mathbf{Spin}(4)$? This is an interesting and non-trivial question in dimension 4. [2] Here follows the answer.

Let $\mathbf{F} \in \bigwedge^2 \mathbb{R}^4$. The antisymmetric function induced by $\mathbf{F}$ is denoted by A, that is, $A(\mathbf{x}) = \mathbf{F} \llcorner \mathbf{x}$ for all $\mathbf{x} \in \mathbb{R}^4$. Write $s = (1+\mathbf{F})(1-\mathbf{F})^{-1}$. The rotation induced by $s \in \mathbf{Spin}(4)$ is denoted by $U \in SO(4)$, that is, $U = (I+A)(I-A)^{-1}$. In other words, $U(\mathbf{x}) = s\mathbf{x}s^{-1}$ for all $\mathbf{x} \in \mathbb{R}^4$. The following cases can be distinguished:

(i) If $\mathbf{F} \in \bigwedge^2 \mathbb{R}^3$ then $U = \left(\frac{I+A}{I-A}\right)^2$.

(ii) If $\mathbf{F} \in \bigwedge^2 \mathbb{R}^4$ is simple, then $U = \left(\frac{I+A}{I-A}\right)^2$.

(iii) If $\mathbf{F} \in \bigwedge^2 \mathbb{R}^4$ is isoclinic, then $U = \frac{I+2A}{I-2A}$.

2 It is also a non-trivial question in dimension 5. In dimension 6, $(1+\mathbf{F})(1-\mathbf{F})^{-1} \notin \mathbf{Spin}(6)$.

(iv) In the case of an arbitrary $\mathbf{F} \in \bigwedge^2 \mathbb{R}^4$ we cannot express U as a rational function of A [although U still has the same eigenplanes as A]. Instead,

$$U = \frac{A^4 + B^4 - 2A^2B^2 + 6A^2 - 2B^2 + I + 4A(A^2 - B^2 + I)}{A^4 + B^4 - 2A^2B^2 - 2A^2 - 2B^2 + I},$$

where $B(\mathbf{x}) = (\mathbf{F}\mathbf{e}_{1234}) \llcorner \mathbf{x}$, the dual of A. The denominator of U is a multiple of the identity I. [3]

Summary

There are three different kinds of rotations in four dimensions depending on the values of the rotation angles α, β satisfying $\pi > \alpha \geq \beta \geq 0$. Let $R : \mathbb{R}^4 \to \mathbb{R}^4$ be a rotation and $\mathbf{a}$ a non-zero vector with iterated images $\mathbf{b} = R(\mathbf{a})$, $\mathbf{c} = R(\mathbf{b})$, $\mathbf{d} = R(\mathbf{c})$. In general, $\mathbf{a}, \mathbf{b}, \mathbf{c}, \mathbf{d}$ are linearly independent, that is, $\mathbf{a} \wedge \mathbf{b} \wedge \mathbf{c} \wedge \mathbf{d} \neq 0$. In the case of a simple rotation with $\beta = 0$, only the vectors $\mathbf{a}, \mathbf{b}, \mathbf{c}$ are linearly independent, that is, $\mathbf{a} \wedge \mathbf{b} \wedge \mathbf{c} \neq 0$ but $\mathbf{a} \wedge \mathbf{b} \wedge \mathbf{c} \wedge \mathbf{d} = 0$. In the case of an isoclinic [4] rotation with $\alpha = \beta$, only the vectors $\mathbf{a}, \mathbf{b}$ are linearly independent, that is, $\mathbf{a} \wedge \mathbf{b} \neq 0$ but $\mathbf{a} \wedge \mathbf{b} \wedge \mathbf{c} = 0$ and $\mathbf{a} \wedge \mathbf{b} \wedge \mathbf{d} = 0$.

In general, a rotation of $\mathbb{R}^4$ has six parameters, computed as

$$(3 + 2 - 1) + 2 = 6.$$

The number 3 comes from picking up a unit vector $\mathbf{a}$; the number 2 comes from picking up a unit vector $\mathbf{b}$ in the orthogonal complement of $\mathbf{a}$; the unit bivector $\mathbf{ab} = \mathbf{a} \wedge \mathbf{b}$ fixes a plane but the same plane is obtained by rotating $\mathbf{a}$ and $\mathbf{b}$ in the plane of $\mathbf{a} \wedge \mathbf{b}$, thus subtract 1; then finally add 2 for the two rotation parameters/angles α and β. On the other hand, an isoclinic rotation has three parameters, computed as

$$(3 - 1) + 1 = 3.$$

The number 3 comes from picking up a unit vector $\mathbf{a}$ in S^3; but in an isoclinic rotation $\mathbf{a}$ stays in a plane or a great circle S^1, so subtract 1; and finally add 1 for the rotation/angle $\alpha = \beta$.

A simple bivector, an exterior product of two vectors, corresponds to simple

3 In dimension 5 the rotation U is given by the same expression, when

$$B(\mathbf{x}) = \left(\mathbf{F}\frac{\mathbf{F} \wedge \mathbf{F}}{|\mathbf{F} \wedge \mathbf{F}|}\right) \llcorner \mathbf{x}.$$

The denominator is no longer a multiple of I, although it still commutes with the numerator by virtue of $AB = BA$.

4 An isoclinic rotation with equal rotation angles corresponds to a multiplication by a quaternion.

rotation turning only one plane. A simple bivector multiplied by one of the idempotents $\frac{1}{2}(1 \pm \mathbf{e}_{1234})$ corresponds to an isoclinic rotation. An isoclinic rotation has an infinity of rotation planes, and in fact, each vector is in some invariant rotation plane of an isoclinic rotation.

The two-fold cover **Spin**(4) of $SO(4)$ has three different subgroups isomorphic to **Spin**(3), each with a Lie algebra

$$\bigwedge^2 \mathbb{R}^3, \quad \frac{1}{2}(1+\mathbf{e}_{1234})\bigwedge^2 \mathbb{R}^4, \quad \frac{1}{2}(1-\mathbf{e}_{1234})\bigwedge^2 \mathbb{R}^4.$$

There is an automorphism of **Spin**(4) which swaps the last two copies of **Spin**(3), but there is no automorphism of **Spin**(4) swapping the first copy of **Spin**(3) with either of the other two copies.

Exercises

1. Compute the squares of $\frac{1}{2}(1+\mathbf{e}_{12}+\mathbf{e}_{34} \pm \mathbf{e}_{1234})$.
2. Take a vector $\mathbf{a} \in \mathbb{R}^4$ and a bivector $\mathbf{B} = \alpha\mathbf{e}_{12} + \beta\mathbf{e}_{34} \in \bigwedge^2 \mathbb{R}^4$. Show that $\mathbf{BaB} \in \mathbb{R}^4$.
3. Compute $\exp(\alpha\mathbf{e}_{12} + \beta\mathbf{e}_{34})$.
4. Let $\mathbf{a} = a_1\mathbf{e}_1 + a_2\mathbf{e}_2 + a_3\mathbf{e}_3$ and $\mathbf{b} = b_1\mathbf{e}_1 + b_2\mathbf{e}_2 + b_3\mathbf{e}_3$. Compute $\mathbf{A} = \mathbf{a}\mathbf{e}_{123}$ and $\mathbf{B} = \mathbf{b}\mathbf{e}_{123}$. Determine $\frac{1}{2}(1+\mathbf{e}_{1234})\mathbf{A}$ and $\frac{1}{2}(1-\mathbf{e}_{1234})\mathbf{B}$, and show that these bivectors commute.
5. Compute $\mathbf{C} = \frac{1}{2}(1+\mathbf{e}_{1234})\mathbf{A} + \frac{1}{2}(1-\mathbf{e}_{1234})\mathbf{B}$, and express $\exp(\mathbf{C})$ using $|\mathbf{a}|$ and $|\mathbf{b}|$. What are the two rotation angles of the rotation $\mathbb{R}^4 \to \mathbb{R}^4$, $\mathbf{x} \to c\mathbf{x}c^{-1}$ where $c = \exp(\mathbf{C})$?
6. Consider the Lie algebra $\bigwedge^2 \mathbb{R}^4$ with the commutator product $[a, b] = ab - ba$, and its three subalgebras spanned by

$$\begin{aligned} \mathcal{V}: &\quad \tfrac{1}{2}\mathbf{e}_{23},\ \tfrac{1}{2}\mathbf{e}_{31},\ \tfrac{1}{2}\mathbf{e}_{12} \\ \mathcal{I}_1: &\quad \tfrac{1}{4}(\mathbf{e}_{23}-\mathbf{e}_{14}),\ \tfrac{1}{4}(\mathbf{e}_{31}-\mathbf{e}_{24}),\ \tfrac{1}{4}(\mathbf{e}_{12}-\mathbf{e}_{34}) \\ \mathcal{I}_2: &\quad \tfrac{1}{4}(\mathbf{e}_{23}+\mathbf{e}_{14}),\ \tfrac{1}{4}(\mathbf{e}_{31}+\mathbf{e}_{24}),\ \tfrac{1}{4}(\mathbf{e}_{12}+\mathbf{e}_{34}), \end{aligned}$$

   each isomorphic to $\bigwedge^2 \mathbb{R}^3$. Show that there is no automorphism of the Lie algebra $\bigwedge^2 \mathbb{R}^4$ which permuts $\mathcal{V}, \mathcal{I}_1, \mathcal{I}_2$ cyclically or swaps $\mathcal{V}$ for $\mathcal{I}_1$ or $\mathcal{I}_2$.
7. In two dimensions we can place 4 circles of radius r inside a square of side $4r$, and put a circle of radius $(\sqrt{2}-1)r$ in the middle of the 4 circles. In three dimensions we can place 8 spheres of radius r inside a cube of side $4r$, and put a sphere of radius $(\sqrt{3}-1)r$ in the middle of the 8 circles. In n dimensions we can place 2^n spheres of radius r inside a hypercube of side $4r$, and put a sphere of radius $(\sqrt{n}-1)r$ in the middle of the 2^n spheres.

Dimensions 2 and 3 differ topologically: in dimension 3 one can see the middle sphere from outside the cube. Let the dimension be progressively increased. In some dimension the middle sphere actually emerges out of the hypercube. In some dimension the middle sphere becomes even bigger than the hypercube, in the sense that its volume is larger than the volume of the hypercube. Determine those dimensions.

Solutions

1. $\mathbf{e}_{1234}$, $\mathbf{e}_{12} + \mathbf{e}_{34}$.
3. $\cos\alpha\cos\beta + \mathbf{e}_{12}\sin\alpha\cos\beta + \mathbf{e}_{34}\cos\alpha\sin\beta + \mathbf{e}_{1234}\sin\alpha\sin\beta$.
5. The rotation angles are $\alpha = (|\mathbf{a}| + |\mathbf{b}|)/2$ and $\beta = (|\mathbf{a}| - |\mathbf{b}|)/2$, and

$$\frac{1}{2}(1+\mathbf{e}_{1234})\Big(\cos|\mathbf{a}| + \frac{\mathbf{A}}{|\mathbf{a}|}\sin|\mathbf{a}|\Big) + \frac{1}{2}(1-\mathbf{e}_{1234})\Big(\cos|\mathbf{b}| + \frac{\mathbf{B}}{|\mathbf{b}|}\sin|\mathbf{b}|\Big)$$
$$= \cos\alpha\cos\beta - \mathbf{e}_{1234}\sin\alpha\sin\beta$$
$$+\mathbf{C}\frac{\alpha - \beta\mathbf{e}_{1234}}{\alpha^2 - \beta^2}(\sin\alpha\cos\beta + \mathbf{e}_{1234}\cos\alpha\sin\beta).$$

7. In dimension 9 the middle sphere touches the surface of the hypercube, and in dimension 10 it emerges out of the hypercube. In dimension 1206 the volume of the middle sphere is larger than the volume of the hypercube.

Bibliography

S.L. Altmann: *Rotations, Quaternions, and Double Groups*, Oxford University Press, Oxford, 1986.

H.S.M. Coxeter: *Regular Polytopes*, Methuen, London, 1948.

P. du Val: *Homographies, Quaternions and Rotations*, Oxford University Press, Oxford, 1964.

D. Hilbert, S. Cohn-Vossen: *Anschauliche Geometrie*, Dover, New York, 1944. *Geometry and the Imagination*, Chelsea, New York, 1952.

7
The Cross Product

The cross product is useful in many physical applications. It measures the angular velocity $\vec{\omega} = \vec{r} \times \vec{v}$ about O of a body moving at velocity $\vec{v}$ at the position P, $\vec{r} = \overrightarrow{OP}$. It is used to describe the torque $\vec{r} \times \vec{F}$ about O of a force $\vec{F}$ acting at $\vec{r}$. It also gives the force $\vec{F} = q\vec{v} \times \vec{B}$ acting on a charge q moving at velocity $\vec{v}$ in a magnetic field $\vec{B}$.

The usefulness of the cross product in three dimensions suggests the following questions: Is there a higher-dimensional analog of the cross product of two vectors in $\mathbb{R}^3$? If an analog exists, is it unique?

The first question is usually responded to by giving an answer to a modified question by explaining that there is a higher-dimensional analog of the cross product of $n-1$ vectors in $\mathbb{R}^n$. However, such a reply not only does not answer the original question, but also gives an incomplete answer to the modified question. In this chapter we will give a complete answer to the above questions and their modifications.

7.1 Scalar product in $\mathbb{R}^3$

The linear space $\mathbb{R}^3$ can be given extra structure by introducing the *scalar product* or *dot product*

$$\mathbf{a} \cdot \mathbf{b} = a_1 b_1 + a_2 b_2 + a_3 b_3$$

for vectors $\mathbf{a} = a_1\mathbf{e}_1 + a_2\mathbf{e}_2 + a_3\mathbf{e}_3$ and $\mathbf{b} = b_1\mathbf{e}_1 + b_2\mathbf{e}_2 + b_3\mathbf{e}_3$ in $\mathbb{R}^3$. The scalar product is scalar valued, $\mathbf{a} \cdot \mathbf{b} \in \mathbb{R}$, and satisfies

$$\left.\begin{aligned} &(\mathbf{a}+\mathbf{b}) \cdot \mathbf{c} = \mathbf{a} \cdot \mathbf{c} + \mathbf{b} \cdot \mathbf{c} \\ &(\lambda \mathbf{a}) \cdot \mathbf{b} = \lambda(\mathbf{a} \cdot \mathbf{b}) \end{aligned}\right\} \quad \text{linear in the first factor}$$

$$\mathbf{a} \cdot \mathbf{b} = \mathbf{b} \cdot \mathbf{a} \qquad \text{symmetric}$$

$$\mathbf{a} \cdot \mathbf{a} > 0 \quad \text{for} \quad \mathbf{a} \neq 0 \qquad \text{positive definite.}$$

Linearity with respect to the first argument together with symmetry implies that the scalar product is linear with respect to both arguments, that is, it is *bilinear*. The symmetric bilinear scalar valued product gives rise to the quadratic form

$$\mathbb{R}^3 \to \mathbb{R}, \quad \mathbf{a} = a_1\mathbf{e}_1 + a_2\mathbf{e}_2 + a_3\mathbf{e}_3 \to \mathbf{a} \cdot \mathbf{a} = a_1^2 + a_2^2 + a_3^2,$$

which makes the linear space $\mathbb{R}^3$ into a *quadratic space* $\mathbb{R}^3$. The quadratic form is positive definite, that is, $\mathbf{a} \cdot \mathbf{a} = 0$ implies $\mathbf{a} = 0$, which allows us to introduce the *length* [1] $|\mathbf{a}| = \sqrt{\mathbf{a} \cdot \mathbf{a}}$ of a vector $\mathbf{a} \in \mathbb{R}^3$. The real linear space $\mathbb{R}^3$ with a positive definite quadratic form on itself is called a *Euclidean space* $\mathbb{R}^3$. The length and the scalar product satisfy

$\|\mathbf{a} + \mathbf{b}\| \leq \|\mathbf{a}\| + \|\mathbf{b}\|$	triangle inequality
$\|\mathbf{a} \cdot \mathbf{b}\| \leq \|\mathbf{a}\|\|\mathbf{b}\|$	Cauchy-Schwarz inequality

where the latter inequality gives rise to the concept of angle. The angle φ between two directions $\mathbf{a}$ and $\mathbf{b}$ is obtained from

$$\cos\varphi = \frac{\mathbf{a} \cdot \mathbf{b}}{|\mathbf{a}||\mathbf{b}|}.$$

Thus, we can write the scalar product in the form

$$\mathbf{a} \cdot \mathbf{b} = |\mathbf{a}||\mathbf{b}|\cos\varphi,$$

a formula which is usually taken as a definition of the scalar product, although this requires prior introduction of the concepts of length and angle.

7.2 Cross product in $\mathbb{R}^3$

In the Euclidean space $\mathbb{R}^3$ it is convenient to introduce a vector valued product, the *cross product* $\mathbf{a} \times \mathbf{b} \in \mathbb{R}^3$ of $\mathbf{a}, \mathbf{b} \in \mathbb{R}^3$, with the following properties:

$(\mathbf{a} \times \mathbf{b}) \perp \mathbf{a}$, $(\mathbf{a} \times \mathbf{b}) \perp \mathbf{b}$	orthogonality
$\|\mathbf{a} \times \mathbf{b}\| = \|\mathbf{a}\|\|\mathbf{b}\|\sin\varphi$	length equals area
$\mathbf{a}, \mathbf{b}, \mathbf{a} \times \mathbf{b}$	right-hand system.

In other words, the vector $\mathbf{a} \times \mathbf{b}$ is perpendicular to $\mathbf{a}$ and $\mathbf{b}$, its length is equal to the area of the parallelogram with $\mathbf{a}$ and $\mathbf{b}$ as edges, and the vectors

1 The function $\mathbb{R}^3 \to \mathbb{R}$, $\mathbf{a} \to |\mathbf{a}|$ is a *norm* satisfying $|\lambda\mathbf{a}| = |\lambda||\mathbf{a}|$, $|\mathbf{a} + \mathbf{b}| \leq |\mathbf{a}| + |\mathbf{b}|$, $|\mathbf{a}| = 0 \Rightarrow \mathbf{a} = 0$. Since this norm can be obtained from a scalar product, it satisfies the parallelogram law $|\mathbf{a} + \mathbf{b}|^2 + |\mathbf{a} - \mathbf{b}|^2 = 2|\mathbf{a}|^2 + 2|\mathbf{b}|^2$.

a, **b** and **a** × **b** are oriented according to the right hand rule.

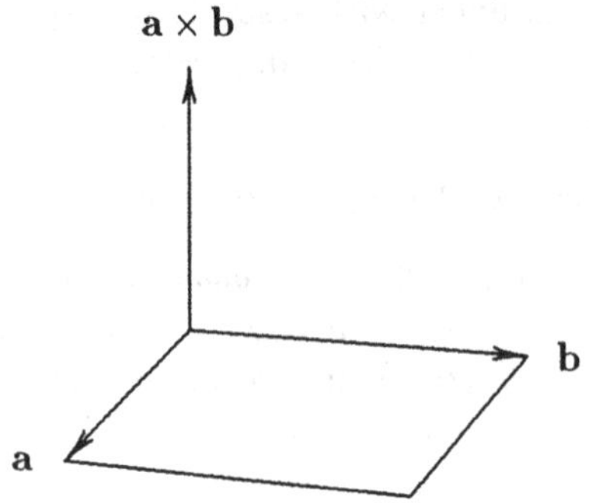

The above definition results in the following multiplication rules:

$$\mathbf{e}_1 \times \mathbf{e}_2 = \mathbf{e}_3 = -\mathbf{e}_2 \times \mathbf{e}_1,$$
$$\mathbf{e}_2 \times \mathbf{e}_3 = \mathbf{e}_1 = -\mathbf{e}_3 \times \mathbf{e}_2,$$
$$\mathbf{e}_3 \times \mathbf{e}_1 = \mathbf{e}_2 = -\mathbf{e}_1 \times \mathbf{e}_3.$$

It is convenient to write the cross product in the form

$$\mathbf{a} \times \mathbf{b} = \begin{vmatrix} \mathbf{e}_1 & \mathbf{e}_2 & \mathbf{e}_3 \\ a_1 & a_2 & a_3 \\ b_1 & b_2 & b_3 \end{vmatrix}.$$

The cross product is uniquely determined by

$$\begin{array}{ll} (\mathbf{a} \times \mathbf{b}) \cdot \mathbf{a} = 0, \ (\mathbf{a} \times \mathbf{b}) \cdot \mathbf{b} = 0 & \text{orthogonality} \\ |\mathbf{a} \times \mathbf{b}|^2 = |\mathbf{a}|^2|\mathbf{b}|^2 - (\mathbf{a} \cdot \mathbf{b})^2 & \text{Pythagorean theorem} \end{array}$$

together with the right hand rule. The Pythagorean theorem can be written using the *Gram determinant* as

$$|\mathbf{a} \times \mathbf{b}|^2 = \begin{vmatrix} \mathbf{a} \cdot \mathbf{a} & \mathbf{a} \cdot \mathbf{b} \\ \mathbf{b} \cdot \mathbf{a} & \mathbf{b} \cdot \mathbf{b} \end{vmatrix}$$

which in coordinate form means *Lagrange's identity*

$$\begin{vmatrix} a_2 & a_3 \\ b_2 & b_3 \end{vmatrix}^2 + \begin{vmatrix} a_3 & a_1 \\ b_3 & b_1 \end{vmatrix}^2 + \begin{vmatrix} a_1 & a_2 \\ b_1 & b_2 \end{vmatrix}^2$$
$$= (a_1^2 + a_2^2 + a_3^2)(b_1^2 + b_2^2 + b_3^2) - (a_1b_1 + a_2b_2 + a_3b_3)^2.$$

The cross product satisfies the following rules for all $\mathbf{a}, \mathbf{b}, \mathbf{c} \in \mathbb{R}^3$:

$$\begin{array}{ll} \mathbf{a} \times \mathbf{b} = -\mathbf{b} \times \mathbf{a} & \text{antisymmetry} \\ (\mathbf{a} \times \mathbf{b}) \cdot \mathbf{c} = \mathbf{a} \cdot (\mathbf{b} \times \mathbf{c}) & \text{interchange rule.} \end{array}$$

The antisymmetry of the cross product has a geometric meaning: the lack of symmetry measures how much the two directions diverge. The cross product is not associative, $\mathbf{a}\times(\mathbf{b}\times\mathbf{c})\neq(\mathbf{a}\times\mathbf{b})\times\mathbf{c}$, which results in an inconvenience in computation, because parentheses cannot be omitted. [2]

The cross product is dual to the exterior product of two vectors:

$$\mathbf{a}\times\mathbf{b}=-(\mathbf{a}\wedge\mathbf{b})\mathbf{e}_{123}.$$

Taking the exterior product of $\mathbf{a}\times(\mathbf{a}\times\mathbf{b})=\mathbf{a}(\mathbf{a}\cdot\mathbf{b})-|\mathbf{a}|^2\mathbf{b}$ and $\mathbf{b}$ one finds that

$$\mathbf{a}\cdot\mathbf{b}=\frac{(\mathbf{a}\times(\mathbf{a}\times\mathbf{b}))\wedge\mathbf{b}}{\mathbf{a}\wedge\mathbf{b}}\quad\text{for}\quad\mathbf{a}\nparallel\mathbf{b},$$

that is, the scalar product can be recaptured from the cross product [you can also replace $\wedge$ by $\times$ in the above formula].

7.3 Cross product of $n-1$ vectors in $\mathbb{R}^n$

We can associate to three given vectors $\mathbf{a},\mathbf{b},\mathbf{c}$ in $\mathbb{R}^4$ a fourth vector

$$\mathbf{a}\times\mathbf{b}\times\mathbf{c}=\begin{vmatrix}\mathbf{e}_1&\mathbf{e}_2&\mathbf{e}_3&\mathbf{e}_4\\a_1&a_2&a_3&a_4\\b_1&b_2&b_3&b_4\\c_1&c_2&c_3&c_4\end{vmatrix}$$

which is orthogonal to the factors $\mathbf{a},\mathbf{b},\mathbf{c}$ and whose length is equal to the volume of the parallelepiped with $\mathbf{a},\mathbf{b},\mathbf{c}$ as edges, that is,

$$|\mathbf{a}\times\mathbf{b}\times\mathbf{c}|^2=\begin{vmatrix}\mathbf{a}\cdot\mathbf{a}&\mathbf{a}\cdot\mathbf{b}&\mathbf{a}\cdot\mathbf{c}\\\mathbf{b}\cdot\mathbf{a}&\mathbf{b}\cdot\mathbf{b}&\mathbf{b}\cdot\mathbf{c}\\\mathbf{c}\cdot\mathbf{a}&\mathbf{c}\cdot\mathbf{b}&\mathbf{c}\cdot\mathbf{c}\end{vmatrix}.$$

The cross product $\mathbf{a}\times\mathbf{b}\times\mathbf{c}$ of three vectors $\mathbf{a},\mathbf{b},\mathbf{c}$ in $\mathbb{R}^4$ is completely antisymmetric and obeys the interchange rule slightly modified:

$$(\mathbf{a}\times\mathbf{b}\times\mathbf{c})\cdot\mathbf{d}=-\mathbf{a}\cdot(\mathbf{b}\times\mathbf{c}\times\mathbf{d})$$

where $\mathbf{d}\in\mathbb{R}^4$. The oriented volume of the 4-dimensional parallelepiped with $\mathbf{a},\mathbf{b},\mathbf{c},\mathbf{d}$ as edges is the scalar

$$\det(\mathbf{a},\mathbf{b},\mathbf{c},\mathbf{d})=(\mathbf{a}\times\mathbf{b}\times\mathbf{c})\cdot\mathbf{d}$$

multiplied by (the unit oriented volume) $\mathbf{e}_{1234}$.

2 The cross product is antisymmetric, $\mathbf{a}\times\mathbf{b}=-\mathbf{b}\times\mathbf{a}$, and satisfies the Jacobi identity $\mathbf{a}\times(\mathbf{b}\times\mathbf{c})+\mathbf{b}\times(\mathbf{c}\times\mathbf{a})+\mathbf{c}\times(\mathbf{a}\times\mathbf{b})=0$, which makes the linear space $\mathbb{R}^3$, with cross product on $\mathbb{R}^3$, a non-associative algebra, called a *Lie algebra*. The Jacobi identity can be verified using $\mathbf{a}\times(\mathbf{b}\times\mathbf{c})=(\mathbf{a}\cdot\mathbf{c})\mathbf{b}-(\mathbf{a}\cdot\mathbf{b})\mathbf{c}$.

The cross product of three vectors in $\mathbb{R}^4$ is dual to the exterior product:

$$\mathbf{a} \times \mathbf{b} \times \mathbf{c} = -(\mathbf{a} \wedge \mathbf{b} \wedge \mathbf{c})\mathbf{e}_{1234}$$

where the latter product is computed in the Clifford algebra $C\ell_4$.

In a similar manner we can introduce in n dimensions a cross product of $n-1$ factors. The result is a vector orthogonal to the factors, and the length of the vector is equal to the hypervolume of the parallelepiped formed by the factors.

7.4 Cross product of two vectors in $\mathbb{R}^7$

Is there a cross product in n dimensions with just two factors? If we require the cross product to be orthogonal to the factors and have length equal to the area of the parallelogram, then the answer is no, unless $n = 3$ or $n = 7$.

The cross product of two vectors in $\mathbb{R}^7$ can be defined in terms of an orthonormal basis $\mathbf{e}_1, \mathbf{e}_2, \ldots, \mathbf{e}_7$ by antisymmetry, $\mathbf{e}_i \times \mathbf{e}_j = -\mathbf{e}_j \times \mathbf{e}_i$, and

$$\begin{array}{lll}
\mathbf{e}_1 \times \mathbf{e}_2 = \mathbf{e}_4, & \mathbf{e}_2 \times \mathbf{e}_4 = \mathbf{e}_1, & \mathbf{e}_4 \times \mathbf{e}_1 = \mathbf{e}_2, \\
\mathbf{e}_2 \times \mathbf{e}_3 = \mathbf{e}_5, & \mathbf{e}_3 \times \mathbf{e}_5 = \mathbf{e}_2, & \mathbf{e}_5 \times \mathbf{e}_2 = \mathbf{e}_3, \\
\vdots & \vdots & \vdots \\
\mathbf{e}_7 \times \mathbf{e}_1 = \mathbf{e}_3, & \mathbf{e}_1 \times \mathbf{e}_3 = \mathbf{e}_7, & \mathbf{e}_3 \times \mathbf{e}_7 = \mathbf{e}_1.
\end{array}$$

The above table can be condensed into the form

$$\mathbf{e}_i \times \mathbf{e}_{i+1} = \mathbf{e}_{i+3}$$

where the indices are permuted cyclically and translated modulo 7.

This cross product of vectors in $\mathbb{R}^7$ satisfies the usual properties, that is,

$$\begin{array}{ll}
(\mathbf{a} \times \mathbf{b}) \cdot \mathbf{a} = 0, \ (\mathbf{a} \times \mathbf{b}) \cdot \mathbf{b} = 0 & \text{orthogonality} \\
|\mathbf{a} \times \mathbf{b}|^2 = |\mathbf{a}|^2|\mathbf{b}|^2 - (\mathbf{a} \cdot \mathbf{b})^2 & \text{Pythagorean theorem}
\end{array}$$

where the second rule can also be written as $|\mathbf{a} \times \mathbf{b}| = |\mathbf{a}||\mathbf{b}| \sin \sphericalangle(\mathbf{a}, \mathbf{b})$. Unlike the 3-dimensional cross product, the 7-dimensional cross product does not satisfy the Jacobi identity, $(\mathbf{a} \times \mathbf{b}) \times \mathbf{c} + (\mathbf{b} \times \mathbf{c}) \times \mathbf{a} + (\mathbf{c} \times \mathbf{a}) \times \mathbf{b} \neq 0$, and so it does not form a Lie algebra. However, the 7-dimensional cross product satisfies the Malcev identity, a generalization of Jacobi, see Ebbinghaus et al. 1991 p. 279.

In $\mathbb{R}^3$ the direction of $\mathbf{a} \times \mathbf{b}$ is unique, up to two alternatives for the orientation, but in $\mathbb{R}^7$ the direction of $\mathbf{a} \times \mathbf{b}$ depends on a 3-vector defining the cross product; to wit,

$$\mathbf{a} \times \mathbf{b} = -(\mathbf{a} \wedge \mathbf{b}) \,\lrcorner\, \mathbf{v} \quad [\neq -(\mathbf{a} \wedge \mathbf{b})\mathbf{v}]$$

depends on

$$\mathbf{v} = \mathbf{e}_{124} + \mathbf{e}_{235} + \mathbf{e}_{346} + \mathbf{e}_{457} + \mathbf{e}_{561} + \mathbf{e}_{672} + \mathbf{e}_{713} \in \bigwedge^3 \mathbb{R}^7.$$

In the 3-dimensional space $\mathbf{a} \times \mathbf{b} = \mathbf{c} \times \mathbf{d}$ implies that $\mathbf{a}, \mathbf{b}, \mathbf{c}, \mathbf{d}$ are in the same plane, but for the cross product $\mathbf{a} \times \mathbf{b}$ in $\mathbb{R}^7$ there are also other planes than the linear span of $\mathbf{a}$ and $\mathbf{b}$ giving the same direction as $\mathbf{a} \times \mathbf{b}$.

The 3-dimensional cross product is invariant under all rotations of $SO(3)$, while the 7-dimensional cross product is not invariant under all of $SO(7)$, but only under the exceptional Lie group G_2, a subgroup of $SO(7)$. When we let $\mathbf{a}$ and $\mathbf{b}$ run through all of $\mathbb{R}^7$, the image set of the simple bivectors $\mathbf{a} \wedge \mathbf{b}$ is a manifold of dimension $2 \cdot 7 - 3 = 11 > 7$ in $\bigwedge^2 \mathbb{R}^7$, $\dim(\bigwedge^2 \mathbb{R}^7) = \frac{1}{2}7(7-1) = 21$, while the image set of $\mathbf{a} \times \mathbf{b}$ is just $\mathbb{R}^7$. So the mapping

$$\mathbf{a} \wedge \mathbf{b} \to \mathbf{a} \times \mathbf{b} = -(\mathbf{a} \wedge \mathbf{b}) \lrcorner \mathbf{v}$$

is not a one-to-one correspondence, but only a method of associating a vector to a bivector.

The 3-dimensional cross product is the pure/vector part of the quaternion product of two pure quaternions, that is,

$$\mathbf{a} \times \mathbf{b} = \mathrm{Im}(\mathbf{ab}) \quad \text{for} \quad \mathbf{a}, \mathbf{b} \in \mathbb{R}^3 \subset \mathbb{H}.$$

In terms of the Clifford algebra $\mathcal{C}\ell_3 \simeq \mathrm{Mat}(2, \mathbb{C})$ of the Euclidean space $\mathbb{R}^3$ the cross product could also be expressed as

$$\mathbf{a} \times \mathbf{b} = -\langle \mathbf{abe}_{123} \rangle_1 \quad \text{for} \quad \mathbf{a}, \mathbf{b} \in \mathbb{R}^3 \subset \mathcal{C}\ell_3.$$

In terms of the Clifford algebra $\mathcal{C}\ell_{0,3} \simeq \mathbb{H} \times \mathbb{H}$ of the negative definite quadratic space $\mathbb{R}^{0,3}$ the cross product can be expressed not only as

$$\mathbf{a} \times \mathbf{b} = -\langle \mathbf{abe}_{123} \rangle_1 \quad \text{for} \quad \mathbf{a}, \mathbf{b} \in \mathbb{R}^{0,3} \subset \mathcal{C}\ell_{0,3}$$

but also as [3]

$$\mathbf{a} \times \mathbf{b} = \langle \mathbf{ab}(1 - \mathbf{e}_{123}) \rangle_1 \quad \text{for} \quad \mathbf{a}, \mathbf{b} \in \mathbb{R}^{0,3} \subset \mathcal{C}\ell_{0,3}.$$

Similarly, the 7-dimensional cross product is the pure/vector part of the octonion product of two pure octonions, that is, $\mathbf{a} \times \mathbf{b} = \langle \mathbf{a} \circ \mathbf{b} \rangle_1$. The octonion algebra $\mathbb{O}$ is a norm-preserving algebra with unity 1, whence its pure/imaginary part is an algebra with cross product, that is, $\mathbf{a} \times \mathbf{b} = \frac{1}{2}(\mathbf{a} \circ \mathbf{b} - \mathbf{b} \circ \mathbf{a})$ for $\mathbf{a}, \mathbf{b} \in \mathbb{R}^7 \subset \mathbb{O} = \mathbb{R} \oplus \mathbb{R}^7$. The octonion product in turn is given by

$$a \circ b = \alpha\beta + \alpha\mathbf{b} + \mathbf{a}\beta - \mathbf{a} \cdot \mathbf{b} + \mathbf{a} \times \mathbf{b}$$

3 This expression is also valid for $\mathbf{a}, \mathbf{b} \in \mathbb{R}^3 \subset \mathcal{C}\ell_3$, but the element $1 - \mathbf{e}_{123}$ does not pick up an ideal of $\mathcal{C}\ell_3$. Recall that $\mathcal{C}\ell_3$ is simple, that is, it has no proper two-sided ideals.

for $a = \alpha + \mathbf{a}$ and $b = \beta + \mathbf{b}$ in $\mathbb{R} \oplus \mathbb{R}^7$. If we replace the Euclidean space $\mathbb{R}^7$ by the negative definite quadratic space $\mathbb{R}^{0,7}$, then not only

$$a \circ b = \alpha\beta + \alpha\mathbf{b} + \mathbf{a}\beta + \mathbf{a} \cdot \mathbf{b} + \mathbf{a} \times \mathbf{b}$$

for $a, b \in \mathbb{R} \oplus \mathbb{R}^{0,7}$, but also

$$a \circ b = \langle ab(1 - \mathbf{v}) \rangle_{0,1}$$

where $\mathbf{v} = \mathbf{e}_{124} + \mathbf{e}_{235} + \mathbf{e}_{346} + \mathbf{e}_{457} + \mathbf{e}_{561} + \mathbf{e}_{672} + \mathbf{e}_{713} \in \bigwedge^3 \mathbb{R}^{0,7}$.

7.5 Cross products of k vectors in $\mathbb{R}^n$

If one reformulates the question about the existence of a cross product of two vectors in $\mathbb{R}^n$, and also allows $n-1$ factors, then one is led to a more general problem on the existence of a cross product of k factors in $\mathbb{R}^n$. If we were looking for a vector valued product of k factors in $\mathbb{R}^n$, then we should first formalize our problem by modifying the Pythagorean theorem, a candidate being the Gram determinant. A natural thing to do is to consider a vector valued product $\mathbf{a}_1 \times \mathbf{a}_2 \times \cdots \times \mathbf{a}_k$ satisfying

$$(\mathbf{a}_1 \times \mathbf{a}_2 \times \cdots \times \mathbf{a}_k) \cdot \mathbf{a}_i = 0 \qquad \text{orthogonality}$$
$$|\mathbf{a}_1 \times \mathbf{a}_2 \times \cdots \times \mathbf{a}_k|^2 = \det(\mathbf{a}_i \cdot \mathbf{a}_j) \qquad \text{Gram determinant}$$

where the second condition means that the length of $\mathbf{a}_1 \times \mathbf{a}_2 \times \ldots \times \mathbf{a}_k$ equals the volume of the parallelepiped with $\mathbf{a}_1, \mathbf{a}_2, \ldots, \mathbf{a}_k$ as edges.

The solution to this problem is that there are vector valued cross products in

3	dimensions with	2	factors
7	dimensions with	2	factors
n	dimensions with	$n-1$	factors
8	dimensions with	3	factors

and no others – except if one allows degenerate solutions, when there would also be in all even dimensions n, $n \in 2\mathbb{Z}$, a vector product with only one factor (and in one dimension an identically vanishing cross product with two factors).

The cross product of three vectors in $\mathbb{R}^8$ can be expressed as

$$\mathbf{a} \times \mathbf{b} \times \mathbf{c} = (\mathbf{a} \wedge \mathbf{b} \wedge \mathbf{c}) \,\lrcorner\, (\mathbf{w} - \mathbf{v}\mathbf{e}_8) = \langle (\mathbf{a} \wedge \mathbf{b} \wedge \mathbf{c})(1 - \mathbf{e}_{12\ldots8})\mathbf{w} \rangle_1$$

where

$$\begin{aligned}\mathbf{w} &= -(\mathbf{e}_{124} + \mathbf{e}_{235} + \mathbf{e}_{346} + \mathbf{e}_{457} + \mathbf{e}_{561} + \mathbf{e}_{672} + \mathbf{e}_{713})\mathbf{e}_{12\ldots7} \\ &= \mathbf{e}_{1236} - \mathbf{e}_{1257} - \mathbf{e}_{1345} + \mathbf{e}_{1467} + \mathbf{e}_{2347} - \mathbf{e}_{2456} - \mathbf{e}_{3567}\end{aligned}$$

and $\mathbf{w} \in \bigwedge^4 \mathbb{R}^7 \subset \bigwedge^4 \mathbb{R}^8$.

The trivial cross product with one factor in an even number of dimensions rotates all vectors by 90°. Thus, let n be even and let $\mathbf{a}$ be the only factor of a trivial cross product with value $\mathbf{b}$, $|\mathbf{b}| = |\mathbf{a}|$, $\mathbf{b} \cdot \mathbf{a} = 0$. This can be accomplished by

$$\mathbf{b} = \mathbf{a} \lrcorner (\mathbf{e}_1\mathbf{e}_2 + \mathbf{e}_3\mathbf{e}_4 + \ldots + \mathbf{e}_{n-1}\mathbf{e}_n).$$

Exercises

1. Show that the cross product $\mathbf{a} \times \mathbf{r}$ can be represented by a matrix multiplication $A\mathbf{r} = \mathbf{a} \times \mathbf{r}$, where

$$A\mathbf{r} = \begin{pmatrix} 0 & -a_3 & a_2 \\ a_3 & 0 & -a_1 \\ -a_2 & a_1 & 0 \end{pmatrix} \begin{pmatrix} x \\ y \\ z \end{pmatrix}.$$

2. Express the rotation matrix e^A in terms of I, A and A^2. Hint: use the Cayley-Hamilton theorem, $A^3 + |\mathbf{a}|^2 A = 0$.
3. Express the rotated vector $e^A\mathbf{r}$ as a linear combination of $\mathbf{r}$, $\mathbf{a} \times \mathbf{r}$ and $(\mathbf{a} \cdot \mathbf{r})\mathbf{a}$. Hint: $A^2\mathbf{r} = (\mathbf{a} \cdot \mathbf{r})\mathbf{a} - \mathbf{a}^2\mathbf{r}$.
4. Compute the square of $\mathbf{w} = -\mathbf{v}\mathbf{e}_{12\ldots7} \in \bigwedge^4 \mathbb{R}^7$.
5. Show that $\frac{1}{8}(1 + \mathbf{w})$ is an idempotent of $\mathcal{C}\ell_7 \simeq \mathrm{Mat}(8, \mathbb{C})$.

Solutions

2. $e^A = I + \dfrac{A}{\alpha}\sin\alpha + \dfrac{A^2}{\alpha^2}(1 - \cos\alpha)$, where $\alpha = |\mathbf{a}|$.
3. $e^A\mathbf{r} = \cos\alpha\,\mathbf{r} + \dfrac{\sin\alpha}{\alpha}\mathbf{a} \times \mathbf{r} + \dfrac{1 - \cos\alpha}{\alpha^2}(\mathbf{a} \cdot \mathbf{r})\mathbf{a}$.
4. $\mathbf{w}^2 = 7 + 6\mathbf{w}$.

Bibliography

H.-D. Ebbinghaus et al. (eds.): *Numbers*. Springer, New York, 1991.
F.R. Harvey: *Spinors and Calibrations*. Academic Press, San Diego, 1990.
W.S. Massey: Cross products of vectors in higher dimensional Euclidean spaces. *Amer. Math. Monthly* **90** (1983), #10, 697-701.
R.D. Schafer: On the algebras formed by the Cayley-Dickson process. *Amer. J. Math.* **76** (1954), 435-446.

8 Electromagnetism

The Maxwell equations can be formulated with vectors or more advanced notation like tensors, differential forms or Clifford bivectors. In these advanced formalisms the Maxwell equations become more uniform and easier to manipulate; for instance, relativistic covariance is more apparent. However, the cost of the convenience is that one has to master new concepts in addition to scalars and vectors; and antisymmetric tensors have to be untangled for physical interpretation.

8.1 The Maxwell equations

The electric field $\vec{E}$ and the magnetic induction $\vec{B}$ act on a charge q moving at velocity $\vec{v}$ by the Lorentz force

$$\vec{F} = q(\vec{E} + \vec{v} \times \vec{B}).$$

The electric displacement $\vec{D}$ and the magnetic intensity $\vec{H}$ are related to $\vec{E}$ and $\vec{B}$ by the constitutive relations

$$\vec{D} = \varepsilon \vec{E}, \qquad \vec{B} = \mu \vec{H}.$$

J. C. Maxwell brought together the following four equations in 1864:

$\nabla \cdot \vec{D} = \rho$	$\oint_S \vec{D} \cdot d\vec{s} = Q$	Gauss' law
$\nabla \times \vec{H} = \vec{J}$	$\oint_C \vec{H} \cdot d\vec{\ell} = I$	Ampère's law
$\nabla \cdot \vec{B} = 0$	$\oint_S \vec{B} \cdot d\vec{s} = 0$	no magnetic sources
$\nabla \times \vec{E} = -\dfrac{\partial \vec{B}}{\partial t}$	$\oint_C \vec{E} \cdot d\vec{\ell} = -\dfrac{d\Phi}{dt}$	Faraday's law

Maxwell also complemented Ampère's law by a new term, which observed time-dependence. Ampère had developed a mathematical formulation for producing magnetism by electricity, a phenomenon detected by Ørsted [1] in 1820, but his law is not valid in a time-varying situation: take the divergence of both sides to obtain

$$\nabla \cdot (\nabla \times \vec{H}) = 0 = \nabla \cdot \vec{J}$$

which violates charge conservation. [2] Maxwell corrected this equation into the form

$$\nabla \cdot (\nabla \times \vec{H}) = 0 = \nabla \cdot \vec{J} + \frac{\partial \rho}{\partial t},$$

applied Gauss' law, and got

$$\nabla \times \vec{H} = \vec{J} + \frac{\partial \vec{D}}{\partial t}.$$

This predicted the existence of a displacement current $\partial \vec{D}/\partial t$, which was first detected experimentally by H. Hertz in 1888, when he radiated electromagnetic waves by a dipole antenna. The electromagnetic field is now described by the *Maxwell equations* [3]

$$\boxed{\begin{array}{ll} \nabla \cdot \vec{D} = \rho & \nabla \times \vec{H} - \dfrac{\partial \vec{D}}{\partial t} = \vec{J} \\ \nabla \cdot \vec{B} = 0 & \nabla \times \vec{E} + \dfrac{\partial \vec{B}}{\partial t} = 0 \end{array}}$$

These equations are linear, and the last two equations with a vanishing right-hand side are *homogeneous*.

If ε, μ are constants, so that they do not depend on position, then the medium is *homogeneous*. If ε, μ are scalars, and not matrices or tensors, then the medium is *isotropic*. [4] In a medium that is uniform in space, i.e. homogeneous and isotropic, and stationary [5] in time, the Maxwell equations can be

1 In the paper of 1820, Ørsted's name is printed as Örsted, because the printer had no Ø.

2 Charge conservation requires that the continuity equation

$$\nabla \cdot \vec{J} = -\frac{\partial \rho}{\partial t}$$

holds for the charge density ρ and the current density $\vec{J}$ in $\mathbb{R}^3$.

3 We use SI units: $[\vec{E}] = \frac{V}{m}$, $[\vec{B}] = \frac{Vs}{m^2}$, $[\vec{D}] = \frac{C}{m^2}$, $[\vec{H}] = \frac{A}{m}$.

4 In the case that the material is non-isotropic, $D_i = \varepsilon_{ij} E_j$, $B_i = \mu_{ij} H_j$, where the matrices are symmetric $\varepsilon_{ij} = \varepsilon_{ji}$, $\mu_{ij} = \mu_{ji}$.

5 Stationary means that ε and μ do not depend on time. In an explosion ε and μ are time dependent.

expressed in terms of $\vec{E}$ and $\vec{B}$ alone:

$$\nabla \cdot \vec{E} = \frac{\rho}{\varepsilon}, \qquad \nabla \times \vec{B} - \frac{1}{c^2}\frac{\partial \vec{E}}{\partial t} = \mu \vec{J},$$
$$\nabla \cdot \vec{B} = 0, \qquad \nabla \times \vec{E} + \frac{\partial \vec{B}}{\partial t} = 0,$$

where $1/c^2 = \varepsilon\mu$. These equations hold *in a vacuum*. In a vacuum it is customary to set $\varepsilon = 1$, $\mu = 1$.

8.2 The Minkowski space-time $\mathbb{R}^{3,1}$

The electromagnetic quantities depend on time $t \in \mathbb{R}$ and position $\mathbf{x} = x_1\mathbf{e}_1 + x_2\mathbf{e}_2 + x_3\mathbf{e}_3 \in \mathbb{R}^3$. Position and time can be united into a single entity

$$x = x_1\mathbf{e}_1 + x_2\mathbf{e}_2 + x_3\mathbf{e}_3 + ct\mathbf{e}_4,$$

a vector in a 4-dimensional real linear space $\mathbb{R}^4 = \mathbb{R}^3 \times \mathbb{R}$. In this linear space we introduce a metric (or a quadratic form)

$$x_1^2 + x_2^2 + x_3^2 - c^2t^2$$

which makes it a quadratic space, called the *Minkowski space-time* $\mathbb{R}^{3,1}$.

In the Minkowski space-time it is customary to set $x^4 = ct = -x_4$ and agree that the indices are raised and lowered as follows:

$$x^1 = x_1,\ x^2 = x_2,\ x^3 = x_3 \quad \text{and} \quad x^4 = -x_4.$$

With this convention the quadratic form $x_1^2 + x_2^2 + x_3^2 - c^2t^2$ becomes

$$x_1^2 + x_2^2 + x_3^2 - x_4^2 = x^1x_1 + x^2x_2 + x^3x_3 + x^4x_4 = x^\alpha x_\alpha$$

where in the last step we have used the summation convention.

Examples. 1. The two densities ρ and $\vec{J}$ can be combined into a single quantity

$$\mathbf{J} = \vec{J} + c\rho\mathbf{e}_4 \quad \text{in} \quad \mathbb{R}^{3,1}$$

with four components $J^1, J^2, J^3, J^4 = c\rho = -J_4$ and the quadratic form $J_1^2 + J_2^2 + J_3^2 - J_4^2$.

2. We can combine the two potentials V and $\vec{A}$ in $\mathbb{R}^3$ into a single quantity with four components A^1, A^2, A^3 and $A^4 = \frac{1}{c}V = -A_4$, a space-time vector

$$\mathbf{A} = \vec{A} + \frac{1}{c}V\mathbf{e}_4 \quad \text{in} \quad \mathbb{R}^{3,1}$$

with a quadratic form $A_1^2 + A_2^2 + A_3^2 - A_4^2$.

8.3 Antisymmetric tensor of the electromagnetic field

H. Minkowski combined the two vectors $\vec{E}$ and $\vec{B}$ into a single quantity, a 4×4-matrix with entries $F^{\alpha\beta}$ given by

$$(F^{14}, F^{24}, F^{34}) = (\tfrac{1}{c}E_1, \tfrac{1}{c}E_2, \tfrac{1}{c}E_3),$$
$$(F^{23}, F^{31}, F^{12}) = (-B_1, -B_2, -B_3)$$

and antisymmetry, $F^{\alpha\beta} = -F^{\beta\alpha}$, so that

$$(F^{\alpha\beta}) = \begin{pmatrix} 0 & -B_3 & B_2 & \frac{1}{c}E_1 \\ B_3 & 0 & -B_1 & \frac{1}{c}E_2 \\ -B_2 & B_1 & 0 & \frac{1}{c}E_3 \\ -\frac{1}{c}E_1 & -\frac{1}{c}E_2 & -\frac{1}{c}E_3 & 0 \end{pmatrix}.$$

The matrix entries $F^{\alpha\beta}$ are coordinates of an antisymmetric tensor of rank 2, namely the electromagnetic field in space-time $\mathbb{R}^{3,1}$.

With this change of notation from $\vec{E}, \vec{B}$ to $F^{\alpha\beta}$ we can write the Maxwell equations in a vacuum:

$$\frac{\partial F^{14}}{\partial x^1} + \frac{\partial F^{24}}{\partial x^2} + \frac{\partial F^{34}}{\partial x^3} = \frac{\rho}{c\varepsilon}, \quad \left(\frac{\partial F^{21}}{\partial x^2} - \frac{\partial F^{13}}{\partial x^3}\right) - \frac{\partial F^{14}}{\partial x^4} = \mu J^1, \ \ldots,$$
$$\frac{\partial F^{32}}{\partial x^1} + \frac{\partial F^{13}}{\partial x^2} + \frac{\partial F^{21}}{\partial x^3} = 0, \quad \left(\frac{\partial F^{34}}{\partial x^2} - \frac{\partial F^{24}}{\partial x^3}\right) + \frac{\partial F^{32}}{\partial x^4} = 0, \ \ldots .$$

The last displayed equation can also be written as

$$\left(\frac{\partial F^{34}}{\partial x_2} + \frac{\partial F^{42}}{\partial x_3}\right) + \frac{\partial F^{23}}{\partial x_4} = 0$$

by employing antisymmetry and the lowering convention $x_4 = -x^4$.

Using the summation convention the Maxwell equations for $F^{\alpha\beta}$ can be condensed to

$$\frac{\partial F^{\alpha\beta}}{\partial x^\alpha} = \mu J^\beta,$$
$$\frac{\partial F^{\beta\gamma}}{\partial x_\alpha} + \frac{\partial F^{\gamma\alpha}}{\partial x_\beta} + \frac{\partial F^{\alpha\beta}}{\partial x_\gamma} = 0,$$

and further adopting the notations $\partial_\alpha = \dfrac{\partial}{\partial x^\alpha}$ and $\partial^\alpha = \dfrac{\partial}{\partial x_\alpha}$ to

$$\partial_\alpha F^{\alpha\beta} = \mu J^\beta,$$
$$\partial^\alpha F^{\beta\gamma} + \partial^\beta F^{\gamma\alpha} + \partial^\gamma F^{\alpha\beta} = 0.$$

Similarly, $\vec{D}$ and $\vec{H}$ can be combined to a second-rank antisymmetric tensor

$$(G^{\alpha\beta}) = \begin{pmatrix} 0 & -H_3 & H_2 & cD_1 \\ H_3 & 0 & -H_1 & cD_2 \\ -H_2 & H_1 & 0 & cD_3 \\ -cD_1 & -cD_2 & -cD_3 & 0 \end{pmatrix}.$$

Using $G^{\alpha\beta}$ the general Maxwell equations (non-homogeneous, non-isotropic, time-varying) can be written in tensor/index form, due to Minkowski:

$$\boxed{\begin{array}{l} \partial_\alpha G^{\alpha\beta} = J^\beta \\ \partial^\alpha F^{\beta\gamma} + \partial^\beta F^{\gamma\alpha} + \partial^\gamma F^{\alpha\beta} = 0 \end{array}}$$

Exercises 1ab,2ab,3a

8.4 Electromagnetic potentials

Because of $\nabla \cdot \vec{B} = 0$ there exists, in a contractible region, a vector-potential $\vec{A}$ such that

$$\vec{B} = \nabla \times \vec{A}.$$

If this equation is substituted into Faraday's law, we get

$$\nabla \times \vec{E} = -\frac{\partial}{\partial t}(\nabla \times \vec{A})$$

or

$$\nabla \times \left(\vec{E} + \frac{\partial \vec{A}}{\partial t}\right) = 0.$$

This curl-free quantity is up to a sign the gradient of a scalar, called the electric potential V,

$$\vec{E} + \frac{\partial \vec{A}}{\partial t} = -\nabla V.$$

We have shown that $\vec{E}$ and $\vec{B}$ can be expressed in terms of the potentials V and $\vec{A}$ as follows:

$$\vec{E} = -\nabla V - \frac{\partial \vec{A}}{\partial t}, \qquad \vec{B} = \nabla \times \vec{A}.$$

Combine the two potentials V and $\vec{A}$ in $\mathbb{R}^3$ into a single quantity with four components A^1, A^2, A^3 and $A^4 = \frac{1}{c}V$, a space-time vector

$$\mathbf{A} = \vec{A} + \frac{1}{c}V\mathbf{e}_4 \in \mathbb{R}^{3,1}.$$

The above equations mean that $F^{\alpha\beta}$ can be expressed in terms of the potential A^α as follows:

$$F^{14} = -\frac{\partial A^4}{\partial x^1} - \frac{\partial A^1}{\partial x^4} = \frac{\partial A^1}{\partial x_4} - \frac{\partial A^4}{\partial x_1}, \dots,$$

$$F^{32} = \frac{\partial A^3}{\partial x^2} - \frac{\partial A^2}{\partial x^3} = \frac{\partial A^3}{\partial x_2} - \frac{\partial A^2}{\partial x_3}, \dots,$$

which can be condensed to

$$F^{\alpha\beta} = -(\partial^\alpha A^\beta - \partial^\beta A^\alpha).$$

We can now verify that $\partial^\alpha F^{\beta\gamma} + \partial^\beta F^{\gamma\alpha} + \partial^\gamma F^{\alpha\beta} = 0$ by computing

$$\partial^\alpha(\partial^\beta A^\gamma - \partial^\gamma A^\beta) + \partial^\beta(\partial^\gamma A^\alpha - \partial^\alpha A^\gamma) + \partial^\gamma(\partial^\alpha A^\beta - \partial^\beta A^\alpha) = 0.$$

Exercises 2c,3b

8.5 Gauge transformations

The vector-potential $\vec{A}$ is not unique, since we can add to it, without changing physics, any vector with a vanishing curl. Adding to $\vec{A}$ a curl-free vector, the gradient of a scalar Φ, gives us $\vec{A}' = \vec{A} + \nabla\Phi$. In order to keep $\vec{E} = -\nabla V - \partial\vec{A}/\partial t$ we also change V to V',

$$\begin{aligned}\vec{E} &= -\nabla V' - \frac{\partial \vec{A}'}{\partial t} \\ &= -\nabla V' - \frac{\partial}{\partial t}(\vec{A} + \nabla\Phi) \\ &= -\nabla\left(V' + \frac{\partial\Phi}{\partial t}\right) - \frac{\partial\vec{A}}{\partial t},\end{aligned}$$

which implies $V' = V - \partial\Phi/\partial t$. The change of potentials

$$\begin{aligned}\vec{A}' &= \vec{A} + \nabla\Phi, \\ V' &= V - \frac{\partial\Phi}{\partial t}\end{aligned}$$

is called a *gauge transformation*. In coordinate form this means

$$A'^\alpha = A^\alpha + \frac{\partial\Phi}{\partial x_\alpha} = A^\alpha + \partial^\alpha\Phi$$

or swapping the sign of the time component

$$A'_\alpha = A_\alpha + \frac{\partial\Phi}{\partial x^\alpha} = A_\alpha + \partial_\alpha\Phi.$$

The fact that $\vec{E}, \vec{B}$ remain unchanged in a gauge transformation is called *gauge invariance*. In quantum electrodynamics gauge invariance is used to deduce the existence of a zero-mass carrier for the electromagnetic field.

8.6 The Lorenz condition for potentials

The two homogeneous Maxwell equations guaranteed existence of potentials for the electromagnetic field. Now we shall find out conditions imposed on the potentials by the remaining Maxwell equations. Substitute $\vec{E} = -\nabla V - \partial\vec{A}/\partial t$ into $\nabla \cdot \vec{E} = \rho/\varepsilon$ to obtain

$$-\nabla^2 V - \frac{\partial}{\partial t}(\nabla \cdot \vec{A}) = \frac{\rho}{\varepsilon}$$

in a vacuum. Substitute $\vec{E} = -\nabla V - \partial\vec{A}/\partial t$ and $\vec{B} = \nabla \times \vec{A}$ into $\nabla \times \vec{B} - \frac{1}{c^2}\partial\vec{E}/\partial t = \mu\vec{J}$, and use the identity $\nabla \times (\nabla \times \vec{A}) = \nabla(\nabla \cdot \vec{A}) - \nabla^2\vec{A}$, to obtain

$$\nabla\left(\nabla \cdot \vec{A} + \frac{1}{c^2}\frac{\partial V}{\partial t}\right) - \nabla^2\vec{A} + \frac{1}{c^2}\frac{\partial^2\vec{A}}{\partial t^2} = \mu\vec{J}.$$

The last two displayed equations couple V and $\vec{A}$.

Although the curl of $\vec{A}$ is designated to $\vec{B}$, we are still at liberty to choose the divergence of $\vec{A}$, which ensures the choice

$$\boxed{\nabla \cdot \vec{A} + \frac{1}{c^2}\frac{\partial V}{\partial t} = 0}$$

called the *Lorenz condition*. [6] In coordinate form,

$$\frac{\partial A^\alpha}{\partial x^\alpha} = 0 \quad \text{or} \quad \partial_\alpha A^\alpha = 0.$$

When the Lorenz condition is satisfied, the above two second-order differential equations, which coupled V and $\vec{A}$, can be decoupled

$$\nabla^2 V - \frac{1}{c^2}\frac{\partial^2 V}{\partial t^2} = -\frac{\rho}{\varepsilon},$$

$$\nabla^2\vec{A} - \frac{1}{c^2}\frac{\partial^2\vec{A}}{\partial t^2} = -\mu\vec{J}$$

into wave equations with the d'Alembert operator $\nabla^2 - \frac{1}{c^2}\frac{\partial^2}{\partial t^2} = \partial^\alpha\partial_\alpha$.

---

6 The Lorenz condition/gauge was discovered by the Danish physicist Ludwig Lorenz in 1867, and not by the Dutch physicist H. A. Lorentz, who demonstrated covariance of the Maxwell equations under Lorentz transformations in 1903. See J. van Bladel: Lorenz or Lorentz? *IEEE Antennas and Propagation Magazine* **33** (1991) p. 69 and *The Radioscientist* **2** (1991) p. 55.

ELECTROMAGNETISM IN CLIFFORD ALGEBRAS

In the rest of this chapter we shall discuss electromagnetism in terms of the Clifford algebras. Clifford algebras automatically take care of the manipulation of indices. The Clifford algebra approach allows various degrees of abstraction which gradually become more and more distant from classical vector analysis.

We reformulate the Maxwell equations first in terms of the Clifford algebras $\mathcal{C}\ell_3 \simeq \mathrm{Mat}(2,\mathbb{C})$ of the Euclidean space $\mathbb{R}^3$ and then in terms of the Clifford algebra $\mathcal{C}\ell_{3,1} \simeq \mathrm{Mat}(4,\mathbb{R})$ of the Minkowski space $\mathbb{R}^{3,1}$. In the Euclidean space $\mathbb{R}^3$ we shall deal with the vector $\vec{E}$ and the bivector $\vec{B}\mathbf{e}_{123}$, and in the Minkowski space $\mathbb{R}^{3,1}$ we shall deal with the bivector

$$\mathbf{F} = \frac{1}{c}\vec{E}\mathbf{e}_4 - \vec{B}\mathbf{e}_{123}.$$

8.7 The vector $\vec{E}$ and the bivector $\vec{B}\mathbf{e}_{123}$

The work W done by an electric field $\vec{E}$ in moving a charge q along a path C is given by the line integral

$$W = q\int_C \vec{E}\cdot d\vec{\ell}.$$

We conclude that the electric field $\vec{E}$ is a vector, because it is integrated along a path.

Similarly, the magnetic induction $\vec{B}$ is integrated over a surface S in order to get the magnetic flux:

$$\Phi = \int_S \vec{B}\cdot d\vec{s}.$$

Since we are integrating over a surface, we conclude that we are actually dealing with the bivector $\vec{B}\mathbf{e}_{123} = B_1\mathbf{e}_{23} + B_2\mathbf{e}_{31} + B_3\mathbf{e}_{12}$, rather than the vector $\vec{B} = B_1\mathbf{e}_1 + B_2\mathbf{e}_2 + B_3\mathbf{e}_3$.

8.8 Differentiating vectors and bivectors

Differentiate the vector $\vec{E}$, in $\mathbb{R}^3 \subset \mathcal{C}\ell_3$, to find

$$\begin{aligned}\nabla\vec{E} &= \nabla\cdot\vec{E} + \nabla\wedge\vec{E}\\ &= \nabla\cdot\vec{E} + \mathbf{e}_{123}(\nabla\times\vec{E})\end{aligned}$$

where $\nabla\times\vec{E} = -\mathbf{e}_{123}(\nabla\wedge\vec{E})$. Differentiate the bivector $\vec{B}\mathbf{e}_{123}$ to find $\nabla(\vec{B}\mathbf{e}_{123}) = \nabla\wedge(\vec{B}\mathbf{e}_{123}) + \nabla\,\lrcorner\,(\vec{B}\mathbf{e}_{123})$ where

$$\nabla\wedge(\vec{B}\mathbf{e}_{123}) = \mathbf{e}_{123}(\nabla\cdot\vec{B}),$$
$$\nabla\,\lrcorner\,(\vec{B}\mathbf{e}_{123}) = \mathbf{e}_{123}(\nabla\wedge\vec{B}) = -\nabla\times\vec{B}.$$

8.9 Single equation in $\mathcal{C}\ell_3$

Recall the Maxwell equations in a vacuum:

$$\nabla\cdot\vec{E} = \rho,$$
$$\frac{\partial\vec{E}}{\partial t} - \nabla\times\vec{B} = -\vec{J},$$
$$\frac{\partial\vec{B}}{\partial t} + \nabla\times\vec{E} = 0,$$
$$\nabla\cdot\vec{B} = 0.$$

Multiply the last two equations by $\mathbf{e}_{123}$, use the following replacements $\nabla\wedge\vec{E} = (\nabla\times\vec{E})\mathbf{e}_{123}$, $\nabla\,\lrcorner\,(\vec{B}\mathbf{e}_{123}) = -\nabla\times\vec{B}$ and $\nabla\wedge(\vec{B}\mathbf{e}_{123}) = (\nabla\cdot\vec{B})\mathbf{e}_{123}$, and you will get

$$\begin{array}{ll} 0 & \nabla\cdot\vec{E} = \rho \\ 1 & \dfrac{\partial\vec{E}}{\partial t} + \nabla\,\lrcorner\,(\vec{B}\mathbf{e}_{123}) = -\vec{J} \\ 2 & \dfrac{\partial}{\partial t}(\vec{B}\mathbf{e}_{123}) + \nabla\wedge\vec{E} = 0 \\ 3 & \nabla\wedge(\vec{B}\mathbf{e}_{123}) = 0. \end{array}$$

The numbers on the left indicate the dimension degrees of the equations. Summing up these four equations we get (use $\nabla\vec{E} = \nabla\cdot\vec{E} + \nabla\wedge\vec{E}$)

$$\frac{\partial}{\partial t}(\vec{E} + \vec{B}\mathbf{e}_{123}) + \nabla\vec{E} + \nabla\,\lrcorner\,(\vec{B}\mathbf{e}_{123}) + \nabla\wedge(\vec{B}\mathbf{e}_{123}) = \rho - \vec{J}.$$

Use $\nabla F = \nabla\,\lrcorner\,F + \nabla\wedge F$ to find

$$(\frac{\partial}{\partial t} + \nabla)(\vec{E} + \vec{B}\mathbf{e}_{123}) = \rho - \vec{J},$$

and we have condensed all the Maxwell equations into a single equation in terms of the Clifford algebra $\mathcal{C}\ell_3$. Taking the grade involute of both sides results in

$$(\frac{\partial}{\partial t} - \nabla)(-\vec{E} + \vec{B}\mathbf{e}_{123}) = \rho + \vec{J}.$$

The potentials V and $\vec{A}$, a scalar and a vector, can be united into a paravector $V+\vec{A}$. Differentiate the paravector $V+\vec{A}$ by the paravector differential operator, $\frac{\partial}{\partial t} + \nabla$,

$$(\frac{\partial}{\partial t} + \nabla)(V + \vec{A}) = \frac{\partial V}{\partial t} + \frac{\partial \vec{A}}{\partial t} + \nabla V + \nabla \vec{A},$$

where $\nabla \vec{A} = \nabla \cdot \vec{A} + (\nabla \times \vec{A})\mathbf{e}_{123}$, and you will get

$$(\frac{\partial}{\partial t} + \nabla)(V + \vec{A}) = -\vec{E} + \vec{B}\mathbf{e}_{123}.$$

Taking the grade involute of both sides results in

$$(\frac{\partial}{\partial t} - \nabla)(V - \vec{A}) = \vec{E} + \vec{B}\mathbf{e}_{123}.$$

8.10 The use of the Clifford algebra $C\ell_{3,1}$

Consider the Clifford algebra $C\ell_3 \simeq \mathrm{Mat}(2, \mathbb{C})$ as a subalgebra of the Clifford algebra $C\ell_{3,1}$ generated by $\mathbf{e}_1, \mathbf{e}_2, \mathbf{e}_3, \mathbf{e}_4$ with the relations

$$\mathbf{e}_1^2 = \mathbf{e}_2^2 = \mathbf{e}_3^2 = 1, \ \mathbf{e}_4^2 = -1,$$
$$\mathbf{e}_\alpha \mathbf{e}_\beta = -\mathbf{e}_\beta \mathbf{e}_\alpha \quad \text{for} \quad \alpha \neq \beta.$$

The Clifford algebra $C\ell_{3,1}$ is isomorphic, as an associative algebra, with the algebra of real 4×4-matrices $\mathrm{Mat}(4, \mathbb{R})$. In the Clifford algebra $C\ell_{3,1}$ we consider the electromagnetic bivector [7]

$$\mathbf{F} = \frac{1}{c}\vec{E}\mathbf{e}_4 - \vec{B}\mathbf{e}_{123} \in \textstyle\bigwedge^2 \mathbb{R}^{3,1}$$

and the space-time current vector

$$\mathbf{J} = \vec{J} + c\rho\mathbf{e}_4 \in \mathbb{R}^{3,1}.$$

From $\mathbf{F}$ we can find $\vec{E}$ by $\vec{E} = c\mathbf{e}_4 \lrcorner \mathbf{F}$, and from $\mathbf{J}$ we can find $\vec{J}$ by $\vec{J} = (\mathbf{J} \wedge \mathbf{e}_4)\mathbf{e}_4^{-1}$.

We introduce the differential operator

$$\partial = \nabla - \mathbf{e}_4 \frac{1}{c}\frac{\partial}{\partial t}.$$

For a function $f : \mathbb{R}^{3,1} \to C\ell_{3,1}$ we have $\partial f = \partial \wedge f + \partial \lrcorner f$, where $\partial \wedge f$ is the *raising differential* and $\partial \lrcorner f$ is the *lowering differential*.

7 If we use the orthonormal basis $\{\mathbf{e}_0, \mathbf{e}_1, \mathbf{e}_2, \mathbf{e}_3\}$ with $\mathbf{e}_1^2 = \mathbf{e}_2^2 = \mathbf{e}_3^2 = 1$ and $\mathbf{e}_0^2 = -1$, that is, $\mathbf{e}_0 = \mathbf{e}_4$, then we find by reordering the indices that $\mathbf{e}_{0123} = -\mathbf{e}_{1234}$ and $\mathbf{F} = \frac{1}{c}\vec{E}\mathbf{e}_4 - (\vec{B}\mathbf{e}_4)\mathbf{e}_{1234} = \frac{1}{c}\vec{E}\mathbf{e}_0 + (\vec{B}\mathbf{e}_0)\mathbf{e}_{0123}$.

Compute the raising differential

$$\partial \wedge \mathbf{F} = \Big(\nabla - \mathbf{e}_4 \frac{1}{c}\frac{\partial}{\partial t}\Big) \wedge \Big(\frac{1}{c}\vec{E}\mathbf{e}_4 - \vec{B}\mathbf{e}_{123}\Big)$$
$$= \frac{1}{c}\mathbf{e}_{123}(\nabla \times \vec{E})\mathbf{e}_4 - \mathbf{e}_{123}(\nabla \cdot \vec{B}) - \mathbf{e}_{1234}\frac{1}{c}\frac{\partial \vec{B}}{\partial t} = 0.$$

Define $\mathbf{G} = c\vec{D}\mathbf{e}_4 - \vec{H}\mathbf{e}_{123}$ and compute the lowering differential

$$\partial \,\lrcorner\, \mathbf{G} = \Big(\nabla - \mathbf{e}_4 \frac{1}{c}\frac{\partial}{\partial t}\Big) \,\lrcorner\, (c\vec{D}\mathbf{e}_4 - \vec{H}\mathbf{e}_{123})$$
$$= c(\nabla \cdot \vec{D})\mathbf{e}_4 + \nabla \times \vec{H} - \frac{\partial \vec{D}}{\partial t} = c\rho\mathbf{e}_4 + \vec{J}.$$

The Maxwell equations now have a particularly succinct form [8]

$$\boxed{\begin{array}{c} \partial \,\lrcorner\, \mathbf{G} = \mathbf{J} \\ \partial \wedge \mathbf{F} = 0 \end{array}}$$

corresponding to

$$\nabla \cdot \vec{D} = \rho, \qquad -\nabla \,\lrcorner\, (\vec{H}\mathbf{e}_{123}) - \frac{\partial \vec{D}}{\partial t} = \vec{J},$$
$$-\nabla \wedge (\vec{B}\mathbf{e}_{123}) = 0, \qquad \nabla \wedge \vec{E} + \frac{\partial}{\partial t}(\vec{B}\mathbf{e}_{123}) = 0.$$

8.11 Single equation in a vacuum, $C\ell_{3,1}$

In a vacuum the Maxwell equations can be further compressed into a single equation

$$\partial \mathbf{F} = \mathbf{J},$$

which decomposes into two parts, $\partial \wedge \mathbf{F} = 0$ and $\partial \,\lrcorner\, \mathbf{F} = \mathbf{J}$. Also, $\partial \wedge \mathbf{A} = -\mathbf{F}$ and the Lorenz condition $\partial \cdot \mathbf{A} = 0$ imply

$$\partial \mathbf{A} = -\mathbf{F}.$$

8.12 The energy-momentum tensor

Marcel Riesz in 1947 wrote the energy-momentum tensor in the form

$$T_{\mu\nu} = -\frac{1}{2}\langle \mathbf{e}_\mu \mathbf{F} \mathbf{e}_\nu \mathbf{F}\rangle_0.$$

8 The 3D formulation differs from this 4D formulation in the sense that $\mathbf{G}$ and $\mathbf{F}$ are bivectors in $\bigwedge^2 \mathbb{R}^{3,1}$.

D. Hestenes 1966 p. 31 introduced the vectors

$$\mathbf{T}_\mu = -\frac{1}{2}\mathbf{F}\mathbf{e}_\mu\mathbf{F}$$

for which $T_{\mu\nu} = \mathbf{T}_\mu \cdot \mathbf{e}_\nu = \mathbf{e}_\mu \cdot \mathbf{T}_\nu$, and also the mapping $T\mathbf{x} = -\frac{1}{2}\mathbf{F}\mathbf{x}\mathbf{F}$ where $(T\mathbf{x})^\mu = T^\mu{}_\nu x^\nu$. [9] The energy-momentum tensor is symmetric, that is, $T_{\mu\nu} = T_{\nu\mu}$ or $gT^\top g^{-1} = T$, where $g_{\mu\nu} = \mathbf{e}_\mu \cdot \mathbf{e}_\nu$, and traceless, that is, $T^\mu{}_\mu = 0$. [10]

Note that the Poynting vector $\mathbf{T}_4 = \vec{E} \times \vec{B} + \frac{1}{2}(\vec{E}^2 + \vec{B}^2)\mathbf{e}_4$ is not a space-time vector, in the sense that it does not transform properly under Lorentz transformations, but rather it is just the last column of the energy-momentum matrix $T = (T^\mu{}_\nu)$ which transforms as $T' = LTL^{-1}$.

Electromagnetism in Differential Forms

Electromagnetism can also be formulated with differential forms, based on Grassmann's exterior algebra. In this context it is customary to invoke the dual space

$$V^* = \{f : V \to \mathbb{R} \mid f \ \text{linear}\}$$

of the real linear space $V = \mathbb{R}^{3,1}$. Instead of vectors and bivectors, in V and $\bigwedge^2 V$, one considers 1-forms and 2-forms, in V^* and $\bigwedge^2 V^*$.

In theoretical physics one applies differential forms to electromagnetism, but in electrical engineering one uses almost exclusively the vector analysis of Gibbs and Heaviside. [11] Electrical engineers are not interested in transformation laws, [12] and so it is convenient for them to place all vectors in $V = \mathbb{R}^{3,1}$ [and disregard the dual space V^*]. However, a theory without the dual space V^* cannot be generalized to curved space-times. In a curved space-time it is not possible to differentiate vector valued functions, only differential forms can be differentiated [in general relativity vectors are differentiated covariantly].

Although differential forms are not of practical value for electrical engineers,

9 Juvet & Schidlof 1932 p. 141 gave $T_{\mu\nu} = F_\mu{}^\alpha F_{\alpha\nu} + \frac{1}{4}g_{\mu\nu}F_{\alpha\beta}F^{\alpha\beta}$ but did not consider $T\mathbf{x} = -\frac{1}{2}\mathbf{F}\mathbf{x}\mathbf{F}$, compare this to Bolinder p. 469 in Chisholm & Common (eds.) 1986.

10 The tracelessness of $T^\mu{}_\nu = -\frac{1}{2}\langle\mathbf{e}^\mu\mathbf{F}\mathbf{e}_\nu\mathbf{F}\rangle_0$ is an accident in dimension 4, since $\mathbf{e}^\mu\mathbf{F}\mathbf{e}_\mu = 0$, and in general $\mathbf{e}^\mu\mathbf{a}\mathbf{e}_\mu = (n-2k)\hat{\mathbf{a}}$ for $\mathbf{a} \in \bigwedge^k \mathbb{R}^n$.

11 As far as the author knows the only university where electrical engineers have used differential forms in teaching is Helsinki University of Technology, see lecture notes Lindell & Lounesto 1995.

12 For instance, the space-time position $\mathbf{x} = \vec{x} + ct\mathbf{e}_4$ and the current density $\mathbf{J} = \vec{J} + c\rho\mathbf{e}_4$ transform differently under the Lorentz group; one transforms contravariantly and the other covariantly. In tensor calculus elements of V are called vectors and elements of the dual space V^* are called covectors.

we shall close this chapter with a short discussion on the formulation of electromagnetism with differential forms, see Lindell 1995. But first some observations about functions with values in $\bigwedge V = \mathcal{C}\ell_{3,1}$.

8.13 Using only raising or lowering differentials

Since the current density $\vec{J}$ integrates over a surface S,

$$\oint_S \vec{J} \cdot d\vec{s} = -I,$$

we can replace it by a bivector $\vec{J}\mathbf{e}_{123}$, and since the charge density ρ integrates over a 3-volume, we can replace it by a 3-vector $\rho\mathbf{e}_{123}$. Similarly, we can regard $\vec{H}$ as a vector, but replace the vector $\vec{D}$ by a bivector $\vec{D}\mathbf{e}_{123}$.

The two Maxwell equations with a source-term on the right hand side can be rewritten in the form

$$\nabla \wedge (\vec{D}\mathbf{e}_{123}) = \rho\mathbf{e}_{123}, \qquad -\nabla \wedge \vec{H} - \frac{\partial}{\partial t}(\vec{D}\mathbf{e}_{123}) = \vec{J}\mathbf{e}_{123}.$$

Take the Hodge dual

$$\begin{aligned} \star\mathbf{G} &= \tilde{\mathbf{G}}\mathbf{e}_{1234} = -c\vec{D}\mathbf{e}_{123} - \vec{H}\mathbf{e}_4 \quad \text{and} \\ \star\mathbf{J} &= \tilde{\mathbf{J}}\mathbf{e}_{1234} = c\rho\mathbf{e}_{123} + (\vec{J}\mathbf{e}_{123})\mathbf{e}_4, \end{aligned}$$

and compute the raising differential

$$\begin{aligned} \partial \wedge \star\mathbf{G} &= \left(\nabla - \mathbf{e}_4 \frac{1}{c}\frac{\partial}{\partial t}\right) \wedge (-c\vec{D}\mathbf{e}_{123} - \vec{H}\mathbf{e}_4) \\ &= -c(\nabla \cdot \vec{D})\mathbf{e}_{123} - \mathbf{e}_{123}(\nabla \times \vec{H})\mathbf{e}_4 + \frac{\partial \vec{D}}{\partial t}\mathbf{e}_{1234}. \end{aligned}$$

The Maxwell equations can now be expressed in terms of the raising differential alone:

$$\begin{aligned} \partial \wedge \star\mathbf{G} &= -\star\mathbf{J}, \\ \partial \wedge \mathbf{F} &= 0. \end{aligned}$$

Dually, we can write down the Maxwell equations using only the lowering differential:

$$\begin{aligned} \partial \lrcorner \mathbf{G} &= \mathbf{J}, \\ \partial \lrcorner \star\mathbf{F} &= 0. \end{aligned}$$

These equations are invariant under the general linear group $GL(4, \mathbb{R})$, and the solutions are independent of the choice of metric. [13]

13 In the absence of a metric it is customary to invoke the dual algebra $\bigwedge V^*$ of the exterior algebra $\bigwedge V$ and take exterior differentials of differential forms rather than differentials of multivector valued functions.

8.14 The constitutive relations

The constitutive relations of the medium are

$$\vec{D} = \varepsilon\vec{E} + \alpha\vec{B},$$
$$\vec{H} = \beta\vec{E} + \mu^{-1}\vec{B}.$$

Here $\varepsilon, \alpha, \beta$ and μ^{-1} are 3×3-matrices. To find the rules imposed on them, write the above equations in coordinate form:

$$G^{\kappa\lambda} = \tfrac{1}{2}\chi^{\kappa\lambda\mu\nu}F_{\mu\nu}.$$

Then, if $\chi^{\kappa\lambda\mu\nu}$ is an irreducible tensor, [14] we must have

$$\chi^{\kappa\lambda\mu\nu} = -\chi^{\lambda\kappa\mu\nu}, \quad \chi^{\kappa\lambda\mu\nu} = -\chi^{\kappa\lambda\nu\mu},$$
$$\chi^{\kappa\lambda\mu\nu} = \chi^{\mu\nu\lambda\kappa},$$
$$\chi^{[\kappa\lambda\mu\nu]} = 0,$$

where the brackets [ ] mean complete alternation of indices. The second relation implies $\varepsilon^{\mathsf{T}} = \varepsilon$, $\mu^{\mathsf{T}} = \mu$ and $\alpha = -\beta^{\mathsf{T}}$ and the third relation implies $\mathrm{trace}(\alpha) = \mathrm{trace}(\beta)$, which together with the former implies $\mathrm{trace}(\alpha) = \mathrm{trace}(\beta) = 0$. These considerations can be condensed into saying that the indices of the constitutive tensor $\chi^{\kappa\lambda\mu\nu} = \chi^{\kappa\mu}_{\lambda\nu}$ can be arranged into a Young tableau

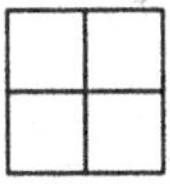

The irreducible tensor χ has 20 components, where $20 = \frac{1}{12}n^2(n^2-1)$ for $n = 4$. In chiral media the tensor χ need not be irreducible, and the number of components may rise to 36.

8.15 The derivative and the exterior differential

Let U and V be real linear spaces with norms. The derivative of $f : U \to V$ at $\mathbf{x} \in U$ is a linear function

$$f'(\mathbf{x}) : U \to V, \; \mathbf{h} \to f'(\mathbf{x})\mathbf{h}$$

such that

$$f(\mathbf{x}+\mathbf{h}) - f(\mathbf{x}) = f'(\mathbf{x})\mathbf{h} + \|\, \mathbf{h} \,\|\, \varepsilon(\mathbf{x},\mathbf{h})$$

where $\varepsilon(\mathbf{x},\mathbf{h}) \to 0$ as $\mathbf{h} \to 0$. The linear function $f'(\mathbf{x}) : U \to V$ can be identified with an element of $U^* \otimes V$.

14 The factor χ need not be a tensor. For instance, magnetic saturation and hysteresis are not expressible with a tensor χ.

Consider now a function $f : V \to \bigwedge V^*$. Its derivative at $\mathbf{x}$,

$$f'(\mathbf{x}) \in V^* \otimes \bigwedge V^*,$$

is no longer an element of the dual exterior algebra $\bigwedge V^* \subset \otimes V^*$. The alternation, which antisymmetrizes tensor product of vectors, is a linear function projecting $\bigwedge V^*$ out of $\otimes V^*$ so that

$$u \wedge u = Alt(u \otimes v) \quad \text{for} \quad u, v \in \bigwedge V^*.$$

We define the *exterior differential* of $f : V \to \bigwedge V^*$ at $\mathbf{x}$ by [15]

$$d \wedge f(\mathbf{x}) = Alt(f'(\mathbf{x})).$$

Next, we will replace vector valued functions $V \to V$ by 1-forms $V \to V^*$, and bivector valued functions $V \to \bigwedge^2 V$ by 2-forms $V \to \bigwedge^2 V^*$. The electromagnetic bivectors $\mathbf{F}$ and $\mathbf{G}$ are replaced by 2-forms F and G. The current vector $\mathbf{J}$ is replaced by a 1-form J.

The exterior differential raises the degree. The dual of the exterior differential, called the *contraction differential* $d \lrcorner f = \star^{-1} d \wedge \star f$, [16] lowers the degree. In differential forms the Maxwell equations look like

$$d \lrcorner G = J,$$
$$d \wedge F = 0.$$

8.16 General linear covariance of the Maxwell equations

Using the differential forms we may find the most general expression of the Maxwell equations:

$$\boxed{\begin{aligned} d \wedge \star G &= - \star J \\ d \wedge F &= 0 \end{aligned}}$$

These equations include only the exterior differential, and no contraction differential, so that a metric is not involved. This makes the equations independent of any coordinate system. The metric gets involved by the constitutive relations of the medium

$$\boxed{G = \chi(F)}$$

and the Hodge dual.

This form of the Maxwell equations is not only relativistically covariant,

15 The exterior differential is usually denoted by df.
16 The contraction differential is commonly called the co-differential and denoted by δf.

under the Lorentz group $O(3,1)$, [17] but also covariant under any linear transformation of space-time coordinates, that is, under the general linear group $GL(4,\mathbb{R})$. This general linear covariance of the Maxwell equations, and their independence of metric/medium, were recognized by Weyl 1921, Cartan 1926 and van Dantzig 1934.

Historical Survey

The Maxwell equations have been condensed into a single equation using complex vectors (Silberstein 1907), complex quaternions (Silberstein 1912/1914, Lanczos 1919), spinors (Laporte & Uhlenbeck 1931, Bleuler & Kustaanheimo 1968) and using Clifford algebras (Juvet & Schidlof 1932, Mercier 1935, M. Riesz 1958). Marcel Riesz 1947 wrote the energy-momentum tensor in the form $T_{\mu\nu} = -\frac{1}{2}\langle \mathbf{e}_\mu \mathbf{F} \mathbf{e}_\nu \mathbf{F}\rangle_0$.

Exercises

Metric $x_1^2 + x_2^2 + x_3^2 - x_4^2$:

1. Recall that $(F^{14}, F^{24}, F^{34}) = (\frac{1}{c}E_1, \frac{1}{c}E_2, \frac{1}{c}E_3)$ and $(F^{23}, F^{31}, F^{12}) = (-B_1, -B_2, -B_3)$. Compute the matrices

a) ${F^\alpha}_\beta$, b) ${F_\alpha}^\beta$, and the vector

c) $v^\alpha {F_\alpha}^\beta$ for $(v^1, v^2, v^3, v^4) = (v_1, v_2, v_3, c)$. [18]

Metric $-x_0^2 + x_1^2 + x_2^2 + x_3^2$:

In this metric $\partial^\alpha = (-\frac{1}{c}\frac{\partial}{\partial t}, \frac{\partial}{\partial x}, \frac{\partial}{\partial y}, \frac{\partial}{\partial z})$.

2. Replace $\vec{E}$ and $\vec{B}$ by $F^{\alpha\beta} = -F^{\beta\alpha}$ so that $(F^{01}, F^{02}, F^{03}) = (-\frac{1}{c}E_1, -\frac{1}{c}E_2, -\frac{1}{c}E_3)$ and $(F^{23}, F^{31}, F^{12}) = (-B_1, -B_2, -B_3)$, and determine

a) the antisymmetric matrix $F^{\alpha\beta}$,

b) the Maxwell equations in terms of $F^{\alpha\beta}$,

c) $F^{\alpha\beta}$ in terms of A^α,

17 The Maxwell equations describe massless particles, photons, and as such they are conformally covariant, as was demonstrated by Cunningham and Bateman in 1910. The conformal transformations are not linear in general, that is, they are not in $GL(4,\mathbb{R})$.

18 For simplicity we have omitted the factor

$$\gamma = \frac{1}{\sqrt{1-\frac{v^2}{c^2}}},$$

which makes both sides of the equation of the Lorentz force, $f^\beta = u^\alpha {F_\alpha}^\beta$, $u^\alpha = \gamma v^\alpha$, properly transforming space-time vectors.

d) $v^\alpha F_\alpha{}^\beta$ for $(v^0, v^1, v^2, v^3) = (c, v_1, v_2, v_3)$.

Metric $x_0^2 - x_1^2 - x_2^2 - x_3^2$:

In this metric $\partial^\alpha = (\frac{1}{c}\frac{\partial}{\partial t}, -\frac{\partial}{\partial x}, -\frac{\partial}{\partial y}, -\frac{\partial}{\partial z})$.

3. Replace $\vec{E}$ and $\vec{B}$ by $F^{\alpha\beta} = -F^{\beta\alpha}$ so that $(F^{01}, F^{02}, F^{03}) = (-\frac{1}{c}E^1, -\frac{1}{c}E^2, -\frac{1}{c}E^3)$ and $(F^{23}, F^{31}, F^{12}) = (-B^1, -B^2, -B^3)$, and determine

a) the Maxwell equations in terms of $F^{\alpha\beta}$,

b) $F^{\alpha\beta}$ in terms of A^α.
[Note that $A^1 = A_x = -A_1$, $A^2 = A_y = -A_2$, $A^3 = A_z = -A_3$ but $A^0 = \frac{1}{c}V = A_0$.]

4. Electrical engineers use the pairs $\vec{E}, \vec{H}$ and $\vec{D}, \vec{B}$. The constitutive relations sending $\vec{E}, \vec{H}$ to $\vec{D}, \vec{B}$ are then

$$\vec{D} = \breve{\varepsilon}\vec{E} + \breve{\alpha}\vec{H},$$
$$\vec{B} = \breve{\beta}\vec{E} + \breve{\mu}\vec{H}.$$

Show that $\breve{\mu} = \mu$, $\breve{\alpha} = \alpha\mu$, $\breve{\beta} = -\mu\beta$ and $\breve{\varepsilon} = \varepsilon - \alpha\mu\beta$.

Solutions

1a.

$$(F^\alpha{}_\beta) = \begin{pmatrix} 0 & -B_3 & B_2 & -\frac{1}{c}E_1 \\ B_3 & 0 & -B_1 & -\frac{1}{c}E_2 \\ -B_2 & B_1 & 0 & -\frac{1}{c}E_3 \\ -\frac{1}{c}E_1 & -\frac{1}{c}E_2 & -\frac{1}{c}E_3 & 0 \end{pmatrix}$$

b.

$$(F_\alpha{}^\beta) = \begin{pmatrix} 0 & -B_3 & B_2 & \frac{1}{c}E_1 \\ B_3 & 0 & -B_1 & \frac{1}{c}E_2 \\ -B_2 & B_1 & 0 & \frac{1}{c}E_3 \\ \frac{1}{c}E_1 & \frac{1}{c}E_2 & \frac{1}{c}E_3 & 0 \end{pmatrix}$$

c. The space-component is $\vec{E} + \vec{v} \times \vec{B}$ and the time-component $\frac{1}{c}\vec{v} \cdot \vec{E}$.

2a.

$$(F^{\alpha\beta}) = \begin{pmatrix} 0 & -\frac{1}{c}E_1 & -\frac{1}{c}E_2 & -\frac{1}{c}E_3 \\ \frac{1}{c}E_1 & 0 & -B_3 & B_2 \\ \frac{1}{c}E_2 & B_3 & 0 & -B_1 \\ \frac{1}{c}E_3 & -B_2 & B_1 & 0 \end{pmatrix}$$

b. $\partial_\alpha F^{\alpha\beta} = \mu J^\beta$, $\partial^\alpha F^{\beta\gamma} + \partial^\beta F^{\gamma\alpha} + \partial^\gamma F^{\alpha\beta} = 0$.

c. $F^{\alpha\beta} = -(\partial^\alpha A^\beta - \partial^\beta A^\alpha)$.

d. The time-component is $\frac{1}{c}\vec{v} \cdot \vec{E}$ and the space-component $\vec{E} + \vec{v} \times \vec{B}$.

3a. $\partial_\alpha F^{\alpha\beta} = \mu J^\beta, \quad \partial^\alpha F^{\beta\gamma} + \partial^\beta F^{\gamma\alpha} + \partial^\gamma F^{\alpha\beta} = 0.$
b. $F^{\alpha\beta} = \partial^\alpha A^\beta - \partial^\beta A^\alpha.$

Bibliography

W.E. Baylis: *Electrodynamics: A Modern Geometric Approach.* Birkäuser, Boston, 1999.

E.F. Bolinder: Unified microwave network theory based on Clifford algebra in Lorentz space, pp. 25-35 in *Conference Proceedings, 12th European Microwave Conference (Helsinki 1982).* Microwave Exhibitions and Publishers, Tunbridge Wells, Kent, 1982.

E.F. Bolinder: Clifford algebra: What is it? *IEEE Antennas and Propagation Society Newsletter* **29** (1987), 18-23.

D.K. Cheng: *Field and Wave Electromagnetics.* Addison-Wesley, Reading, MA, 1989, 1991.

G.A. Deschamps: Electromagnetics and differential forms. *Proc. IEEE*, **69** (1981), 676-696.

D. Hestenes: *Space-Time Algebra.* Gordon and Breach, Philadelphia, PA, 1966, 1987, 1992.

J.D. Jackson: *Classical Electrodynamics.* Wiley, New York, 1962, 1975.

B. Jancewicz: *Multivectors and Clifford Algebra in Electrodynamics.* World Scientific, Singapore, 1988.

G. Juvet, A. Schidlof: Sur les nombres hypercomplexes de Clifford et leurs applications à l'analyse vectorielle ordinaire, à l'électromagnetisme de Minkowski et à la théorie de Dirac. *Bull. Soc. Neuchat. Sci. Nat.*, **57** (1932), 127-147.

I.V. Lindell: *Methods for Electromagnetics Field Analysis.* Clarendon Press, Oxford, 1992.

I.V. Lindell, A. Sihvola, S. Tretyakov, A. Viitanen: *Electromagnetic Waves in Chiral and Bi-Isotropic Media.* Artech House, Boston, MA, 1994.

I. Lindell, P. Lounesto: *Differentiaalimuodot sähkömagnetiikassa.* Helsinki University of Technology, Electromagnetics Laboratory, 1995.

P. Lorrain, D.R. Corson: *Electromagnetic Fields and Waves: Principles and Applications.* W.H. Freeman, San Francisco, 1990.

A. Mercier: *Expression des équations de l'électromagnétisme au moyen des nombres de Clifford.* Thesis, Université de Genève, 1935.

E.J. Post: *Formal Structure of Electromagnetics.* North-Holland, Amsterdam, 1962.

M. Riesz: *Clifford Numbers and Spinors.* The Institute for Fluid Dynamics and Applied Mathematics, Lecture Series No. **38**, University of Maryland, 1958. Reprinted with comments as facsimile by E.F. Bolinder, P. Lounesto (eds.), Kluwer, Dordrecht, The Netherlands, 1993.

9
Lorentz Transformations

According to the *Galilean principle of relativity* the laws of classical mechanics are the same for all observers (moving at constant velocity with respect to each other). More precisely, the laws of classical mechanics remain the same under *Galilean transformations*

direct	inverse
$x' = x - vt$	$x = x' + vt'$
$t' = t$	$t = t'$

relating two frames (x, t) and (x', t') moving at relative velocity v. The equations on the left show that the origin of the second frame $x' = 0$ corresponds to uniform motion $x = vt$ in the first frame. There is no privileged inertial frame or absolute rest for moving bodies, but time is preserved, that is, *time is absolute*.

The Galilean principle or invariance does not govern all of physics, most notably electromagnetism and in particular light propagation. For instance, the wave equation

$$\frac{\partial^2 f}{\partial x^2} - \frac{1}{c^2}\frac{\partial^2 f}{\partial t^2} = 0$$

is not preserved in a Galilean change of variables $(x, t) \to (x', t')$. The wave equation is instead invariant under another transformation, named after H.A. Lorentz. In 1887, Michelson & Morley carried out an experiment which indicated that light travels at the same velocity independent of the motion of the source. In 1905, Einstein took the constancy of the velocity of light as a postulate, and showed that this postulate, together with the principle of relativity, is sufficient for deriving the kinematical formulas of Lorentz. In so doing, Einstein had to revise the notion of time, and abandon the concept of absolute time.

9.1 Lorentz transformations in one space dimension

The simplest modification of the Galilean transformation, preserving linearity and the implication $x' = 0 \Rightarrow x = vt$, is obtained by multiplying with a factor γ:

$$x' = \gamma(x - vt), \qquad x = \gamma(x' + vt')$$

where γ is independent of x and t but may depend on v. We require that γ is the same in both equations since the inverse transformation should be identical to the direct one except for a change of v to $-v$.

In computing γ we use the observation of equal velocity of light. Consider a light-signal travelling at velocity c in both frames, so that $x = ct$ and $x' = ct'$, which substituted into the right-hand side of the previous equations results in

$$x' = \gamma(ct - vt), \qquad x = \gamma(ct' + vt')$$

or, substituting $x' = ct'$ and $x = ct$ also into the left-hand side,

$$ct' = \gamma(c - v)t, \qquad ct = \gamma(c + v)t',$$

a formula admitting explicitly a transformation of time. Divide the two equations

$$\frac{c}{\gamma(c+v)} = \frac{\gamma(c-v)}{c},$$

which gives the factor

$$\gamma = \frac{1}{\sqrt{1 - \frac{v^2}{c^2}}}.$$

Next, compute the transformation of the time coordinate of events. Substitute $x' = \gamma(x - vt)$ into $x = \gamma(x' + vt')$,

$$x = \gamma^2(x - vt) + \gamma vt',$$

use the explicit form of γ, and solve for t',

$$t' = \gamma(t - \frac{v}{c^2}x).$$

Similarly, compute the inverse transformation,

$$t = \gamma(t' + \frac{v}{c^2}x').$$

The fact that time is also transformed is referred to as the *relativity of time*.

Summarizing, we have the following transformation laws for the space and time coordinates

direct	inverse
$x' = \dfrac{x - vt}{\sqrt{1 - \frac{v^2}{c^2}}}$	$x = \dfrac{x' + vt'}{\sqrt{1 - \frac{v^2}{c^2}}}$
$t' = \dfrac{t - \frac{v}{c^2}x}{\sqrt{1 - \frac{v^2}{c^2}}}$	$t = \dfrac{t' + \frac{v}{c^2}x'}{\sqrt{1 - \frac{v^2}{c^2}}}$

known as the **Lorentz transformation**. Lorentz transformations preserve the quadratic form $x^2 - c^2t^2 = x'^2 - c^2t'^2$ and orthogonality of events; two events x_1, ct_1 and x_2, ct_2 are said to be *orthogonal* if $x_1x_2 - c^2t_1t_2 = 0$. In particular, time and space are orthogonal.

It should be noted that time and space do not diverge by 90°, that is, they are not 'perpendicular' or 'rectangular'. If we draw space-time coordinates x, ct on paper so that the time-axis is 'perpendicular' to the space and perform a Lorentz transformation, then the transformed coordinate-axes x', ct' are no longer 'rectangular' (but they are orthogonal, by definition).

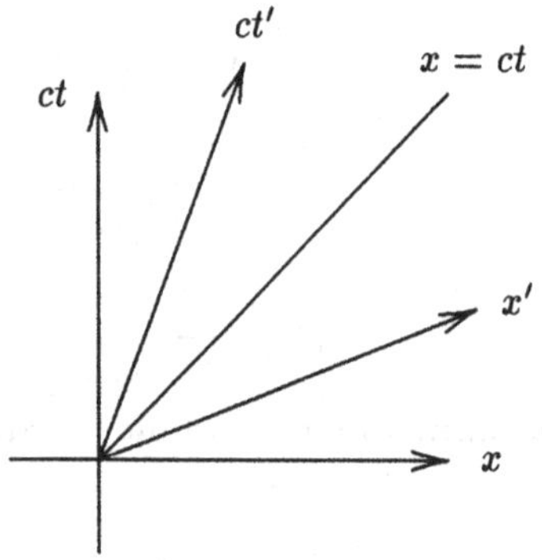

Write the direct Lorentz transformation in matrix form:

$$\begin{pmatrix} x' \\ ct' \end{pmatrix} = \frac{1}{\sqrt{1 - \frac{v^2}{c^2}}} \begin{pmatrix} 1 & -\frac{v}{c} \\ -\frac{v}{c} & 1 \end{pmatrix} \begin{pmatrix} x \\ ct \end{pmatrix}.$$

This resumes the composition of Lorentz transformations into multiplication of matrices:

$$L_v = \frac{1}{\sqrt{1 - \frac{v^2}{c^2}}} \begin{pmatrix} 1 & -\frac{v}{c} \\ -\frac{v}{c} & 1 \end{pmatrix}.$$

The composition of two Lorentz transformations at velocities v_1 and v_2 results in a Lorentz transformation $L_v = L_{v_2} L_{v_1}$ at velocity

$$v = \frac{v_1 + v_2}{1 + \frac{v_1 v_2}{c^2}},$$

a formula known as the relativistic composition of parallel velocities.

9.2 The Minkowski space-time $\mathbb{R}^{3,1}$

Space-time events can be labelled by points (x^1, x^2, x^3, x^4) or vectors $\mathbf{x} = x^1\mathbf{e}_1 + x^2\mathbf{e}_2 + x^3\mathbf{e}_3 + x^4\mathbf{e}_4$ in the Minkowski space-time $\mathbb{R}^{3,1}$. Indices are raised and lowered according to

$$x^1 = x_1,\ x^2 = x_2,\ x^3 = x_3,\ x^4 = ct = -x_4.$$

The Minkowski space-time $\mathbb{R}^{3,1}$ has a quadratic form sending a vector $\mathbf{x} \in \mathbb{R}^{3,1}$ to a scalar which we shall denote by $\mathbf{x}^2$,

$$\mathbf{x}^2 = x_1^2 + x_2^2 + x_3^2 - x_4^2.$$

Solutions to the equation

$$x_1^2 + x_2^2 + x_3^2 = x_4^2 \quad \text{or} \quad \mathbf{x}^2 = 0$$

form the *null-cone* or *light-cone*.

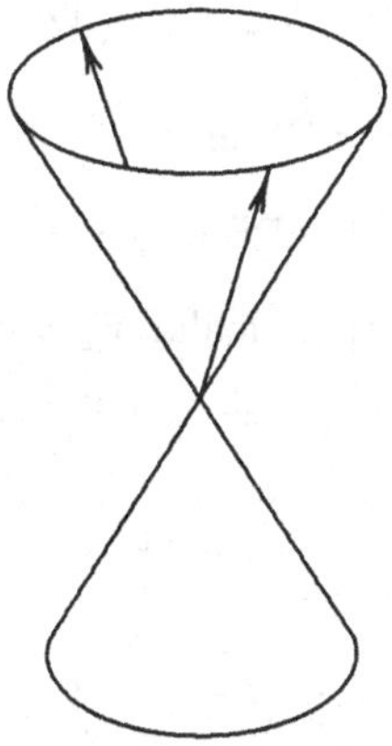

Light-cone and light-like vectors

The set of non-zero vectors, or *space-time intervals*, $\mathbf{x} \in \mathbb{R}^{3,1}$ can be divided into

$\mathbf{x}^2 > 0$, space-like vectors,
$\mathbf{x}^2 < 0$, time-like vectors,
$\mathbf{x}^2 = 0$, null vectors or light-like vectors.

The set of time-like and light-like vectors can be divided into *future oriented* $x^4 > 0$, and *past oriented* $x^4 < 0$ [recall that $x^4 = ct = -x_4$].

Planes passing through the origin can be divided into time-like, light-like and space-like according as they intersect the light-cone along two, one or zero light-like vectors.

Space-like unit vectors $\mathbf{x}^2 = 1$ form a connected hyperboloid, and time-like unit vectors $\mathbf{x}^2 = -1$ form a two-sheeted hyperboloid. Future oriented time-like unit vectors correspond to *observers*; [1] an observer travelling at velocity $\vec{v} \in \mathbb{R}^3$ is associated to the *time-axis*

$$\mathbf{v} = \frac{\vec{v} + c\mathbf{e}_4}{\sqrt{c^2 - \vec{v}^2}} \quad \text{where} \quad \mathbf{v}^2 = -1.$$

9.3 Lorentz boost at velocity $\vec{v} \in \mathbb{R}^3$

Let us review how a space-time event $(\vec{r}, t)$ of an observer O is seen by another, O', moving at velocity $\vec{v}$ with respect to O.

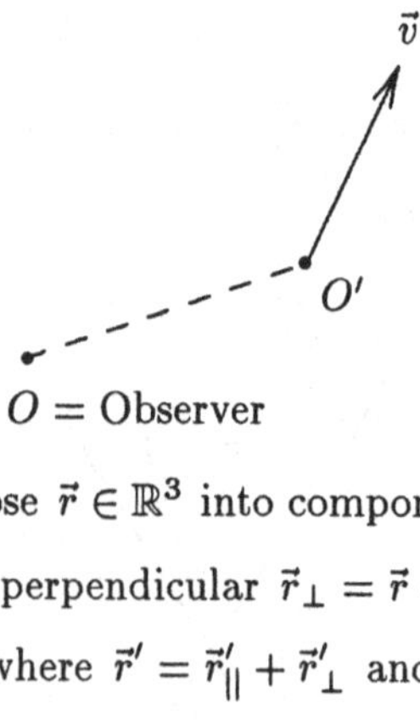

To do this we first decompose $\vec{r} \in \mathbb{R}^3$ into components $\vec{r} = \vec{r}_{||} + \vec{r}_{\perp}$ which are parallel $\vec{r}_{||} = (\vec{r} \cdot \vec{v})\dfrac{\vec{v}}{\vec{v}^2}$ and perpendicular $\vec{r}_{\perp} = \vec{r} - \vec{r}_{||}$ to $\vec{v}$. The transformed space-time event is $(\vec{r}\,', t')$ where $\vec{r}\,' = \vec{r}\,'_{||} + \vec{r}\,'_{\perp}$ and

$$\vec{r}\,'_{||} = \frac{1}{\sqrt{1 - \frac{\vec{v}^2}{c^2}}}(\vec{r}_{||} - \vec{v}t), \qquad \vec{r}\,'_{\perp} = \vec{r}_{\perp},$$

$$t' = \frac{1}{\sqrt{1 - \frac{\vec{v}^2}{c^2}}}\left(t - \frac{\vec{v} \cdot \vec{r}}{c^2}\right).$$

This transformation is called a *boost* at velocity $\vec{v}$.

The scalar $\vec{r}^2 - c^2t^2$, where $\vec{r}^2 = |\vec{r}|^2$, remains invariant under a boost. A boost leaves untouched the perpendicular component $\vec{r}_{\perp}$, but alters the parallel component $\vec{r}_{||}$.

1 We consider only inertial observers; inertial = free of forces (no acceleration).

A boost at velocity $\vec{v} = v_1\mathbf{e}_1 + v_2\mathbf{e}_2 + v_3\mathbf{e}_3$ can be represented by matrix multiplication $x'^\alpha = L^\alpha{}_\beta x^\beta$ or $x' = Lx$ where

$$L_{\vec{v}} = I - \frac{1}{\sqrt{1-\frac{\vec{v}^2}{c^2}}}\frac{V}{c} + \Big(\frac{1}{\sqrt{1-\frac{\vec{v}^2}{c^2}}} - 1\Big)\frac{V^2}{\vec{v}^2}$$

and

$$V = \begin{pmatrix} 0 & 0 & 0 & v_1 \\ 0 & 0 & 0 & v_2 \\ 0 & 0 & 0 & v_3 \\ v_1 & v_2 & v_3 & 0 \end{pmatrix}.$$

A boost is a special case of a Lorentz transformation, which can in general also rotate the space $\mathbb{R}^3$.

9.4 Lorentz transformations of the electromagnetic field

The electromagnetic field $\vec{E}, \vec{B}$ experiences a boost at velocity $\vec{v}$ as

$$\vec{E}'_\perp = \frac{1}{\sqrt{1-\frac{\vec{v}^2}{c^2}}}(\vec{E}_\perp + \vec{v}\times\vec{B}), \qquad \vec{E}'_\parallel = \vec{E}_\parallel,$$

$$\vec{B}'_\perp = \frac{1}{\sqrt{1-\frac{\vec{v}^2}{c^2}}}\Big(\vec{B}_\perp - \frac{\vec{v}\times\vec{E}}{c^2}\Big), \quad \vec{B}'_\parallel = \vec{B}_\parallel.$$

A boost of the electromagnetic field leaves invariant two scalars

$$\vec{E}^2 - c^2\vec{B}^2 \quad \text{and} \quad \vec{E}\cdot\vec{B},$$

called *Lorentz invariants.* [2]

A Lorentz transformation of the electromagnetic field can be written in coordinate form as $F'^\alpha{}_\beta = L^\alpha{}_\mu F^\mu{}_\nu (L^{-1})^\nu{}_\beta$ or concisely as matrix multiplication $F' = LFL^{-1}$. The matrix $F = (F^\alpha{}_\beta)$ satisfies $gF^\top g^{-1} = -F$, where

$$g = \begin{pmatrix} 1 & 0 & 0 & 0 \\ 0 & 1 & 0 & 0 \\ 0 & 0 & 1 & 0 \\ 0 & 0 & 0 & -1 \end{pmatrix}.$$

F is said to be Minkowski-antisymmetric.

2 The Lorentz invariants remain the same also under rotations of $\mathbb{R}^3$, and therefore under the special Lorentz group $SO(3,1)$. This can be seen by squaring $\mathbf{F} = \vec{E}\mathbf{e}_4 - \vec{B}\mathbf{e}_{123}$: $\mathbf{F}^2 = \mathbf{F}\,\lrcorner\,\mathbf{F} + \mathbf{F}\wedge\mathbf{F} = \vec{E}^2 - \vec{B}^2 - 2(\vec{E}\cdot\vec{B})\mathbf{e}_{1234}$. The scalar part remains invariant under $L \in O(3,1)$ and the 4-volume part remains invariant under $L \in SL(4,\mathbb{R})$. Note that $O(3,1) \cap SL(4,\mathbb{R}) = SO(3,1)$, $SL(4,\mathbb{R})/\mathbb{Z}_2 \simeq SO_+(3,3)$ and $\bigwedge^2 L \in SO_+(3,3)$ acts on $\bigwedge^2 \mathbb{R}^{3,1} \simeq \mathbb{R}^{3,3}$.

9.5 The Lorentz group $O(3,1)$

A matrix L satisfying $LgL^\top = g$ [3] is said to be Minkowski-orthogonal or a *Lorentz transformation*. The Lorentz transformations form the *Lorentz group*

$$O(3,1) = \{L \in \mathrm{Mat}(4,\mathbb{R}) \mid LgL^\top = g\}.$$

A Lorentz transformation has a unit determinant: $\det L = \pm 1$. The subgroup with positive determinant,

$$SO(3,1) = O(3,1) \cap SL(4,\mathbb{R}),$$

is called the *special Lorentz group*. The special Lorentz group $SO(3,1)$ has two components. The component connected to the identity I is denoted by $SO_+(3,1)$; it preserves orientations of both space and time. The other component $SO(3,1) \setminus SO_+(3,1)$ reverses orientations of both space and time.

The Lorentz group $O(3,1)$ has four components; these form three two-component subgroups preserving space orientation, time orientation or space-time orientation. Time-orientation–preserving Lorentz transformations form the *orthochronous Lorentz group* $O_\uparrow(3,1)$. A *restricted* or special orthochronous Lorentz transformation $L \in SO_+(3,1)$ preserves space-time orientation (orientation of both space and time); its opposite $-L \in SO(3,1) \setminus SO_+(3,1)$ reverses space-time orientation, while gL reverses time orientation, and $-gL \in O_\uparrow(3,1)$ reverses space orientation, $-gL \in O_\uparrow(3,1) \setminus SO_+(3,1)$.

The Lorentz transformations, which stabilize a time-like vector, form a subgroup $O(3)$, the orthogonal group of $\mathbb{R}^3 = \mathbb{R}^{3,0}$. The Lorentz transformations, which stabilize a space-like vector, form a subgroup $O(2,1)$, the small Lorentz group of $\mathbb{R}^{2,1}$. The Lorentz transformations, which stabilize a light-like vector, form a subgroup isomorphic to the group of rigid movements of the Euclidean plane $\mathbb{R}^2$.

Any special orthochronous Lorentz transformation $L \in SO_+(3,1)$ can be written as an exponential $L = e^A$ of a Minkowski-antisymmetric matrix

$$A = \begin{pmatrix} 0 & b_3 & -b_2 & a_1 \\ -b_3 & 0 & b_1 & a_2 \\ b_2 & -b_1 & 0 & a_3 \\ a_1 & a_2 & a_3 & 0 \end{pmatrix},$$

which satisfies $gA^\top g^{-1} = -A$. The matrix A can be characterized by two vectors $\vec{a} = a_1\mathbf{e}_1 + a_2\mathbf{e}_2 + a_3\mathbf{e}_3$ and $\vec{b} = b_1\mathbf{e}_1 + b_2\mathbf{e}_2 + b_3\mathbf{e}_3$ in $\mathbb{R}^3$. If $\vec{b} = 0$, then L is a boost at velocity

$$|\vec{v}| = c\tanh|\vec{a}|.$$

3 Note the resemblance between $LgL^\top g^{-1} = I$ and the condition of orthogonality $RR^\top = I$ of a matrix R, $R \in O(n)$.

If $\vec{a} = 0$, then $L \in SO(3)$ is a rotation of the Euclidean space $\mathbb{R}^3$ around the axis $\vec{b}$ by the angle $|\vec{b}|$. Boosts and rotations are special cases of simple Lorentz transformations.

9.6 Simple Lorentz transformations

A special orthochronous Lorentz transformation $L \in SO_+(3,1)$, $L \neq I$, has one or two light-like vectors as eigenvectors. If there are two light-like eigenvectors, then they span a time-like eigenplane, which is preserved by the Lorentz transformation; there is also a space-like eigenplane, which is completely orthogonal to the time-like eigenplane. [4] A special orthochronous Lorentz transformation is called *simple*, if it turns vectors only in one eigenplane, leaving the other eigenplane point-wise invariant. Disregarding the case $L = I$, a special orthochronous Lorentz transformation $L = e^A$, where A is characterized as before by $\vec{a}, \vec{b} \in \mathbb{R}^3$, is simple if and only if $\vec{a} \cdot \vec{b} = 0$. A simple Lorentz transformation is called

$$\begin{array}{ll} \textit{hyperbolic,} & |\vec{a}| > |\vec{b}|, \\ \textit{elliptic,} & |\vec{a}| < |\vec{b}|, \\ \textit{parabolic,} & |\vec{a}| = |\vec{b}|. \end{array}$$

A hyperbolic Lorentz transformation is a boost for an observer, whose time-axis is in the time-like eigenplane of the Lorentz transformation. An elliptic Lorentz transformation is a rotation of the Euclidean space $\mathbb{R}^3$, which is orthogonal to an observer, whose time-axis is orthogonal to the space-like eigenplane of the Lorentz transformation. A parabolic Lorentz transformation has only one light-like eigenvector; it is of the form

$$L = I + A + \frac{1}{2}A^2, \quad \text{since} \quad A^3 = 0,$$

and has only one eigenplane, which is light-like and tangent to the light-cone along the light-like eigenvector. A non-parabolic Lorentz transformation can be written as a product of two commuting simple transformations, one hyperbolic and the other elliptic.

Lorentz Transformations in Clifford Algebras

Lorentz transformations can be described within the Clifford algebras $\mathcal{C}\ell_3 \simeq \mathrm{Mat}(2,\mathbb{C})$, $\mathcal{C}\ell_{3,1} \simeq \mathrm{Mat}(4,\mathbb{R})$ and $\mathcal{C}\ell_{1,3} \simeq \mathrm{Mat}(2,\mathbb{H})$.

4 Completely orthogonal planes have only one point in common, the origin O. For two vectors $\mathbf{x}, \mathbf{y}$ in completely orthogonal planes, the scalar product $\mathbf{x} \cdot \mathbf{y} = x_1y_1 + x_2y_2 + x_3y_3 - x_4y_4$ vanishes: $\mathbf{x} \cdot \mathbf{y} = 0$.

9.7 In the Clifford algebra $\mathcal{C}\ell_3 \simeq \mathrm{Mat}(2,\mathbb{C})$

Events in time and space can be labelled by sums of scalars and vectors,

$$x = ct + \vec{x},$$

in $\mathbb{R} \oplus \mathbb{R}^3 \subset \mathcal{C}\ell_3$. A paravector $x = x^0 + x^1\mathbf{e}_1 + x^2\mathbf{e}_2 + x^3\mathbf{e}_3$ can be provided with a quadratic form

$$x\bar{x} = c^2t^2 - \vec{x}^2 = x_0^2 - x_1^2 - x_2^2 - x_3^2$$

making $\mathbb{R} \oplus \mathbb{R}^3$ isometric to the Minkowski time-space $\mathbb{R}^{1,3}$. [5] This quadratic form is preserved in a special orthochronous Lorentz transformation $L \in SO_+(1,3)$,

$$L : \mathbb{R} \oplus \mathbb{R}^3 \to \mathbb{R} \oplus \mathbb{R}^3, \quad x \to L(x) = sx\hat{s}^{-1},$$

where s is in the spin group [6]

$$\$pin_+(1,3) = \{s \in \mathcal{C}\ell_3 \mid s\bar{s} = 1\} \simeq SL(2,\mathbb{C}).$$

The time-space event $x = t + \vec{x} \in \mathbb{R} \oplus \mathbb{R}^3$ and the electromagnetic field $F = \vec{E} - \vec{B}\mathbf{e}_{123} \in \mathbb{R}^3 \oplus \bigwedge^2 \mathbb{R}^3$ behave slightly differently under restricted Lorentz transformations:

$$\begin{aligned} x' &= L(x) = sx\hat{s}^{-1}, \\ F' &= sFs^{-1}. \end{aligned}$$

The spin group $\$pin_+(1,3)$ is a two-fold covering of the special orthochronous Lorentz group $SO_+(1,3)$. In other words, there are two elements $\pm s$ in the group $\$pin_+(1,3)$ inducing the same Lorentz transformation L in $SO_+(1,3)$. This can be expressed by saying that the kernel of the group homomorphism

$$\rho : \$pin_+(1,3) \to SO_+(1,3), \quad s \to L = \rho(s)$$

consists of two elements $\{\pm 1\} \in \$pin_+(1,3)$ [the kernel is the pre-image of the identity element $I \in SO_+(1,3)$].

Every element s in the spin group $\$pin_+(1,3)$ is of the form

$$s = \pm \exp \frac{1}{2}(\vec{a} + \vec{b}\mathbf{e}_{123})$$

where $\vec{a}$ and $\vec{b}$ are vectors in $\mathbb{R}^3$. The minus sign in front of the exponential

5 The raising and lowering conventions are different in $\mathbb{R} \oplus \mathbb{R}^3$ and $\mathbb{R}^{1,3}$. In $\mathbb{R}^{1,3}$ $(x^0, x^1, x^2, x^3) = (x_0, -x_1, -x_2, -x_3)$ whereas in $\mathbb{R}^3 \subset \mathbb{R} \oplus \mathbb{R}^3$ there is a prescribed metric such that $(x^1, x^2, x^3) = (x_1, x_2, x_3)$.

6 The groups $\$pin_+(1,3) \subset \mathcal{C}\ell_3$ and $\mathbf{Spin}_+(1,3) \subset \mathcal{C}\ell_{1,3}$ are isomorphic, and so are their Lie algebras $\mathbb{R}^3 \oplus \bigwedge^2 \mathbb{R}^3$ and $\bigwedge^2 \mathbb{R}^{1,3}$.

is needed, [7] because not all the elements in the two-fold cover $\$pin_+(1,3)$ of $SO_+(1,3)$ can be written as exponentials of para-bivectors in $\mathbb{R}^3 \oplus \bigwedge^2 \mathbb{R}^3$. In the case $\vec{a} = 0$ the Lorentz transformation is a rotation of $\mathbb{R}^3$, and in the case $\vec{b} = 0$ we have a boost at velocity [8]

$$\vec{v} = \tanh \vec{a} = \frac{\vec{a}}{|\vec{a}|} \tanh|\vec{a}|.$$

For any $s \in \$pin_+(1,3)$ the product $s\tilde{s}$ is a boost, that is, $s\tilde{s} \in \mathbb{R} \oplus \mathbb{R}^3$. Since $\langle s\tilde{s}\rangle_0 > 0$, there is a unique square root of $u = s\tilde{s}$, a boost such that $\sqrt{u} = \alpha(1+u)$, $\alpha > 0$. Squaring both sides and using $u\bar{u} = 1$ results in

$$\sqrt{u} = \frac{1+u}{\sqrt{2(1+\langle u\rangle_0)}}.$$

Write $b_1 = \sqrt{s\tilde{s}}$. The product $r = b_1^{-1}s$ satisfies $\tilde{r}r = 1$ and $\bar{r}r = 1$, and so it is a rotation, $r \in \mathbf{Spin}(3)$. A special orthochronous Lorentz transformation can be uniquely decomposed into a product of a boost and a rotation,

$$s = b_1 r,$$

called the *polar decomposition*. Similarly computing $b_2 = \sqrt{\tilde{s}s}$ and $r = sb_2^{-1}$, we find that $s = rb_2$ with the same rotation r, that is,

$$s = b_1 r = r b_2.$$

Both the decompositions have as a factor the same rotation $r \in \mathbf{Spin}(3) = \{s \in C\ell_3^+ \mid s\tilde{s} = 1\}$, but the boosts are different: $b_1 \neq b_2$.

9.8 In the Clifford algebra $C\ell_{3,1} \simeq \mathrm{Mat}(4,\mathbb{R})$

A boost $b \in \mathbb{R} \oplus \mathbb{R}^3\mathbf{e}_4$, at velocity $\vec{v} \in \mathbb{R}^3$, can be computed by

$$b = \exp(\vec{a}\mathbf{e}_4/2), \quad \vec{a} = \operatorname{artanh}(\vec{v}/c),$$

and results in

$$b = \frac{1+\gamma(1+\vec{v}\mathbf{e}_4)}{\sqrt{2(1+\gamma)}}, \quad \gamma = \frac{1}{\sqrt{1-\frac{\vec{v}^2}{c^2}}},$$

obtained also by taking a square root of $b^2 = \gamma(1+\vec{v}\mathbf{e}_4)$.

The restricted Lorentz group $SO_+(3,1)$ has a double cover

$$\mathbf{Spin}_+(3,1) = \{s \in C\ell_{3,1}^+ \mid s\tilde{s} = 1\}.$$

7 If $|\vec{a}| = |\vec{b}|$ and $\vec{a} \cdot \vec{b} = 0$ then $(\vec{a} + \vec{b}\mathbf{e}_{123})^2 = 0$, and for a non-zero $F = \vec{a} + \vec{b}\mathbf{e}_{123}$ there is no para-bivector $B \in \mathbb{R}^3 \oplus \bigwedge^2 \mathbb{R}^3$ such that $e^B = -e^F$.

8 The first tanh-function is evaluated in the Clifford algebra $C\ell_3$.

Under a Lorentz transformation induced by $s \in \mathbf{Spin}_+(3,1)$ the space-time vector $\mathbf{x}$ transforms according to $\mathbf{x}' = s\mathbf{x}s^{-1}$ and the electromagnetic bivector $\mathbf{F} = \frac{1}{c}\vec{E}\mathbf{e}_4 - \vec{B}\mathbf{e}_{123}$ transforms according to $\mathbf{F}' = s\mathbf{F}s^{-1}$.

9.9 In the Clifford algebra $C\ell_{1,3} \simeq \mathrm{Mat}(2,\mathbb{H})$

Consider the Lorentz group of the Minkowski time-space $\mathbb{R}^{1,3}$ in the Clifford algebra $C\ell_{1,3}$ which is isomorphic, as an associative algebra, to the real algebra of 2×2-matrices $\mathrm{Mat}(2,\mathbb{H})$ with quaternions as entries. The Clifford algebra $C\ell_{1,3}$ is generated as a real algebra by the Dirac gamma-matrices $\gamma_0, \gamma_1, \gamma_2, \gamma_3$ satisfying

$$\gamma_0^2 = I,\ \gamma_1^2 = \gamma_2^2 = \gamma_3^2 = -I, \quad \text{and} \quad \gamma_\mu\gamma_\nu = -\gamma_\nu\gamma_\mu \quad \text{for} \quad \mu \neq \nu.$$

In this case the Lorentz groups $O(1,3)$, $SO(1,3)$, $SO_+(1,3)$ are doubly covered by

$$\begin{aligned}
&\mathbf{Pin}(1,3) = \{s \in C\ell_{1,3}^+ \cup C\ell_{1,3}^- \mid s\tilde{s} = \pm 1\}, \\
&\mathbf{Spin}(1,3) = \{s \in C\ell_{1,3}^+ \mid s\tilde{s} = \pm 1\}, \\
&\mathbf{Spin}_+(1,3) = \{s \in C\ell_{1,3}^+ \mid s\tilde{s} = 1\} \simeq SL(2,\mathbb{C}).
\end{aligned}$$

A Lorentz transformation $L \in O(1,3)$ is given by $L(\mathbf{x}) = s\mathbf{x}\hat{s}^{-1}$ in general, but a special Lorentz transformation $L \in SO(1,3)$ corresponds to an even s and can also be written as $L(\mathbf{x}) = s\mathbf{x}s^{-1}$. The group homomorphism $\rho : \mathbf{Pin}(1,3) \to O(1,3)$ is fixed by $L = \rho(s)$, $L(\mathbf{x}) = s\mathbf{x}\hat{s}^{-1}$, and its kernel is $\{\pm 1\}$, that is, each $L \in O(1,3)$ has two pre-images $\pm s$ in $\mathbf{Pin}(1,3)$.

An element $s \in \mathbf{Spin}_+(1,3)$ has a unique polar decomposition

$$s = b_1 r = r b_2,$$

where the boosts are different,

$$b_1 = \sqrt{s\gamma_0\tilde{s}\gamma_0^{-1}} \quad \text{and} \quad b_2 = \sqrt{\gamma_0\tilde{s}\gamma_0^{-1}s},$$

but the rotation is the same, $r = b_1^{-1}s = sb_2^{-1}$.

Penrose & Rindler 1984. On pp. 31-32 the authors give a geometric interpretation for Lorentz transformations, reviewed here in terms of the Clifford algebra $C\ell_{1,3}$. Take four distinct light-like vectors $\mathbf{a}$, $\mathbf{b}$, $\mathbf{c}$, $\mathbf{d}$ such that $\mathbf{a}\cdot\mathbf{b} = 1$ and $\mathbf{c}\cdot\mathbf{d} = 1$. The bivector $\mathbf{a}\wedge\mathbf{b}$ represents a time-like plane, since $(\mathbf{a}\wedge\mathbf{b})^2 = (\mathbf{a}\cdot\mathbf{b})^2 - \mathbf{a}^2\mathbf{b}^2 = 1$; the bivector $\mathbf{a}\wedge\mathbf{b}$ belongs to $\mathbf{Spin}(1,3)\setminus\mathbf{Spin}_+(1,3)$ and represents a Lorentz transformation, which reverses the space-time orientation. Therefore, the product

$$s = (\mathbf{a}\wedge\mathbf{b})(\mathbf{c}\wedge\mathbf{d})$$

is in $\mathbf{Spin}_+(1,3)$. Let the light-like eigenvectors of the corresponding Lorentz transformation be $\mathbf{l}_1$ and $\mathbf{l}_2$, and choose $\mathbf{l}_1 \cdot \mathbf{l}_2 = 1$ so that $(\mathbf{l}_1 \wedge \mathbf{l}_2)^2 = 1$. The bivector $\mathbf{l}_1 \wedge \mathbf{l}_2$ anticommutes with $\mathbf{a} \wedge \mathbf{b}$ and $\mathbf{c} \wedge \mathbf{d}$, that is, it is the unique 'normal' to $\mathbf{a}\wedge\mathbf{b}$ and $\mathbf{c}\wedge\mathbf{d}$. The bivector $\mathbf{F} = \log(s)$ in $\bigwedge^2 \mathbb{R}^{1,3}$ is determined up to a multiple of $2\pi\gamma_{0123}(\mathbf{l}_1 \wedge \mathbf{l}_2)$. The square root $\phi + \psi\gamma_{0123} = \sqrt{\mathbf{F}^2}$ is such that $\mathbf{F} = \pm(\phi + \psi\gamma_{0123})\mathbf{l}_1 \wedge \mathbf{l}_2$; it is determined up to a sign; choosing $\phi \geq 0$, the Lorentz transformation $L = \rho(s)$ has velocity $v = \tanh(2\phi)$ and eigenvalues

$$e^{\pm 2\phi} = \sqrt{\frac{1 \pm v}{1 \mp v}}.$$

The planes $\mathbf{a} \wedge \mathbf{b}$ and $\mathbf{c} \wedge \mathbf{d}$ 'differ' in the sense that $(\mathbf{a} \wedge \mathbf{b})s = \mathbf{c} \wedge \mathbf{d}$ and $s(\mathbf{c} \wedge \mathbf{d}) = \mathbf{a} \wedge \mathbf{b}$ by a sum of an elliptic angle ψ about the plane $\mathbf{l}_1 \wedge \mathbf{l}_2$ and a hyperbolic angle ϕ in the plane $\mathbf{l}_1 \wedge \mathbf{l}_2$. Indicating the transformed light-like vectors by primes,

$$\mathbf{a}' = s\mathbf{a}s^{-1}, \quad \mathbf{b}' = s\mathbf{b}s^{-1}, \quad \mathbf{c}' = s\mathbf{c}s^{-1}, \quad \mathbf{d}' = s\mathbf{d}s^{-1},$$

we find that $\mathbf{c}' \wedge \mathbf{d}' = s(\mathbf{c} \wedge \mathbf{d})s^{-1} = (\mathbf{a} \wedge \mathbf{b})(\mathbf{c} \wedge \mathbf{d})(\mathbf{a} \wedge \mathbf{b})^{-1}$, that is, the Lorentz transformation reflects the plane $\mathbf{c} \wedge \mathbf{d}$ across the plane $\mathbf{a} \wedge \mathbf{b}$. But, $s(\mathbf{a}\wedge\mathbf{b})s^{-1} = (\mathbf{c}'\wedge\mathbf{d}')(\mathbf{a}\wedge\mathbf{b})(\mathbf{c}'\wedge\mathbf{d}')^{-1}$ and $(\mathbf{c}\wedge\mathbf{d})(\mathbf{a}\wedge\mathbf{b})(\mathbf{c}\wedge\mathbf{d})^{-1} = s^{-1}(\mathbf{a}\wedge\mathbf{b})s$, that is, the inverse of $s(\mathbf{a} \wedge \mathbf{b})s^{-1}$. Take a square root of the inverse of s,

$$u = \pm\frac{1}{\sqrt{s}},$$

within $\mathbf{Spin}_+(1,3)$, and find that

$$u\mathbf{a}u^{-1} = \mathbf{c} \quad \text{and} \quad u\mathbf{b}u^{-1} = \mathbf{d},$$

a kind of 'half' of the reflection above. ∎

Jancewicz 1988. On pp. 252-256 the author shows how to decompose a non-simple bivector $\mathbf{F}$ into simple components. He defines

$$\alpha + \beta\gamma_{0123} = (\phi + \psi\gamma_{0123})^2 = \mathbf{F}^2$$

and sets $\delta^2 = \alpha^2 + \beta^2$. Then he gives the simple components

$$\mathbf{F}_{1,2} = \frac{\mathbf{F}}{2\delta}(\delta \pm \alpha \mp \beta\gamma_{0123})$$

for $\mathbf{F} = \mathbf{F}_1 + \mathbf{F}_2$ so that $\mathbf{F}_1^2 > 0$ and $\mathbf{F}_2^2 < 0$, that is, $\mathbf{F}_1$ is hyperbolic or time-like and $\mathbf{F}_2$ is elliptic or space-like. Observing that

$$\left(\frac{\mathbf{F}}{\phi + \psi\gamma_{0123}}\right)^2 = 1$$

enables us to work out the decomposition in another way:

$$\mathbf{F}_1 = \frac{\mathbf{F}\phi}{\phi + \psi\gamma_{0123}} \quad \text{and} \quad \mathbf{F}_2 = \frac{\mathbf{F}\psi\gamma_{0123}}{\phi + \psi\gamma_{0123}}.$$

Hestenes & Sobczyk 1984 p. 81 note that $\phi^2 = \mathbf{F}_1^2$, $\psi^2 = -\mathbf{F}_2^2$ and

$$\mathbf{F}_{1,2}^2 = \frac{1}{2}[\mathbf{F} \rfloor \mathbf{F} \pm \sqrt{(\mathbf{F} \rfloor \mathbf{F})^2 - (\mathbf{F} \wedge \mathbf{F})^2}]$$

(their formula 4.16 concerns only the positive definite case). ∎

Hestenes 1966. The author gives on pp. 52-53 a method to find out $s \in$ $\mathbf{Spin}_+(1,3)$ from the coordinates $L^\mu{}_\nu$ of a special orthochronous Lorentz transformation $L(\mathbf{x}) = s\mathbf{x}s^{-1}$, $L \in SO_+(1,3)$. Recall that $L(\gamma_\nu) = \gamma_\mu L^\mu{}_\nu$, and deduce

$$L^\mu{}_\nu = \gamma^\mu \cdot L(\gamma_\nu) = \langle \gamma^\mu s \gamma_\nu \tilde{s} \rangle_0.$$

To compute s in terms of $L^\mu{}_\nu$ define first

$$\mathcal{L} \equiv L^\mu{}_\nu \gamma_\mu \gamma^\nu = L^\mu{}_\mu + L^\mu{}_\nu \gamma_\mu \wedge \gamma^\nu \in \mathbb{R} \oplus \bigwedge^2 \mathbb{R}^{1,3}. \tag{1}$$

It follows that

$$\mathcal{L} = L(\gamma_\nu)\gamma^\nu = s\gamma_\nu \tilde{s} \gamma^\nu.$$

In computing $\gamma_\nu \tilde{s} \gamma^\nu$, note that in general $\mathbf{e}_\nu u \mathbf{e}^\nu = (n-2k)\hat{u}$ for $u \in \bigwedge^k \mathbb{R}^n$, and deduce that for $s = \langle s \rangle_0 + \langle s \rangle_2 + \langle s \rangle_4$

$$\gamma_\nu \tilde{s} \gamma^\nu = 4[\langle s \rangle_0 - \langle s \rangle_4].$$

Therefore,

$$\mathcal{L} = 4s[\langle s \rangle_0 - \langle s \rangle_4].$$

Since $\tilde{s}s = 1$, $\tilde{\mathcal{L}}\mathcal{L} = 16[\langle s \rangle_0 - \langle s \rangle_4]^2$, and

$$s = \pm \frac{\mathcal{L}}{\sqrt{\tilde{\mathcal{L}}\mathcal{L}}}.$$

Substituting (1) this gives s explicitly as a function of $L^\mu{}_\nu$. This construction is an accident in dimension $n = 4$, because only then does the sum $\mathbf{e}_\nu u \mathbf{e}^\nu$ vanish for a bivector $u \in \bigwedge^2 \mathbb{R}^n$. ∎

Historical survey

In 1881, A.A. Michelson carried out, for the first time, measurements intended to determine the motion of the Earth relative to an absolute, imaginary 'light medium'. For this purpose he measured the velocity of light in different directions. Michelson & Morley repeated the experiment in 1887 and came to the conclusion that light travels at the same velocity independent of the motion of the source with respect to the 'light medium'.

Voigt in 1887 was the first to recognize that the wave equation

$$\frac{\partial^2 f}{\partial x^2} - \frac{1}{c^2}\frac{\partial^2 f}{\partial t^2} = 0$$

is invariant with respect to the change of variables

$$x' = x - vt,$$
$$t' = t - \tfrac{vx}{c^2},$$

where also time is transformed. Voigt's formulas are not identical for direct and inverse transformations; symmetry was restored later by introducing the factor $\sqrt{1-v^2/c^2}$. This factor was first encountered in another connection: FitzGerald and Lorentz 1892 gave independently an explanation of the Michelson & Morley experiment by suggesting that moving bodies are contracted in the direction of motion by the ratio $\sqrt{1-v^2/c^2}$.

The Lorentz transformations of space-time events were introduced by Larmor in 1900, while the relativistic covariance of the Maxwell equations was demonstrated by H.A. Lorentz 1903 [9] (and conformal covariance by Cunningham 1909/1910 and Bateman 1910).

In 1905 [10] Einstein supplemented the principle of relativity by postulating the principle of independence of the velocity of light (of the motion of the source). These two principles led Einstein to a revision of the notion of time and enabled him to deduce the kinematical transformation laws of Lorentz; his predecessors had obtained the transformation laws by considering transformations which do not change the form of the Maxwell equations.

---

9 Poincaré noticed that restricted Lorentz transformations of space-dimension 1 form a group $SO_+(1,1)$ consisting of the elements

$$\begin{pmatrix} \cosh\chi & \sinh\chi \\ \sinh\chi & \cosh\chi \end{pmatrix}$$

where $\chi \in \mathbb{R}$.

10 A. Einstein: Zur Elektrodynamik bewegter Körper. *Ann. Physik* **17** (1905), 891-921. In this paper Einstein compared the same phenomenon when observed in two different frames: a magnet moving near a closed conductor and a closed conductor moving near a magnet. In another paper of 1905 Einstein gave a relation between mass and energy, which was later popularized as the formula $E = mc^2$, written today as $E = mc^2/\sqrt{1-v^2/c^2}$ or $E^2 = m^2c^4 + p^2c^2$.

Later Einstein reformulated the principle of relativity so that it embraces not only mechanical but also electromagnetic phenomena:

the laws of physics have the same form in all reference frames

When this *Einsteinian principle of relativity* is applied to the Maxwell equations, one is compelled to conclude that the velocity of light is the same in all reference frames. In other words, the principle of constancy of the velocity of light becomes superfluous as an amendment to the principle of relativity. The principle of relativity and knowledge of the Maxwell equations are enough to deduce the transformation laws of Lorentz.

Nowadays the terms 'relativistic' and 'relativity' almost invariably refer to the Einsteinian principle.

Questions

1. How many light-like eigenvectors does a Lorentz transformation have?
2. Are all $L \in SO_+(3,1)$ of the form $L = \exp(A)$, $gA^\top g^{-1} = -A$?
3. Are all $s \in \mathbf{Spin}_+(3,1)$ of the form $s = \exp(\mathbf{B}/2)$, $\mathbf{B} \in \bigwedge^2 \mathbb{R}^{3,1}$?
4. A special orthochronous Lorentz transformation can be written as a product of a boost and a rotation, in two different orders. In the two expressions, which factor is the same: the boost or the rotation?
5. Are all the special orthochronous Lorentz transformations products of two commuting transformations, one hyperbolic and one elliptic?

Let $\mathbf{B} \in \bigwedge^2 \mathbb{R}^{3,1}$.

6. Is $\mathbb{R}^{3,1} \ni \mathbf{x} \to u\mathbf{x}u^{-1}$, $u = 1 + \mathbf{B} + \frac{1}{2}\mathbf{B} \wedge \mathbf{B}$, a Lorentz transformation?
7. Do the Lorentz transformations induced by $\exp(\mathbf{B}/2)$ and $1 + \mathbf{B} + \frac{1}{2}\mathbf{B} \wedge \mathbf{B}$, $\mathbf{B}^2 \neq 1$, have the same eigenvectors?
8. Does $(1 + \mathbf{B})(1 - \mathbf{B})^{-1}$ represent a Lorentz transformation?
9. Do the Lorentz transformations induced by $\exp(\mathbf{B}/2)$ and $(1 + \mathbf{B})(1 - \mathbf{B})^{-1}$, $\mathbf{B}^2 \neq 1$, have the same eigenvectors?

Answers

1. In general two, parabolic has one, $\pm I$ have all of them.
2. Yes. 3. No. 4. Rotation. 5. No (parabolic are not).
6. Yes, if $\mathbf{B}^2 \neq 1$.
7. Yes, because both the Lorentz transformations are functions (= power series with real coefficients) of A, $A(\mathbf{x}) = \mathbf{B} \lrcorner \mathbf{x}$; namely e^A and $(I + A)(I - A)^{-1}$, respectively.

8. Yes, if $\mathbf{B}^2 \neq 1$ (but this is no longer true in dimension 6).
9. No, because the latter Lorentz transformation is not a rational function of A alone (but also of $A^\top$).

Exercises

1. Derive the composition rule for non-parallel velocities,
$$\vec{v}_2' = \frac{1}{1+\frac{\vec{v}_1\cdot\vec{v}_2}{c^2}}[\vec{v}_1 + \vec{v}_{2\|} + \sqrt{1-\tfrac{\vec{v}_1^2}{c^2}}\,\vec{v}_{2\perp}].$$
Hint: use the inverse Lorentz transformation
$$\vec{r}_2' = \frac{1}{\sqrt{1-\frac{\vec{v}_1^2}{c^2}}}(\vec{r}_{2\|} + \vec{v}_1 t_2) + \vec{r}_{2\perp},$$
$$t_2' = \frac{1}{\sqrt{1-\frac{\vec{v}_1^2}{c^2}}}(t_2 + \tfrac{\vec{v}_1\cdot\vec{r}_2}{c^2}).$$
2. Show that $\vec{v}_2' = \tanh(2\log(b_2'))$ where $b_2' = \sqrt{s_2'\bar{s}_2'}$, $s_2' = s_1 s_2 \in \mathbb{R}\oplus\mathbb{R}^3$ and $s_1 = \exp(\frac{1}{2}\vec{a}_1)$, $\vec{v}_1 = \tanh(\vec{a}_1)$.
3. Show that the composite of two boosts is a hyperbolic transformation.
4. Consider a time-space event $x = ct + \vec{x}$ in $\mathbb{R}\oplus\mathbb{R}^3$ corresponding to $y = \vec{x} + ct\mathbf{e}_4$ in $\mathbb{R}^{3,1}$. Define $s = \exp(\vec{a}/2)$ and $u = \exp(\vec{a}\mathbf{e}_4/2)$ for $\vec{a}\in\mathbb{R}^3$. Show that the boost $\hat{s}xs^{-1} = s^{-1}x\hat{s}$ corresponds to the boost uyu^{-1}.
5. Show that for $u \in \mathbf{Spin}_+(3,1)$, when decomposed into a product of a boost and a rotation, $u = b_1 r = r b_2$, the rotation-factor $r \in \mathbf{Spin}(3)$ can be obtained by normalizing $(u\wedge\mathbf{e}_4)\mathbf{e}_4^{-1}$.
6. Take a bivector $\mathbf{F} = \vec{a}\mathbf{e}_4 + \vec{b}\mathbf{e}_{123} \in \bigwedge^2\mathbb{R}^{3,1}$ such that $|\vec{a}| = |\vec{b}|$. Consider the antisymmetric linear transformation $\mathbb{R}^{3,1}\to\mathbb{R}^{3,1}$, $\mathbf{x}\to A\mathbf{x} = \langle\mathbf{F}\mathbf{x}\rangle_1$. Show that $(A^3\mathbf{x}) \,\|\, \mathbf{x}$.
7. Take a non-simple bivector $\mathbf{F}\in\bigwedge^2\mathbb{R}^{3,1}$ with simple components $\mathbf{F} = \mathbf{F}_1 + \mathbf{F}_2$, $\mathbf{F}_1^2 > 0$, $\mathbf{F}_2^2 < 0$. Show that
$$\mathbb{R}^{3,1}\to\mathbb{R}^{3,1}, \quad \mathbf{x}\to\frac{\mathbf{F}\mathbf{x}\mathbf{F}}{\mathbf{F}_1^2 - \mathbf{F}_2^2}$$
is a Lorentz transformation, a reflection across the plane of $\mathbf{F}_1$.
8. Show that $(\phi + \psi\gamma_{0123})(\delta + \alpha - \beta\gamma_{0123}) = 2\delta\phi$.
9. Show that as topological spaces $\mathbf{Spin}_+(1,3) \simeq \mathbb{R}^3\times S^3$.
10. Show that as groups $\mathcal{S}pin_+(3,1) \simeq \mathbf{Spin}_+(3,1) \simeq SL(2,\mathbb{C})$ and $SO_+(3,1) \simeq SO(3,\mathbb{C}) = \{R\in\mathrm{Mat}(3,\mathbb{C}) \mid RR^\top = I,\ \det R = 1\}$.
11. Show that for $u\in\mathbf{Spin}_+(3,1)$ there is a square root in $\mathbf{Spin}_+(3,1)$ given

by

$$\sqrt{u} = \frac{u+1}{\sqrt{2(1+\langle u\rangle_0+\langle u\rangle_4)}}.$$

Hint: for $s \in \mathbf{Spin}_+(3,1)$, $s^2+1 = s^2+\tilde{s}s = (s+\tilde{s})s = 2(\langle s\rangle_0+\langle s\rangle_4)s$. Therefore, $(\alpha+\beta\mathbf{e}_{1234})\sqrt{u} = u+1$ with $\alpha,\beta \in \mathbb{R}$.

Bibliography

V. Bargmann, E.P. Wigner: Group theoretical discussion of relativistic wave equations. *Proc. Nat. Acad. Sci. U.S.A.* **34** (1948), 211-223. Discussion on unitary representations of the Poincaré group.

W. Baylis: *Theoretical Methods in the Physical Sciences: an Introduction to Problem Solving with MAPLE V.* Birkhäuser, Boston, MA, 1994.

A. Einstein: Zur Elektrodynamik bewegter Körper. *Ann. Physik.* (4) **17** (1905), 891-921.

I.M. Gelfand, R.A. Minlos, Z.Ya. Shapiro: *Representations of the Rotation and Lorentz Groups and their Applications.* Pergamon, New York, 1963.

D. Hestenes: *Space-Time Algebra.* Gordon and Breach, Philadelphia, PA, 1966, 1987, 1992.

B. Jancewicz: *Multivectors and Clifford Algebra in Electrodynamics.* World Scientific, Singapore, 1988.

W. Kopczyński, A. Trautman: *Spacetime and Gravitation.* PWN, Warszawa; John Wiley, Chichester; 1992. Includes a historical survey on relativity.

M.A. Naimark: *Linear Representations of the Lorentz Group.* Macmillan, New York, 1964.

R. Penrose, W. Rindler: *Spinors and Space-Time.* Vol. 1. Cambridge University Press, Cambridge, 1984.

M. Riesz: *Clifford Numbers and Spinors.* The Institute for Fluid Dynamics and Applied Mathematics, Lecture Series No. **38**, University of Maryland, 1958. Reprinted with comments as facsimile by E.F. Bolinder, P. Lounesto (eds.), Kluwer, Dordrecht, The Netherlands, 1993.

L. Silberstein: *The Theory of Relativity.* Macmillan, London, 1914.

R.F. Streater, A.S. Wightman: *PCT, Spin and Statistics, and all that.* Benjamin, New York, 1964. Contains a discussion on the two-fold cover of the Lorentz group.

E.P. Wigner: On unitary representations of the inhomogeneous Lorentz group. *Ann. of Math.* **40** (1939), 149-204. A historical article in which is discussed, for the first time, the physical role of the *Poincaré group*.

10
The Dirac Equation

The Schrödinger equation describes all atomic phenomena except those involving magnetism and relativity. The Schrödinger-Pauli equation takes care of magnetism by including the spin of the electron.

The relativistic phenomena can be taken into consideration by starting from the equation $E^2/c^2 - \vec{p}^2 = m^2c^2$. Inserting energy and momentum operators into this equation, results in the *Klein-Gordon equation*

$$\hbar^2\Big(-\frac{1}{c^2}\frac{\partial^2}{\partial t^2}+\frac{\partial^2}{\partial x_1^2}+\frac{\partial^2}{\partial x_2^2}+\frac{\partial^2}{\partial x_3^2}\Big)\psi = m^2c^2\psi,$$

which treats time and space on an equal footing. Dirac 1928 linearized the Klein-Gordon equation, or replaced it by a first-order equation,

$$i\hbar\Big(\gamma_0\frac{1}{c}\frac{\partial}{\partial t}+\gamma_1\frac{\partial}{\partial x_1}+\gamma_2\frac{\partial}{\partial x_2}+\gamma_3\frac{\partial}{\partial x_3}\Big)\psi = mc\psi.$$

The above *Dirac equation* implies the Klein-Gordon equation provided the symbols γ_μ satisfy the relations

$$\gamma_0^2 = I, \quad \gamma_1^2 = \gamma_2^2 = \gamma_3^2 = -I,$$
$$\gamma_\mu\gamma_\nu = -\gamma_\nu\gamma_\mu \quad \text{for} \quad \mu \neq \nu.$$

Dirac found a set of 4×4-matrices satisfying these relations, namely, the following *Dirac matrices*:

$$\gamma_0 = \begin{pmatrix} 1 & 0 & 0 & 0 \\ 0 & 1 & 0 & 0 \\ 0 & 0 & -1 & 0 \\ 0 & 0 & 0 & -1 \end{pmatrix},$$

$$\gamma_1 = \begin{pmatrix} 0 & 0 & 0 & -1 \\ 0 & 0 & -1 & 0 \\ 0 & 1 & 0 & 0 \\ 1 & 0 & 0 & 0 \end{pmatrix}, \quad \gamma_2 = \begin{pmatrix} 0 & 0 & 0 & i \\ 0 & 0 & -i & 0 \\ 0 & -i & 0 & 0 \\ i & 0 & 0 & 0 \end{pmatrix}, \quad \gamma_3 = \begin{pmatrix} 0 & 0 & -1 & 0 \\ 0 & 0 & 0 & 1 \\ 1 & 0 & 0 & 0 \\ 0 & -1 & 0 & 0 \end{pmatrix}.$$

In terms of the Pauli spin-matrices σ_k the Dirac gamma-matrices γ_μ can be expressed as [1]

$$\gamma_0 = \gamma^0 = \begin{pmatrix} I & 0 \\ 0 & -I \end{pmatrix}, \qquad \gamma_k = -\gamma^k = \begin{pmatrix} 0 & -\sigma_k \\ \sigma_k & 0 \end{pmatrix}.$$

Writing $x_0 = ct$, the Dirac equation can be condensed into the form

$$i\hbar\gamma^\mu \partial_\mu \psi = mc\psi$$

where $\partial_\mu = \dfrac{\partial}{\partial x^\mu}$. [2] An interaction with the electromagnetic field $F^{\mu\nu}$ is included via the space-time potential $(A^0, A^1, A^2, A^3) = (\frac{1}{c}V, A_x, A_y, A_z)$ of $F^{\mu\nu}$ by employing the replacement $i\hbar\partial^\mu \to i\hbar\partial^\mu - eA^\mu$. This leads to the conventional *Dirac equation*

$$\boxed{\gamma_\mu(i\hbar\partial^\mu - eA^\mu)\psi = mc\psi}$$

where the wave function is a *column spinor*, that is,

$$\psi(x) = \begin{pmatrix} \psi_1 \\ \psi_2 \\ \psi_3 \\ \psi_4 \end{pmatrix} \in \mathbb{C}^4 \quad \text{with} \quad \psi_\alpha \in \mathbb{C}.$$

The Dirac equation takes into account the relativistic phenomena and also spin; it describes spin-$\frac{1}{2}$ particles, like the electron.

10.1 Bilinear covariants

The *Dirac adjoint* [3] of a column spinor $\psi \in \mathbb{C}^4$ is a row matrix

$$\psi^\dagger\gamma_0 = (\,\psi_1^* \quad \psi_2^* \quad -\psi_3^* \quad -\psi_4^*\,).$$

A column spinor $\psi(\mathbf{x})$ and its Dirac adjoint $\psi^\dagger(\mathbf{x})\gamma_0$ can be used to define four real valued functions

$$J^\mu(\mathbf{x}) = \psi^\dagger(\mathbf{x})\gamma_0\gamma^\mu\psi(\mathbf{x})$$

which are components of a space-time vector, the Dirac current,

$$\mathbf{J}(\mathbf{x}) = \gamma_\mu J^\mu(\mathbf{x}).$$

Under a Lorentz transformation

$$\mathbf{x}' = s\mathbf{x}s^{-1}, \quad s \in \mathbf{Spin}_+(1,3),$$

1 The above matrix representation is called the *Pauli-Dirac representation* (although it should be called the Pauli-Dirac basis).

2 Note that $\gamma^\mu\partial_\mu = \gamma_\mu\partial^\mu$ where $\partial^\mu = \frac{\partial}{\partial x_\mu}$.

3 The Dirac adjoint of ψ is often denoted by $\bar{\psi}$, but we have reserved this bar-notation for the Clifford-conjugation.

the Dirac spinor transforms according to

$$\psi' = s^{-1}\psi \quad \text{or} \quad \psi'(\mathbf{x}') = s^{-1}\psi(s\mathbf{x}s^{-1})$$

and the Dirac current according to

$$\mathbf{J}' = s^{-1}\mathbf{J}s \quad \text{or} \quad \mathbf{J}'(\mathbf{x}') = s^{-1}\mathbf{J}(s\mathbf{x}s^{-1})s.$$

Thus, the Dirac current is covariant under the Lorentz transformations. The components $J^\mu = \psi^\dagger\gamma_0\gamma^\mu\psi$ are called *bilinear*[4] *covariants.*

The physical state of the electron is determined by the following 16 bilinear covariants:

$$\begin{aligned}
&\Omega_1 = \psi^\dagger\gamma_0\psi = \psi_1^*\psi_1 + \psi_2^*\psi_2 - \psi_3^*\psi_3 - \psi_4^*\psi_4, \\
&J^\mu = \psi^\dagger\gamma_0\gamma^\mu\psi, \\
&S^{\mu\nu} = \psi^\dagger\gamma_0 i\gamma^{\mu\nu}\psi, \qquad\qquad \gamma^{\mu\nu} = \gamma^\mu\gamma^\nu \neq i\gamma^\mu\gamma^\nu, \\
&K^\mu = \psi^\dagger\gamma_0 i\gamma^{0123}\gamma_\mu\psi, \\
&\Omega_2 = \psi^\dagger\gamma_0\gamma^{0123}\psi, \qquad\qquad \gamma^{0123} = \gamma^0\gamma^1\gamma^2\gamma^3.
\end{aligned}$$

Their integrals over space give expectation values of the physical observables.

The quantity $J^0 = \psi^\dagger\psi$, $J^0 \geq 0$, integrated over a space-like domain gives the probability of finding the electron in that domain.[5] The quantities $J^k = \psi^\dagger\gamma_0\gamma^k\psi$ $(k = 1, 2, 3)$ give the current of probability $\vec{J} = \gamma_k J^k$; they satisfy the continuity equation

$$\frac{1}{c}\frac{\partial J^0}{\partial t} + \frac{\partial J^k}{\partial x^k} = 0.$$

The Dirac current $\mathbf{J}$ is a future-oriented vector, $\mathbf{J}^2 \geq 0$.[6] The time-component $u_0 = \gamma_0 \cdot \mathbf{u}$ of the unit vector $\mathbf{u} = \mathbf{J}/\sqrt{\mathbf{J}^2}$, $\mathbf{J}^2 \neq 0$, gives the probable velocity of the electron,

$$u_0 = \frac{1}{\sqrt{1 - \frac{v^2}{c^2}}}.$$

The bivector $\mathbf{S} = \frac{1}{2}S^{\mu\nu}\gamma_{\mu\nu}$[7] is usually interpreted as the electromagnetic moment density, while it gives the probability density of the electromagnetic moment of the electron.

---

4 The quantities $\psi^\dagger\gamma_0\gamma^\mu\psi$ are actually quadratic in ψ. Also their polarized forms $\psi^\dagger\gamma_0\gamma^\mu\varphi$ are not bilinear but rather sesquilinear, while anti-linear in ψ.

5 Or rather the probability multiplied by the (negative of the) charge of the electron. In the case of a large number of particles J^0 can be interpreted as the charge density.

6 Recall that $\mathbf{J}^2 = (J^0)^2 - (J^1)^2 - (J^2)^2 - (J^3)^2$.

7 This is a shorthand notation for $\mathbf{S} = \frac{1}{2}\sum_{\mu,\nu} S^{\mu\nu}\gamma_{\mu\nu} = \sum_{\mu<\nu} S^{\mu\nu}\gamma_{\mu\nu}$.

The vector $\mathbf{K} = K^\mu \gamma_\mu$ is space-like, and such that $\mathbf{K}^2 = -\mathbf{J}^2$. It is orthogonal to $\mathbf{J}$, $\mathbf{K} \cdot \mathbf{J} = 0$, and gives the direction of the spin of the electron, the spin vector $\frac{1}{2}\hbar\mathbf{K}/\sqrt{-\mathbf{K}^2}$, $\mathbf{K}^2 \neq 0$. Note that $K^\mu = \psi^\dagger \gamma_0 \gamma^\mu i\gamma_{0123}\psi$.

The first and last of the bilinear covariants were combined into a single quantity by de Broglie:

$$\boxed{\Omega = \Omega_1 + \Omega_2 \gamma_{0123}}$$

Note that $\Omega_2 = -\psi^\dagger \gamma_0 \gamma_{0123} \psi$.

Spinors in Ideals

Here we shall take a new view on spinors and regard them as elements of minimal left ideals, [8] first in matrix algebras, then in complexified Clifford algebras, and finally in real Clifford algebras.

10.2 Square matrix spinors

Usually the wave function is a column spinor $\psi \in \mathbb{C}^4$, but we shall also regard it as a 4×4-matrix with only the first column being non-zero; that is, $\psi \in \mathrm{Mat}(4, \mathbb{C})f$ where f is the primitive idempotent [9]

$$f = \frac{1}{2}(1 + \gamma_0)\frac{1}{2}(1 + i\gamma_1\gamma_2) = \begin{pmatrix} 1 & 0 & 0 & 0 \\ 0 & 0 & 0 & 0 \\ 0 & 0 & 0 & 0 \\ 0 & 0 & 0 & 0 \end{pmatrix}.$$

More explicitly, a Dirac spinor might appear as a *column spinor* or as a *square matrix spinor*: [10]

$$\psi = \begin{pmatrix} \psi_1 \\ \psi_2 \\ \psi_3 \\ \psi_4 \end{pmatrix} \in \mathbb{C}^4 \quad \text{or} \quad \psi = \begin{pmatrix} \psi_1 & 0 & 0 & 0 \\ \psi_2 & 0 & 0 & 0 \\ \psi_3 & 0 & 0 & 0 \\ \psi_4 & 0 & 0 & 0 \end{pmatrix} \in \mathrm{Mat}(4, \mathbb{C})f$$

8 We shall reject ideal spinors later in favor of spinor operators.

9 The factors $\frac{1}{2}(1 + \gamma_0)$ and $\frac{1}{2}(1 + i\gamma_{12})$ are energy and spin projection operators.

10 We replace column spinors by square matrix spinors in order to be able to get everything – vectors, rotations and spinors – represented within one mathematical system, namely the Clifford algebra.

where $\psi_\alpha \in \mathbb{C}$. In the latter case $\psi = \psi_1 f_1 + \psi_2 f_2 + \psi_3 f_3 + \psi_4 f_4$ expressed in a basis of the complex linear spinor space $S = (\mathbb{C} \otimes C\ell_{1,3})f$,

$$\begin{aligned} f_1 &= \tfrac{1}{4}(1 + \gamma_0 + i\gamma_{12} + i\gamma_{012}) &&= f, \\ f_2 &= \tfrac{1}{4}(-\gamma_{13} + i\gamma_{23} - \gamma_{013} + i\gamma_{023}) &&= -\gamma_{13} f, \\ f_3 &= \tfrac{1}{4}(\gamma_3 - \gamma_{03} + i\gamma_{123} - i\gamma_{0123}) &&= -\gamma_{03} f, \\ f_4 &= \tfrac{1}{4}(\gamma_1 - i\gamma_2 - \gamma_{01} + i\gamma_{02}) &&= -\gamma_{01} f. \end{aligned}$$

We write $\gamma_{12} = \gamma_1\gamma_2$ $[\neq i\gamma_1\gamma_2]$ and $\gamma_{0123} = \gamma_0\gamma_1\gamma_2\gamma_3$.

10.3 Real structures and involutions

Although we have an isomorphism of real algebras $\mathbb{C} \otimes \mathrm{Mat}(4, \mathbb{R}) \simeq \mathbb{C} \otimes C\ell_{1,3}$, the complex conjugations are not the same in $\mathbb{C} \otimes C\ell_{1,3}$ and $\mathbb{C} \otimes \mathrm{Mat}(4, \mathbb{R}) \simeq \mathrm{Mat}(4, \mathbb{C})$. In the matrix algebra $\mathrm{Mat}(4, \mathbb{C})$ we take complex conjugates of the matrix entries $u^* = (u_{jk})^* = (u_{jk}^*)$, whereas in the complexified Clifford algebra $\mathbb{C} \otimes C\ell_{1,3}$ complex conjugation has no effect on the real part $C\ell_{1,3}$, and we have $u^* = (a + ib)^* = a - ib$ for $a, b \in C\ell_{1,3}$. Thus there are two different complex conjugations (real parts) in the algebra $\mathbb{C} \otimes C\ell_{1,3} \simeq \mathrm{Mat}(4, \mathbb{C})$. This is referred to by saying that there are two different *real structures* [11] in the same complex algebra.

To make this point more explicit, the following table lists some correspondences of involutions.

	$\mathbb{C} \otimes C\ell_{1,3}$	$\mathrm{Mat}(4, \mathbb{C})$	
complex conjugate	u^*	$\gamma_{013} u^* \gamma_{013}^{-1}$	
	$\gamma_{013} u^* \gamma_{013}^{-1}$	u^*	complex conjugate
grade involute	$\hat{u}$	$\gamma_{0123} u \gamma_{0123}^{-1}$	
reverse	$\tilde{u}$	$\gamma_{13} u^\top \gamma_{13}^{-1}$	
Clifford-conjugate	$\bar{u}$	$\gamma_{02} u^\top \gamma_{02}^{-1}$	
	$\gamma_{13} \tilde{u} \gamma_{13}^{-1}$	$u^\top$	transpose
	$\gamma_0 \tilde{u}^* \gamma_0^{-1}$	$u^\dagger = u^{*\top}$	Hermitian conjugate
	$\tilde{u}^*$	$\gamma_0 u^\dagger \gamma_0^{-1}$	Dirac adjoint

An element $u = \langle u \rangle_0 + \langle u \rangle_1 + \langle u \rangle_2 + \langle u \rangle_3 + \langle u \rangle_4 \in C\ell_{1,3}$, decomposed in

11 Not to be confused with the *complex structure* of an even-dimensional real linear space, a real linear transformation J such that $J^2 = -I$.

dimension degrees $\langle u\rangle_k \in \bigwedge^k \mathbb{R}^{1,3}$, has three important involutions:

$$\hat{u} = \langle u\rangle_0 - \langle u\rangle_1 + \langle u\rangle_2 - \langle u\rangle_3 + \langle u\rangle_4, \qquad \text{grade involution,}$$
$$\tilde{u} = \langle u\rangle_0 + \langle u\rangle_1 - \langle u\rangle_2 - \langle u\rangle_3 + \langle u\rangle_4, \qquad \text{reversion,}$$
$$\bar{u} = \langle u\rangle_0 - \langle u\rangle_1 - \langle u\rangle_2 + \langle u\rangle_3 + \langle u\rangle_4, \qquad \text{Clifford-conjugation.}$$

The reversion and Clifford-conjugation are anti-automorphisms satisfying $\widetilde{uv} = \tilde{v}\tilde{u}$, $\overline{uv} = \bar{v}\bar{u}$, whereas the grade involution is an automorphism $\widehat{uv} = \hat{u}\hat{v}$. These three involutions are extended to $\mathbb{C} \otimes \mathcal{C}\ell_{1,3}$ as complex linear functions, that is, for $\lambda \in \mathbb{C}$ and $u \in \mathcal{C}\ell_{1,3}$ we have $(\lambda u)\hat{} = \lambda\hat{u}$, $(\lambda u)\tilde{} = \lambda\tilde{u}$, $(\lambda u)^- = \lambda\bar{u}$, whereas the complex conjugation is by definition anti-linear: $(\lambda u)^* = \lambda^* u$. Complex conjugation is of course an automorphism, $(uv)^* = u^* v^*$ for $u, v \in \mathbb{C} \otimes \mathcal{C}\ell_{1,3}$.

10.4 Comparison of real parts/structures

Note that the real part and the complex conjugate of a Dirac spinor depend on the decomposition (in the real structure) singling out the real part. For $\psi \in \mathrm{Mat}(4, \mathbb{C})f$:

$$\mathrm{Re}(\psi) = \begin{pmatrix} \mathrm{Re}(\psi_1) & 0 & 0 & 0 \\ \mathrm{Re}(\psi_2) & 0 & 0 & 0 \\ \mathrm{Re}(\psi_3) & 0 & 0 & 0 \\ \mathrm{Re}(\psi_4) & 0 & 0 & 0 \end{pmatrix}, \qquad \psi^* = \begin{pmatrix} \psi_1^* & 0 & 0 & 0 \\ \psi_2^* & 0 & 0 & 0 \\ \psi_3^* & 0 & 0 & 0 \\ \psi_4^* & 0 & 0 & 0 \end{pmatrix}.$$

For $\psi \in (\mathbb{C} \otimes \mathcal{C}\ell_{1,3})f$ [viewed as a matrix]:

$$\mathrm{Re}(\psi) = \frac{1}{2}\begin{pmatrix} \psi_1 & -\psi_2^* & 0 & 0 \\ \psi_2 & \psi_1^* & 0 & 0 \\ \psi_3 & \psi_4^* & 0 & 0 \\ \psi_4 & -\psi_3^* & 0 & 0 \end{pmatrix}, \qquad \psi^* = \begin{pmatrix} 0 & -\psi_2^* & 0 & 0 \\ 0 & \psi_1^* & 0 & 0 \\ 0 & \psi_4^* & 0 & 0 \\ 0 & -\psi_3^* & 0 & 0 \end{pmatrix}.$$

The Dirac spinor ψ might appear as a column spinor $\psi \in \mathbb{C}^4$ or else as a square matrix spinor $\psi \in \mathrm{Mat}(4, \mathbb{C})f$ or as a *Clifford algebraic spinor* $\psi \in (\mathbb{C} \otimes \mathcal{C}\ell_{1,3})f$ where the last two differ in their real structures.

Important Note. To indicate in what real structure the real part and the complex conjugate are taken we write

$$\mathrm{Re}(\psi) \text{ in } \mathrm{Mat}(4, \mathbb{C})f \qquad \text{and} \qquad \psi^* \text{ in } \mathrm{Mat}(4, \mathbb{C})f$$

or

$$\mathrm{Re}(\psi) \text{ in } \mathbb{C} \otimes \mathcal{C}\ell_{1,3} \qquad \text{and} \qquad \psi^* \text{ in } \mathbb{C} \otimes \mathcal{C}\ell_{1,3}.$$

Other contextual indicators are the Hermitian conjugation [either $\psi^\dagger\gamma_0$ is a row spinor or else it is in $\mathrm{Mat}(4, \mathbb{C})$] and for instance the reversion [the composite of the reversion and the complex conjugation $\tilde{\psi}^*$ in $\mathbb{C}\otimes C\ell_{1,3}$ corresponds to $\psi^\dagger\gamma_0$ in $\mathrm{Mat}(4,\mathbb{C})$].

The reader should also observe that the real part $\mathrm{Re}(\psi)$ in $\mathbb{C}\otimes C\ell_{1,3}$ carries the same information as the original Dirac spinor $\psi\in\mathbb{C}^4$ [in contrast to $\mathrm{Re}(\psi)$ in $\mathrm{Mat}(4,\mathbb{C})f$]. ∎

Exercises 1,2,3,4,5

10.5 Bilinear covariants via algebraic spinors

For a column spinor $\psi\in\mathbb{C}^4$ the Dirac adjoint is a row matrix

$$\psi^\dagger\gamma_0 = (\,\psi_1^*\quad \psi_2^*\quad -\psi_3^*\quad -\psi_4^*\,)$$

but for a square matrix spinor $\psi\in\mathrm{Mat}(4,\mathbb{C})f$ the Dirac adjoint is a square matrix

$$\psi^\dagger\gamma_0 = \gamma_0\psi^\dagger\gamma_0^{-1} = \begin{pmatrix} \psi_1^* & \psi_2^* & -\psi_3^* & -\psi_4^* \\ 0 & 0 & 0 & 0 \\ 0 & 0 & 0 & 0 \\ 0 & 0 & 0 & 0 \end{pmatrix}$$

with only the first row being non-zero.

The components of the Dirac current can be computed as follows for column spinors, square matrix spinors and Clifford algebraic spinors

$$\begin{aligned} J_\mu &= \psi^\dagger\gamma_0\gamma_\mu\psi && \psi\in\mathbb{C}^4 \\ &= \mathrm{trace}(\psi^\dagger\gamma_0\gamma_\mu\psi) && \psi\in\mathrm{Mat}(4,\mathbb{C})f \\ &= \mathrm{trace}(\gamma_\mu\psi\psi^\dagger\gamma_0) = 4\langle\gamma_\mu\psi\psi^\dagger\gamma_0\rangle_0 && \\ &= 4\langle\gamma_\mu\psi\tilde{\psi}^*\rangle_0 && \psi\in(\mathbb{C}\otimes C\ell_{1,3})f \end{aligned}$$

where the factor 4 appeared because

$$f = \frac{1}{4}(1+\gamma_0+i\gamma_{12}+i\gamma_{012}) = \begin{pmatrix} 1 & 0 & 0 & 0 \\ 0 & 0 & 0 & 0 \\ 0 & 0 & 0 & 0 \\ 0 & 0 & 0 & 0 \end{pmatrix}$$

has scalar part $\frac{1}{4}$, that is, $\langle f\rangle_0 = \frac{1}{4}$, while $\mathrm{trace}(f)=1$. The current vector is the resultant

$$\begin{aligned} \mathbf{J} = \gamma^\mu J_\mu &= \gamma^\mu 4\langle\gamma_\mu\psi\psi^\dagger\gamma_0\rangle_0 && \psi\in\mathrm{Mat}(4,\mathbb{C})f \\ &= \gamma^\mu\langle\gamma_\mu \lrcorner (4\psi\psi^\dagger\gamma_0)\rangle_0 = \gamma^\mu <\gamma_\mu, 4\psi\psi^\dagger\gamma_0> && \\ &= \langle 4\psi\psi^\dagger\gamma_0\rangle_1 && J_\mu = \gamma_\mu\cdot\mathbf{J} \\ &= \langle 4\psi\tilde{\psi}^*\rangle_1 && \psi\in(\mathbb{C}\otimes C\ell_{1,3})f. \end{aligned}$$

Similarly $\psi \in \mathbb{C}^4$ carries a real bivector $\mathbf{S}$ with components

$$S_{\mu\nu} = \psi^\dagger \gamma_0 i\gamma_{\mu\nu}\psi \qquad (\gamma_{\mu\nu} = \gamma_\mu\gamma_\nu \neq i\gamma_\mu\gamma_\nu)$$

for which $S_{\mu\nu} = -\gamma_{\mu\nu} \,\lrcorner\, \mathbf{S}$ and $\mathbf{S} = \frac{1}{2}S_{\mu\nu}\gamma^{\mu\nu}$. In various formalisms

$$\begin{aligned}
S_{\mu\nu} &= \psi^\dagger\gamma_0 i\gamma_{\mu\nu}\psi && \psi \in \mathbb{C}^4\\
&= \operatorname{trace}(\psi^\dagger\gamma_0 i\gamma_{\mu\nu}\psi) && \psi \in \mathrm{Mat}(4,\mathbb{C})f\\
&= \operatorname{trace}(i\gamma_{\mu\nu}\psi\psi^\dagger\gamma_0) = 4\langle i\gamma_{\mu\nu}\psi\psi^\dagger\gamma_0\rangle_0 &&\\
&= 4\langle i\gamma_{\mu\nu}\psi\tilde{\psi}^*\rangle_0 && \psi \in (\mathbb{C}\otimes C\ell_{1,3})f
\end{aligned}$$

$$\begin{aligned}
\mathbf{S} = \tfrac{1}{2}\gamma^{\mu\nu}S_{\mu\nu} &= \tfrac{1}{2}\gamma^{\mu\nu}4\langle i\gamma_{\mu\nu}\psi\psi^\dagger\gamma_0\rangle_0 && \psi \in \mathrm{Mat}(4,\mathbb{C})f\\
&= \tfrac{1}{2}\gamma^{\mu\nu}\langle i\gamma_{\mu\nu} \,\lrcorner\, (4\psi\psi^\dagger\gamma_0)\rangle_0 && S_{\mu\nu} = -\gamma_{\mu\nu} \,\lrcorner\, \mathbf{S}\\
&= \tfrac{1}{2}\gamma^{\mu\nu}{<}{-i\gamma_{\mu\nu}}, 4\psi\psi^\dagger\gamma_0{>} && {<}u,v{>} = \langle\tilde{u} \,\lrcorner\, v\rangle_0\\
&= \langle -i4\psi\psi^\dagger\gamma_0\rangle_2 = -i\langle 4\psi\psi^\dagger\gamma_0\rangle_2 &&\\
&= -i\langle 4\psi\tilde{\psi}^*\rangle_2 && \psi \in (\mathbb{C}\otimes C\ell_{1,3})f.
\end{aligned}$$

The Dirac adjoint $\psi^\dagger\gamma_0$ of a column spinor $\psi \in \mathbb{C}^4$ corresponds to $\tilde{\psi}^*$ of an algebraic spinor $\psi \in (\mathbb{C}\otimes C\ell_{1,3})f$. The current vector $\mathbf{J}$ and the bivector $\mathbf{S}$ are examples of *bilinear covariants* listed below for a column spinor $\psi \in \mathbb{C}^4$ and for an algebraic spinor $\psi \in (\mathbb{C}\otimes C\ell_{1,3})f$.

$$\begin{aligned}
\Omega_1 &= \psi^\dagger\gamma_0\psi = 4\langle\tilde{\psi}^*\psi\rangle_0 = 4\langle\psi\tilde{\psi}^*\rangle_0 &&\\
J_\mu &= \psi^\dagger\gamma_0\gamma_\mu\psi = 4\langle\tilde{\psi}^*\gamma_\mu\psi\rangle_0 &&\\
S_{\mu\nu} &= \psi^\dagger\gamma_0 i\gamma_{\mu\nu}\psi = 4\langle\tilde{\psi}^* i\gamma_{\mu\nu}\psi\rangle_0 && \gamma_{\mu\nu} = \gamma_\mu\gamma_\nu \neq i\gamma_\mu\gamma_\nu\\
K_\mu &= \psi^\dagger\gamma_0 i\gamma_{0123}\gamma_\mu\psi = 4\langle\tilde{\psi}^* i\gamma_{0123}\gamma_\mu\psi\rangle_0 && \mathbf{K} = K_\mu\gamma^\mu\\
\Omega_2 &= -\psi^\dagger\gamma_0\gamma_{0123}\psi = -4\langle\tilde{\psi}^*\gamma_{0123}\psi\rangle_0 && \gamma_{0123} = \gamma_0\gamma_1\gamma_2\gamma_3.
\end{aligned}$$

Later we shall need the following aggregate of bilinear covariants $Z = \Omega_1 + \mathbf{J} + i\mathbf{S} + i\mathbf{K}\gamma_{0123} + \Omega_2\gamma_{0123}$.

Spinors as Operators

Here we shall view spinors as new kinds of objects: rather than being something which are operated upon they are regarded as active operators. The big advantage is that the physical observables, which were earlier calculated component-wise, can now be obtained at one stroke.

10.6 Spinor operators $\Psi \in \mathcal{C}\ell^+_{1,3}$

We will associate to a Clifford algebraic spinor $\psi \in (\mathbb{C} \otimes \mathcal{C}\ell_{1,3})f$ [viewed here as a matrix] the *mother spinor* [this will be the mother of all real spinors]

$$\Phi = 4\,\mathrm{Re}(\psi) = 2\begin{pmatrix} \psi_1 & -\psi_2^* & 0 & 0 \\ \psi_2 & \psi_1^* & 0 & 0 \\ \psi_3 & \psi_4^* & 0 & 0 \\ \psi_4 & -\psi_3^* & 0 & 0 \end{pmatrix}$$

and the *spinor operator*

$$\Psi = \mathrm{even}(\Phi) = \begin{pmatrix} \psi_1 & -\psi_2^* & \psi_3 & \psi_4^* \\ \psi_2 & \psi_1^* & \psi_4 & -\psi_3^* \\ \psi_3 & \psi_4^* & \psi_1 & -\psi_2^* \\ \psi_4 & -\psi_3^* & \psi_2 & \psi_1^* \end{pmatrix}.$$

From the mother spinor $\Phi \in \mathcal{C}\ell_{1,3}\frac{1}{2}(1+\gamma_0)$ we may reobtain the original Dirac spinor

$$\psi = \Phi\frac{1}{4}(1+i\gamma_{12}) \in (\mathbb{C} \otimes \mathcal{C}\ell_{1,3})f$$

[that is, the square matrix spinor, not the column spinor], and from the spinor operator Ψ we may reobtain the mother spinor $\Phi = \Psi(1+\gamma_0)$ and the original Dirac spinor

$$\psi = \Psi\frac{1}{2}(1+\gamma_0)\frac{1}{2}(1+i\gamma_{12}) \in (\mathbb{C} \otimes \mathcal{C}\ell_{1,3})f.$$

Note that the spinor operator is invertible if

$$|\psi_1|^2 + |\psi_2|^2 - |\psi_3|^2 - |\psi_4|^2 \neq 0 \quad \text{and} \quad 2\,\mathrm{Im}(\psi_1^*\psi_3 + \psi_2^*\psi_4) \neq 0,$$

or equivalently $\Psi\tilde{\Psi} \neq 0$, the inverse being

$$\Psi^{-1} = \frac{\tilde{\Psi}}{\Psi\tilde{\Psi}}.$$

Multiplication by $i = \sqrt{-1}$ corresponds to right multiplication by the bivector

$$\gamma_2\gamma_1 = \begin{pmatrix} i & 0 & 0 & 0 \\ 0 & -i & 0 & 0 \\ 0 & 0 & i & 0 \\ 0 & 0 & 0 & -i \end{pmatrix},$$

that is, $i\psi = \psi\gamma_2\gamma_1$ for $\psi \in \mathbb{C}^4$. In other words, the real part of $i\psi$, $\psi \in (\mathbb{C}\otimes\mathcal{C}\ell_{1,3})f$, is the mother spinor $\Phi\gamma_2\gamma_1$ whose even part is the spinor operator $\Psi\gamma_2\gamma_1$, $4\,\mathrm{Re}(\mathrm{even}(i\psi)) = \Psi\gamma_2\gamma_1$.

Decompose the mother spinor $\Phi \in \mathcal{C}\ell_{1,3}\frac{1}{2}(1+\gamma_0)$ into even and odd parts

$\Phi = \Phi_0 + \Phi_1 = (\Phi_0 + \Phi_1)\frac{1}{2}(1+\gamma_0) = \frac{1}{2}(\Phi_0 + \Phi_1\gamma_0) + \frac{1}{2}(\Phi_1 + \Phi_0\gamma_0)$. It follows that $\Phi_0 = \Phi_1\gamma_0$ and $\Phi_1 = \Phi_0\gamma_0$. Taking the real part [in $\mathbb{C}\otimes C\ell_{1,3}$] of the Dirac equation $(i\partial - e\mathbf{A})\psi = m\psi$ results in

$$\partial\Phi\gamma_2\gamma_1 - e\mathbf{A}\Phi = m\Phi,$$

which decomposes into even and odd parts [$\Phi_0 = \text{even}(\Phi)$, $\Phi_1 = \text{odd}(\Phi)$]

$$\partial\Phi_0\gamma_{21} - e\mathbf{A}\Phi_0 = m\Phi_1 \qquad [\Phi_1 = \Phi_0\gamma_0],$$
$$\partial\Phi_1\gamma_{21} - e\mathbf{A}\Phi_1 = m\Phi_0 \qquad [\Phi_0 = \Phi_1\gamma_0].$$

Therefore, the even part of the mother spinor, the spinor operator, satisfies the equation [12]

$$\boxed{\partial\Psi\gamma_{21} - e\mathbf{A}\Psi = m\Psi\gamma_0}$$

where $\Psi : \mathbb{R}^{1,3} \to C\ell^+_{1,3}$. In this *Dirac-Hestenes equation* the role of the Dirac column spinors is taken over by real even multivectors, which are not in any proper left ideal of the Clifford algebra $C\ell_{1,3}$.

Comments. 1. Under a Lorentz transformation $\mathbf{x} \to s\mathbf{x}\hat{s}^{-1}$, $s \in \mathbf{Pin}(1,3)$, $\mathbf{x} \in \mathbb{R}^{1,3}$, a Dirac spinor $\psi \in \text{Mat}(4,\mathbb{C})f \simeq (\mathbb{C}\otimes C\ell_{1,3})f$ transforms according to $\psi \to s\psi$, and a spinor operator $\Psi \in C\ell^+_{1,3}$ transforms like this:

$$\Psi \to s\Psi \qquad \text{when} \quad s \in \mathbf{Spin}(1,3),$$
$$\Psi \to s\Psi\gamma_0 \qquad \text{when} \quad s \in \mathbf{Pin}(1,3)\backslash\mathbf{Spin}(1,3).$$

This can be seen by the definition $\Psi = 4\,\text{Re}(\text{even}(\psi))$ and using $\psi = \psi f$, $f = \frac{1}{2}(1+\gamma_0)\frac{1}{2}(1+i\gamma_{12})$.

Note that the so-called Wigner time-reversal is not represented by any $s \in \mathbf{Pin}(1,3)\backslash\mathbf{Spin}(1,3)$.

2. The Dirac-Hestenes equation has been criticized on the basis that it is not Lorentz covariant because of an explicit appearance of the two basis elements γ_0 and γ_{12}. This criticism does not hold. The Dirac-Hestenes equation is Lorentz covariant in two different ways: first, we can regard γ_0 and γ_{12} as constants and transform Ψ to $s\Psi$; secondly, we can transform γ_0, γ_{12} to $s\gamma_0 s^{-1}$, $s\gamma_{12}s^{-1}$ and Ψ to $s\Psi s^{-1}$, $s \in \mathbf{Spin}_+(1,3)$.

3. In curved space-times spinor fields/bundles [functions with values in a minimal left ideal of a Clifford algebra] exist globally only under certain topological conditions: the space-time must be a spin manifold [have a spinor structure]. It has been argued that since even multivector functions exist on all oriented manifolds, the theory of spin manifolds is superfluous. This argument is misplaced since γ_0 and γ_{12} do not exist globally.

---

12 Note that $4\,\text{Re}(\text{even}(\mathbf{x}\psi)) = \mathbf{x}\Psi\gamma_0$ for $\mathbf{x} \in \mathbb{R}^{1,3}$.

However, the physical justification of the theory of spin manifolds could be questioned on the following basis: why should we need to know the global properties of the universe if we want to explore the local properties of a single electron?

4. The explicit occurrence of γ_0 and γ_{12} is due to our injection $\mathbb{C}^4 \to S = (\mathbb{C} \otimes \mathcal{C}\ell_{1,3})f$. In curved manifolds it is more appropriate to use abstract representation modules as spinor spaces and not minimal left ideals [nor the even subalgebras] of Clifford algebras. The injection ties these spaces together in a manner that singles out special directions in $\mathbb{R}^{1,3}$. ∎

10.7 Bilinear covariants via spinor operators

Write as before $\Psi = 4\,\mathrm{Re}(\mathrm{even}(\psi))$ [real part taken in $\mathbb{C} \otimes \mathcal{C}\ell_{1,3}$]. Because of the identities [13]

$$\Psi\tilde{\Psi} = \Omega_1 + \Omega_2\gamma_{0123},$$

$$\Psi\gamma_0\tilde{\Psi} = \mathbf{J},$$

$$\Psi\gamma_{12}\tilde{\Psi} = \mathbf{S}, \qquad \Psi\gamma_{03}\tilde{\Psi} = -\mathbf{S}\gamma_{0123},$$

$$\Psi\gamma_3\tilde{\Psi} = \mathbf{K}, \qquad \Psi\gamma_{012}\tilde{\Psi} = \mathbf{K}\gamma_{0123},$$

we call Ψ a *spinor operator*. In the non-null case $\Psi\tilde{\Psi} \neq 0$ the element Ψ operates like a Lorentz transformation composed with a dilation [and a duality transformation]. In coordinate form

$$\Omega_1 + \Omega_2\gamma_{0123} = \Psi\tilde{\Psi} = \tilde{\Psi}\Psi,$$

$$J_\mu = \langle\tilde{\Psi}\gamma_\mu\Psi\gamma_0\rangle_0 = (\tilde{\Psi}\gamma_\mu\Psi)\cdot\gamma_0,$$

$$S_{\mu\nu} = -\langle\tilde{\Psi}\gamma_{\mu\nu}\Psi\gamma_{12}\rangle_0 = (\tilde{\Psi}\gamma_{\mu\nu}\Psi)\cdot\gamma_{12},$$

$$K_\mu = \langle\tilde{\Psi}\gamma_\mu\Psi\gamma_3\rangle_0 = (\tilde{\Psi}\gamma_{0123}\gamma_\mu\Psi)\cdot\gamma_{012}.$$

For later convenience we introduce $P = \Omega + \mathbf{J}$, $\Omega = \Omega_1 + \Omega_2\gamma_{0123}$, and $Q = \mathbf{S} + \mathbf{K}\gamma_{0123}$. We have the following identities:

$$\Psi(1+\gamma_0)\tilde{\Psi} = P, \qquad \Psi(1+\gamma_0)\gamma_{12}\tilde{\Psi} = Q,$$

$$\Psi(1+\gamma_0)\gamma_{ij}\tilde{\Psi} = Q_k \qquad (ijk\ \text{cycl.},\ Q_3 = Q),$$

$$\Psi(1+\gamma_0)(1+i\gamma_{12})\tilde{\Psi} = Z \quad [= P + iQ].$$

Hestenes 1986 p. 334 gives $P, -Q$ and Z in (2.26), (2.27) and (2.28).

Exercises 6,7,8

---

13 The Dirac-Hestenes equation $\partial\Psi\gamma_{21} - e\mathbf{A}\Psi = m\Psi\gamma_0$ contains γ_{21}, γ_0 explicitly. It follows that γ_0, γ_{12} must be explicit in $\mathbf{J} = \Psi\gamma_0\tilde{\Psi}$, $\mathbf{S} = \Psi\gamma_{12}\tilde{\Psi}$.

10.8 Higher-dimensional analogies for spinor operators

In the case of the Minkowski space-time $\mathbb{R}^{1,3}$ the spinor space is a minimal left ideal $(\mathbb{C}\otimes C\ell_{1,3})f$ induced by the primitive idempotent $f = \frac{1}{2}(1+\gamma_0)\frac{1}{2}(1+i\gamma_{12})$. In the primitive idempotent we have projection operators for energy $\frac{1}{2}(1+\gamma_0)$ and spin $\frac{1}{2}(1+i\gamma_{12})$. In other words, spin is quantized in the γ_3-direction or more precisely in the $\gamma_1\gamma_2$-plane.

In the case of a higher-dimensional space-time, say $\mathbb{R}^{1,5}$ with an orthonormal basis $\{\gamma_0, \gamma_1, \ldots, \gamma_5\}$, the spinor space is a minimal left ideal $(\mathbb{C}\otimes C\ell_{1,5})f$ induced, for instance, by the primitive idempotent

$$f = \frac{1}{2}(1+\gamma_0)\frac{1}{2}(1+i\gamma_{12})\frac{1}{2}(1+i\gamma_{34}).$$

The spin is quantized in the $\gamma_1\gamma_2$-plane and the $\gamma_3\gamma_4$-plane. The procedure of taking the real part and the even part does not result in an invertible operator, since $4\,\mathrm{Re}(\mathrm{even}(f)) = \frac{1}{2}(1-\gamma_{1234})$. In other words, for a spinor in a minimal left ideal $\psi \in (\mathbb{C}\otimes C\ell_{1,5})f$ the 'spinor operator' is also in a left ideal, $\Psi = 4\,\mathrm{Re}(\mathrm{even}(\psi)) \in C\ell^+_{1,5}\frac{1}{2}(1-\gamma_{1234})$. We conclude that there is no analogy for spinor operators in higher dimensions.

Appendix 1: Discussion on the role of $i = \sqrt{-1}$ in QM

Are there superfluous complex numbers in the present formulation of quantum mechanics? Is it possible to get rid of *some* complex numbers in QM? To answer these questions, we present analogies which become step by step closer to the present situation in quantum mechanics.

Analogy # 1. Consider someone who uses only the line $y = x$ in the complex plane $\mathbb{C}$, that is, someone who does not use all the complex numbers $z = x+iy$, but instead restricts himself to complex numbers of the form $x+ix$. This person could equally well restrict himself to the real axis and consider instead only the real part $x = \mathrm{Re}(x+ix)$. In terms of the picture

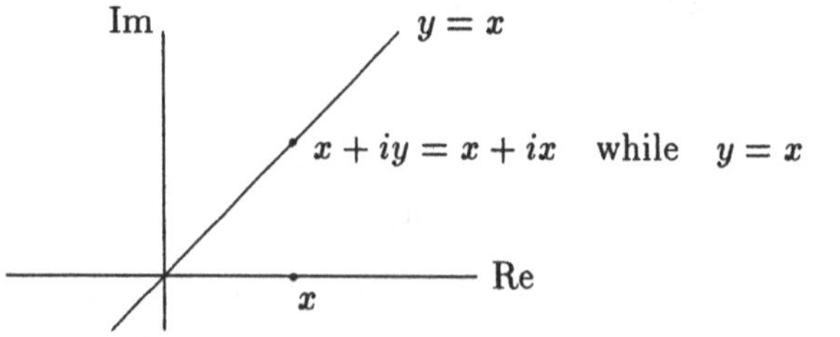

this would mean a projection from the line $y = x$ onto the real axis $y = 0$ with no information lost.

This analogy/picture could be criticized by arguing that the product w of two complex numbers of the form $x + ix$ is not of the same type, that is, $\text{Re}(w) \neq \text{Im}(w)$. ∎

Analogy # 2. The sums and products of matrices of type

$$X = \frac{1}{2}\begin{pmatrix} x & x \\ x & x \end{pmatrix}$$

provide an isomorphic image for addition and multiplication of the reals $x \in \mathbb{R}$. It is of course more economical to use just the real numbers $x \in \mathbb{R}$ instead of the real 2×2-matrices $X \in \text{Mat}(2, \mathbb{R})$. ∎

Analogy # 3. If we have a complex matrix

$$S = \frac{1}{2}\begin{pmatrix} x + iy & -y + ix \\ y - ix & x + iy \end{pmatrix}$$

then the real part, multiplied by two, $Z = 2\,\text{Re}(S)$, that is,

$$Z = \begin{pmatrix} x & -y \\ y & x \end{pmatrix},$$

obeys the same addition and multiplication rules as S and carries the same information as S [contained in the pair (x, y)]. Note that for a complex number $z = x + iy$ we have $S = \frac{z}{2}(I - \sigma_2)$, where the matrix $f = \frac{1}{2}(I - \sigma_2)$ is an idempotent satisfying $f^2 = f$. ∎

Situation in QM. In the present formulation of quantum mechanics one uses column spinors $\psi \in \mathbb{C}^4$, which could be replaced without loss of generality by spinors in a minimal left ideal of the complex Clifford algebra $\psi \in (\mathbb{C} \otimes C\ell_{1,3})f$, $f = \frac{1}{2}(1 + \gamma_0)\frac{1}{2}(1 + i\gamma_{12})$. Spinors in minimal left ideals $\psi \in (\mathbb{C} \otimes C\ell_{1,3})f$ can be replaced without reduction of information by spinor operators $\Psi = 4\,\text{Re}(\text{even}(\psi)) \in C\ell^+_{1,3}$. No information is lost in this replacement, because the original spinor can be recovered as $\psi = \Psi f$, $f = \frac{1}{2}(1 + \gamma_0)\frac{1}{2}(1 + i\gamma_{12})$. ∎

Appendix 2: Real ideal spinors $\phi \in C\ell_{1,3}\,\frac{1}{2}(1 - \gamma_{03})$

This appendix is included mainly for historical reasons. The concept of a spinor operator $\Psi \in C\ell^+_{1,3}$ was introduced by Hestenes 1966. In his invention he used as an intermediate step the *real ideal spinor*

$$\phi = \Phi\,\frac{1}{2}(1 - \gamma_{03}) \in C\ell_{1,3}\,\frac{1}{2}(1 - \gamma_{03})$$

and not the mother spinor $\Phi \in C\ell_{1,3}\,\frac{1}{2}(1 + \gamma_0)$, $\Phi = 4\,\text{Re}(\psi)$, $\psi \in (\mathbb{C} \otimes C\ell_{1,3})f$. The ideal spinor contains the same information as the mother spinor, since

$\Phi = \phi(1+\gamma_0)$. Note that $\phi = \Phi\frac{1}{2}(1-\gamma_{03})$ or $\phi = 2\,\mathrm{Re}(\psi)(1-\gamma_{03})$ which implies $\phi\gamma_2\gamma_1 = 2\,\mathrm{Re}(i\psi)(1-\gamma_{03})$, and so the Dirac equation has the form

$$\partial\phi\gamma_{21} = (e\mathbf{A}+m)\phi, \qquad \phi \in \mathcal{C}\ell_{1,3}\frac{1}{2}(1-\gamma_{03}).$$

In contrast to the mother spinor Φ, the real ideal spinor ϕ satisfies $\phi\gamma_{0123} = \phi\gamma_{21}$, and so we could rewrite the Dirac equation in the same way as Hestenes 1966:

$$\partial\phi\gamma_{0123} = (e\mathbf{A}+m)\phi.$$

Comments. The ideal spinors might be useful in conjunction with conformal transformations of the Dirac equation. Decompose the ideal spinor $\phi = \phi_0+\phi_1$ into its even and odd parts and separate the parts,

$$\begin{aligned} \partial\phi_0\gamma_{0123} &= e\mathbf{A}\phi_0 + m\phi_1, \quad \phi_0 = \mathrm{even}(\phi) \in \mathcal{C}\ell^+_{1,3}\tfrac{1}{2}(1-\gamma_{03}),\\ \partial\phi_1\gamma_{0123} &= e\mathbf{A}\phi_1 + m\phi_0, \quad \phi_1 = \mathrm{odd}(\phi) \in \mathcal{C}\ell^-_{1,3}\tfrac{1}{2}(1-\gamma_{03}),\end{aligned}$$

which can be put into the matrix form

$$\begin{pmatrix}\gamma_{0123} & 0\\ 0 & -\gamma_{0123}\end{pmatrix}\begin{pmatrix}0 & \partial\\ -\partial & 0\end{pmatrix}\begin{pmatrix}\phi_0 & 0\\ \phi_1 & 0\end{pmatrix}$$
$$= e\begin{pmatrix}0 & \mathbf{A}\\ -\mathbf{A} & 0\end{pmatrix}\begin{pmatrix}\phi_0 & 0\\ \phi_1 & 0\end{pmatrix} + m\begin{pmatrix}1 & 0\\ 0 & -1\end{pmatrix}\begin{pmatrix}\phi_0 & 0\\ \phi_1 & 0\end{pmatrix},$$

where we have used the fact that the matrix

$$\begin{pmatrix}\gamma_{0123} & 0\\ 0 & -\gamma_{0123}\end{pmatrix} \text{ commutes with } \begin{pmatrix}0 & \partial\\ -\partial & 0\end{pmatrix} \text{ and } \begin{pmatrix}\phi_0 & 0\\ \phi_1 & 0\end{pmatrix}$$

and takes the role of an overall commuting imaginary unit $\sqrt{-1}$. ▮

Exercises 9,10

Historical survey

Pauli 1927 and Dirac 1928 presented their spinor equations for the description of the electron spin. Juvet 1930 and Sauter 1930 replaced column spinors by square matrix spinors, where only the first column was non-zero. Marcel Riesz 1947 was the first one to consider spinors as elements in a minimal left ideal of a Clifford algebra (although the special case of pure spinors had been considered earlier by Cartan in 1938).

Gürsey 1956-58 rewrote the Dirac equation with 2×2 quaternion matrices in $\mathrm{Mat}(2,\mathbb{H})$ [Lanczos 1929 had used pairs of quaternions, see Gsponer & Hurni 1993]. Kustaanheimo 1964 presented the spinor regularization of the Kepler motion, the KS-transformation, which emphasized the operator aspect of spinors. This led David Hestenes 1966-74 to a reformulation of the Dirac theory, where the role of spinors [in columns $\mathbb{C}^4$ or in minimal left ideals of the complex Clifford algebra $\mathbb{C}\otimes\mathcal{C}\ell_{1,3}\simeq \mathrm{Mat}(4,\mathbb{C})$] was taken over by operators in the even subalgebra $\mathcal{C}\ell_{1,3}^{+}$ of the real Clifford algebra $\mathcal{C}\ell_{1,3}\simeq\mathrm{Mat}(2,\mathbb{H})$.

Exercises

1. Show that for $u\in\mathbb{C}\otimes\mathcal{C}\ell_{1,3}$ the real part $\mathrm{Re}(u)$ corresponds to $\frac{1}{2}(u+\gamma_{013}u^*\gamma_{013}^{-1})\in\mathrm{Mat}(4,\mathbb{C})$.
2. Show that if $u\in\mathbb{C}\otimes\mathcal{C}\ell_{1,3}$ satisfies the condition $u=u\,\frac{1}{2}(1+i\gamma_{12})$ then $u=\mathrm{Re}(u)(1+i\gamma_{12})$ and $iu=u\gamma_2\gamma_1$.
3. Show that $\mathrm{Im}(\psi)=\mathrm{Re}(\psi)\gamma_{12}$ in $\mathbb{C}\otimes\mathcal{C}\ell_{1,3}$.
4. Show that the charge conjugate $\psi_C=-i\gamma_2\psi^*$ of the Dirac spinor $\psi\in\mathrm{Mat}(4,\mathbb{C})f$ corresponds to $\psi_C=\hat{\psi}^*\gamma_1\in(\mathbb{C}\otimes\mathcal{C}\ell_{1,3})f$.
5. Show that although for $\psi\in\mathrm{Mat}(4,\mathbb{C})f$, $\mathrm{Re}(\psi)\in\mathrm{Mat}(4,\mathbb{C})f$, for a non-zero $\psi\in(\mathbb{C}\otimes\mathcal{C}\ell_{1,3})f$, $\mathrm{Re}(\psi)\notin(\mathbb{C}\otimes\mathcal{C}\ell_{1,3})f$.
6. Show that in terms of the ideal spinor $\phi=\Phi\,\frac{1}{2}(1-\gamma_{03})\in\mathcal{C}\ell_{1,3}\,\frac{1}{2}(1-\gamma_{03})$, where $\Phi=4\,\mathrm{Re}(\psi)$, the bilinear covariants can be expressed as

$$
\begin{aligned}
\Omega_1 &= \langle\tilde{\phi}\phi\gamma_0\rangle_0 = (\tilde{\phi}\phi)\cdot\gamma_3,\\
J_\mu &= \langle\tilde{\phi}\gamma_\mu\phi\gamma_0\rangle_0 = (\tilde{\phi}\gamma_\mu\phi)\cdot\gamma_3,\\
S_{\mu\nu} &= -\langle\tilde{\phi}\gamma_{\mu\nu}\phi\gamma_{123}\rangle_0 = -\langle\tilde{\phi}\gamma_{\mu\nu}\phi\rangle_3\cdot\gamma_{123},\\
K_\mu &= -\langle\tilde{\phi}\gamma_{0123}\gamma_\mu\phi\gamma_{123}\rangle_0,\\
\Omega_2 &= -\langle\tilde{\phi}\gamma_{0123}\phi\gamma_0\rangle_0
\end{aligned}
$$

   and the aggregates $P=\Omega+\mathbf{J}$ and $Q=\mathbf{S}+\mathbf{K}\gamma_{0123}$ as

$$
\phi\gamma_0\tilde{\phi}=\phi\gamma_3\tilde{\phi}=P,
$$
$$
\phi\gamma_0\bar{\phi}=\phi\gamma_3\bar{\phi}=-Q\gamma_{0123},\qquad Q=\phi\gamma_{123}\tilde{\phi}.
$$

   [Hestenes 1986 p. 334 gives P in (2.26) and $-Q$ in (2.27).]
7. Show that $\phi\tilde{\phi}$, $\phi\bar{\phi}$, $\phi\gamma_1\tilde{\phi}$, $\phi\gamma_1\bar{\phi}$, $\phi\gamma_2\tilde{\phi}$, $\phi\gamma_2\bar{\phi}$ all vanish.
8. Show that for a different choice of sign in $\phi=\Phi\,\frac{1}{2}(1-\gamma_{03})$, namely $\varphi=\Phi\,\frac{1}{2}(1+\gamma_{03})$, we have $\varphi\gamma_0\tilde{\varphi}=P=-\varphi\gamma_3\tilde{\varphi}$ and $\varphi\gamma_3\bar{\varphi}=-Q\gamma_{0123}=-\varphi\gamma_0\bar{\varphi}$.
9. Show that $\phi=\Phi\,\frac{1}{2}(1-\gamma_3)$.
10. Show that $\phi=\Psi\,\frac{1}{2}(1+\gamma_0)(1-\gamma_{03})$.

Bibliography

W. Baylis: *Theoretical Methods in the Physical Sciences: an Introduction to Problem Solving with MAPLE V*. Birkhäuser, Boston, MA, 1994.

J.D. Bjorken, S.D. Drell: *Relativistic Quantum Mechanics*. McGraw-Hill, New York, 1964.

R. Boudet: Les algèbres de Clifford et les transformations des multivecteurs, pp. 343-352 in A. Micali et al. (eds.): *Proceedings of the Second Workshop on 'Clifford Algebras and their Applications in Mathematical Physics' (Montpellier, 1989)*. Kluwer, Dordrecht, The Netherlands, 1992.

R. Brauer, H. Weyl: Spinors in n dimensions. *Amer. J. Math.* **57** (1935), 425-449. Reprinted in *Selecta Hermann Weyl*, Birkhäuser, Basel, 1956, pp. 431-454.

A. Charlier, A. Bérard, M.-F. Charlier, D. Fristot: *Tensors and the Clifford Algebra, Applications to the Physics of Bosons and Fermions*. Marcel Dekker, New York, 1992.

P.A.M. Dirac: The quantum theory of the electron. *Proc. Roy. Soc. A***117** (1928), 610-624.

A. Gsponer, J.-P. Hurni: Lanczos' equation to replace Dirac's equation? pp. 509-512 in J.D. Brown (ed.): *Proceedings of the Cornelius Lanczos Centenary Conference (Raleigh, NC, 1993)*.

F. Gürsey: Correspondence between quaternions and four-spinors. *Rev. Fac. Sci. Univ. Istanbul* **A21** (1956), 33-54.

F. Gürsey: Relation of charge independence and baryon conservation to Pauli's transformation. *Nuovo Cimento* **7** (1958), 411-415.

J.D. Hamilton: The Dirac equation and Hestenes' geometric algebra. *J. Math. Phys.* **25** (1984), 1823-1832.

D. Hestenes: *Space-Time Algebra*. Gordon and Breach, New York, 1966, 1987, 1992.

D. Hestenes: Real spinor fields. *J. Math. Phys.* **8** (1967), 798-808.

D. Hestenes: Local observables in the Dirac theory. *J. Math. Phys.* **14** (1973), 893-905.

D. Hestenes: Proper dynamics of a rigid point particle. *J. Math. Phys.* **15** (1974), 1778-1786.

D. Hestenes: Clifford algebra and the interpretation of quantum mechanics, pp. 321-346 in J.S.R. Chisholm, A.K. Common (eds.): *Proceedings of the Workshop on 'Clifford Algebras and their Applications in Mathematical Physics' (Canterbury 1985)*. Reidel, Dordrecht, The Netherlands, 1986.

P.R. Holland: Minimal ideals and Clifford algebras in the phase space representation of spin-$\frac{1}{2}$ fields, pp. 273-283 in J.S.R. Chisholm, A.K. Common (eds.): *Proceedings of the Workshop on 'Clifford Algebras and their Applications in Mathematical Physics' (Canterbury 1985)*. Reidel, Dordrecht, 1986.

P.R. Holland: Relativistic algebraic spinors and quantum motions in phase space. *Found. Phys.* **16** (1986), 708-709.

G. Juvet: Opérateurs de Dirac et équations de Maxwell. *Comment. Math. Helv.* **2** (1930), 225-235.

J. Keller, S. Rodríguez-Romo: A multivectorial Dirac equation. *J. Math. Phys.* **31** (1990), 2501-2510.

J. Keller, F. Viniegra: The multivector structure of the matter and interaction field theories. pp. 437-445 in A. Micali et al. (eds.): *Proceedings of the Second Workshop on 'Clifford Algebras and their Applications in Mathematical Physics' (Montpellier 1989)*. Reidel, Dordrecht, The Netherlands, 1992.

P. Kustaanheimo, E. Stiefel: Perturbation theory of Kepler motion based on spinor regularization. *J. Reine Angew. Math.* **218** (1965), 204-219.

R. Penrose, W. Rindler: *Spinors and Space-Time*, Vol. 1. Cambridge University Press, Cambridge, 1984.

M. Riesz: Sur certaines notions fondamentales en théorie quantique relativiste; pp. 123-148 in *C. R. 10e Congrès Math. Scandinaves (Copenhagen, 1946)*. Jul. Gjellerups Forlag, Copenhagen, 1947. Reprinted in L. Gårding, L. Hörmander (eds.): *Marcel Riesz, Collected Papers*. Springer, Berlin, 1988, pp. 545-570.

F. Sauter: Lösung der Diracschen Gleichungen ohne Spezialisierung der Diracschen Operatoren. *Z. Phys.* **63** (1930), 803-814.

11

Fierz Identities and Boomerangs

Fierz identities are quadratic relations between the bilinear covariants (or physical observables) of a Dirac spinor. They are used to recover the original Dirac spinor from its bilinear covariants, up to a phase. The Fierz identities are sufficient to examine the non-null case, when either $\psi^\dagger\gamma_0\psi \neq 0$ or $\psi^\dagger\gamma_0\gamma_{0123}\psi \neq 0$. However, they are insufficient for the null case when both $\psi^\dagger\gamma_0\psi = 0$ and $\psi^\dagger\gamma_0\gamma_{0123}\psi = 0$. In this chapter, we introduce a new object called the **boomerang**, which enables us to study also the null case.

11.1 Fierz identities

The bilinear covariants satisfy certain quadratic relations called **Fierz identities** [see Holland 1986 p. 276 (2.8)]

$$\mathbf{J}^2 = \Omega_1^2 + \Omega_2^2, \qquad \mathbf{K}^2 = -\mathbf{J}^2$$

$$\mathbf{J}\cdot\mathbf{K} = 0, \qquad \mathbf{J}\wedge\mathbf{K} = -(\Omega_2 + \Omega_1\gamma_{0123})\mathbf{S}.$$

In coordinate form the Fierz identities are as follows [see Crawford 1985 p. 1439 (1.2)]

$$J_\mu J^\mu = \Omega_1^2 + \Omega_2^2, \qquad J_\mu J^\mu = -K_\mu K^\mu$$

$$J_\mu K^\mu = 0, \qquad J_\mu K_\nu - K_\mu J_\nu = -\Omega_2 S_{\mu\nu} + \Omega_1(\star S)_{\mu\nu}$$

where $(\star S)_{\mu\nu} = -\frac{1}{2}\varepsilon_{\mu\nu\alpha\beta}S^{\alpha\beta}$ (with $\varepsilon_{0123} = 1$) or $\star\mathbf{S} = \tilde{\mathbf{S}}\gamma_{0123}$ [in general, $\star v = \tilde{v}\gamma_{0123}$ given by $u\wedge\star v = \langle u, v\rangle\gamma_0\wedge\gamma_1\wedge\gamma_2\wedge\gamma_3$].

In the non-null case $\Omega \neq 0$ the Fierz identities result in [Crawford 1985 p. 1439 (1.3) and 1986 p. 356 (2.14)]

$$\mathbf{S}\,\llcorner\,\mathbf{J} = \Omega_2\mathbf{K}, \qquad \mathbf{S}\,\llcorner\,\mathbf{K} = \Omega_2\mathbf{J}$$

$$(\gamma_{0123}\mathbf{S})\,\llcorner\,\mathbf{J} = \Omega_1\mathbf{K}, \qquad (\gamma_{0123}\mathbf{S})\,\llcorner\,\mathbf{K} = \Omega_1\mathbf{J}$$

$$\mathbf{S} \,\lrcorner\, \mathbf{S} = \Omega_2^2 - \Omega_1^2, \qquad (\gamma_{0123}\mathbf{S}) \,\lrcorner\, \mathbf{S} = 2\Omega_1\Omega_2$$

and

$$\begin{aligned}
&\mathbf{JS} = -(\Omega_2 + \Omega_1\gamma_{0123})\mathbf{K}, \quad \mathbf{KS} = -(\Omega_2 + \Omega_1\gamma_{0123})\mathbf{J} \\
&\mathbf{SJ} = (\Omega_2 - \Omega_1\gamma_{0123})\mathbf{K}, \quad \mathbf{SK} = (\Omega_2 - \Omega_1\gamma_{0123})\mathbf{J} \\
&\mathbf{S}^2 = (\Omega_2 - \Omega_1\gamma_{0123})^2 = \Omega_2^2 - \Omega_1^2 - 2\Omega_1\Omega_2\gamma_{0123} \\
&\mathbf{S}^{-1} = -\mathbf{S}(\Omega_1 - \Omega_2\gamma_{0123})^2/(\Omega_1^2 + \Omega_2^2)^2 = \mathbf{KSK}/(\Omega_1^2 + \Omega_2^2)^2.
\end{aligned}$$

In the index-notation some of these identities look like

$$\begin{aligned}
&J_\mu S^{\mu\nu} = -\Omega_2 K^\nu, \qquad J_\mu(\star S)^{\mu\nu} = \Omega_1 K^\nu \\
&\mathbf{S} \,\lrcorner\, \mathbf{S} = -\tfrac{1}{2}S_{\mu\nu}S^{\mu\nu} = \Omega_2^2 - \Omega_1^2 \\
&(\star\mathbf{S}) \,\lrcorner\, \mathbf{S} = -\tfrac{1}{2}(\star S)_{\mu\nu}S^{\mu\nu} = \tfrac{1}{4}\varepsilon_{\mu\nu\alpha\beta}S^{\mu\nu}S^{\alpha\beta} = -2\Omega_1\Omega_2.
\end{aligned}$$

Note also that in general $\mathbf{S} \,\llcorner\, \mathbf{K} = -\mathbf{K} \,\lrcorner\, \mathbf{S}$, $\gamma_{0123}(\mathbf{S} \wedge \mathbf{K}) = (\gamma_{0123}\mathbf{S}) \,\llcorner\, \mathbf{K} = -(\star\mathbf{S}) \,\llcorner\, \mathbf{K} = \mathbf{K} \,\lrcorner\, (\star\mathbf{S})$, $\gamma_{0123}(\mathbf{S} \wedge \mathbf{S}) = (\gamma_{0123}\mathbf{S}) \,\lrcorner\, \mathbf{S} = -(\star\mathbf{S}) \,\lrcorner\, \mathbf{S}$ and that $(\mathbf{J} \,\lrcorner\, \mathbf{S}) \wedge \mathbf{S} = \frac{1}{2}\mathbf{J} \,\lrcorner\, (\mathbf{S} \wedge \mathbf{S})$.

Fierz identities via spinor operators. By direct computation we can see that

$$\begin{aligned}
\mathbf{J}^2 &= (\Psi\gamma_0\tilde{\Psi})(\Psi\gamma_0\tilde{\Psi}) = \Psi\gamma_0\Psi\tilde{\Psi}\gamma_0\tilde{\Psi} = \Psi\gamma_0(\Omega_1 + \Omega_2\gamma_{0123})\gamma_0\tilde{\Psi} \\
&= \Psi(\Omega_1 - \Omega_2\gamma_{0123})\gamma_0\gamma_0\tilde{\Psi} = (\Omega_1 - \Omega_2\gamma_{0123})\Psi\tilde{\Psi} \\
&= (\Omega_1 - \Omega_2\gamma_{0123})(\Omega_1 + \Omega_2\gamma_{0123}) = \Omega_1^2 + \Omega_2^2
\end{aligned}$$

which gives one of the Fierz identities. Computing in a similar manner we find

$$\begin{aligned}
\mathbf{JK} &= (\Psi\gamma_0\tilde{\Psi})(\Psi\gamma_3\tilde{\Psi}) = \Psi\gamma_0\Psi\tilde{\Psi}\gamma_3\tilde{\Psi} = \Psi\gamma_0(\Omega_1 + \Omega_2\gamma_{0123})\gamma_3\tilde{\Psi} \\
&= \Psi(\Omega_1 - \Omega_2\gamma_{0123})\gamma_0\gamma_3\tilde{\Psi} = -(\Omega_1 - \Omega_2\gamma_{0123})\Psi\gamma_{0123}\gamma_{12}\tilde{\Psi} \\
&= -(\Omega_1 - \Omega_2\gamma_{0123})\gamma_{0123}\Psi\gamma_{12}\tilde{\Psi} = -(\Omega_2 + \Omega_1\gamma_{0123})\mathbf{S}.
\end{aligned}$$

Since the result is a bivector, we find that $\mathbf{J} \wedge \mathbf{K} = -(\Omega_2 + \Omega_1\gamma_{0123})\mathbf{S}$ and $\mathbf{J} \cdot \mathbf{K} = 0$. ∎

Exercise 1

11.2 Recovering a spinor from its bilinear covariants

Let the spinor ψ have bilinear covariants Ω_1, $\mathbf{J}$, $\mathbf{S}$, $\mathbf{K}$, Ω_2 [a scalar, a vector, a bivector, a vector, a scalar]. Take an arbitrary spinor η such that $\tilde{\eta}^*\psi \neq 0$

in $\mathbb{C}\otimes \mathcal{C}\ell_{1,3}$ or equivalently $\eta^\dagger\gamma_0\psi \neq 0$ in $\mathrm{Mat}(4,\mathbb{C})$. Then the spinor ψ is proportional to

$$\psi \simeq Z\eta \quad \text{where} \quad \boxed{Z = \Omega_1 + \mathbf{J} + i\mathbf{S} + i\mathbf{K}\gamma_{0123} + \Omega_2\gamma_{0123}}$$

that is, ψ and $Z\eta$ differ only by a complex factor. The original spinor ψ can be recovered by the algorithm [see Takahashi 1983, and Crawford 1985, who also gave a proof in the non-null case $\Omega \neq 0$]

$$N = \sqrt{\langle\tilde{\eta}^* Z\eta\rangle_0} = \frac{1}{2}\sqrt{\eta^\dagger\gamma_0 Z\eta}$$
$$e^{-i\alpha} = \frac{4}{N}\langle\tilde{\eta}^*\psi\rangle_0 = \frac{1}{N}\eta^\dagger\gamma_0\psi$$
$$\psi = \frac{1}{4N}e^{-i\alpha}Z\eta.$$

[For the choice $\eta = f$ we get simply

$$N = \sqrt{\langle Zf\rangle_0} = \frac{1}{2}\sqrt{\Omega_1 + \mathbf{J}\cdot\gamma_0 - \mathbf{S}\,\lrcorner\,\gamma_{12} - \mathbf{K}\cdot\gamma_3}$$
$$e^{-i\alpha} = \frac{\psi_1}{|\psi_1|},$$

which are not the same N, $e^{-i\alpha}$ as those for an arbitrary η.] Once the spinor ψ has been recovered, we may also write

$$N = 4|\langle\tilde{\eta}^*\psi\rangle_0| = |\eta^\dagger\gamma_0\psi|$$
$$e^{-i\alpha} = \frac{\langle\tilde{\eta}^*\psi\rangle_0}{|\langle\tilde{\eta}^*\psi\rangle_0|} = \frac{\eta^\dagger\gamma_0\psi}{|\eta^\dagger\gamma_0\psi|}.$$

A spinor ψ is determined by its bilinear covariants Ω_1, $\mathbf{J}$, $\mathbf{S}$, $\mathbf{K}$, Ω_2 up to a phase-factor $e^{-i\alpha}$, and

$$Z = \Omega_1 + \mathbf{J} + i\mathbf{S} + i\mathbf{K}\gamma_{0123} + \Omega_2\gamma_{0123}$$

projects/extracts out of η the relevant part parallel to ψ.

Recovery via mother spinors $\Phi \in \mathcal{C}\ell_{1,3}\frac{1}{2}(1+\gamma_0)$. Take two arbitrary elements in the real Clifford algebra, $a, b \in \mathcal{C}\ell_{1,3}$, in such a way that $\psi = (a+ib)f$, $f = \frac{1}{2}(1+\gamma_0)\frac{1}{2}(1+\gamma_{12})$. Then $\psi\bar{\psi}^* = 0$ and

$$\begin{aligned}\psi\tilde{\psi}^* &= (a+ib)f(\tilde{a}-i\tilde{b}) = af\tilde{a} + bf\tilde{b} + i(bf\tilde{a} - af\tilde{b})\\ &= \tfrac{1}{2}(ag\tilde{a} + bg\tilde{b} - bg\gamma_{12}\tilde{a} + ag\gamma_{12}\tilde{b} + i(ag\gamma_{12}\tilde{a} + bg\gamma_{12}\tilde{b} + bg\tilde{a} - ag\tilde{b}))\end{aligned}$$

where we have written $g = \frac{1}{2}(1+\gamma_0)$. Next, we introduce a real spinor, called the *mother spinor* [for all real spinors]

$$\Phi = (a - b\gamma_{12})(1+\gamma_0) \in \mathcal{C}\ell_{1,3}\frac{1}{2}(1+\gamma_0).$$

Compute $\Phi\bar{\Phi} = 0$ and

$$\Phi\tilde{\Phi} = 4(a - b\gamma_{12})g(\tilde{a} + \gamma_{12}\tilde{b}) = 4(ag\tilde{a} + ag\gamma_{12}\tilde{b} - bg\gamma_{12}\tilde{a} + bg\tilde{b})$$

to find

$$\frac{1}{2}\Phi\tilde{\Phi} = 4\,\mathrm{Re}(\psi\tilde{\psi}^*), \quad \text{and similarly} \quad \frac{1}{2}\Phi\gamma_{12}\tilde{\Phi} = 4\,\mathrm{Im}(\psi\tilde{\psi}^*).$$

Recall that $Z = 4\psi\tilde{\psi}^*$ is sufficient to reconstruct the original Dirac spinor ψ and conclude that the real mother spinor $\Phi \in \mathcal{C}\ell_{1,3}\,\frac{1}{2}(1+\gamma_0)$ carries all the physically relevant information of the Dirac spinor ψ. In fact,

$$\psi = \Phi\,\frac{1}{4}(1 + i\gamma_{12}) \quad \text{and} \quad \Phi = 4\,\mathrm{Re}(\psi)$$

where the real part is taken in the decomposition $\mathbb{C} \otimes \mathcal{C}\ell_{1,3}$ [and *not* in the decomposition $\mathbb{C} \otimes \mathrm{Mat}(4, \mathbb{R})$].

Write as before $Z = P + iQ$ where $P = \Omega + \mathbf{J}$ and $Q = \mathbf{S} + \mathbf{K}\gamma_{0123}$. We will show how to recover the real mother spinor Φ from its bilinear covariants $[\tilde{g} = g = \frac{1}{2}(1+\gamma_0)]$:

$$N = \sqrt{\tfrac{1}{2}\langle\tilde{g}(P - Q\gamma_{12})g\rangle_0}, \qquad e^{\gamma_{12}\alpha} = \frac{1}{N}((\tilde{g}\phi) \wedge \gamma_{03})\gamma_{03}^{-1},$$

$$\Phi = \frac{1}{2N}(P - Q\gamma_{12})e^{\gamma_{12}\alpha}g$$

[the same N as for the choice $f \in \mathbb{C} \otimes \mathcal{C}\ell_{1,3}$] or for an arbitrary spinor $\eta \in \mathcal{C}\ell_{1,3}g$, $\tilde{\eta}\Phi \neq 0$,

$$N = \sqrt{\tfrac{1}{2}\langle\tilde{\eta}(P\eta - Q\eta\gamma_{12})\rangle_0}, \qquad e^{\gamma_{12}\alpha} = \frac{1}{N}((\tilde{\eta}\Phi) \wedge \gamma_{03})\gamma_{03}^{-1},$$

$$\Phi = \frac{1}{2N}(P\eta - Q\eta\gamma_{12})e^{\gamma_{12}\alpha}g$$

[$\eta' \in (\mathbb{C} \otimes \mathcal{C}\ell_{1,3})f$ and $\eta = 2\,\mathrm{Re}(\eta')$ result in the same numerical value for N]. Note that the role of $i = \sqrt{-1}$ is played by multiplication by $\gamma_2\gamma_1$ on the *right* hand side, that is, $\Phi\gamma_2\gamma_1 = 4\,\mathrm{Re}(i\psi)$.

Exercises 2,3,4

11.3 Fierz identities and the recovery of spinors

It might be interesting to know if given multivectors Ω_1, $\mathbf{J}$, $\mathbf{S}$, $\mathbf{K}$, Ω_2 [a scalar, a vector, a bivector, a vector, a scalar] are bilinear covariants for some spinor. The answer is postponed to the next section in the case $\Omega_1 = 0 = \Omega_2$. Writing $\Omega = \Omega_1 + \Omega_2\gamma_{0123}$, we are left with the remaining case $\Omega \neq 0$, in which we can say that the multivectors are bilinear covariants essentially if they satisfy the Fierz identities.

If the multivectors Ω_1, $\mathbf{J}$, $\mathbf{S}$, $\mathbf{K}$, Ω_2 satisfy the Fierz identities, then their aggregate

$$Z = \Omega_1 + \mathbf{J} + i\mathbf{S} + i\mathbf{K}\gamma_{0123} + \Omega_2\gamma_{0123}$$

can be factored as [see Crawford 1985 p. 1439 (2.2)] [1]

$$Z = (\Omega_1 + \mathbf{J} + \Omega_2\gamma_{0123})(1 + i(\Omega_1 + \Omega_2\gamma_{0123})^{-1}\mathbf{K}\gamma_{0123}).$$

This factorization is valid only in the non-null case $\Omega \neq 0$. Using this factorization Crawford proved that if the multivectors Ω_1, $\mathbf{J}$, $\mathbf{S}$, $\mathbf{K}$, Ω_2 satisfy the Fierz identities [and $J^0 > 0$ with $4\langle\tilde{\eta}^* Z\eta\rangle_0 = \eta^\dagger\gamma_0 Z\eta > 0$ for all non-zero spinors η], then Ω_1, $\mathbf{J}$, $\mathbf{S}$, $\mathbf{K}$, Ω_2 are bilinear covariants for some spinor ψ, for instance,

$$\psi = \frac{1}{4N}Z\eta, \qquad N = \sqrt{\langle\tilde{\eta}^* Z\eta\rangle_0} = \frac{1}{2}\sqrt{\eta^\dagger\gamma_0 Z\eta}$$

[and two such spinors ψ obtained by distinct choices of η differ only in their phases].

Hamilton 1984 p. 1827 (4.2) mentioned how ψ determines $Z = 4\psi\psi^\dagger\gamma_0$, see also Holland 1986 p. 276 (2.9), Keller & Rodríguez-Romo 1990 p. 2502 (2.3b) and Hestenes 1986 p. 334 (2.28).

11.4 Boomerangs

Definition. If the multivectors Ω_1, $\mathbf{J}$, $\mathbf{S}$, $\mathbf{K}$, Ω_2 [a scalar, a vector, a bivector, a vector, a scalar] satisfy the Fierz identities, then their aggregate $Z = \Omega_1 + \mathbf{J} + i\mathbf{S} + i\mathbf{K}\gamma_{0123} + \Omega_2\gamma_{0123}$ is called a *Fierz aggregate*. ∎

Definition. A multivector $Z = \Omega_1 + \mathbf{J} + i\mathbf{S} + i\mathbf{K}\gamma_{0123} + \Omega_2\gamma_{0123}$, which is Dirac self-adjoint $\tilde{Z}^* = Z$, is called a *boomerang*, if its components Ω_1, $\mathbf{J}$, $\mathbf{S}$, $\mathbf{K}$, Ω_2 are bilinear covariants for some spinor $\psi \in \mathbb{C}^4$. ∎

Both in the non-null case $\Omega \neq 0$ and in the null case $\Omega = 0$ a spinor ψ is determined up to a phase-factor by its aggregate of bilinear covariants $Z = \Omega_1 + \mathbf{J} + i\mathbf{S} + i\mathbf{K}\gamma_{0123} + \Omega_2\gamma_{0123}$ [as $\psi = \frac{1}{4N}e^{-i\alpha}Z\eta$], which in turn is determined by the original spinor ψ via the formula $Z = 4\psi\tilde{\psi}^* = 4\psi\psi^\dagger\gamma_0$ [thus we have a boomerang, which comes back].

If Z is a boomerang so that $Z = 4\psi\psi^\dagger\gamma_0$ then $Z^2 = 4\Omega_1 Z$ where $\Omega_1 = \langle Z\rangle_0$, because

---

1 In Crawford's factorization $Z = (\Omega + \mathbf{J})(1 + i\Omega^{-1}\mathbf{K}\gamma_{0123})$ the first factor $P = \Omega + \mathbf{J}$ is Dirac self-adjoint, $\tilde{P}^* = P$. Writing $\Gamma = 1 + i\mathbf{K}\Omega^{-1}\gamma_{0123}$, we can write Crawford's factorization as $Z = P\tilde{\Gamma}^*$, and note that $P\tilde{\Gamma}^* = \Gamma P \neq P\Gamma$. Crawford 1985 posed an open problem of decomposing Z into a product of two commuting Dirac self-adjoint factors. This problem is solved at the end of this chapter.

$$\begin{aligned} Z^2 &= 4\psi\psi^\dagger\gamma_0\, 4\psi\psi^\dagger\gamma_0 = 16\psi(\psi^\dagger\gamma_0\psi)\psi^\dagger\gamma_0 \\ &= 16\,\mathrm{trace}(\psi^\dagger\gamma_0\psi)\psi\psi^\dagger\gamma_0 \qquad [\text{since } \psi^\dagger\gamma_0\psi = \mathrm{trace}(\psi^\dagger\gamma_0\psi)f] \\ &= 16\,\mathrm{trace}(\psi\psi^\dagger\gamma_0)\psi\psi^\dagger\gamma_0 = \mathrm{trace}(4\psi\psi^\dagger\gamma_0)\, 4\psi\psi^\dagger\gamma_0. \end{aligned}$$

Conversely if $\Omega_1 \neq 0$ then $Z^2 = 4\Omega_1 Z$ ensures a boomeranging Z. If Z is a Fierz aggregate and $\Omega \neq 0$, then it boomerangs back to Z. Crawford's results say that in the non-null case $\Omega \neq 0$ we have a boomeranging Z if and only if Z is a Fierz aggregate. However, in the null case $\Omega = 0$, there are such Z which are Fierz aggregates but still do not boomerang [for instance $Z = \mathbf{J}$, $\mathbf{J}^2 = 0$, $\mathbf{J} \neq 0$].

If $\Omega = 0$ and $\mathbf{J}$, $\mathbf{S}$, $\mathbf{K}$ satisfy the Fierz identities, then for a spinor constructed by

$$\psi = \frac{1}{4N} Z\eta \quad \text{where} \quad Z = \mathbf{J} + i\mathbf{S} + i\mathbf{K}\gamma_{0123}$$

we have in general $Z \neq 4\psi\tilde{\psi}^*$ (the Fierz identities are reduced to $\mathbf{J}^2 = \mathbf{K}^2 = 0$, $\mathbf{J} \cdot \mathbf{K} = \mathbf{J} \wedge \mathbf{K} = 0$ which impose no restriction on $\mathbf{S}$). Even if the Fierz identities were supplemented by all the conditions presented in section 12.2 (in the non-null case these conditions are consequences of the Fierz identities) these extended identities would not result in a boomeranging Z. To handle also the null case $\Omega = 0$ we could replace the Fierz identities by the more restrictive conditions

$$\begin{aligned} &Z^2 = 4\Omega_1 Z, \qquad Z\gamma_\mu Z = 4J_\mu Z, \qquad Zi\gamma_{\mu\nu} Z = 4S_{\mu\nu} Z, \\ &Zi\gamma_{0123}\gamma_\mu Z = 4K_\mu Z, \qquad Z\gamma_{0123} Z = -4\Omega_2 Z \end{aligned}$$

[see Crawford 1986 p. 357 (2.16)], but this would result in a tedious checking process. If $Z = \mathbf{J} + i\mathbf{S} + i\mathbf{K}\gamma_{0123}$ is a boomerang, then $Z^2 = 0$, and so each dimension degree vanishes,

$\langle Z^2\rangle_0 = \mathbf{J}^2 - \mathbf{S} \lrcorner \mathbf{S} - \mathbf{K}^2$	
$\langle Z^2\rangle_1 = +2\gamma_{0123}(\mathbf{S} \wedge \mathbf{K})$	$\mathbf{K}$ in the plane of $\mathbf{S}$
$\langle Z^2\rangle_2 = +i2\gamma_{0123}(\mathbf{J} \wedge \mathbf{K})$	$\mathbf{J}$ and $\mathbf{K}$ are parallel
$\langle Z^2\rangle_3 = +i2\mathbf{J} \wedge \mathbf{S}$	$\mathbf{J}$ in the plane of $\mathbf{S}$
$\langle Z^2\rangle_4 = -\mathbf{S} \wedge \mathbf{S}$	$\mathbf{S}$ is simple.

The bivector part implies that $\mathbf{J}$ and $\mathbf{K}$ are parallel, the 4-vector part implies that $\mathbf{S}$ is simple, and the vector and 3-vector parts imply that $\mathbf{J}$ and $\mathbf{K}$ are in the plane of $\mathbf{S}$. Altogether we must have

$$\boxed{Z = \mathbf{J}(1 + is + ih\gamma_{0123})}$$

where h is a real number and $\mathbf{s}$ is a space-like vector orthogonal to $\mathbf{J}$, $\mathbf{J}\cdot\mathbf{s}=0$. We again compute $Z^2=\mathbf{J}^2(1+(\mathbf{s}+h\gamma_{0123})^2)=0$ and conclude that either

1. $\mathbf{J}^2=0$ or else
2. $(\mathbf{s}+h\gamma_{0123})^2=-1$.

Neither condition alone is sufficient to force Z to become a boomerang [Z is not even a Fierz aggregate if $\mathbf{J}^2\neq 0$]. However, such a Z is a boomerang if both conditions are satisfied simultaneously.

Counter-examples. 1. In the case $\Omega_1=0$, the element $Z=\mathbf{J}-\Omega_2\gamma_{0123}$, $\mathbf{J}^2=\Omega_2^2>0$, is such that $Z^2=0$, but Z is not a Fierz aggregate.

2. $Z=\mathbf{J}+i\mathbf{S}$ with $\mathbf{J}^2>0$, $\mathbf{S}=\gamma_{0123}\mathbf{J}\mathbf{s}$, $\mathbf{J}\cdot\mathbf{s}=0$, $\mathbf{s}^2=-1$, is not aFierz aggregate, and $Z^2\neq 0$, but we have $Z\gamma_{0123}Z=0$.

3. $Z=\mathbf{J}+i\mathbf{S}+i\mathbf{K}\gamma_{0123}$ where $\mathbf{J}^2=\mathbf{K}^2=0$, $\mathbf{J}\cdot\mathbf{K}=0$, $\mathbf{J}\wedge\mathbf{K}=0$, $\mathbf{S}\wedge\mathbf{S}\neq 0$, is a Fierz aggregate but does not satisfy $Z^2=0$, $Z\gamma_{0123}Z=0$.

4. $Z=\mathbf{J}(1+i\mathbf{s}+ih\gamma_{0123})$ with $\mathbf{J}^2=0$, $\mathbf{J}\cdot\mathbf{s}=0$, $(\mathbf{s}+h\gamma_{0123})^2\neq -1$, is a Fierz aggregate and satisfies $Z^2=0$ and $Z\gamma_{0123}Z=0$, but still we do not have a boomeranging Z. ∎

Throughout this chapter we assume that Ω_1, $\mathbf{J}$, $\mathbf{S}$, $\mathbf{K}$, Ω_2 are real multivectors or equivalently that $Z=\Omega_1+\mathbf{J}+i\mathbf{S}+i\mathbf{K}\gamma_{0123}+\Omega_2\gamma_{0123}$ is Dirac self-adjoint [$\tilde{Z}^*=Z$ or in matrix notation $\gamma_0 Z^\dagger\gamma_0=Z$]. This implies that $\eta^\dagger\gamma_0 Z\eta$ [$=4\langle\tilde{\eta}^* Z\eta\rangle_0$] is a real number for all spinors η.

For a boomerang Z we have $\eta^\dagger\gamma_0 Z\eta\geq 0$, for all spinors η, and also $J^0>0$ [the grade involute $\hat{Z}$ of Z is such that $\langle\hat{Z}\rangle_0\cdot\gamma_0<0$ and $4\langle\tilde{\eta}^*\hat{Z}\eta\rangle_0=\eta^\dagger\gamma_0\hat{Z}\eta\leq 0$].

Theorem. Let Z be such that $\eta^\dagger\gamma_0 Z\eta\geq 0$ for all spinors η, and that $J^0>0$. Then the following statements hold.

1. Z is a boomerang if and only if $Z\gamma^0\tilde{Z}^*=4J^0Z$ or equivalently $ZZ^\dagger\gamma^0=4J^0Z$.
2. In the non-null case $\Omega\neq 0$, Z is a boomerang if and only if it is a Fierz aggregate.
3. In the null case $\Omega=0$, Z **is a boomerang if and only if** $Z=\mathbf{J}(1+i\mathbf{s}+ih\gamma_{0123})$ where $\mathbf{J}$ is a null-vector, $\mathbf{J}^2=0$, $\mathbf{s}$ is a space-like vector, $\mathbf{s}^2<0$ or $\mathbf{s}=0$, orthogonal to $\mathbf{J}$, $\mathbf{J}\cdot\mathbf{s}=0$, and h is a real number such that $h=\pm\sqrt{1+\mathbf{s}^2}$, $|h|\leq 1$. ∎

The condition $Z\gamma^0\tilde{Z}^*=4J^0Z$ could also be written with an arbitrary time-like vector $\mathbf{v}$ as follows: $Z\mathbf{v}\tilde{Z}^*=4(\mathbf{v}\cdot\mathbf{J})Z$.

11.5 Decomposition and factorization of boomerangs

Write

$$P = \Omega_1 + \mathbf{J} + \Omega_2\gamma_{0123}, \qquad Q = \mathbf{S} + \mathbf{K}\gamma_{0123} = \mathbf{S} - \gamma_{0123}\mathbf{K}$$
$$\Sigma = 1 - i\mathbf{J}\mathbf{K}^{-1}\gamma_{0123} = 1 - i\mathbf{K}\mathbf{J}^{-1}\gamma_{0123} \quad [\text{when } \Omega \neq 0]$$

so that $Z = P + iQ = P\Sigma$. Then $P\Sigma = \Sigma P$ and we have found a solution to the open problem posed by Crawford 1985 p. 1441 ref. (10). [Crawford's second factor in (4.1),

$$1 + i(\Omega_1 + \Omega_2\gamma_{0123})^{-1}\mathbf{K}\gamma_{0123} = 1 - i(\Omega_2 - \Omega_1\gamma_{0123})^{-1}\mathbf{K}$$
$$= 1 - i\mathbf{S}^{-1}\mathbf{J} = 1 + i\mathbf{J}^{-1}\mathbf{S},$$

did not commute with P unless $\Omega_2 = 0$.] In the case $\Omega_1 \neq 0$ there is another factorization

$$Z = P\Big(1 + i\frac{1}{2\Omega_1}Q\Big) = P\frac{1}{2}\Big(1 + i\frac{1}{2\Omega_1}Q\Big)^2$$

where the factors commute and are Dirac self-adjoint, but in this factorization the second factor is not an idempotent (even though it behaves like one when multiplied by P).

11.6 Multiplication by the imaginary unit $i = \sqrt{-1}$

We have found that $i\psi = \psi\gamma_2\gamma_1$ $(\neq \psi\gamma_{0123})$ corresponds to $\Phi\gamma_2\gamma_1 = 4\,\mathrm{Re}(i\psi)$ and further to $\phi\gamma_2\gamma_1 = \Phi\gamma_2\gamma_1\frac{1}{2}(1-\gamma_{03})$. In the non-null case $\Omega \neq 0$ write

$$s = \gamma_{0123}\mathbf{J}\mathbf{K}^{-1} = \mathbf{J}(\mathbf{K}\gamma_{0123})^{-1} = -\Omega^{-1}\mathbf{S} = \Omega\mathbf{S}^{-1}$$
$$k = -(\Omega_2 + \Omega_1\gamma_{0123})^{-1}\mathbf{K} = \mathbf{J}\mathbf{S}^{-1} = -\mathbf{S}\mathbf{J}^{-1}.$$

[Using $\rho e_\mu = \Psi\gamma_\mu\tilde{\Psi}$, $\rho^2 = \Omega_1^2 + \Omega_2^2$ Hestenes 1986 p. 333 (2.23) gives $s = \gamma_{0123}e_3e_0$, Boudet 1985 p. 719 (2.6) gives $-s = e_1e_2$.] Note that

$$s = \Psi(-\gamma_{12})\Psi^{-1} = \Psi\gamma_2\gamma_1\Psi^{-1} \in \textstyle\bigwedge^2\mathbb{R}^{1,3} \qquad [\text{simple bivector}]$$
$$k = \Psi(-\gamma_{012})\Psi^{-1} = \Psi\gamma_{0123}\gamma_3\Psi^{-1} \in \mathbb{R}^{1,3} \oplus \textstyle\bigwedge^3\mathbb{R}^{1,3}$$

and $s^2 = -1$, $k^2 = -1$ and $sk = ks$. Both s and k play the role of the imaginary unit (multiplication on the left side):

$i\psi = s\psi = k\psi = \psi\gamma_2\gamma_1 \neq \psi\gamma_{0123}$	Dirac spinor
$s\Phi = k\Phi = \Phi\gamma_2\gamma_1$	mother spinor
$s\phi = k\phi = \phi\gamma_2\gamma_1 = \phi\gamma_{0123} \neq \phi(-\gamma_{012})$	ideal spinor
$s\Psi = k\Psi\gamma_0 = \Psi\gamma_2\gamma_1$	spinor operator

and $P = sQ = kQ$. [Hestenes 1986 p. 334 (2.24) reports $s\Psi = \Psi\gamma_2\gamma_1$ and also $s\phi = \phi\gamma_2\gamma_1 = \phi\gamma_{0123}$.]

Question

Do the conditions $Z^2 = 0$ and $Z\gamma_{0123}Z = 0$ imply that Z is a Fierz aggregate?

Exercises

1. Compute $\mathbf{K}^2$, when $\mathbf{K} = \Psi\gamma_3\tilde{\Psi}$.

Show that

2. $\mathrm{Im}(\psi) = \mathrm{Re}(\psi)\gamma_{12}$ in $\mathbb{C}\otimes \mathcal{C}\ell_{1,3}$.
3. $\frac{1}{2}\Phi\gamma_1\bar{\Phi} = 4\,\mathrm{Re}(\psi\gamma_1\bar{\psi}) = -4\,\mathrm{Im}(\psi\gamma_2\bar{\psi})$ [no complex conjugation]
   $\frac{1}{2}\Phi\gamma_2\bar{\Phi} = 4\,\mathrm{Re}(\psi\gamma_2\bar{\psi}) = 4\,\mathrm{Im}(\psi\gamma_1\bar{\psi})$.
4. $Q_k^2 = -2\Omega_1 P$, $Q_iQ_j = 2\Omega_1 Q_k$ (ijk cycl.) for $Q_k = \frac{1}{2}(\Phi\gamma_k\bar{\Phi})\gamma_{0123}$
   [$P, Q_1, Q_2, Q_3 = Q$ span a quaternion algebra when $\Omega_1 \neq 0$].

Show that for $Z = 4\psi\tilde{\psi}^*$, where $\psi \in (\mathbb{C}\otimes \mathcal{C}\ell_{1,3})f$, the following hold:

5. $Z\gamma^0 Z = 4J^0 Z$ ($\neq 0$ for $\psi \neq 0$).
6. $P\gamma^0 P = 2J^0 P = -Q\gamma^0 Q$.
7. $P\gamma_0 Q = Q\gamma_0 P$.
8. $\mathrm{even}(\tilde{Z})Z = 0$, $\mathrm{odd}(\tilde{Z})Z = 0$.
9. $\tilde{Z}Z = 0 \Rightarrow P^2 = -Q^2$, $PQ = QP$ [no complex conjugation].
10. $\tilde{Z}Z = 0$, $Z^2 = 4\Omega_1 Z \Rightarrow P^2 = 2\Omega_1 P$, $PQ = 2\Omega_1 Q$.
11. $\bar{Z}Z = 0 \Rightarrow \bar{P}P = \bar{Q}Q$, $\bar{P}Q = -\bar{Q}P$.
12. $\bar{Z}Z = 0$, $\bar{Z}^*Z = 4\Omega_2\gamma_{0123}Z \Rightarrow \bar{P}P = 2\Omega_2\gamma_{0123}P$, $\bar{P}Q = 2\Omega_2\gamma_{0123}Q$.

Write

$$K = \Omega_1 + i\mathbf{K}\gamma_{0123} + \Omega_2\gamma_{0123}, \qquad S = \Omega_1 + i\mathbf{S} + \Omega_2\gamma_{0123}$$
$$\Pi = P(\Omega_1 + \Omega_2\gamma_{0123})^{-1}, \qquad \Gamma = K(\Omega_1 + \Omega_2\gamma_{0123})^{-1}$$

and show that:

13. $\Sigma = S(\Omega_1 + \Omega_2\gamma_{0123})^{-1} = 1 - i\mathbf{s}$.
14. $\Gamma = 1 - i\mathbf{k}$, $\tilde{\Gamma}^* = 1 + i\tilde{\mathbf{k}} = 1 - i(\Omega_1 + \Omega_2\gamma_{0123})^{-1}\gamma_{0123}\mathbf{K}$.
15. $Z = K\Sigma = \Sigma K = \Gamma S = S\tilde{\Gamma}^*$.
16. $Z = \Pi S = S\tilde{\Pi} = \Pi K = K\tilde{\Pi}$ [no complex conjugation needed]
   $= \Gamma P = P\tilde{\Gamma}^*$ [this is Crawford's factorization]
   [$= P\Gamma$ only if $\Omega_2 = 0$ since then $\Gamma = \tilde{\Gamma}^*$].

Bibliography

R. Boudet: Conservation laws in the Dirac theory. *J. Math. Phys.* **26** (1985), 718-724.

R. Boudet: Les algèbres de Clifford et les transformations des multivecteurs; pp. 343-352 in A. Micali et al. (eds.): *Proceedings of the Second Workshop on 'Clifford Algebras and their Applications in Mathematical Physics' (Montpellier, 1989).* Kluwer, Dordrecht, The Netherlands, 1992.

J.P. Crawford: On the algebra of Dirac bispinor densities: Factorization and inversion theorems. *J. Math. Phys.* **26** (1985), 1439-1441.

J.P. Crawford: Dirac equation for bispinor densities; pp. 353-361 in J.S.R. Chisholm, A.K. Common (eds.): *Proceedings of the Workshop on 'Clifford Algebras and their Applications in Mathematical Physics' (Canterbury 1985).* Reidel, Dordrecht, 1986.

J.D. Hamilton: The Dirac equation and Hestenes' geometric algebra. *J. Math. Phys.* **25** (1984), 1823-1832.

D. Hestenes: Clifford algebra and the interpretation of quantum mechanics; pp. 321-346 in J.S.R. Chisholm, A.K. Common (eds.): *Proceedings of the Workshop on 'Clifford Algebras and their Applications in Mathematical Physics' (Canterbury 1985).* Reidel, Dordrecht, The Netherlands, 1986.

P.R. Holland: Minimal ideals and Clifford algebras in the phase space representation of spin-$\frac{1}{2}$ fields; pp. 273-283 in J.S.R. Chisholm, A.K. Common (eds.): *Proceedings of the Workshop on 'Clifford Algebras and their Applications in Mathematical Physics' (Canterbury 1985).* Reidel, Dordrecht, 1986.

P.R. Holland: Relativistic algebraic spinors and quantum motions in phase space. *Found. Phys.* **16** (1986), 708-709.

J. Keller, S. Rodríguez-Romo: A multivectorial Dirac equation. *J. Math. Phys.* **31** (1990), 2501-2510.

J. Keller, F. Viniegra: The multivector structure of the matter and interaction field theories; pp. 437-445 in A. Micali et al. (eds.): *Proceedings of the Second Workshop on 'Clifford Algebras and their Applications in Mathematical Physics' (Montpellier 1985).* Reidel, Dordrecht, The Netherlands, 1992.

Y. Takahashi: A passage between spinors and tensors. *J. Math. Phys.* **24** (1983), 1783-1790.

12

Flags, Poles and Dipoles

The classification of spinors is commonly based on representation theory, irreducible representations of the Lorentz group $SO_+(1,3)$. Thus, one customarily speaks about Dirac, Majorana and Weyl spinors. In this chapter spinors are classified in a new way by their bilinear covariants, multivectors of observables. The new classification is geometric, since it is based on multivectors, and physical, since it is based on observables. The classification reveals new spinors, called *flag-dipole spinors*, which reside between the Weyl, Majorana and Dirac spinors.

Dirac spinors describe the electron, and for them $\Omega \neq 0$. Weyl and Majorana spinors describe the neutrino. Weyl spinors are eigenspinors of the helicity projection operators $\frac{1}{2}(1 \pm \gamma_{0123})$, and their bilinear covariants satisfy $\Omega = 0$, $\mathbf{S} = 0$, $\mathbf{K} \neq 0$. Majorana spinors are eigenspinors of the charge conjugation operator, [1] with eigenvalues ± 1, and their bilinear covariants satisfy $\Omega = 0$, $\mathbf{S} \neq 0$, $\mathbf{K} = 0$. [Weyl and Majorana spinors are usually introduced by properties of matrices, see Benn & Tucker 1987 and Crumeyrolle 1990.]

The flag-dipole spinors satisfy $\Omega = 0$ [and cannot be Dirac spinors] and $\mathbf{S} \neq 0$, $\mathbf{K} \neq 0$ [and so they are neither Weyl nor Majorana spinors]. Unlike Weyl and Majorana spinors, the flag-dipole spinors do not form a real linear subspace, because they are characterized by a quadratic constraint. Therefore the superposition principle is violated, and the flag-dipole spinors cannot describe fermions. It has been conjectured (G. Trayling, Windsor) that the flag-dipole spinors are related to the quark confinement.

1 The charge conjugation operator C is conventionally defined by $C(\psi) = -i\gamma_2\psi^*$ for $\psi \in \mathbb{C}^4$.

12.1 Classification of spinors by their bilinear covariants

In the following we shall present a classification of spinors ψ based on properties of their bilinear covariants Ω_1, $\mathbf{J}$, $\mathbf{S}$, $\mathbf{K}$, Ω_2, collected as

$$Z = \Omega_1 + \mathbf{J} + i\mathbf{S} + i\mathbf{K}\gamma_{0123} + \Omega_2\gamma_{0123}.$$

In other words, we classify the boomerangs $Z = 4\psi\tilde{\psi}^*$.

Recall that $Z = P + iQ$, $P = \Omega_1 + \mathbf{J} + \Omega_2\gamma_{0123}$, $Q = \mathbf{S} + \mathbf{K}\gamma_{0123}$.

Dirac spinors of the electron:

1. $\underline{\Omega_1 \neq 0,\ \Omega_2 \neq 0}$: Using $P^2 = 2\Omega_1 P = -Q^2$ we find the relationship $P = \pm(-\frac{1}{2}Q^2)/\sqrt{\langle -\frac{1}{2}Q^2\rangle_0}$ where the sign is given by $J^0 > 0$ (and coincides with the sign of Ω_1). $P = kQ$, where $k = -(\Omega_2 + \Omega_1\gamma_{0123})^{-1}\mathbf{K}$, $i\psi = k\psi$.

2. $\underline{\Omega_1 \neq 0,\ \Omega_2 = 0}$: P is a multiple of $\frac{1}{2\Omega_1}(\Omega_1 + \mathbf{J})$ which looks like a proper *energy projection operator* and which commutes with the *spin projection operator* $\frac{1}{2}(1 - i\mathbf{K}\gamma_{0123}/\Omega_1)$. $Z = \Omega_1 + \mathbf{J} + i\mathbf{S} + i\mathbf{K}\gamma_{0123} = (\Omega_1 + \mathbf{J})(1 - i\gamma_{0123}\mathbf{K}/\Omega_1)$, $\mathbf{S} = \gamma_{0123}\mathbf{J}\mathbf{K}/\Omega_1$. $P = \gamma_{0123}\frac{1}{\Omega_1}\mathbf{K}Q$, $k = \gamma_{0123}\mathbf{K}/\Omega_1$. In this class the Yvon-Takabayasi angle β gets only two values, 0 and π; and the charge superselection rule applies.

3. $\underline{\Omega_1 = 0,\ \Omega_2 \neq 0}$: Using $P^2 = 2\Omega_1 P$ we find that P is nilpotent: $P^2 = 0$. $Z = \mathbf{J} + i\mathbf{S} + i\mathbf{K}\gamma_{0123} + \Omega_2\gamma_{0123}$, $\mathbf{S} = -\mathbf{J}\mathbf{K}/\Omega_2$. $P = -\frac{1}{\Omega_2}\mathbf{K}Q = \pm\mathbf{K}Q/\sqrt{-\mathbf{K}^2}$ (opposite to the sign of Ω_2), $k = -\mathbf{K}/\Omega_2$.

Singular spinors with a light-like pole/current:

4. $\underline{\Omega_1 = \Omega_2 = 0,\ \mathbf{K} \neq 0,\ \mathbf{S} \neq 0}$: *Flag-dipole spinors.* $Z = \mathbf{J} + i\mathbf{J}\mathbf{s} - ih\gamma_{0123}\mathbf{J}$, $\mathbf{J}^2 = 0$, $\mathbf{s}$ is a space-like vector, $\mathbf{s}^2 < 0$, orthogonal to $\mathbf{J}$, $\mathbf{J}\cdot\mathbf{s} = 0$, $\mathbf{S} = \mathbf{J}\mathbf{s}$ $(= \mathbf{J}\wedge\mathbf{s})$, $\mathbf{K} = h\mathbf{J}$, $h^2 = 1 + \mathbf{s}^2 < 1$ (h real, $h \neq 0$). $P = \mathbf{J}$, $Q = \mathbf{J}(\mathbf{s} + h\gamma_{0123})$, $(1 + i\mathbf{s} + ih\gamma_{0123})Z = 0$. Note that $\frac{1}{2}(1 - i\mathbf{s} - ih\gamma_{0123})\psi = \psi$ and $(1 + i\mathbf{s} + ih\gamma_{0123})\psi = 0$. $\tilde{Z}^* = Z$ and $Z^2 = 0$ imply $Z = \mathbf{J}(1 + i\mathbf{s} + ih\gamma_{0123})$ etc. Let $\psi = \frac{1}{4N}Z\eta$, then $Z = 4\psi\tilde{\psi}^*$ implies $(\mathbf{s} + h\gamma_{0123})^2 = -1$. $P = (\mathbf{s} + h\gamma_{0123})Q$, $i\psi = (\mathbf{s} + h\gamma_{0123})\psi$.

5. $\underline{\Omega_1 = \Omega_2 = \mathbf{K} = 0,\ \mathbf{S} \neq 0}$: *Flag-pole spinors* for which $Z = \mathbf{J} + i\mathbf{J}\mathbf{s}$ is a pole $\mathbf{J}$ plus a flag $\mathbf{S} = \mathbf{J}\mathbf{s}$ $(= \mathbf{J}\wedge\mathbf{s})$, $\mathbf{J}\cdot\mathbf{s} = 0$, $\mathbf{s}^2 = -1$. $P = \mathbf{s}Q$, $i\psi = \mathbf{s}\psi$. The flag-pole spinors are eigenspinors of the charge conjugation operator with eigenvalues $\lambda \in U(1)$, thus $C(\psi) = \lambda\psi$, $|\lambda| = 1$.
— Write $\mathbf{K}_k = \Psi\gamma_k\tilde{\Psi}$, $\mathbf{S}_k = \Psi\gamma_{ij}\tilde{\Psi}$ (ijk cycl.) with $\mathbf{K}_3 = \mathbf{K}$, $\mathbf{S}_3 = \mathbf{S}$. Then $\mathbf{K}_1 = \mathbf{J}$, $\mathbf{K}_2 = \mathbf{K}_3 = 0$ and $\mathbf{S}_1 = 0$, $\mathbf{S}_2 = \mathbf{J}\mathbf{s}_2$ $(= \mathbf{J}\wedge\mathbf{s}_2)$, $\mathbf{S}_3 = \mathbf{J}\mathbf{s}_3$ where $\mathbf{s}_3 = \mathbf{s}$, $\mathbf{s}_2^2 = -1$, $\mathbf{s}_2\cdot\mathbf{s}_3 = 0$.
— Given an arbitrary Dirac spinor ψ with covariants $\mathbf{J}$, $\mathbf{K}$ (and with $\mathbf{K}_1$, $\mathbf{S}_2$) we may construct, as special cases of flag-pole spinors, two *Majorana spinors*

$\psi_\pm = \frac{1}{2}(\psi \pm \psi_C)$, which are seen to be eigenspinors of the charge conjugation $C(\psi_\pm) = \pm\psi_\pm$, and whose bilinear covariants $\mathbf{J}_\pm$, $\mathbf{S}_\pm$ satisfy $\mathbf{K}\cdot\mathbf{J}_\pm = 0$, $\mathbf{K}\cdot\mathbf{S}_\pm = -\Omega_2\mathbf{J}_\pm$ and $\mathbf{J} = \mathbf{J}_+ + \mathbf{J}_-$, $\mathbf{S} = \mathbf{S}_+ + \mathbf{S}_-$. [Note that $\mathbf{J}_\pm = \frac{1}{2}(\mathbf{J} \pm \mathbf{K}_1)$ and $\mathbf{S}_\pm = \frac{1}{2}(\mathbf{S} \mp \mathbf{S}_2\gamma_{0123})$.] The charge conjugations of $i\psi_\pm$ are $C(i\psi_+) = -i\psi_+ \neq \pm\psi_-$ and $C(i\psi_-) = i\psi_- \neq \pm\psi_+$.

6. $\underline{\Omega_1 = \Omega_2 = \mathbf{S} = 0,\ \mathbf{K} \neq 0}$: *Weyl spinors* (of massless neutrinos) are eigenspinors of the chirality operator $\gamma_{0123}\psi_\pm = \pm i\psi_\pm$. $Z = \mathbf{J} \mp i\gamma_{0123}\mathbf{J}$, $\mathbf{J} = \pm\mathbf{K}$, $h = \pm 1$, $\psi_\pm = \frac{1}{2}(1 \mp i\gamma_{0123})\psi_\pm$. Note that $\text{even}(\psi_\pm) = \psi_\pm\frac{1}{2}(1 \mp \gamma_{03})$, $\text{odd}(\psi_\pm) = \psi_\pm\frac{1}{2}(1 \pm \gamma_{03})$. $P = \pm\gamma_{0123}Q$.
— Write $\mathbf{K}_k = \Psi\gamma_k\tilde{\Psi}$, $\mathbf{S}_k = \Psi\gamma_{ij}\tilde{\Psi}$ as before. Then $\mathbf{K}_1 = \mathbf{K}_2 = 0$, $\mathbf{S}_1 = \mathbf{J}\mathbf{s}_1$ $(= \mathbf{J}\wedge\mathbf{s}_1)$, $\mathbf{S}_2 = \mathbf{J}\mathbf{s}_2$ $(= \mathbf{J}\wedge\mathbf{s}_2)$ where $\mathbf{s}_1^2 = \mathbf{s}_2^2 = -1$, $\mathbf{s}_1\cdot\mathbf{s}_2 = 0$.
— Given an arbitrary Dirac spinor ψ with covariants $\mathbf{J}$, $\mathbf{K}$ we may construct two Weyl spinors $\psi_\pm = \frac{1}{2}(1 \mp i\gamma_{0123})\psi$ with covariants $\mathbf{J}_\pm = \frac{1}{2}(\mathbf{J} \pm \mathbf{K})$, $\mathbf{K}_\pm = \frac{1}{2}(\mathbf{K} \pm \mathbf{J})$. Weyl spinors are *pure*: $\tilde{\psi}_\pm\gamma_{0123}\gamma_\mu\psi_\pm = 0$ [no complex conjugation; for arbitrary Dirac spinors $\tilde{\psi}\psi = 0$, $\tilde{\psi}\gamma_\mu\psi = 0$, $\tilde{\psi}\gamma_{0123}\psi = 0$ though $\tilde{\psi}\gamma_{\mu\nu}\psi \neq 0$, $\tilde{\psi}\gamma_{0123}\gamma_\mu\psi \neq 0$ (and also $\bar{\psi}\psi = 0$, $\bar{\psi}\gamma_{0123}\gamma_\mu\psi = 0$, $\bar{\psi}\gamma_{0123}\psi = 0$, though $\bar{\psi}\gamma_\mu\psi \neq 0$, $\bar{\psi}\gamma_{\mu\nu}\psi \neq 0$)]. $C(\psi_\pm)$ is of helicity $h = \mp 1$ with covariants $\mathbf{J}_\pm$, $-\mathbf{K}_\pm$.

In addition to the above six classes *there are no other classes* based on distinctions between bilinear covariants. This can be seen by the following reasoning. First, we always have $\mathbf{J} \neq 0$, because $J^0 > 0$. Secondly, $\Omega \neq 0$ implies $\mathbf{S} \neq 0$ and $\mathbf{K} \neq 0$. Thirdly, $\Omega = 0$ implies $Z = \mathbf{J}(1 + i(\mathbf{s} + h\gamma_{0123}))$ where $(\mathbf{s} + h\gamma_{0123})^2 = -1$ so that we have a non-vanishing $\mathbf{J}(\mathbf{s} + h\gamma_{0123}) = \mathbf{S} + \mathbf{K}\gamma_{0123}$.

Comments:

For classes 1, 2 the element $\frac{1}{4\Omega_1}Z$ is a primitive idempotent in $\mathbb{C}\otimes\mathcal{C}\ell_{1,3}$.

Classes 1, 2, 3 are *Dirac spinors for the electron*. A spinor operator Ψ has a unique (up to a sign) *polar decomposition* $\Psi = \sqrt{\Omega}u$, $u \in \mathbf{Spin}_+(1,3)$. In particular, writing $\mathbf{K}_k = \Psi\gamma_k\tilde{\Psi}$ we have an orthogonal basis $\{\mathbf{J}, \mathbf{K}_1, \mathbf{K}_2, \mathbf{K}_3\}$ $(\mathbf{K}_3 = \mathbf{K})$ of $\mathbb{R}^{1,3}$.

Class 4 consists of *flag-dipole spinors* with a flag $\mathbf{S}$ on a dipole of two poles $\mathbf{J}$ and $\mathbf{K}$. Class 5 consists of *flag-pole spinors* with a flag $\mathbf{S}$ on a pole $\mathbf{J}$. Class 6 consists of *dipole spinors* with two poles $\mathbf{J}$ and $\mathbf{K}$.

In classes 4, 5, 6 the vectors $\mathbf{J}$, $\mathbf{K}_1$, $\mathbf{K}_2$, $\mathbf{K}_3$ no longer form a basis but collapse into a null-line $\mathbf{J}$ (also $\mathbf{S}_1$, $\mathbf{S}_2$, $\mathbf{S}_3$ intersect along $\mathbf{J}$). The even elements $\mathbf{J}(\gamma_0 - \mathbf{s}\gamma_{12} - h\gamma_3)$ and Ψ differ only up to a complex factor $x - y\gamma_{12}$ (on the right).

In addition to the electron (classes 1, 2, 3) the massless neutrino (class 6) has also been discussed by Hestenes [1967 p. 808 (8.13) and 1986 p. 343] who quite correctly observed that $\mathbf{J}_\pm = \frac{1}{2}\Psi(\gamma_0 \pm \gamma_3)\tilde{\Psi}$; note also that $\tilde{\Psi}_\pm \gamma_\mu \Psi_\pm = (\gamma_\mu \cdot \mathbf{J}_\pm)(\gamma_0 \mp \gamma_3)$. Hestenes has not discussed classes 4 and 5. Holland in *Found. Phys.* 1986, pp. 708-709, does not discuss classes 3, 4, 5, 6 with a nilpotent Z, $Z^2 = 0$, but focuses on a nilpotent ψ, $\psi^2 = 0$.

Majorana spinors $\Psi \in \mathcal{C}\ell^+_{1,3}\,\frac{1}{2}(1 \mp \gamma_{01})$ are not stable under the $U(1)$-gauge transformation $\Psi \to \Psi e^{\alpha\gamma_{12}} \notin \mathcal{C}\ell^+_{1,3}\,\frac{1}{2}(1 \mp \gamma_{01})$.

Given a Weyl spinor ψ with bilinear covariants $\mathbf{J}$, $\mathbf{K}$ we can associate to it two Majorana spinors $\psi_\pm = \frac{1}{2}(\psi \pm \psi_C)$ with *Penrose flags* $Z_\pm = \frac{1}{2}(\mathbf{J} \mp i\mathbf{S}_2\gamma_{0123})$. [2]

The number of parameters in the sets of bilinear covariants (or spinors without $U(1)$-gauge) is seen to be

class no.	1	2	3	4	5	6
parameters	7	6	6	5	4	3

If the $U(1)$-gauge is taken into consideration, then the number of parameters will be raised by one unit in all classes except in class 5 of Majorana spinors [Weyl spinors with $U(1)$-gauge and Majorana spinors both have four parameters and can be mapped bijectively onto each other – which enables Penrose flags also to be attached to Weyl spinors].

The Weyl and Majorana spinors can be written with spinor operators in the form

$$\Psi\frac{1}{2}(1 + \gamma_0\mathbf{u}) \qquad [\Psi \in \mathcal{C}\ell^+_{1,3}]$$

where $\mathbf{u} = \pm\gamma_3$ for Weyl spinors and $\mathbf{u} = \pm\gamma_1$ for Majorana spinors. The flag-pole spinors can be written in a similar form with $\mathbf{u} = \gamma_1 \cos\phi + \gamma_2 \sin\phi$. It is easy to see that all elements of the form $\Psi\frac{1}{2}(1 + \gamma_0\mathbf{u})$, $\Psi \in \mathcal{C}\ell^+_{1,3}$ are flag-dipole spinors, when $\mathbf{u}$ is a spatial unit vector, $\mathbf{u} \cdot \gamma_0 = 0$, $\mathbf{u}^2 = -1$, which is not on the γ_3-axis or in the $\gamma_1\gamma_2$-plane. About the converse the following has been presented:

Conjecture (C. Doran, 1995): All the flag-dipole spinors can be written in the form $\Psi\frac{1}{2}(1 + \gamma_0\mathbf{u})$, where $\Psi \in \mathcal{C}\ell^+_{1,3}$, $\mathbf{u} \in \mathbb{R}^3$, $\mathbf{u}^2 = -1$. ∎

When $\mathbf{u}$ varies in the unit sphere S^2 in $\mathbb{R}^3$ (orthogonal to γ_0), the flag-dipole spinor sweeps around the 'paraboloid' $\Psi\tilde{\Psi} = 0$. If the conjecture is true, it would be nice to know the relation between $\mathbf{s}, h$ and $\mathbf{u}$. [Clearly, $h = \mathbf{u} \cdot \gamma_3$.]

2 Our flag-pole $Z = \mathbf{J} + i\mathbf{S}$ is invariant under rotations $\Psi \to \Psi e^{\alpha\gamma_{12}}$, whereas the Penrose flags $Z_\pm = \frac{1}{2}(\mathbf{J} \mp i\mathbf{S}_2\gamma_{0123})$ make a 720° turn under a rotation of 360°.

12.2 Projection operators in $\mathrm{End}(\mathcal{C}\ell_{1,3})$

Write as before $P = \Omega_1 + \mathbf{J} + \Omega_2\gamma_{0123}$, $Q = \mathbf{S} + \mathbf{K}\gamma_{0123}$, $Z = P + iQ = P\Sigma = \Sigma P$, $\Sigma = 1 - i\gamma_{0123}\mathbf{J}\mathbf{K}^{-1}$. Then

$$\frac{1}{4\Omega_1}Z\psi = \psi, \quad \frac{1}{2\Omega_1}P\psi = \psi, \quad \text{when } \Omega_1 \neq 0$$
$$\frac{1}{2}\Sigma\psi = \psi, \quad \text{when } \Omega \neq 0.$$

Define for $u \in \mathcal{C}\ell_{1,3}$ (or $u \in \mathbb{C} \otimes \mathcal{C}\ell_{1,3}$)

$$P_\pm(u) = \frac{1}{2\Omega_1}(\Omega_1 \pm \mathbf{J} \pm \Omega_2\gamma_{0123})u, \quad \Omega_1 \neq 0$$
$$\Sigma_\pm(u) = \frac{1}{2}(u \pm \gamma_{0123}\mathbf{J}\mathbf{K}^{-1}u\gamma_{12}), \quad \Sigma_\pm \in \mathrm{End}(\mathcal{C}\ell_{1,3}).$$

Then

$$\begin{array}{llll} P_+(\psi) = \psi & P_+(\Phi) = \Phi & P_+(\phi) = \phi & \\ P_-(\psi) = 0 & P_-(\Phi) = 0 & P_-(\phi) = 0 & \\ \Sigma_+(\psi) = \psi & \Sigma_+(\Phi) = \Phi & \Sigma_+(\phi) = \phi & \Sigma_+(\Psi) = \Psi \\ \Sigma_-(\psi) = 0 & \Sigma_-(\Phi) = 0 & \Sigma_-(\phi) = 0 & \Sigma_-(\Psi) = 0. \end{array}$$

In general, for $u \in \mathcal{C}\ell_{1,3}$, $P_\pm^2(u) = P_\pm(u)$, $\Sigma_\pm^2(u) = \Sigma_\pm(u)$ and $P_\pm(\Sigma_\pm(u)) = \Sigma_\pm(P_\pm(u))$, that is, $P_\pm$ and $\Sigma_\pm$ are commuting projection operators. For an arbitrary η in $\mathcal{C}\ell_{1,3}\,\frac{1}{2}(1+\gamma_0)$ [or in $\mathcal{C}\ell_{1,3}\,\frac{1}{2}(1-\gamma_{03})$] the spinor $P_+(\Sigma_+(\eta))$ is parallel to Φ [or to ϕ], that is, the bilinear covariants of $P_+(\Sigma_+(\eta))$ are proportional to P, Q. However, for an arbitrary $u \in \mathcal{C}\ell_{1,3}$, $P_+(\Sigma_+(u)) \notin \mathcal{C}\ell_{1,3}\,\frac{1}{2}(1+\gamma_0)$ [or $P_+(\Sigma_+(u)) \notin \mathcal{C}\ell_{1,3}\,\frac{1}{2}(1-\gamma_{03})$].

Define

$$\Sigma_\pm^I(u) = \frac{1}{2}(u \mp \gamma_{0123}\mathbf{J}\mathbf{K}^{-1}u\gamma_{0123})$$

where I stands for ideal spinor. Then for an arbitrary $u \in \mathcal{C}\ell_{1,3}$ we have $\Sigma_+(\Sigma_+^I(u)) \in \mathcal{C}\ell_{1,3}\,\frac{1}{2}(1-\gamma_{03})$, and $P_+(\Sigma_+(\Sigma_+^I(u)))$ is an ideal spinor parallel to ϕ (with bilinear covariants proportional to P, Q). Furthermore, $\Sigma_+^I(\phi) = \phi$, $\Sigma_-^I(\phi) = 0$, and $\Sigma_\pm^I$ are projection operators commuting with $P_\pm$, $\Sigma_\pm$.

Define (O stands for spinor operator)

$$P_\pm^O(u) = \frac{1}{2\Omega_1}((\Omega_1 \pm \Omega_2\gamma_{0123})u \pm \mathbf{J}u\gamma_0), \quad u \in \mathcal{C}\ell_{1,3}^+,$$

which are projection operators commuting with $\Sigma_\pm$.

Exercise. (Inspired by Crawford 1985) Define

$$\Gamma_\pm(u) = \frac{1}{2}(u \mp (\Omega_2 + \Omega_1\gamma_{0123})^{-1}\mathbf{K}u\gamma_{12})$$

and show that $\Gamma_\pm$ are projection operators commuting with $\Sigma_\pm$ [but not with $P_\pm$ unless $\Omega_2 = 0$; recall here the factorization of Crawford]. Show that $P_+(\Gamma_+(\Phi)) = \Phi$, $P_+(\Gamma_+(\phi)) = \phi$. How would you define $\Gamma_\pm$ for a spinor operator Ψ?
[Answer: $\Gamma^O_\pm(u) = \frac{1}{2}(u \mp (\Omega_2 + \Omega_1\gamma_{0123})^{-1}\mathbf{K}u\gamma_{012})$ for $u \in \mathcal{C}\ell^+_{1,3}$.] ▮

Remark. Define $\Gamma_\pm$ for an ideal spinor ϕ (I stands for ideal):

$$\Gamma^I_\pm(u) = \frac{1}{2}(u \pm (\Omega_2 + \Omega_1\gamma_{0123})^{-1}\mathbf{K}u\gamma_{0123}), \qquad u \in \mathcal{C}\ell_{1,3}\,\frac{1}{2}(1 - \gamma_{03}).$$

In the special case $\Omega_2 = 0$ of type 2 these take the form

$$\Gamma^I_\pm(u) = \frac{1}{2}(u \mp \gamma_{0123}\frac{1}{\Omega_1}\mathbf{K}u\gamma_{0123}), \qquad u \in \mathcal{C}\ell_{1,3}\,\frac{1}{2}(1 - \gamma_{03}),$$

and commute with $P_\pm$ [this special case was also observed by Hestenes 1986 p. 336 (2.32)]. ▮

12.3 Projection operators for Majorana and Weyl spinors

Treat first the general case (class 4) $\Omega_1 = 0 = \Omega_2$, $\mathbf{K} \neq 0 \neq \mathbf{S}$. Recall that $(1 + i\mathbf{s} + ih\gamma_{0123})\psi = 0$ or $i\psi = (\mathbf{s} + h\gamma_{0123})\psi$. Define

$$\Sigma^G_\pm(u) = \frac{1}{2}(u \pm (\mathbf{s} + h\gamma_{0123})u\gamma_{12}).$$

Then $\Sigma^G_+(\Phi) = \Phi$, $\Sigma^G_+(\phi) = \phi$. Majorana and Weyl spinors are now the limiting cases

$$\Sigma^M_\pm(u) = \frac{1}{2}(u \pm \mathbf{s}u\gamma_{12}), \qquad \Sigma^W_\pm(u) = \frac{1}{2}(u \pm \gamma_{0123}u\gamma_{12}).$$

Exercises 1,2,3,4,5

12.4 Charge conjugate $\psi_C = \mathcal{C}(\psi)$

The charge conjugate spinor $\psi_C = -i\gamma_2\psi^*$ sits in $\mathbb{C}^4$ [3] or in the same minimal left ideal $\mathrm{Mat}(4,\mathbb{C})f$; it satisfies

$$\gamma^\mu(i\partial_\mu + eA_\mu)\psi_C = m\psi_C.$$

Charge conjugation is an anti-linear operation, that is, $\mathcal{C}(i\psi) = -i\mathcal{C}(\psi)$. Other characteristics of charge conjugation are $\psi_C^\dagger\gamma_0\psi_C = -\psi^\dagger\gamma_0\psi$ and $\psi_C^\dagger\gamma_0\gamma_\mu\psi_C = +\psi^\dagger\gamma_0\gamma_\mu\psi$.

In the notation of $\mathbb{C}\otimes\mathcal{C}\ell_{1,3}$ we have $\psi_C = -i\gamma_2\gamma_{013}\psi^*\gamma_{013}^{-1} = \hat{\psi}^*\gamma_1$, which also sits in the minimal left ideal $(\mathbb{C}\otimes\mathcal{C}\ell_{1,3})f$, while γ_1 swaps the signs of both factors of the primitive idempotent $f = \frac{1}{2}(1+\gamma_0)\frac{1}{2}(1+i\gamma_1\gamma_2)$. The bilinear covariants are transformed as follows under the charge conjugation

$$\begin{aligned}
&\langle\tilde{\psi}_C^*\psi_C\rangle_0 &&= -\langle\tilde{\psi}^*\psi\rangle_0 &&\text{(as above)}\\
&\langle\tilde{\psi}_C^*\gamma_\mu\psi_C\rangle_0 &&= +\langle\tilde{\psi}^*\gamma_\mu\psi\rangle_0 &&\text{(as above)}\\
&\langle\tilde{\psi}_C^*i\gamma_{\mu\nu}\psi_C\rangle_0 &&= +\langle\tilde{\psi}^*i\gamma_{\mu\nu}\psi\rangle_0 &&\\
&\langle\tilde{\psi}_C^*i\gamma_{0123}\gamma_\mu\psi_C\rangle_0 &&= -\langle\tilde{\psi}^*i\gamma_{0123}\gamma_\mu\psi\rangle_0 &&\\
&\langle\tilde{\psi}_C^*\gamma_{0123}\psi_C\rangle_0 &&= -\langle\tilde{\psi}^*\gamma_{0123}\psi\rangle_0 &&
\end{aligned}$$

and $4\psi_C\tilde{\psi}_C^* = -\hat{Z}^* = -\hat{P} + i\hat{Q} = -\Omega_1 + \mathbf{J} + i\mathbf{S} + i\gamma_{0123}\mathbf{K} - \gamma_{0123}\Omega_2$, since $4\psi_C\tilde{\psi}_C^* = 4\hat{\psi}^*\gamma_1(\hat{\psi}^*\gamma_1)^{\tilde{}*} = 4\hat{\psi}^*\gamma_1\gamma_1\bar{\psi} = -4(\psi\tilde{\psi}^*)^{\hat{}*}$

The charge conjugate of the mother spinor is

$$\Phi_C = \hat{\Phi}\gamma_1 = 4\,\mathrm{Re}(\hat{\psi}^*\gamma_1);$$

it satisfies $\Phi_C \in \mathcal{C}\ell_{1,3}\,\frac{1}{2}(1+\gamma_0)$, and has the same properties as were listed above for the charge conjugation. The charge conjugate of the ideal spinor is $\phi_C = \hat{\phi}\gamma_1 = \hat{\Phi}\gamma_1\frac{1}{2}(1-\gamma_{03})$. Its bilinear covariants are (as above)

$$\begin{aligned}
\phi_C\gamma_0\tilde{\phi}_C &= -\hat{P}, & \phi_C\gamma_3\bar{\phi}_C &= -\hat{Q}\gamma_{0123}\\
\phi_C\gamma_3\tilde{\phi}_C &= -P, & \phi_C\gamma_0\bar{\phi}_C &= -\hat{Q}\gamma_{0123}.
\end{aligned}$$

The charge conjugate of the spinor operator is $\Psi_C = \Psi\gamma_1\gamma_0 = \mathrm{even}(\hat{\Phi}\gamma_1)$, where $\hat{\Phi}\gamma_1 = (\Phi_0 - \Phi_0\gamma_0)\gamma_1$.

Exercise. Show that the operator form of a Majorana spinor $\frac{1}{2}(\psi \pm \psi_C)$ is

---

3 In this case the complex conjugate is

$$\psi^* = \begin{pmatrix}\psi_1^*\\ \psi_2^*\\ \psi_3^*\\ \psi_4^*\end{pmatrix} \quad \text{for} \quad \psi = \begin{pmatrix}\psi_1\\ \psi_2\\ \psi_3\\ \psi_4\end{pmatrix}.$$

$\Psi\frac{1}{2}(1 \mp \gamma_{01}) \in C\ell^+_{1,3}\,\frac{1}{2}(1 \mp \gamma_{01})$. ∎

The Wigner time-reversal is $\psi_T = -i\gamma_{13}\psi^* \in \mathrm{Mat}(4,\mathbb{C})f$ or $\psi_T = \gamma_{123}\hat{\psi}^*\gamma_1 \in (\mathbb{C} \otimes C\ell_{1,3})f$ and the parity involution of space is $\psi_{\mathcal{P}} = \gamma_0\psi$. So

$\psi_{\mathcal{T}} = \gamma_{123}\hat{\psi}^*\gamma_1$	$\psi_{\mathcal{P}} = \gamma_0\psi$	$\psi_{\mathcal{C}} = \hat{\psi}^*\gamma_1$	Dirac
$\Phi_{\mathcal{T}} = \gamma_{123}\hat{\Phi}\gamma_1$	$\Phi_{\mathcal{P}} = \gamma_0\Phi$	$\Phi_{\mathcal{C}} = \hat{\Phi}\gamma_1$	mother
$\phi_{\mathcal{T}} = \gamma_{123}\hat{\phi}\gamma_1$	$\phi_{\mathcal{P}} = \gamma_0\phi$	$\phi_{\mathcal{C}} = \hat{\phi}\gamma_1$	ideal
$\Psi_{\mathcal{T}} = \gamma_{123}\Psi\gamma_1$	$\Psi_{\mathcal{P}} = \gamma_0\Psi\gamma_0^{-1}$	$\Psi_{\mathcal{C}} = \Psi\gamma_1\gamma_0$	operator

and $\psi_{\mathcal{TPC}} = \gamma_{0123}\psi$. Note that charge conjugation $\mathcal{C}$ anticommutes with both parity involution $\mathcal{P}$ and time-reversal $\mathcal{T}$.

Exercises 6,7,8,9,10

Appendix: Crumeyrolle's spinoriality transformation

Crumeyrolle introduced a number of *spinoriality groups* to be able to treat the complicated situations with spinors. However, one relevant problem remained unsolved: how can the usual bilinear covariants be obtained from Crumeyrolle's spinors? The bilinear covariants of Crumeyrolle's spinors mix the Dirac current vector **J** and the electromagnetic moment bivector **S**. A solution to this problem can be given by a variation of Crumeyrolle's spinoriality group. In this appendix it is shown how to extract the standard bilinear covariants (see the standard textbooks on quantum mechanics, like Bjorken & Drell 1964) from Crumeyrolle's or Cartan's pure spinors.

Crumeyrolle considered the complexification $\mathbb{C} \otimes C\ell_{1,3}$ of the Clifford algebra $C\ell_{1,3}$ of the Minkowski space $\mathbb{R}^{1,3}$. In the complex linear space $\mathbb{C} \otimes \mathbb{R}^{1,3}$ Crumeyrolle picked up a maximal totally null subspace spanned by the orthogonal null vectors

$$\frac{1}{2}(\gamma_0 - \gamma_3) \quad \text{and} \quad \frac{1}{2}(\gamma_1 - i\gamma_2).$$

Denote the product of these vectors by

$$\mathbf{v} = \frac{1}{2}(\gamma_0 - \gamma_3)\frac{1}{2}(\gamma_1 - i\gamma_2)$$

which is the volume element of the totally null subspace. Crumeyrolle chose as his spinor space the minimal left ideal $(\mathbb{C} \otimes C\ell_{1,3})\mathbf{v}$ of the complex Clifford algebra $\mathbb{C}\otimes C\ell_{1,3}$. The difficulty with this choice is that the bilinear covariants of such a spinor are not directly related to those of a column spinor in standard textbooks on quantum mechanics. [4] To overcome this difficulty, first note that the element

$$g = \gamma_{01}\mathbf{v} = \frac{1}{2}(1-\gamma_{03})\frac{1}{2}(1+i\gamma_{12})$$

is a primitive idempotent generating Crumeyrolle's spinor space, that is,

$$(\mathbb{C}\otimes \mathcal{C}\ell_{1,3})\mathbf{v} = (\mathbb{C}\otimes \mathcal{C}\ell_{1,3})g.$$

Unfortunately, $g = \frac{1}{2}(1-\gamma_{03})\frac{1}{2}(1+i\gamma_{12})$ is an even element and does not contain as a factor the 'energy projection' operator $\frac{1}{2}(1+\gamma_0)$. The physical observables are obtained from column spinors sitting in $\mathbb{C}^4$ which are related to Clifford algebraic spinors sitting in $(\mathbb{C}\otimes \mathcal{C}\ell_{1,3})f$, where

$$f = \frac{1}{2}(1+\gamma_0)\frac{1}{2}(1+i\gamma_{12}).$$

In order to move from the spinor space $(\mathbb{C}\otimes \mathcal{C}\ell_{1,3})g$ to the spinor space $(\mathbb{C}\otimes \mathcal{C}\ell_{1,3})f$ we must find a transformation law for spinors $\psi_g \in (\mathbb{C}\otimes \mathcal{C}\ell_{1,3})g$ and $\psi_f \in (\mathbb{C}\otimes \mathcal{C}\ell_{1,3})f$. [5] This transformation law is a slight variation of Crumeyrolle's spinoriality transformation. [6]

Before giving our variation of the spinoriality transformation let us recall that ψ_g is a sum of two Weyl spinors

$$\frac{1}{2}(1+i\gamma_{0123})\psi_g \in \mathcal{C}\ell^{-}_{1,3}\otimes\mathbb{C}$$
$$\frac{1}{2}(1-i\gamma_{0123})\psi_g \in \mathcal{C}\ell^{+}_{1,3}\otimes\mathbb{C}$$

so that the components are of homogeneous parity [the correspondence between the even/odd parts and the negative/positive helicities is irrelevant, since it could be swapped by a different choice of g, for instance, by $g = \frac{1}{2}(1+\gamma_{03})\frac{1}{2}(1+i\gamma_{12})$ for which $g = \gamma_{01}\mathbf{v}$ with $\mathbf{v} = \frac{1}{2}(\gamma_0+\gamma_3)\frac{1}{2}(\gamma_1 - i\gamma_2)$].

Our variation of the spinoriality transformation is carried out by the element

$$z = \frac{1}{\sqrt{2}}(1+\gamma_3)$$

4 The bilinear covariants of Crumeyrolle's spinors either vanish identically or else, as in Crumeyrolle 1990 p. 229 formula 24, mix the Dirac current $\mathbf{J}$ and the electromagnetic moment $\mathbf{S}$.

5 Recall that, for instance, $J_\mu = 4\langle\tilde{\psi}_f^*\gamma_\mu\psi_f\rangle_0$. In contrast, these bilinear covariants of ψ_g vanish: $4\langle\tilde{\psi}_g^*\gamma_\mu\psi_g\rangle_0 = 0$. However, ψ_g does carry all the information of $\mathbf{J}$: $J_\mu = 4\langle\tilde{\psi}_g^*\gamma_\mu\psi_g\rangle_1\cdot\gamma_0$.

6 For Crumeyrolle the spinoriality group meant a number of things, with different adjectives added as specification. First, it is the subgroup of those $s\in\mathbf{Spin}(1,3)$ for which $s\mathbf{v} = \pm\mathbf{v}$, see Crumeyrolle 1990 p. 145. Secondly, it is the group of invertible elements z in $\mathbb{C}\otimes\mathcal{C}\ell_{1,3}$ such that the primitive idempotents g and zgz^{-1} determine the same minimal left ideal $(\mathbb{C}\otimes\mathcal{C}\ell_{1,3})g = (\mathbb{C}\otimes\mathcal{C}\ell_{1,3})g'$, see p. 277. Thirdly, it is, if normalized, the intersection of the previous group with $\mathbf{Spin}(1,3)$, see p. 281.

for which $g = zfz^{-1}$ or $f = z^{-1}gz$. The latter rule gives us a relation between Crumeyrolle's nilpotent induced spinors ψ_g and idempotent induced spinors ψ_f (directly related to the standard column spinors like those in Bjorken & Drell 1964),

$$\psi_f = \psi_g z.$$

(Earlier we wrote $\psi = \psi_f$ but here it is necessary to indicate to which minimal left ideal the spinor belongs.)

Now we can compute the spinor operator $\Psi = \text{even}(4\,\text{Re}(\psi_f))$ and the bilinear covariants, for instance, $\mathbf{J} = \Psi\gamma_0\tilde{\Psi}$. For later convenience note that $\Psi = \text{Oper}(\psi_f)$ where

$$\text{Oper}(\psi_f) = \frac{1}{2}(\text{even}(4\,\text{Re}(\psi_f)) + \text{odd}(4\,\text{Re}(\psi_f))\gamma_0).$$

Recall the aggregate of bilinear covariants

$$Z = 4\psi_f\tilde{\psi}_f^*$$

and note that $4\psi_f\tilde{\psi}_f^* = 4\psi_f\gamma_0\tilde{\psi}_f^*$. Our variation of the spinoriality group is the group of those elements z in $\mathcal{C}\ell_{1,3}$ or $\mathbb{C}\otimes\mathcal{C}\ell_{1,3}$ which preserve the aggregate Z under the transformation $\psi_f \to \psi_f z^{-1}$. Crumeyrolle's spinoriality groups preserve the ideals whereas our spinoriality groups preserve the physical observables. The spinoriality groups are seen to be the following (see Lounesto 1981 p. 733):

	$\mathcal{C}\ell_{1,3}$	$\mathbb{C}\otimes\mathcal{C}\ell_{1,3}$
$Z = 4\psi_f\tilde{\psi}_f^*$	$Sp(2,2)$	$U(2,2)$
$Z = 4\psi_f\gamma_0\tilde{\psi}_f^*$	$Sp(4)$	$U(4)$

where, as an example, the Lie algebra of $Sp(4) \simeq \mathbf{Spin}(5)$ is spanned by the elements

$$\begin{aligned}&\gamma_1,\ \gamma_2,\ \gamma_3\\&\gamma_{12},\ \gamma_{13},\ \gamma_{23}\\&\gamma_{012},\ \gamma_{013},\ \gamma_{023}\\&\gamma_{0123}.\end{aligned}$$

For those z in $\mathcal{C}\ell_{1,3}$ which preserve $Z = 4\psi_f\gamma_0\tilde{\psi}_f^*$, under the replacement $\psi_f \to \psi_f z^{-1}$, that is $z \in Sp(4)$, we may find that the spinor operator is preserved under the following transformations:

$$\Psi = \frac{1}{2\langle w\rangle_0}(\text{even}(4\,\text{Re}(\psi_f w^{-1})) + \text{odd}(4\,\text{Re}(\psi_f w^{-1}))\gamma_0)$$

where $w = z\tilde{z}$, that is,

$$\Psi = \frac{1}{\langle w\rangle_0}\mathrm{Oper}(\psi_f w^{-1}).$$

To put all this in a nutshell: our variation of

spinoriality transformation preserves bilinear covariants

However, this preservation should be distinguished from our use of the particular element $z = \frac{1}{\sqrt{2}}(1+\gamma_3) \in Sp(4)$ to retrieve the aggregate of bilinear covariants $Z = 4\psi_f\tilde{\psi}_f^*$ by sending ψ_g to $\psi_g z = \psi_f$.

Exercises

1. Recall that $\Psi\gamma_2\gamma_1 = \mathrm{s}\Psi\gamma_0 + h\gamma_{0123}\Psi$. How would you define $\Sigma_\pm^G$ for a spinor operator Ψ?
2. Recall that $\phi\gamma_{0123} = \phi\gamma_2\gamma_1$. How would you define another pair $\Sigma_\pm^G$ for an ideal spinor ϕ?
3. Show that up to a unit complex factor $e^{\gamma_{12}\alpha}$:
   $\Psi \simeq \frac{1}{4N}(\Omega_1 + \mathbf{J}\gamma_0 - \mathbf{S}\gamma_{12} - \mathbf{K}\gamma_3 + \Omega_2\gamma_{0123})$, when $N = \sqrt{\langle Zf\rangle_0} \neq 0$.
4. Show that the operator form of a Weyl spinor is $\Psi\frac{1}{2}(1 \mp \gamma_{03})$.
5. Show that Weyl spinors $\frac{1}{2}(1 \mp i\gamma_{0123})\psi$ correspond to even and odd parts of the ideal spinor $\phi = \phi_0 + \phi_1$.

Write $W = 4\psi\tilde{\psi}_{\mathcal{C}}^*$, [7] note that $\psi\tilde{\psi}_{\mathcal{C}}^* = 0$, and show that

6. $W = -(Q_1 + iQ_2)\gamma_{0123}$ where $Q_k = \frac{1}{2}(\Phi\gamma_k\bar{\Phi})\gamma_{0123}$ or $Q_k = \Psi(1+\gamma_0)\gamma_{ij}\tilde{\Psi}$ (ijk cycl.).
7. $W = \mathcal{K} - \mathcal{S}\gamma_{0123}$ where $\mathcal{K} = \mathbf{K}_1 + i\mathbf{K}_2$ and $\mathcal{S} = \mathbf{S}_1 + i\mathbf{S}_2$, where as before $\mathbf{K}_k = \Psi\gamma_k\tilde{\Psi}$, $\mathbf{S}_k = \Psi\gamma_{ij}\tilde{\Psi}$ (ijk cycl.).
8. $W^2 = 0$.
9. $WZ = 0$.
10. $ZW = 4\Omega_1 W$ and so the 3-vector part vanishes:
    $\langle ZW\rangle_3 = -\mathbf{J}\wedge(\gamma_{0123}\mathcal{S}) + i\mathbf{S}\wedge\mathcal{K} + \gamma_{0123}\Omega_2\mathcal{K} + i\mathbf{K}\wedge\mathcal{S} = 0$.
11. Show that $4\psi_g\tilde{\psi}_g^* = 0$.
12. Write $\Psi_f = \mathrm{Oper}(\psi_f)$ and $\Psi_g = \mathrm{Oper}(\psi_g)$. Show that
    $\Psi_f\gamma_0\tilde{\Psi}_f = 2\Psi_g\gamma_0\tilde{\Psi}_g$.
13. Write $\Psi_f = \mathrm{even}(4\,\mathrm{Re}(\psi_f)) = \mathrm{Oper}(\psi_f)$ and
    $\Psi_g = \mathrm{even}(4\,\mathrm{Re}(\psi_g)) \neq \mathrm{Oper}(\psi_g)$. Show that $\Psi_g\tilde{\Psi}_g = 0$, $\Psi_g\gamma_{12}\tilde{\Psi}_g = 0$ and $\Psi_g\gamma_0\tilde{\Psi}_g = \mathbf{J} + \mathbf{K}$ where $\mathbf{J} = \Psi_f\gamma_0\tilde{\Psi}_f$ and $\mathbf{K} = \Psi_f\gamma_3\tilde{\Psi}_f$.

---

7 $\tilde{\psi}_{\mathcal{C}}^* = (\psi_{\mathcal{C}})^{\tilde{}\,*} \neq (\tilde{\psi}^*)_{\mathcal{C}}$.

Solutions

1. $\Sigma_{\pm}^{GO}(u) = \frac{1}{2}(u \pm su\gamma_{012} \pm h\gamma_{0123}u\gamma_{12})$ for $u \in \mathcal{C}\ell_{1,3}^{+}$.
2. $\Sigma_{\pm}^{GI}(u) = \frac{1}{2}(u \mp (s + h\gamma_{0123})u\gamma_{0123}),\ u \in \mathcal{C}\ell_{1,3}\,\frac{1}{2}(1-\gamma_{03})$.
4. Hint: compute the even part of $4\,\mathrm{Re}(\frac{1}{2}(1 \mp i\gamma_{0123})\psi)$ in the decomposition $\mathbb{C} \otimes \mathcal{C}\ell_{1,3}$.

Bibliography

I.M. Benn, R.W. Tucker: *An Introduction to Spinors and Geometry with Applications in Physics*. Adam Hilger, Bristol, 1987.

J.D. Bjorken, S.D. Drell: *Relativistic Quantum Mechanics*. McGraw-Hill, New York, 1964.

J.P. Crawford: On the algebra of Dirac bispinor densities: factorization and inversion theorems. *J. Math. Phys.* **26** (1985), 1439-1441.

J.P. Crawford: Dirac equation for bispinor densities; pp. 353-361 in J.S.R. Chisholm, A.K. Common (eds.): *Proceedings of the Workshop on 'Clifford Algebras and their Applications in Mathematical Physics' (Canterbury 1985)*. Reidel, Dordrecht, The Netherlands, 1986.

A. Crumeyrolle: Groupes de spinorialité. *Ann. Inst. H. Poincaré* **14** (1971), 309-323.

A. Crumeyrolle: *Orthogonal and Symplectic Clifford Algebras, Spinor Structures*. Kluwer, Dordrecht, The Netherlands, 1990.

C. Doran: *Geometric Algebra and its Applications to Mathematical Physics*. Thesis, Univ. Cambridge, 1994.

J.D. Hamilton: The Dirac equation and Hestenes' geometric algebra. *J. Math. Phys.* **25** (1984), 1823-1832.

P.R. Holland: Minimal ideals and Clifford algebras in the phase space representation of spin-$\frac{1}{2}$ fields; pp. 273-283 in J.S.R. Chisholm, A.K. Common (eds.): *Proceedings of the Workshop on 'Clifford Algebras and their Applications in Mathematical Physics' (Canterbury 1985)*. Reidel, Dordrecht, 1986.

P.R. Holland: Relativistic algebraic spinors and quantum motions in phase space. *Found. Phys.* **16** (1986), 708-709.

J. Keller, S. Rodríguez-Romo: A multivectorial Dirac equation. *J. Math. Phys.* **31** (1990), 2501-2510.

J. Keller, F. Viniegra: The multivector structure of the matter and interaction field theories; pp. 437-445 in A. Micali et al. (eds.): *Proceedings of the Second Workshop on 'Clifford Algebras and their Applications in Mathematical Physics' (Montpellier 1985)*. Reidel, Dordrecht, The Netherlands, 1992.

R. Penrose, W. Rindler: *Spinors and Space-Time*. Vol. 1. Cambridge University Press, Cambridge, 1984.

13

Tilt to the Opposite Metric

Physicists usually go from $C\ell_{1,3} \simeq \mathrm{Mat}(2,\mathbb{H})$ to its opposite algebra $C\ell_{3,1} \simeq \mathrm{Mat}(4,\mathbb{R})$ by replacing γ_μ by $i\gamma_\mu$ [within $\mathrm{Mat}(4,\mathbb{C})$]. However, such a transition to the opposite metric does not make sense within the space-time $\mathbb{R}^4$, because it calls for $i\mathbb{R}^4$ which is outside of $\mathbb{R}^4$. We will instead regard the linear space $\mathbb{R}^4$ as one and the same space-time, endowed with two different metrics or quadratic structures, $\mathbb{R}^{1,3}$ and $\mathbb{R}^{3,1}$.

Let the basis $\{\gamma_0, \gamma_1, \gamma_2, \gamma_3\}$ of the space-time $\mathbb{R}^{1,3}$ generating $C\ell_{1,3}$ correspond to the basis $\{\mathbf{e}_0, \mathbf{e}_1, \mathbf{e}_2, \mathbf{e}_3\}$ of the space-time $\mathbb{R}^{3,1}$ generating $C\ell_{3,1}$,

$$\mathbf{e}_0^2 = -1, \ \mathbf{e}_1^2 = \mathbf{e}_2^2 = \mathbf{e}_3^2 = 1 \qquad [\mathbf{e}_\mu \neq \pm i\gamma_\mu].$$

So the vectors $A^0\gamma_0 + A^1\gamma_1 + A^2\gamma_2 + A^3\gamma_3$ and $A^0\mathbf{e}_0 + A^1\mathbf{e}_1 + A^2\mathbf{e}_2 + A^3\mathbf{e}_3$ correspond to each other but have opposite squares

$$(A^0\gamma_0 + A^1\gamma_1 + A^2\gamma_2 + A^3\gamma_3)^2 = (A^0)^2 - (A^1)^2 - (A^2)^2 - (A^3)^2,$$
$$(A^0\mathbf{e}_0 + A^1\mathbf{e}_1 + A^2\mathbf{e}_2 + A^3\mathbf{e}_3)^2 = -(A^0)^2 + (A^1)^2 + (A^2)^2 + (A^3)^2.$$

We shall go further and regard $A^0\gamma_0 + A^1\gamma_1 + A^2\gamma_2 + A^3\gamma_3 \in \mathbb{R}^{1,3} \subset C\ell_{1,3}$ and $A^0\mathbf{e}_0 + A^1\mathbf{e}_1 + A^2\mathbf{e}_2 + A^3\mathbf{e}_3 \in \mathbb{R}^{3,1} \subset C\ell_{3,1}$ as one and the same vector $\mathbf{A} \in \mathbb{R}^4$ embedded in two non-isomorphic algebras $C\ell_{1,3}$ and $C\ell_{3,1}$ which are identified as linear spaces by the correspondences

$C\ell_{1,3}$	$C\ell_{3,1}$
1	1
γ_μ	$\mathbf{e}_\mu$
$\gamma_\mu \wedge \gamma_\nu$	$\mathbf{e}_\mu \wedge \mathbf{e}_\nu$ $\quad (= \mathbf{e}_\mu\mathbf{e}_\nu$ for $\mu \neq \nu)$

[Note that $\gamma_0 = \gamma^0$ and $A_0 = A^0$ in $C\ell_{1,3}$ whereas in $C\ell_{3,1}$ we have $\mathbf{e}_0 = -\mathbf{e}^0$ and $A_0 = -A^0$ and that the numerical values of A^0 are the same in $C\ell_{1,3}$ and $C\ell_{3,1}$.]

The products in the Clifford algebras $\mathcal{C}\ell_{1,3}$ and $\mathcal{C}\ell_{3,1}$ are related to each other by (all the terms in this table are computed in $\mathcal{C}\ell_{1,3}$)

$\mathcal{C}\ell_{1,3}$	$\mathcal{C}\ell_{3,1}$
ab	$b_0a_0 + b_0a_1 + b_1a_0 - b_1a_1$

where $a_0 = \text{even}(a)$ and $a_1 = \text{odd}(a)$. For $a \in \mathcal{C}\ell_{1,3}$ we may sometimes emphasize that $\text{opp}[a] \in \mathcal{C}\ell_{3,1}$ (or the other way round). In this notation, the products in the (graded) opposite algebras are related by [1]

$$\text{opp}[ab] = b_0a_0 + b_0a_1 + b_1a_0 - b_1a_1.$$

In this chapter we shall study the Maxwell equations and the Dirac equation in opposite metrics, in the quadratic spaces $\mathbb{R}^{1,3}$ and $\mathbb{R}^{3,1}$. In particular, we do not consider curved space-times, only flat space-times. In a flat space-time it is also possible to differentiate multivector fields, not only differential forms; we will focus on differentiating multivector fields.

The Maxwell Equations in Opposite Metrics

There are a few changes of sign in the Maxwell equations in the quadratic spaces $\mathbb{R}^{3,1}$ and $\mathbb{R}^{1,3}$.

13.1 The Maxwell equations in $\mathbb{R}^{3,1}$

We use the following definitions for the potential $\mathbf{A}$, the current $\mathbf{J}$, the differential operator ∂, and the electromagnetic field $\mathbf{F}$:

$$\begin{aligned}
&A^\alpha = (\tfrac{1}{c}V, A_x, A_y, A_z), \qquad A_\alpha = (-\tfrac{1}{c}V, A_x, A_y, A_z)\\
&J^\alpha = (\rho c, J_x, J_y, J_z), \qquad J_\alpha = (-\rho c, J_x, J_y, J_z)\\
&\partial^\alpha = (-\tfrac{1}{c}\tfrac{\partial}{\partial t}, \tfrac{\partial}{\partial x}, \tfrac{\partial}{\partial y}, \tfrac{\partial}{\partial z}), \qquad \partial_\alpha = (\tfrac{1}{c}\tfrac{\partial}{\partial t}, \tfrac{\partial}{\partial x}, \tfrac{\partial}{\partial y}, \tfrac{\partial}{\partial z})\\
&(F^{01}, F^{02}, F^{03}) = (-\tfrac{1}{c}E_x, -\tfrac{1}{c}E_y, -\tfrac{1}{c}E_z)\\
&(F^{23}, F^{31}, F^{12}) = (-B_x, -B_y, -B_z).
\end{aligned}$$

This leads to the d'Alembert operator

$$\partial^\alpha\partial_\alpha = \nabla^2 - \frac{1}{c^2}\frac{\partial^2}{\partial t^2}$$

and the equation

$$\partial^\alpha\partial_\alpha A^\beta = -\mu J^\beta.$$

1 The symbol 'opp' is not a function of one or two variables; it is rather an indicator signaling that all the computations in the brackets will be computed in the opposite algebra.

The equations

$$\partial^0 A^1 - \partial^1 A^0 = -\tfrac{1}{c}\tfrac{\partial}{\partial t}A_x - \tfrac{\partial}{\partial x}\tfrac{V}{c} = \tfrac{1}{c}E_x, \ \dots$$
$$\partial^1 A^2 - \partial^2 A^1 = \tfrac{\partial}{\partial x}A_y - \tfrac{\partial}{\partial y}A_x = B_z, \ \dots$$

give

$$F^{\alpha\beta} = -(\partial^\alpha A^\beta - \partial^\beta A^\alpha),$$

and

$$\partial_\alpha F^{\alpha 0} = \tfrac{\partial}{\partial x}\tfrac{E_x}{c} + \tfrac{\partial}{\partial y}\tfrac{E_y}{c} + \tfrac{\partial}{\partial z}\tfrac{E_z}{c} = \tfrac{1}{c}\nabla\cdot\vec{E} = \tfrac{\rho}{c\varepsilon} = \mu J^0$$
$$\partial_\alpha F^{\alpha 1} = -\tfrac{1}{c}\tfrac{\partial}{\partial t}\tfrac{E_x}{c} + \tfrac{\partial}{\partial y}B_z - \tfrac{\partial}{\partial z}B_y = \mu J^1, \dots$$

give the Maxwell equations

$$\partial_\alpha F^{\alpha\beta} = \mu J^\beta.$$

In the Clifford algebra $\mathcal{C}\ell_{3,1}$ we have

$$\partial = -\mathbf{e}_0\tfrac{1}{c}\tfrac{\partial}{\partial t} + \mathbf{e}_1\tfrac{\partial}{\partial x} + \mathbf{e}_2\tfrac{\partial}{\partial y} + \mathbf{e}_3\tfrac{\partial}{\partial z}$$
$$\mathbf{F} = -\tfrac{1}{c}E_x\mathbf{e}_{01} - \tfrac{1}{c}E_y\mathbf{e}_{02} - \tfrac{1}{c}E_z\mathbf{e}_{03} - B_x\mathbf{e}_{23} - B_y\mathbf{e}_{31} - B_z\mathbf{e}_{12}$$
$$= \tfrac{1}{c}\vec{E}\mathbf{e}_0 - \vec{B}\mathbf{e}_{123}$$
$$\mathbf{A} = \tfrac{1}{c}V\mathbf{e}_0 + A_x\mathbf{e}_1 + A_y\mathbf{e}_2 + A_z\mathbf{e}_3$$
$$\mathbf{J} = c\rho\mathbf{e}_0 + J_x\mathbf{e}_1 + J_y\mathbf{e}_2 + J_z\mathbf{e}_3$$

and the equations

$$\partial\mathbf{A} = \mathbf{e}_{01}(-\tfrac{1}{c}\tfrac{\partial}{\partial t}A_x - \tfrac{\partial}{\partial x}\tfrac{V}{c}) + \dots + \mathbf{e}_{12}(\tfrac{\partial}{\partial x}A_y - \tfrac{\partial}{\partial y}A_x)$$
$$\partial\mathbf{F} = \mathbf{e}_0(\tfrac{\partial}{\partial x}\tfrac{E_x}{c} + \tfrac{\partial}{\partial y}\tfrac{E_y}{c} + \tfrac{\partial}{\partial z}\tfrac{E_z}{c}) + \mathbf{e}_1(-\tfrac{1}{c^2}\tfrac{\partial}{\partial t}E_x - \tfrac{\partial}{\partial z}B_y + \tfrac{\partial}{\partial y}B_z) + \dots$$

which lead to

$$\boxed{\partial\mathbf{A} = -\mathbf{F}, \quad \partial\mathbf{F} = \mu\mathbf{J}, \quad \partial^2\mathbf{A} = -\mu\mathbf{J}}$$

The computations can be related to Gibbs' vector algebra as follows (here $c = 1,\ \mu = 1$):

$$\partial\mathbf{F} = (\nabla\cdot\vec{E})\mathbf{e}_0 + (\nabla\wedge\vec{E})\mathbf{e}_0 + \nabla\times\vec{B} - (\nabla\cdot\vec{B})\mathbf{e}_{123} - \mathbf{e}_0\tfrac{\partial\vec{E}}{\partial t}\mathbf{e}_0 - \mathbf{e}_0\tfrac{\partial\vec{B}}{\partial t}\mathbf{e}_{123}$$
$$= (\nabla\cdot\vec{E})\mathbf{e}_0 + \nabla\times\vec{B} - \tfrac{\partial\vec{E}}{\partial t} - (\nabla\cdot\vec{B})\mathbf{e}_{123} + (\nabla\times\vec{E})\mathbf{e}_{0123} + \tfrac{\partial\vec{B}}{\partial t}\mathbf{e}_{0123}$$
$$= \rho\mathbf{e}_0 + \vec{J} = \mathbf{J}$$
$$\partial\mathbf{A} = \nabla\cdot\vec{A} + (\nabla\times\vec{A})\mathbf{e}_{123} + (\nabla V)\mathbf{e}_0 - \mathbf{e}_0\tfrac{\partial V}{\partial t}\mathbf{e}_0 - \mathbf{e}_0\tfrac{\partial\vec{A}}{\partial t}$$
$$= -\vec{E}\mathbf{e}_0 + \vec{B}\mathbf{e}_{123} = -\mathbf{F}$$

where in the last step we used $-\mathbf{e}_0\frac{\partial\vec{A}}{\partial t} = \frac{\partial\vec{A}}{\partial t}\mathbf{e}_0$.

13.2 The Maxwell equations in $\mathbb{R}^{1,3}$

We use the following definitions for the potential **A**, the current **J**, the differential operator ∂, and the electromagnetic field **F**:

$$\begin{aligned}
&A^\alpha = (\tfrac{1}{c}V, A_x, A_y, A_z), \qquad A_\alpha = (\tfrac{1}{c}V, -A_x, -A_y, -A_z)\\
&J^\alpha = (\rho c, J_x, J_y, J_z), \qquad J_\alpha = (\rho c, -J_x, -J_y, -J_z)\\
&\partial^\alpha = (\tfrac{1}{c}\tfrac{\partial}{\partial t}, -\tfrac{\partial}{\partial x}, -\tfrac{\partial}{\partial y}, -\tfrac{\partial}{\partial z}), \qquad \partial_\alpha = (\tfrac{1}{c}\tfrac{\partial}{\partial t}, \tfrac{\partial}{\partial x}, \tfrac{\partial}{\partial y}, \tfrac{\partial}{\partial z})\\
&(F^{01}, F^{02}, F^{03}) = (-\tfrac{1}{c}E_x, -\tfrac{1}{c}E_y, -\tfrac{1}{c}E_z)\\
&(F^{23}, F^{31}, F^{12}) = (-B_x, -B_y, -B_z).
\end{aligned}$$

This leads to the d'Alembert operator

$$\partial^\alpha \partial_\alpha = \frac{1}{c^2}\frac{\partial^2}{\partial t^2} - \nabla^2$$

and the equations

$$\partial^\alpha \partial_\alpha A^\beta = \mu J^\beta.$$

The equations

$$\begin{aligned}
\partial^0 A^1 - \partial^1 A^0 &= \tfrac{1}{c}\tfrac{\partial}{\partial t}A_x + \tfrac{\partial}{\partial x}\tfrac{V}{c} = -\tfrac{1}{c}E_x, \ \ldots\\
\partial^1 A^2 - \partial^2 A^1 &= -\tfrac{\partial}{\partial x}A_y + \tfrac{\partial}{\partial y}A_x = -B_z, \ \ldots
\end{aligned}$$

give

$$F^{\alpha\beta} = \partial^\alpha A^\beta - \partial^\beta A^\alpha,$$

and

$$\begin{aligned}
\partial_\alpha F^{\alpha 0} &= \tfrac{\partial}{\partial x}\tfrac{E_x}{c} + \tfrac{\partial}{\partial y}\tfrac{E_y}{c} + \tfrac{\partial}{\partial z}\tfrac{E_z}{c} = \tfrac{1}{c}\nabla\cdot\vec{E} = \tfrac{\rho}{c\varepsilon} = \mu J^0\\
\partial_\alpha F^{\alpha 1} &= -\tfrac{1}{c}\tfrac{\partial}{\partial t}\tfrac{E_x}{c} + \tfrac{\partial}{\partial y}B_z - \tfrac{\partial}{\partial z}B_y = \mu J^1, \ \ldots
\end{aligned}$$

give the Maxwell equations

$$\partial_\alpha F^{\alpha\beta} = \mu J^\beta.$$

In the Clifford algebra $\mathcal{C}\ell_{1,3}$ we have

$$\begin{aligned}
&\partial = \gamma_0\tfrac{1}{c}\tfrac{\partial}{\partial t} - \gamma_1\tfrac{\partial}{\partial x} - \gamma_2\tfrac{\partial}{\partial y} - \gamma_3\tfrac{\partial}{\partial z}\\
&\mathbf{F} = -\tfrac{1}{c}E_x\gamma_{01} - \tfrac{1}{c}E_y\gamma_{02} - \tfrac{1}{c}E_z\gamma_{03} - B_x\gamma_{23} - B_y\gamma_{31} - B_z\gamma_{12}\\
&\mathbf{A} = \tfrac{1}{c}V\gamma_0 + A_x\gamma_1 + A_y\gamma_2 + A_z\gamma_3\\
&\mathbf{J} = c\rho\gamma_0 + J_x\gamma_1 + J_y\gamma_2 + J_z\gamma_3
\end{aligned}$$

and the equations

$$\begin{aligned}
&\partial\mathbf{A} = \gamma_{01}(\tfrac{1}{c}\tfrac{\partial}{\partial t}A_x + \tfrac{\partial}{\partial x}\tfrac{V}{c}) + \ldots + \gamma_{12}(-\tfrac{\partial}{\partial x}A_y + \tfrac{\partial}{\partial y}A_x)\\
&\partial\mathbf{F} = \gamma_0(\tfrac{\partial}{\partial x}\tfrac{E_x}{c} + \tfrac{\partial}{\partial y}\tfrac{E_y}{c} + \tfrac{\partial}{\partial z}\tfrac{E_z}{c}) + \gamma_1(-\tfrac{1}{c^2}\tfrac{\partial}{\partial t}E_x - \tfrac{\partial}{\partial z}B_y + \tfrac{\partial}{\partial y}B_z) + \ldots
\end{aligned}$$

which lead to

$$\boxed{\partial \mathbf{A} = \mathbf{F}, \quad \partial \mathbf{F} = \mu \mathbf{J}, \quad \partial^2 \mathbf{A} = \mu \mathbf{J}}$$

13.3 Comparison of $\mathbb{R}^{3,1}$ and $\mathbb{R}^{1,3}$

The equation $\partial \mathbf{A} = \pm \mathbf{F}$ has opposite signs in opposite metrics. This means that the raising part $\partial \wedge \mathbf{A} = \pm \mathbf{F}$ has opposite signs in opposite metrics, whereas the lowering part $\partial \,\lrcorner\, \mathbf{A} = 0$ is independent of metric. The Maxwell equations $\partial \mathbf{F} = \mu \mathbf{J}$ have the same signs in both metrics. This means that the lowering part $\partial \,\lrcorner\, \mathbf{F} = \mu \mathbf{J}$ is invariant under the metric swap (and that in both metrics $\partial \wedge \mathbf{F} = 0$). The above unexpected results are consequences of our definition: the differential operator ∂ experiences a sign change under the metric swap, that is,

$$\boxed{\text{opp}[\partial] = -\partial}$$

Spinors and Observables in $\mathbb{R}^{3,1}$

In the rest of this chapter we shall study the Dirac equation, spinors and observables in $\mathbb{R}^{3,1}$. Our special concern will be the behavior of spinors under the transition to the opposite metric.

Since going to the opposite algebra interchanges left and right ideals, we will study real ideal spinors $\phi \in \mathcal{C}\ell_{1,3}\frac{1}{2}(1 - \gamma_{03})$ in conjunction with their *opposite-reverses*

$$\underline{\phi} = \text{opp}[\tilde{\phi}] \in \mathcal{C}\ell_{3,1}\frac{1}{2}(1 + \mathbf{e}_{03})$$

(both ϕ and the opposite of its reverse $\underline{\phi}$ are in left ideals). Clearly, $\text{opp}[\tilde{\phi}] = \text{opp}[\phi]^{\sim}$.

For instance, the Dirac equation for the real ideal spinors, $\partial \phi \gamma_{21} = e\mathbf{A}\phi + m\phi$, $\phi \in \mathcal{C}\ell_{1,3}\frac{1}{2}(1 - \gamma_{03})$, is transformed by the opposite-reversion to $\partial \hat{\underline{\phi}} \mathbf{e}_{21} = e\mathbf{A}\hat{\underline{\phi}} + m\underline{\phi}$, $\underline{\phi} \in \mathcal{C}\ell_{3,1}\frac{1}{2}(1 + \mathbf{e}_{03})$, [2] and further by grade involution to

$$\partial \underline{\phi} \mathbf{e}_{21} = e\mathbf{A}\underline{\phi} - m\hat{\underline{\phi}}.$$

However, this is not a nice formula, because we have to explain the occurrence of the grade involution in the last term. There are even more interpretational difficulties for the opposite-reverses $\underline{\psi} = \text{opp}[\tilde{\psi}]$ [of Dirac spinor $\psi = \psi\frac{1}{2}(1 +$

2 Note that $\phi \in \mathcal{C}\ell_{1,3}\frac{1}{2}(1 - \gamma_{03})$ is in a graded minimal left ideal of $\mathcal{C}\ell_{1,3} \simeq \text{Mat}(2, \mathbb{H})$, which is also an ungraded minimal left ideal, while $\underline{\phi} \in \mathcal{C}\ell_{3,1}\frac{1}{2}(1 + \mathbf{e}_{03})$ is in a graded minimal left ideal of $\mathcal{C}\ell_{3,1} \simeq \text{Mat}(4, \mathbb{R})$, which is not an ungraded minimal left ideal [the minimal left ideals of $\mathcal{C}\ell_{3,1}$ are not graded].

$\gamma_0)\frac{1}{2}(1+i\gamma_{12})]$ and $\underline{\Phi} = \text{opp}[\tilde{\Phi}]$ [of mother spinor $\Phi = \Phi\frac{1}{2}(1+\gamma_0)$], since $\underline{\psi} \notin (\mathbb{C}\otimes C\ell_{3,1})\frac{1}{2}(1\pm i\mathbf{e}_0)$ and $\underline{\Phi}$ is not in any proper left ideal of $C\ell_{3,1}$. An obvious attempt for a possible solution would be to study $\underline{\eta} = \underline{\psi}\frac{1}{2}(1-i\mathbf{e}_0)$, but then (like evaluating $4\psi\tilde{\psi}^* = Z$ in the case $\mathbb{R}^{1,3}$) for the aggregate of observables

$$4\underline{\eta}\tilde{\underline{\eta}}^* = \frac{1}{2}(\Omega_1 - i\mathbf{J} - i\mathbf{S} + \mathbf{K}\mathbf{e}_{0123} + \Omega_2\mathbf{e}_{0123}), \qquad \underline{\eta}\tilde{\underline{\eta}}^* = 0,$$

and we would have the inconvenience of an extra factor $\frac{1}{2}$. This shortcoming could be circumvented by multiplying $\underline{\eta}$ by $\sqrt{2}$, but then the relation to the original Dirac spinor ψ would be irrational. Again there is an obvious solution: multiply $\underline{\eta}$ by $1-i$, which has absolute value $\sqrt{2}$, and study instead the *flip* of the opposite-reverse, that is, the *tilted* spinor

$$\underset{\sim}{\psi} = (1-i)\underline{\psi}\frac{1}{2}(1-i\mathbf{e}_0) = \underline{\psi}\frac{1}{2}(1-\mathbf{e}_{012})(1-\mathbf{e}_{12}), \qquad \underline{\psi} = \text{opp}[\tilde{\psi}],$$

for which

$$4\underset{\sim}{\psi}\tilde{\underset{\sim}{\psi}}^* = \Omega_1 - i\mathbf{J} - i\mathbf{S} + \mathbf{K}\mathbf{e}_{0123} + \Omega_2\mathbf{e}_{0123}, \qquad \underset{\sim}{\psi}\underset{\sim}{\psi}^* = 0.$$

The opposite-reverse of $\psi\gamma_0 = \psi$ is $\hat{\underline{\psi}}\mathbf{e}_0 = \underline{\psi}$. Therefore, we find

$$(1+i)\underset{\sim}{\psi} = \underline{\psi}(1-i\mathbf{e}_0) = \underline{\psi} + i\hat{\underline{\psi}} = (1+i)\underline{\psi}_0 + (1-i)\underline{\psi}_1$$

$[\underline{\psi}_0 = \text{even}(\underline{\psi}),\ \underline{\psi}_1 = \text{odd}(\underline{\psi})]$ which implies $\underset{\sim}{\psi} = \underline{\psi}_0 - i\underline{\psi}_1$, or since $i\underline{\psi} = \underline{\psi}\mathbf{e}_{12}$,

$$\underset{\sim}{\psi} = \underline{\psi}_0 - \underline{\psi}_1\mathbf{e}_{12} \qquad (\underline{\psi} = \text{opp}[\tilde{\psi}], \quad \psi \in \mathbb{C}\otimes C\ell_{1,3}).$$

Similarly, we define the *tilted* spinor for the mother of all real spinors Φ and for the real ideal spinor ϕ:

$$\underset{\sim}{\Phi} = \underline{\Phi}_0 - \underline{\Phi}_1\mathbf{e}_{12} = \underline{\Phi}\frac{1}{2}(1-\mathbf{e}_{012})(1-\mathbf{e}_{12}), \qquad \underline{\Phi} = \text{opp}[\tilde{\Phi}],$$

$$\underset{\sim}{\phi} = \underline{\phi}_0 - \underline{\phi}_1\mathbf{e}_{12} \neq \underline{\phi}\frac{1}{2}(1-\mathbf{e}_{012})(1-\mathbf{e}_{12}), \qquad \underline{\phi} = \text{opp}[\tilde{\phi}].$$

Of course, for the spinor operator $\underset{\sim}{\Psi} = \underline{\Psi}$ $(= \text{opp}[\tilde{\Psi}])$.

The transition back from $\underset{\sim}{\psi} \in (\mathbb{C}\otimes C\ell_{3,1})\frac{1}{2}(1-i\mathbf{e}_0)\frac{1}{2}(1-i\mathbf{e}_{12})$ to $\psi \in (\mathbb{C}\otimes C\ell_{1,3})\frac{1}{2}(1+\gamma_0)\frac{1}{2}(1+i\gamma_{12})$ is given by tilting again,

$$\psi = \text{opp}[\tilde{\underset{\sim}{\psi}}_0] - \text{opp}[\tilde{\underset{\sim}{\psi}}_1]\gamma_{12}, \qquad \underset{\sim}{\psi} \in \mathbb{C}\otimes C\ell_{3,1},$$

since

$$\begin{aligned}\text{opp}[\tilde{\underset{\sim}{\psi}}_0] - \text{opp}[\tilde{\underset{\sim}{\psi}}_1]\gamma_{12} &= (\tilde{\psi}_0)^{\tilde{}} - \text{opp}[(-\underline{\psi}_1\mathbf{e}_{12})^{\tilde{}}]\gamma_{12} \\ &= \psi_0 - (-\gamma_{12}\tilde{\psi}_1)^{\tilde{}}\gamma_{12} = \psi_0 - \psi_1\gamma_{12}\gamma_{12}.\end{aligned}$$

13.4 The Dirac equation in $\mathbb{R}^{3,1}$

The opposite-reverse Dirac equation

$$\partial \underline{\phi} \mathbf{e}_{21} = e\mathbf{A}\underline{\phi} - m\hat{\underline{\phi}}, \qquad \underline{\phi} \in C\ell_{3,1}\,\frac{1}{2}(1+\mathbf{e}_{03}),$$

splits into even and odd parts $[\underline{\phi}_0 = \mathrm{even}(\underline{\phi}),\ \underline{\phi}_1 = \mathrm{odd}(\underline{\phi})]$

$$\partial \underline{\phi}_1 \mathbf{e}_{21} = e\mathbf{A}\underline{\phi}_1 - m\underline{\phi}_0$$
$$\partial \underline{\phi}_0 \mathbf{e}_{21} = e\mathbf{A}\underline{\phi}_0 + m\underline{\phi}_1.$$

Recalling that $\underset{\sim}{\phi}_0 = \underline{\phi}_0$ and $\underset{\sim}{\phi}_1 = \underline{\phi}_1 \mathbf{e}_{21}$ we find

$$\partial\, \underset{\sim}{\phi}_0 = e\mathbf{A}\, \underset{\sim}{\phi}_0 \mathbf{e}_{12} - m\, \underset{\sim}{\phi}_1$$
$$\partial\, \underset{\sim}{\phi}_1 = e\mathbf{A}\, \underset{\sim}{\phi}_1 \mathbf{e}_{12} - m\, \underset{\sim}{\phi}_0$$

which added together result in

$$\partial\, \underset{\sim}{\phi} = e\mathbf{A}\, \underset{\sim}{\phi} \mathbf{e}_{12} - m\, \underset{\sim}{\phi}, \qquad \underset{\sim}{\phi} \in C\ell_{3,1}\,\frac{1}{2}(1+\mathbf{e}_{03}).$$

Similarly, the flip of the opposite-reverse, or simply tilted, Dirac spinor $\underset{\sim}{\psi}$ obeys

$$\partial\, \underset{\sim}{\psi} = ie\mathbf{A}\, \underset{\sim}{\psi} - m\, \underset{\sim}{\psi}, \qquad \underset{\sim}{\psi} \in (\mathbb{C} \otimes C\ell_{3,1})\frac{1}{2}(1-i\mathbf{e}_0)\frac{1}{2}(1-i\mathbf{e}_{12}),$$

a formula found essentially in Benn & Tucker 1987 p. 284 (and p. 256). So the tilted mother spinor $\underset{\sim}{\Phi} = 4\,\mathrm{Re}(\underset{\sim}{\psi}) = \underset{\sim}{\Psi}(1-\mathbf{e}_{012})$ [3] obeys

$$\partial\, \underset{\sim}{\Phi} = e\mathbf{A}\, \underset{\sim}{\Phi} \mathbf{e}_{12} - m\, \underset{\sim}{\Phi}, \qquad \underset{\sim}{\Phi} \in C\ell_{3,1}\,\frac{1}{2}(1-\mathbf{e}_{012}),$$

which decomposed into even and odd parts (and recalling that $\underset{\sim}{\Phi}_0 = -\underset{\sim}{\Phi}_1 \mathbf{e}_{012}$, $\underset{\sim}{\Phi}_1 = -\underset{\sim}{\Phi}_0 \mathbf{e}_{012}$) results in the Dirac-Hestenes equation in the opposite metric,

$$\boxed{\partial\, \underset{\sim}{\Psi} \mathbf{e}_{21} - e\mathbf{A}\, \underset{\sim}{\Psi} = m\, \underset{\sim}{\Psi} \mathbf{e}_0}$$

where $\underset{\sim}{\Psi} : \mathbb{R}^{3,1} \to C\ell^+_{3,1}$.

13.5 Bilinear covariants in $\mathbb{R}^{3,1}$

Recall that for $\underset{\sim}{\psi} \in (\mathbb{C} \otimes C\ell_{3,1})\frac{1}{2}(1-i\mathbf{e}_0)\frac{1}{2}(1-i\mathbf{e}_{12})$

$$4\,\underset{\sim}{\psi}\,\underset{\sim}{\bar{\psi}}^* = \Omega_1 - i\mathbf{J} - i\mathbf{S} + \mathbf{K}\mathbf{e}_{0123} + \Omega_2 \mathbf{e}_{0123}, \qquad \underset{\sim}{\psi}\,\underset{\sim}{\tilde{\psi}}^* = 0.$$

Compute the bilinear covariants in the opposite metric $\mathbb{R}^{3,1}$:

3 Note that $\underset{\sim}{\Phi} = \underset{\sim}{\phi}(1-\mathbf{e}_{123})$ and $\underset{\sim}{\phi} = \underset{\sim}{\Phi}\frac{1}{2}(1+\mathbf{e}_{03})$.

$$\Omega_1 = 4\langle \bar{\underset{\sim}{\psi}}^* \underset{\sim}{\psi}\rangle_0$$
$$J_\mu = 4\langle \bar{\underset{\sim}{\psi}}^* i\mathbf{e}_\mu \underset{\sim}{\psi}\rangle_0 \qquad (J^0 = -J_0)$$
$$S_{\mu\nu} = -4\langle \bar{\underset{\sim}{\psi}}^* i\mathbf{e}_{\mu\nu} \underset{\sim}{\psi}\rangle_0$$
$$K_\mu = -4\langle \bar{\underset{\sim}{\psi}}^* \mathbf{e}_{0123}\mathbf{e}_\mu \underset{\sim}{\psi}\rangle_0$$
$$\Omega_2 = -4\langle \bar{\underset{\sim}{\psi}}^* \mathbf{e}_{0123} \underset{\sim}{\psi}\rangle_0.$$

(Observe that the coordinates J^μ of $\mathbf{J}$ have the same numerical values in $\mathcal{C}\ell_{1,3}$ and $\mathcal{C}\ell_{3,1}$. In contrast, the coordinates J_μ are opposite in $\mathcal{C}\ell_{1,3}$ and $\mathcal{C}\ell_{3,1}$.)

13.6 Fierz identities in $\mathbb{R}^{3,1}$

The Fierz identities in the opposite metric are

$$\mathbf{J}^2 = -(\Omega_1^2 + \Omega_2^2), \qquad \mathbf{K}^2 = -\mathbf{J}^2$$
$$\mathbf{J} \cdot \mathbf{K} = 0, \qquad \mathbf{J} \wedge \mathbf{K} = -(\Omega_2 + \mathbf{e}_{0123}\Omega_1)\mathbf{S}.$$

Note also that $\mathbf{S}^2 = (\Omega_2 - \mathbf{e}_{0123}\Omega_1)^2$.

Exercise. Derive the real theory from the mother of all real tilted spinors $4\,\mathrm{Re}(\underset{\sim}{\psi}) = \underset{\approx}{\Phi} = \underset{\approx}{\Phi}\frac{1}{2}(1 - \mathbf{e}_{012}) \in \mathcal{C}\ell_{3,1}\,\frac{1}{2}(1 - \mathbf{e}_{012})$. ∎

13.7 Decomposition of boomerangs in $\mathbb{R}^{3,1}$

Write for $\underset{\sim}{\psi} \in (\mathbb{C} \otimes \mathcal{C}\ell_{3,1})\frac{1}{2}(1 - i\mathbf{e}_0)\frac{1}{2}(1 - i\mathbf{e}_{12})$

$$Z = 4\,\underset{\sim}{\psi}\,\bar{\underset{\sim}{\psi}}^*$$
$$K = \Omega_1 + \mathbf{K}\mathbf{e}_{0123} + \Omega_2\mathbf{e}_{0123}, \qquad L = -\mathbf{J} - \mathbf{S}$$
$$\underset{\sim}{S} = \Omega_1 - i\mathbf{S} + \Omega_2\mathbf{e}_{0123}, \qquad \underset{\sim}{\Sigma} = 1 - i\mathbf{e}_{0123}\mathbf{J}\mathbf{K}^{-1}$$
$$P = \Omega_1 - i\mathbf{J} + \Omega_2\mathbf{e}_{0123}, \qquad \Pi = 1 - i\mathbf{J}(\Omega_1 + \Omega_2\mathbf{e}_{0123})^{-1}$$
$$\Gamma = K(\Omega_1 + \Omega_2\mathbf{e}_{0123})^{-1}$$

[Z, K, P, Π, Γ are not the same as those in the case of $\mathbb{R}^{1,3}$ but instead as a sample Z in $\mathbb{C} \otimes \mathcal{C}\ell_{3,1}$ is obtained by sending Z in $\mathbb{C} \otimes \mathcal{C}\ell_{1,3}$ to $\mathrm{even}(\tilde{Z}) - i\,\mathrm{odd}(\tilde{Z})$]. Then

$$Z = K + iL = K\underset{\sim}{\Sigma} = \underset{\sim}{\Sigma}K = \Pi K = K\bar{\Pi}^* \quad (\neq K\Pi \text{ unless } \Omega_2 = 0)$$
$$= P\underset{\sim}{\Sigma} = \underset{\sim}{\Sigma}P = \Pi\underset{\sim}{S} = \underset{\sim}{S}\bar{\Pi}^*$$
$$= \Gamma P = P\bar{\Gamma} \quad \text{(no complex conjugation)}$$
$$= K\left(1 + i\frac{1}{2\Omega_1}L\right),$$

$$K^2 = 2\Omega_1 K = -L^2, \qquad KL = LK = 2\Omega_1 L$$
$$\Pi = P(\Omega_1 + \Omega_2 \mathbf{e}_{0123})^{-1}, \qquad \utilde{\Sigma} = \utilde{S}(\Omega_1 + \Omega_2 \mathbf{e}_{0123})^{-1}.$$

13.8 Multiplication by $i = \sqrt{-1}$ in $\mathbb{C} \otimes C\ell_{3,1}$

Write

$$j = (\Omega_1 - \Omega_2 \mathbf{e}_{0123})^{-1}\mathbf{J} = -\mathbf{e}_{0123}\mathbf{S}\mathbf{K}^{-1} = \utilde{\Psi}\mathbf{e}_0\,\utilde{\Psi}^{-1}$$
$$\utilde{s} = \mathbf{e}_{0123}\mathbf{J}\mathbf{K}^{-1} = (\Omega_1 + \Omega_2 \mathbf{e}_{0123})^{-1}\mathbf{S} = \utilde{\Psi}\mathbf{e}_{12}\,\utilde{\Psi}^{-1}.$$

Then $K = jL = \utilde{s}L$, $\Pi = 1 - ij$ and $\utilde{\Sigma} = 1 - i\utilde{s}$. Also

$$i\utilde{\psi} = j\utilde{\psi} = \utilde{s}\utilde{\psi} = \utilde{\psi}\mathbf{e}_{12} = \utilde{\psi}\mathbf{e}_0 \neq \utilde{\psi}\mathbf{e}_{0123} \qquad \text{tilted Dirac}$$
$$j\utilde{\Phi} = \utilde{s}\utilde{\Phi} = \utilde{\Phi}\mathbf{e}_{12} = \utilde{\Phi}\mathbf{e}_0 \qquad \text{tilted mother}$$
$$j\utilde{\phi} = \utilde{s}\utilde{\phi} = \utilde{\phi}\mathbf{e}_{12} = \utilde{\phi}\mathbf{e}_{0123} \neq \utilde{\phi}\mathbf{e}_0 \qquad \text{tilted ideal}$$
$$\utilde{s}\utilde{\Psi} = \utilde{\Psi}\mathbf{e}_{12} \quad (\text{but } j\utilde{\Psi}\mathbf{e}_0 = -\utilde{\Psi}) \qquad \text{operator.}$$

13.9 Some differences between $C\ell_{1,3}$ and $C\ell_{3,1}$

Compute as a sample in $C\ell_{3,1}$

$$\utilde{\Phi}\tilde{\utilde{\Phi}} = 0, \qquad \frac{1}{2}\utilde{\Phi}\bar{\utilde{\Phi}} = K, \qquad \frac{1}{2}\utilde{\Phi}\mathbf{e}_0\bar{\utilde{\Phi}} = -L$$
$$\frac{1}{2}\utilde{\Phi}\mathbf{e}_3\bar{\utilde{\Phi}} = 0, \qquad \frac{1}{2}\utilde{\Phi}\mathbf{e}_3\tilde{\utilde{\Phi}} = -K\mathbf{e}_{0123}.$$

Note that

$$\langle P\gamma_0\tilde{P}\gamma_0\rangle_0 = \langle Q\gamma_0\tilde{Q}\gamma_0\rangle_0 > 0 \quad \text{for non-zero} \quad \psi \in \mathbb{C} \otimes C\ell_{1,3}$$

while

$$\langle K\mathbf{e}_0\tilde{K}\mathbf{e}_0\rangle_0 = \langle L\mathbf{e}_0\tilde{L}\mathbf{e}_0\rangle_0 \gtrless 0 \quad \text{for non-zero} \quad \utilde{\psi} \in \mathbb{C} \otimes C\ell_{3,1}$$
$$\langle K\mathbf{e}_0\bar{K}\mathbf{e}_0^{-1}\rangle_0 \neq \langle L\mathbf{e}_0\bar{L}\mathbf{e}_0^{-1}\rangle_0 > 0$$

so that it is possible that $K = 0$ while $L \neq 0$ (this happens in the case $\Omega_1 = \Omega_2 = 0$, $\mathbf{K} = 0$ of Majorana spinors).

13.10 Charge conjugate in $\mathbb{R}^{3,1}$

The charge conjugate of the tilted Dirac spinor is obtained as follows:

$$
\begin{aligned}
&\psi_C = \hat{\psi}^* \gamma_1 \ \text{ take opposite-reverse } \ \underline{\psi}_C = \underline{\psi}^* \mathbf{e}_1 \ \text{ and tilt} \\
&\underset{\sim}{\psi}_C = (1-i)\underline{\psi}^* \mathbf{e}_1 \tfrac{1}{2}(1 - i\mathbf{e}_0) = (1-i)\underline{\psi}^* \tfrac{1}{2}(1 + i\mathbf{e}_0)\mathbf{e}_1 \\
&\quad = [(1+i)\underline{\psi}\tfrac{1}{2}(1 - i\mathbf{e}_0)]^* \mathbf{e}_1 = -i[(1-i)\underline{\psi}\tfrac{1}{2}(1 - i\mathbf{e}_0)]^* \mathbf{e}_1 \\
&\quad = -i\,\underset{\sim}{\psi}^* \mathbf{e}_1 = -\underset{\sim}{\psi}^* \mathbf{e}_1 \mathbf{e}_{12} = -\underset{\sim}{\psi}^* \mathbf{e}_2 \quad \text{ or} \\
&\quad = -\underset{\sim}{\psi}^* \mathbf{e}_1 \mathbf{e}_0 = \underset{\sim}{\psi}^* \mathbf{e}_{01} \in (\mathbb{C} \otimes \mathcal{C}\ell_{3,1})\tfrac{1}{2}(1 - i\mathbf{e}_0)\tfrac{1}{2}(1 - i\mathbf{e}_{12}).
\end{aligned}
$$

Numerical Example

Start from $\mathcal{C}\ell_{1,3}$. Take a column spinor

$$
\psi = \begin{pmatrix} 4+3i \\ 1+6i \\ 5+2i \\ 2+i \end{pmatrix} \in \mathbb{C}^4.
$$

Then $Z = 4\psi\psi^\dagger \gamma_0$

$$
\begin{aligned}
&= 4\begin{pmatrix} 25 & 22-21i & -26-7i & -11-2i \\ 22+21i & 37 & -17-28i & -8-11i \\ 26-7i & 17-28i & -29 & -12+i \\ 11-2i & 8-11i & -12-i & -5 \end{pmatrix} \\
&= \Omega_1 + \mathbf{J} + i\mathbf{S} + i\mathbf{K}\gamma_{0123} + \Omega_2\gamma_{0123}
\end{aligned}
$$

where

$$
\begin{aligned}
&\Omega_1 = 28 \\
&\mathbf{J} = 96\gamma_0 + 56\gamma_1 + 52\gamma_2 + 36\gamma_3 \\
&\mathbf{S} = 60\gamma_{01} - 12\gamma_{02} - 8\gamma_{03} - 36\gamma_{12} - 40\gamma_{13} + 20\gamma_{23} \\
&\mathbf{K} = 68\gamma_0 + 68\gamma_1 + 44\gamma_2 + 12\gamma_3 \\
&\Omega_2 = -36.
\end{aligned}
$$

In the opposite algebra $\mathcal{C}\ell_{3,1}$ we must first fix the matrix representation, for instance, using the Pauli spin matrices σ_k,

$$
\mathbf{e}_0 = \begin{pmatrix} i & 0 \\ 0 & -i \end{pmatrix}, \qquad \mathbf{e}_k = \begin{pmatrix} 0 & \sigma_k \\ \sigma_k & 0 \end{pmatrix}
$$

corresponding to the tilted primitive idempotent

$$\underset{\sim}{f} = \frac{1}{2}(1 - i\mathbf{e}_0)\,\frac{1}{2}(1 - i\mathbf{e}_{12})$$

and the tilted spinor basis

$$\begin{aligned}
\underset{\sim}{f}_1 &= \frac{1}{4}(1 - i\mathbf{e}_0 - i\mathbf{e}_{12} - \mathbf{e}_{012}) && = \underset{\sim}{f} \\
\underset{\sim}{f}_2 &= \frac{1}{4}(\mathbf{e}_{13} - i\mathbf{e}_{23} - i\mathbf{e}_{013} - \mathbf{e}_{023}) && = \mathbf{e}_{13}\underset{\sim}{f} \\
\underset{\sim}{f}_3 &= \frac{1}{4}(\mathbf{e}_3 + i\mathbf{e}_{03} - i\mathbf{e}_{123} + \mathbf{e}_{0123}) && = \mathbf{e}_3\underset{\sim}{f} \\
\underset{\sim}{f}_4 &= \frac{1}{4}(\mathbf{e}_1 - i\mathbf{e}_2 + i\mathbf{e}_{01} + \mathbf{e}_{02}) && = \mathbf{e}_1\underset{\sim}{f}.
\end{aligned}$$

Then the tilted spinor is

$$\underset{\sim}{\psi} = \begin{pmatrix} 4+3i \\ 1+6i \\ 2-5i \\ 1-2i \end{pmatrix}$$

and the boomerang $Z = 4\underset{\sim}{\psi}\,\underset{\sim}{\psi}^\dagger(-i\mathbf{e}_0) = 4\underset{\sim}{\psi}\,\underset{\sim}{\psi}^\dagger i\mathbf{e}^0$

$$\begin{aligned}
&= 4\begin{pmatrix} 25 & 22-21i & 7-26i & 2-11i \\ 22+21i & 37 & 28-17i & 11-8i \\ -7-26i & -28-17i & -29 & -12+i \\ -2-11i & -11-8i & -12-i & -5 \end{pmatrix} \\
&= \Omega_1 - i\mathbf{J} - i\mathbf{S} + \mathbf{K}\mathbf{e}_{0123} + \Omega_2\mathbf{e}_{0123}
\end{aligned}$$

where

$$\begin{aligned}
\Omega_1 &= 28 \\
\mathbf{J} &= 96\mathbf{e}_0 + 56\mathbf{e}_1 + 52\mathbf{e}_2 + 36\mathbf{e}_3 \\
\mathbf{S} &= 60\mathbf{e}_{01} - 12\mathbf{e}_{02} - 8\mathbf{e}_{03} - 36\mathbf{e}_{12} - 40\mathbf{e}_{13} + 20\mathbf{e}_{23} \\
\mathbf{K} &= 68\mathbf{e}_0 + 68\mathbf{e}_1 + 44\mathbf{e}_2 + 12\mathbf{e}_3 \\
\Omega_2 &= -36.
\end{aligned}$$

Note that for $u \in \mathbb{C} \otimes C\ell_{3,1}$, $\bar{u}^* = \mathbf{e}_0 u^\dagger \mathbf{e}_0^{-1}$.

Note that Z for $\mathbb{R}^{1,3}$ and Z for $\mathbb{R}^{3,1}$ are related via a similarity transformation by $\frac{1}{\sqrt{2}}(I + i\gamma_0)$.

Summary

To realize the transition to the opposite metric we use the rules

$$\text{opp}[\partial] = -\partial \quad \text{and} \quad \text{opp}[ab] = b_0a_0 + b_0a_1 + b_1a_0 - b_1a_1$$

and apply reversion to get the tilted spinors also in left ideals. Thus, the two sides of $\partial \mathbf{A} = \mathbf{F}$ in $\mathbb{R}^{1,3}$ are transformed as

$$\text{opp}[\partial \mathbf{A}]^{\tilde{}} = -(-\partial)\tilde{\mathbf{A}} = \partial \mathbf{A}$$
$$\text{opp}[\mathbf{F}]^{\tilde{}} = \tilde{\mathbf{F}} = -\mathbf{F}$$

and so we have $\partial \mathbf{A} = -\mathbf{F}$ in $\mathbb{R}^{3,1}$. The two sides of $\partial \mathbf{F} = \mathbf{J}$ in $\mathbb{R}^{1,3}$ are transformed as

$$\text{opp}[\partial \mathbf{F}]^{\tilde{}} = (-\partial)\tilde{\mathbf{F}} = \partial \mathbf{F}$$
$$\text{opp}[\mathbf{J}]^{\tilde{}} = \tilde{\mathbf{J}} = \mathbf{J}$$

and so we have $\partial \mathbf{F} = \mathbf{J}$ in $\mathbb{R}^{3,1}$. The terms of $\partial \Psi \gamma_{21} - e\mathbf{A}\Psi = m\Psi\gamma_0$ in $\mathbb{R}^{1,3}$ are transformed as

$$\text{opp}[\partial \Psi \gamma_{21}]^{\tilde{}} = (-\partial)\tilde{\Psi}\tilde{\mathbf{e}}_{21} = \partial \tilde{\Psi}\mathbf{e}_{21}$$
$$\text{opp}[\mathbf{A}\Psi]^{\tilde{}} = \tilde{\mathbf{A}}\tilde{\Psi} = \mathbf{A}\tilde{\Psi}$$
$$\text{opp}[\Psi\gamma_0]^{\tilde{}} = \tilde{\Psi}\tilde{\mathbf{e}}_0 = \tilde{\Psi}\mathbf{e}_0$$

and so we have $\partial\tilde{\Psi}\mathbf{e}_{21} - e\mathbf{A}\tilde{\Psi} = m\tilde{\Psi}\mathbf{e}_0$ in $\mathbb{R}^{3,1}$. [Earlier in this chapter we wrote $\underset{\sim}{\Psi}$ for $\tilde{\Psi}$.] Note in particular that the Dirac-Hestenes equation has the same form in both metrics, only the spinor operators are reversed. For complex ideal spinors the situation is more complicated, an extra flip is needed to complete the metric tilt.

In our differential operator

$$\partial = \mathbf{e}^0 \frac{\partial}{\partial x^0} + \mathbf{e}^1 \frac{\partial}{\partial x^1} + \mathbf{e}^2 \frac{\partial}{\partial x^2} + \mathbf{e}^3 \frac{\partial}{\partial x^3}$$

we have used an orthonormal basis, but this formula gives the same ∂ for any basis $\{\mathbf{e}_0, \mathbf{e}_1, \mathbf{e}_2, \mathbf{e}_3\}$ for $\mathbb{R}^{3,1}$ when $\mathbf{e}^\mu \cdot \mathbf{e}_\nu = \delta^\mu_\nu$, that is, when $\{\mathbf{e}_0, \mathbf{e}_1, \mathbf{e}_2, \mathbf{e}_3\}$ and $\{\mathbf{e}^0, \mathbf{e}^1, \mathbf{e}^2, \mathbf{e}^3\}$ are reciprocal. In this sense our differential operator is not only Lorentz covariant but also invariant under all of $GL(4, \mathbb{R})$.

Note that the raising differential $\partial \wedge \mathbf{f}$ is metric dependent and therefore it is *not* related to the exterior differential $d \wedge f$ [in a metric inpendent way]. In general, in dimension n, not necessarily $n = 4$, the lowering differential $\partial \lrcorner \mathbf{f}$ is metric independent and related to the exterior differential by

$$\partial \lrcorner \mathbf{f} = \pm[d \wedge (\mathbf{f} \lrcorner w^*)] \lrcorner \mathbf{w}$$

for an n-volume $\mathbf{w} \in \bigwedge^n V$ such that $w^* \lrcorner \mathbf{w} = 1$ for $w^* \in \bigwedge^n V^*$. The relation $(\partial \lrcorner \mathbf{f}) \lrcorner \mathbf{w} = d \wedge (\mathbf{f} \lrcorner w^*)$ requires identification of multivector-valued functions with differential forms, which is possible only in flat spaces, while multivectors cannot be differentiated on curved spaces. Such an identification can be carried out by lowering the coordinate-indices of a multivector by means of the metric tensor $g_{\mu\nu}$ (or raising the indices of a differential form by the inverse $g^{\mu\nu}$).

Exercises

Show that

1. $\underset{\sim}{\Psi}\mathbf{e}_0\,\bar{\underset{\sim}{\Psi}} = \mathbf{J}$. 2. $\underset{\sim}{\Psi}\mathbf{e}_{12}\,\bar{\underset{\sim}{\Psi}} = \mathbf{S}$.
3. $\underset{\sim}{\Psi}(1 - i\mathbf{e}_0)(1 - i\mathbf{e}_{12})\,\bar{\underset{\sim}{\Psi}} = K + iL$. 4. $\underset{\sim}{\phi}\mathbf{e}_0\,\bar{\underset{\sim}{\phi}} = -L$.
5. $\underset{\sim}{\psi}_0 = -\underset{\sim}{\psi}_1\mathbf{e}_{012}$, $\underset{\sim}{\psi}_1 = -\underset{\sim}{\psi}_0\mathbf{e}_{012}$.
6. $\underset{\sim}{\psi}_0 + \underset{\sim}{\psi}_1\mathbf{e}_{12} \in (\mathbb{C}\otimes C\ell_{3,1})\frac{1}{2}(1 + i\mathbf{e}_0)$.

Write $\text{flip}(u) = u_0 - u_1\mathbf{e}_{12}$ and recall that the opposite-reverse of $\mathbf{A}\psi$ is $\mathbf{A}\underset{\sim}{\hat{\psi}}$ (for a vector $\mathbf{A}$). Show that

7. $\text{flip}(\mathbf{A}\hat{\psi}) = -\underset{\sim}{\mathbf{A}}\,\underset{\sim}{\psi}\mathbf{e}_{12}$ $[\Rightarrow \partial\,\underset{\sim}{\psi} = e\mathbf{A}\,\underset{\sim}{\psi}\mathbf{e}_{12} - m\,\underset{\sim}{\psi}]$.
8. $\mathbf{A}\psi$ and $\mathbf{A}\Phi = 4\,\text{Re}(\mathbf{A}\psi)$ correspond to $\mathbf{A}\Psi\gamma_0 = \text{even}(\mathbf{A}\Phi)$, and in the opposite algebra to $-\mathbf{A}\,\underset{\sim}{\Psi}\mathbf{e}_0$.
9. $4\,\underset{\sim}{\psi}_C\,\bar{\underset{\sim}{\psi}}_C^* = -4(\,\underset{\sim}{\psi}\,\bar{\underset{\sim}{\psi}}^*)^*$. [This means that $\mathbf{J}$, $\mathbf{S}$ are preserved under charge conjugation while Ω_1, $\mathbf{K}$, Ω_2 swap their signs – as in $\mathbb{R}^{1,3}$. This should be contrasted with Crumeyrolle (1990 p. 135, l. -9), who considers charge conjugation in conjunction with a scalar product of spinors induced by the reversion (composed with complex conjugation), in which case $\mathbf{S}$, $\mathbf{K}$ are preserved and Ω_1, $\mathbf{J}$, Ω_2 swap signs. Crumeyrolle's numerical results are not directly related to the Bjorken & Drell formulation of the Dirac theory, as he induces spinor spaces by totally isotropic subspaces of $\mathbb{C}\otimes\mathbb{R}^{3,1}$. To relate the results one must permute the primitive idempotents by an algebra automorphism of the Clifford algebra in such a way that dimension grades are mixed. The next exercise gives a hint on how the scalar product of spinors induced by the reversion (composed with complex conjugation) can be used to find the bilinear covariants.]
10. $K_\mu = 4\langle\,\tilde{\underset{\sim}{\psi}}^*\mathbf{e}_\mu\,\underset{\sim}{\psi}\rangle_1\cdot\mathbf{e}_3$ (find similar formulas for J_μ, $S_{\mu\nu}$).

In the next exercise we have a scalar product of spinors induced by the reversion alone without composing it with complex conjugation.

11. Take a Majorana spinor $\underset{\sim}{\psi} = \underset{\sim}{\psi}_C$ with bilinear covariants $\mathbf{J}$, $\mathbf{S} = \mathbf{J}\wedge\mathbf{s}$. Then the Weyl spinor $u = \frac{1}{2}(1 + i\mathbf{e}_{0123})\,\underset{\sim}{\psi}$ has charge conjugate $u_C = \frac{1}{2}(1 - i\mathbf{e}_{0123})\,\underset{\sim}{\psi}$ so that $\underset{\sim}{\psi} = u + u_C$. Show that

$$\mathbf{J}(u + u_C) = 0 \qquad (3.1.21)$$

$$\mathbf{s}(u + u_C) = -(u + u_C) \qquad (3.1.25)$$

$$4\langle iu\mathbf{e}_{13}\tilde{u}_C\rangle_1 = \tfrac{1}{2}\,\mathbf{J} \qquad (3.1.28)$$

$$\text{Re}(4iu\mathbf{e}_{13}\tilde{u}) = -\tfrac{1}{2}\,\mathbf{S} \qquad (3.1.29)$$

$$4\,\underset{\sim}{\psi}\mathbf{e}_{13}\,\tilde{\underset{\sim}{\psi}}\mathbf{e}_{0123} = \mathbf{J} + \mathbf{S} \qquad (3.1.30/31)$$

The numbering on the right refers to Benn & Tucker 1987 pp. 113-116. Try to work out a translation to their notation, and discuss the physical relevance of the connection between the Majorana and Weyl spinors. Hint: $u_C = \hat{u}^* \mathbf{e}_{13}$ while $u = \frac{1}{2}(1 + i\mathbf{e}_{0123})\,\underset{\sim}{\psi}$. Benn & Tucker use the scalar product of spinors

$$(\underset{\sim}{\psi}, \underset{\sim}{\varphi}) \to \mathbf{e}_{13}\,\tilde{\underset{\sim}{\psi}}\,\underset{\sim}{\varphi} \in \underset{\sim}{f}(\mathbb{C} \otimes C\ell_{3,1})\,\underset{\sim}{f} \simeq \mathbb{C}.$$

Bibliography

This material grew out of discussions with Bill Pezzaglia at a meeting in Banff, 1995.

W. M. Pezzaglia Jr.: Classification of multivector theories and the modification of the postulates of physics; pp. 317-323 in F. Brackx, R. Delanghe, H. Serras (eds.): *Proceedings on the Third Conference on 'Clifford Algebras and their Applications in Mathematical Physics' (Deinze 1993).* Kluwer, Dordrecht, The Netherlands, 1993.

W. M. Pezzaglia Jr.: Multivector solutions to the hyperholomorphic massive Dirac equation; pp. 345-360 in J. Ryan (ed.): *'Clifford Algebras in Analysis and Related Topics' (Fayetteville, AR, 1993).* CRC Press, Boca Raton, FL, 1996.

I.R. Porteous: *Clifford Algebras and the Classical Groups.* Cambridge University Press, Cambridge, U.K., 1995.

14
Definitions of the Clifford Algebra

In this chapter we shall for the first time give a formal definition of the Clifford algebra. There are several definitions, suitable for different purposes. In mathematics, definitions serve as premises for deductions; in physics, however, definitions are more or less secondary and serve as characterizations. We shall review Clifford's original definition, its basis-free variation given as a deformation of the exterior algebra, definition by the universal property, which does not guarantee existence, and the definition as an ideal of the tensor algebra. The construction of Chevalley, where Clifford algebra is regarded as a subalgebra of the endomorphism algebra of the exterior algebra, is postponed till the discussion on characteristic 2. The definitions by the multiplication table of the basis elements, and by index sets, are postponed till the chapter on the Walsh functions. The definition of Clifford algebras as group algebras of extra-special groups will be omitted.

14.1 Clifford's original definition

Grassmann's exterior algebra $\bigwedge \mathbb{R}^n$ of the linear space $\mathbb{R}^n$ is an associative algebra of dimension 2^n. In terms of a basis $\{\mathbf{e}_1, \mathbf{e}_2, \ldots, \mathbf{e}_n\}$ for $\mathbb{R}^n$ the exterior algebra $\bigwedge \mathbb{R}^n$ has a basis

$$\begin{aligned}
&1\\
&\mathbf{e}_1, \mathbf{e}_2, \ldots, \mathbf{e}_n\\
&\mathbf{e}_1 \wedge \mathbf{e}_2,\ \mathbf{e}_1 \wedge \mathbf{e}_3, \ldots, \mathbf{e}_1 \wedge \mathbf{e}_n,\ \mathbf{e}_2 \wedge \mathbf{e}_3, \ldots, \mathbf{e}_{n-1} \wedge \mathbf{e}_n\\
&\vdots\\
&\mathbf{e}_1 \wedge \mathbf{e}_2 \wedge \ldots \wedge \mathbf{e}_n.
\end{aligned}$$

The exterior algebra has a unity 1 and satisfies the multiplication rules

$$\mathbf{e}_i \wedge \mathbf{e}_j = -\mathbf{e}_j \wedge \mathbf{e}_i \quad \text{for} \quad i \neq j,$$
$$\mathbf{e}_i \wedge \mathbf{e}_i = 0.$$

Clifford 1882 kept the first rule but altered the second rule, and arrived at the multiplication rules

$$\mathbf{e}_i\mathbf{e}_j = -\mathbf{e}_j\mathbf{e}_i \quad \text{for} \quad i \neq j,$$
$$\mathbf{e}_i\mathbf{e}_i = 1.$$

This time $\{\mathbf{e}_1, \mathbf{e}_2, \ldots, \mathbf{e}_n\}$ is an orthonormal basis for the positive definite Euclidean space $\mathbb{R}^n$. An associative algebra of dimension 2^n so defined is the Clifford algebra $\mathcal{C}\ell_n$.

Clifford had earlier, in 1878, considered the multiplication rules

$$\mathbf{e}_i\mathbf{e}_j = -\mathbf{e}_j\mathbf{e}_i \quad \text{for} \quad i \neq j,$$
$$\mathbf{e}_i\mathbf{e}_i = -1$$

of the Clifford algebra $\mathcal{C}\ell_{0,n}$ of the negative definite space $\mathbb{R}^{0,n}$.

14.2 Basis-free version of Clifford's definition

Here we consider as an example the exterior algebra $\bigwedge\mathbb{R}^4$ of the 4-dimensional real linear space $\mathbb{R}^4$. Provide the linear space $\mathbb{R}^4$ with a *quadratic form*

$$Q(\mathbf{x}) = x_0^2 - x_1^2 - x_2^2 - x_3^2$$

and associate to Q the *symmetric* bilinear form

$$<\mathbf{x}, \mathbf{y}> = \frac{1}{2}[Q(\mathbf{x}+\mathbf{y}) - Q(\mathbf{x}) - Q(\mathbf{y})].$$

This makes $\mathbb{R}^4$ isometric with the Minkowski space-time $\mathbb{R}^{1,3}$. Then define the **left contraction** $u \rfloor v \in \bigwedge\mathbb{R}^{1,3}$ by

(*a*) $\quad \mathbf{x} \rfloor \mathbf{y} = <\mathbf{x}, \mathbf{y}>$

(*b*) $\quad \mathbf{x} \rfloor (u \wedge v) = (\mathbf{x} \rfloor u) \wedge v + \hat{u} \wedge (\mathbf{x} \rfloor v)$

(*c*) $\quad (u \wedge v) \rfloor w = u \rfloor (v \rfloor w)$

for $\mathbf{x}, \mathbf{y} \in \mathbb{R}^{1,3}$ and $u, v, w \in \bigwedge\mathbb{R}^{1,3}$. [1] The identity (*b*) says that $\mathbf{x}$ operates like a *derivation* and the identity (*c*) makes $\bigwedge\mathbb{R}^{1,3}$ a left module over $\bigwedge\mathbb{R}^{1,3}$. Then introduce the *Clifford product* of $\mathbf{x} \in \mathbb{R}^{1,3}$ and $u \in \bigwedge\mathbb{R}^{1,3}$ by the formula

$$\mathbf{x}u = \mathbf{x} \rfloor u + \mathbf{x} \wedge u$$

1 Recall that $\hat{u}$ is the grade involute of $u \in \bigwedge V$, defined for a k-vector $u \in \bigwedge^k V$ by $\hat{u} = (-1)^k u$.

and extend this product by linearity and associativity to all of $\bigwedge\mathbb{R}^{1,3}$. Provided with the Clifford product (the linear space underlying) the exterior algebra $\bigwedge\mathbb{R}^{1,3}$ becomes the **Clifford algebra** $\mathcal{C}\ell_{1,3}$.

14.3 Definition by generators and relations

The following definition is favored by physicists. It is suitable for non-degenerate quadratic forms, especially the real quadratic spaces $\mathbb{R}^{p,q}$.

Definition. An associative algebra over $\mathbb{F}$ with unity 1 is the Clifford algebra $\mathcal{C}\ell(Q)$ of a non-degenerate Q on V if it contains V and $\mathbb{F} = \mathbb{F}\cdot 1$ as distinct subspaces so that

(1) $\mathbf{x}^2 = Q(\mathbf{x})$ for any $\mathbf{x} \in V$

(2) V generates $\mathcal{C}\ell(Q)$ as an algebra over $\mathbb{F}$

(3) $\mathcal{C}\ell(Q)$ is not generated by any proper subspace of V. ∎

The third condition (3) guarantees the universal property [see below], and dimension 2^n. Using an orthonormal basis $\{\mathbf{e}_1, \mathbf{e}_2, \ldots, \mathbf{e}_n\}$ for $\mathbb{R}^{p,q}$, generating $\mathcal{C}\ell_{p,q}$, the condition (1) can be expressed as

(1.*a*) $\mathbf{e}_i^2 = 1,\ 1 \le i \le p, \quad \mathbf{e}_i^2 = -1,\ p < i \le n, \quad \mathbf{e}_i\mathbf{e}_j = -\mathbf{e}_j\mathbf{e}_i,\ i < j,$

while condition (3) becomes $\mathbf{e}_1\mathbf{e}_2\ldots\mathbf{e}_n \neq \pm 1$, as in Porteous 1969. Condition (3) is needed only in signatures $p - q = 1 \bmod 4$ where $(\mathbf{e}_1\mathbf{e}_2\ldots\mathbf{e}_n)^2 = 1$. The relations (1.*a*) without (3) also generate a lower-dimensional non-universal algebra of dimension 2^{n-1} in any signature $p - q = 1 \bmod 4$ in which all the basis elements $\mathbf{e}_i$ commute with $\mathbf{e}_{12\ldots n} = \mathbf{e}_1\mathbf{e}_2\ldots\mathbf{e}_n$. No similar non-universal algebra exists in even dimensions, and so it is correct to introduce the Clifford algebra of the Minkowski space-time without condition (3). However, in arbitrary dimensions it is controversial to omit condition (3).

The above definition gives a unique algebra only for non-degenerate (non-singular) quadratic forms Q. In particular, the definition is not good for a degenerate Q, for which $\mathbf{e}_1\mathbf{e}_2\ldots\mathbf{e}_n = 0$, as is shown by the following two counter-examples where $Q = 0$.

1. Define for $\mathbf{x}, \mathbf{y} \in V$, $\dim V = n$, the product $\mathbf{xy} = 0$. This makes the direct sum $\mathbb{F} \oplus V$ an associative algebra with unity 1. It is of dimension $n + 1$.

2. Introduce a product in $\bigwedge\mathbb{R}^3$ by $\mathbf{e}_i\mathbf{e}_j = \mathbf{e}_i \wedge \mathbf{e}_j$ for all $i, j = 1, 2, 3$ and $\mathbf{e}_1\mathbf{e}_2\mathbf{e}_3 = 0$. Thus the subspace $\mathbb{R} \oplus \mathbb{R}^3 \oplus \bigwedge^2\mathbb{R}^3$ of $\bigwedge\mathbb{R}^3$ is a 7-dimensional associative algebra with unity, generated by $\mathbb{R}$ and $\mathbb{R}^3$.

This shows that it is not possible to replace condition (3) by the requirement that only parallel vectors commute. We could include arbitrary quadratic forms Q by requiring instead of condition (3) that the product of any set of linearly free vectors in V should not belong to $\mathbb{F}$. However, even this would leave some 'ambiguity' in the definition by generators and relations. The above definition results in a unique algebra only 'up to isomorphism'. Here are two more examples to clarify the meaning of this statement:

3. The multiplication table of the exterior algebra $\bigwedge\mathbb{R}^2$ with respect to the basis $\{1, \mathbf{e}_1, \mathbf{e}_2, \mathbf{e}_1 \wedge \mathbf{e}_2\}$ is

$\wedge$	$\mathbf{e}_1$	$\mathbf{e}_2$	$\mathbf{e}_1 \wedge \mathbf{e}_2$
$\mathbf{e}_1$	0	$\mathbf{e}_1 \wedge \mathbf{e}_2$	0
$\mathbf{e}_2$	$-\mathbf{e}_1 \wedge \mathbf{e}_2$	0	0
$\mathbf{e}_1 \wedge \mathbf{e}_2$	0	0	0

Introduce a second product on $\bigwedge\mathbb{R}^2$ with multiplication table

$\dot\wedge$	$\mathbf{e}_1$	$\mathbf{e}_2$	$\mathbf{e}_1 \wedge \mathbf{e}_2$
$\mathbf{e}_1$	0	$\mathbf{e}_1 \wedge \mathbf{e}_2 + b$	$-b\mathbf{e}_1$
$\mathbf{e}_2$	$-\mathbf{e}_1 \wedge \mathbf{e}_2 - b$	0	$-b\mathbf{e}_2$
$\mathbf{e}_1 \wedge \mathbf{e}_2$	$-b\mathbf{e}_1$	$-b\mathbf{e}_2$	$-b^2 - 2b\mathbf{e}_1 \wedge \mathbf{e}_2$

where $b > 0$. Denote the second product by $u \dot\wedge v$. Rearrange the multiplication table of the second product into the form

$\dot\wedge$	$\mathbf{e}_1$	$\mathbf{e}_2$	$\mathbf{e}_1 \wedge \mathbf{e}_2 + b$
$\mathbf{e}_1$	0	$\mathbf{e}_1 \wedge \mathbf{e}_2 + b$	0
$\mathbf{e}_2$	$-\mathbf{e}_1 \wedge \mathbf{e}_2 - b$	0	0
$\mathbf{e}_1 \wedge \mathbf{e}_2 + b$	0	0	0

which shows that we have generated a new exterior algebra $\dot\bigwedge\mathbb{R}^2$ on $\mathbb{R}^2$, different from $\bigwedge\mathbb{R}^2$ but isomorphic with $\bigwedge\mathbb{R}^2$. In other words, we have introduced a linear mapping $\alpha : \bigwedge\mathbb{R}^2 \to \dot\bigwedge\mathbb{R}^2$ for which $\alpha(\mathbf{e}_i) = \mathbf{e}_i$, $i = 1, 2$, and $\alpha(\mathbf{e}_1 \wedge \mathbf{e}_2) = \mathbf{e}_1 \dot\wedge \mathbf{e}_2 = \mathbf{e}_1 \wedge \mathbf{e}_2 + b$ so that it is the identity on $\mathbb{R}^2$, preserves even-odd grading and gives an isomorphism between the two products, $\alpha(u \wedge v) = \alpha(u) \dot\wedge \alpha(v)$.

4. An orthonormal basis $\mathbf{e}_1, \mathbf{e}_2$ for $\mathbb{R}^2$ satisfying $\mathbf{e}_i\mathbf{e}_j + \mathbf{e}_j\mathbf{e}_i = 2\delta_{ij}$ generates the Clifford algebra $\mathcal{C}\ell_2 = \mathcal{C}\ell_{2,0}$ with basis $\{1, \mathbf{e}_1, \mathbf{e}_2, \mathbf{e}_{12}\}$ where $\mathbf{e}_{12} =$

$\mathbf{e}_1\mathbf{e}_2\ (= \mathbf{e}_1 \wedge \mathbf{e}_2)$. We have the following multiplication table for $C\ell_2$:

	$\mathbf{e}_1$	$\mathbf{e}_2$	$\mathbf{e}_{12}$
$\mathbf{e}_1$	1	$\mathbf{e}_{12}$	$\mathbf{e}_2$
$\mathbf{e}_2$	$-\mathbf{e}_{12}$	1	$-\mathbf{e}_1$
$\mathbf{e}_{12}$	$-\mathbf{e}_2$	$\mathbf{e}_1$	-1

Introduce a second product on $C\ell_2$ with multiplication table

	$\mathbf{e}_1$	$\mathbf{e}_2$	$\mathbf{e}_{12}$
$\mathbf{e}_1$	1	$\mathbf{e}_{12} + b$	$\mathbf{e}_2 - b\mathbf{e}_1$
$\mathbf{e}_2$	$-\mathbf{e}_{12} - b$	1	$-\mathbf{e}_1 - b\mathbf{e}_2$
$\mathbf{e}_{12}$	$-\mathbf{e}_2 - b\mathbf{e}_1$	$\mathbf{e}_1 - b\mathbf{e}_2$	$-1 - b^2 - 2b\mathbf{e}_{12}$

The anticommutation relations $\mathbf{e}_i\mathbf{e}_j + \mathbf{e}_j\mathbf{e}_i = 2\delta_{ij}$ are also satisfied by the new product, and one may directly verify associativity. As the real number b varies we have a family of different but isomorphic Clifford algebras on $\mathbb{R}^2$.

14.4 Universal object of quadratic algebras

The Clifford algebra $C\ell(Q)$ is the universal associative algebra over $\mathbb{F}$ generated by V with the relations $\mathbf{x}^2 = Q(\mathbf{x})$, $\mathbf{x} \in V$.

Let Q be the quadratic form on a linear space V over a field $\mathbb{F}$, and let A be an associative algebra over $\mathbb{F}$ with unity 1_A. A linear mapping $V \to A$, $\mathbf{x} \to \varphi_\mathbf{x}$ such that

$$(\varphi_\mathbf{x})^2 = Q(\mathbf{x}) \cdot 1_A \quad \text{for all} \quad \mathbf{x} \in V$$

is called a *Clifford map*. The subalgebra of A generated by $\mathbb{F} = \mathbb{F} \cdot 1_A$ and V (or more precisely by the images of $\mathbb{F}$ and V in A) will be called a *quadratic algebra*. [2] The **Clifford algebra** $C\ell(Q)$ is a quadratic algebra with a Clifford map $V \to C\ell(Q)$, $\mathbf{x} \to \gamma_\mathbf{x}$ such that for any Clifford map $\varphi : V \to A$ there exists a unique algebra homomorphism $\psi : C\ell(Q) \to A$ making the following diagram commutative:

$$\begin{array}{ccc} V & \xrightarrow{\gamma} & C\ell(Q) \\ & \varphi \searrow & \downarrow \psi \\ & & A \end{array} \qquad\qquad \varphi_\mathbf{x} = \psi(\gamma_\mathbf{x})$$

This definition says that **all** Clifford maps may be obtained from $\gamma : V \to C\ell(Q)$ which is thereby **universal**.

2 The term quadratic algebra is commonly used for something else: in a quadratic algebra each square x^2 is linearly dependent on x and 1.

The definition by the universal property is meaningful for an algebraist who knows categories and morphisms up to the theory of universal objects. A category contains objects and morphisms between the objects. Invertible morphisms are called isomorphisms. In a category there is an initial (resp. final) universal object U, if for any object A, there is a unique morphism $\alpha : U \to A$ (resp. $A \to U$). The universal objects are unique up to isomorphism. In many categories there exists trivially the final universal object, which often reduces to 0. The Clifford algebra is the initial universal object in the category of quadratic algebras.

Example. Consider the category of quadratic algebras on $\mathbb{R}^{p,q}$. In this category the initial universal object is the Clifford algebra $\mathcal{C}\ell_{p,q}$ of dimension 2^n and the final universal object is 0. Between these two objects there are no other objects, when $p - q \neq 1 \bmod 4$. However, there are four objects in this category, when $p - q = 1 \bmod 4$; between $\mathcal{C}\ell_{p,q}$ and 0 there are two algebras both of dimension 2^{n-1}; in one we have the relation $\mathbf{e}_1\mathbf{e}_2 \ldots \mathbf{e}_n = 1$ and in the other $\mathbf{e}_1\mathbf{e}_2 \ldots \mathbf{e}_n = -1$; these two algebras are not isomorphic in the category of quadratic algebras (the identity mapping on $\mathbb{R}^{p,q}$ does not extend to an isomorphism from one algebra to the other); however, they are isomorphic as associative algebras (in the category of all real algebras). ∎

The above definition of Clifford algebras is most suitable for an algebraist who wants to study Clifford algebras over commutative rings (and who does not insist on injectivity of mappings $\mathbb{F} \to A$ and $V \to A$). However, this approach does not guarantee existence, which is given by constructing the Clifford algebra as the quotient algebra of the tensor algebra (which in turn is regarded by algebraists as the mother of all algebras).

14.5 Clifford algebra as a quotient of the tensor algebra

Chevalley 1954 p. 37 constructs the Clifford algebra $\mathcal{C}\ell(Q)$ as the quotient algebra $\otimes V/I(Q)$ of the tensor algebra $\otimes V$ with respect to the two-sided ideal $I(Q)$ generated by the elements $\mathbf{x}\otimes\mathbf{x}-Q(\mathbf{x})$ where $\mathbf{x} \in V$. See also N. Bourbaki 1959 p. 139 and T.Y. Lam 1973 p. 103. The tensor algebra approach gives a proof of existence by construction – suitable for an algebraist who is interested in rapid access to the main properties of Clifford algebras over commutative rings.

In characteristic zero we may avoid quotient structures by making the exterior algebra $\bigwedge V$ concrete as the subspace of antisymmetric tensors in $\otimes V$. For example, if $\mathbf{x}, \mathbf{y} \in V$, then $\mathbf{x} \wedge \mathbf{y} = \frac{1}{2}(\mathbf{x} \otimes \mathbf{y} - \mathbf{y} \otimes \mathbf{x}) \in \bigwedge^2 V$. More generally, a simple k-vector $\mathbf{x}_1 \wedge \mathbf{x}_2 \wedge \ldots \wedge \mathbf{x}_k$ is identified with [3]

$$Alt(\mathbf{x}_1 \otimes \mathbf{x}_2 \otimes \ldots \otimes \mathbf{x}_k) = \frac{1}{k!} \sum_{\pi} \operatorname{sign}(\pi)\, \mathbf{x}_{\pi(1)} \otimes \mathbf{x}_{\pi(2)} \otimes \ldots \otimes \mathbf{x}_{\pi(k)}\,,$$

where the linear operator $Alt : \otimes V \to \bigwedge V$, called **alternation**, is a projection operator $Alt(\otimes V) = \bigwedge V$ satisfying $u \wedge v = Alt(u \otimes v)$.

Similarly, we may obtain an isomorphism of linear spaces $\bigwedge V \to C\ell(Q)$ by identifying simple k-vectors with antisymmetrized Clifford products

$$\mathbf{x}_1 \wedge \mathbf{x}_2 \wedge \ldots \wedge \mathbf{x}_k \to \mathbf{x}_1 \dot{\wedge} \mathbf{x}_2 \dot{\wedge} \ldots \dot{\wedge} \mathbf{x}_k = \frac{1}{k!} \sum_{\pi} \operatorname{sign}(\pi)\, \mathbf{x}_{\pi(1)} \mathbf{x}_{\pi(2)} \ldots \mathbf{x}_{\pi(k)}$$

thus splitting the Clifford algebra $C\ell(Q)$ into fixed subspaces of k-vectors $\bigwedge^k V \subset C\ell(Q)$. Any orthogonal basis $\mathbf{e}_1, \mathbf{e}_2, \ldots, \mathbf{e}_n$ of V gives a correspondence

$$\mathbf{e}_{i_1} \wedge \mathbf{e}_{i_2} \wedge \ldots \wedge \mathbf{e}_{i_k} \to \mathbf{e}_{i_1} \dot{\wedge} \mathbf{e}_{i_2} \dot{\wedge} \ldots \dot{\wedge} \mathbf{e}_{i_k} = \mathbf{e}_{i_1} \mathbf{e}_{i_2} \ldots \mathbf{e}_{i_k}$$

of bases for $\bigwedge V$ and $C\ell(Q)$.

Exercises

1. Show that the subspace $Alt(\otimes V)$ of $\otimes V$ is not closed under the tensor product.
2. Show that $\mathbf{A} \otimes \mathbf{B} - \mathbf{B} \otimes \mathbf{A} = \frac{1}{2}(\mathbf{AB} - \mathbf{BA})$ for bivectors $\mathbf{A}, \mathbf{B} \in \bigwedge^2 V$.

Bibliography

E. Artin: *Geometric Algebra*. Interscience, New York, 1957, 1988.

N. Bourbaki: *Algèbre, Chapitre 9, Formes sesquilinéaires et formes quadratiques*. Hermann, Paris, 1959.

C. Chevalley: *Theory of Lie Groups*. Princeton University Press, Princeton, NJ, 1946.

C. Chevalley: *The Algebraic Theory of Spinors*. Columbia University Press, New York, 1954.

W.K. Clifford: Applications of Grassmann's extensive algebra. *Amer. J. Math.* **1** (1878), 350-358.

W.K. Clifford: On the classification of geometric algebras; pp. 397-401 in R. Tucker (ed.): *Mathematical Papers by William Kingdon Clifford*, Macmillan, London, 1882. Reprinted by Chelsea, New York, 1968. Title of talk announced already in *Proc. London Math. Soc.* **7** (1876), p. 135.

J. Helmstetter: Algèbres de Clifford et algèbres de Weyl. *Cahiers Math.* **25**, Montpellier, 1982.

I.R. Porteous: *Clifford Algebras and the Classical Groups*. Cambridge University Press, Cambridge, 1995.

---

3 Another alternative is to omit the factor $\frac{1}{k!}$. This gives in all characteristics a correspondence between the exterior product and the antisymmetrized tensor product.

15

Witt Rings and Brauer Groups

Quadratic forms can be classified by their Witt classes in Witt rings (of concerned fields). This is a slightly coarser classification than the one given by the Clifford algebras (of quadratic forms).

Associative algebras with unity can be studied by means of Brauer groups (of fields); for this one needs to know tensor products of algebras. These topics will be discussed in this chapter.

15.1 Quadratic forms

A *quadratic form* on a linear space V over a field $\mathbb{F}$ is a map $Q : V \to \mathbb{F}$ such that for any $\lambda \in \mathbb{F}$ and $\mathbf{x} \in V$

$$Q(\lambda \mathbf{x}) = \lambda^2 Q(\mathbf{x})$$

and such that the map

$$V \times V \to \mathbb{F}, \quad (\mathbf{x}, \mathbf{y}) \to Q(\mathbf{x} + \mathbf{y}) - Q(\mathbf{x}) - Q(\mathbf{y})$$

is bilinear, that is, linear in both arguments. A linear space with a quadratic form on itself is called a *quadratic space*. A quadratic form obeys the parallelogram law

$$Q(\mathbf{x} + \mathbf{y}) + Q(\mathbf{x} - \mathbf{y}) = 2Q(\mathbf{x}) + 2Q(\mathbf{y}).$$

In characteristic $\neq 2$ the quadratic form may be recaptured from its symmetric bilinear form

$$B(\mathbf{x}, \mathbf{y}) = \frac{1}{2}[Q(\mathbf{x} + \mathbf{y}) - Q(\mathbf{x}) - Q(\mathbf{y})]$$

since $Q(\mathbf{x}) = B(\mathbf{x}, \mathbf{x})$. In characteristic $\neq 2$ the theory of quadratic forms is the same as the theory of symmetric bilinear forms, but in characteristic 2 there are quadratic forms which are not induced by any *symmetric* bilinear form.

Example. Consider the 2-dimensional space $\mathbb{F}_2^2$ over $\mathbb{F}_2 = \{0, 1\}$. For the quadratic form $Q : \mathbb{F}_2^2 \to \mathbb{F}_2$, $\mathbf{x} = (x_1, x_2) \to x_1 x_2$ there is no symmetric bilinear form B such that $Q(\mathbf{x}) = B(\mathbf{x}, \mathbf{x})$. However, there is a bilinear form B, not symmetric, such that $Q(\mathbf{x}) = B(\mathbf{x}, \mathbf{x})$, namely $B(\mathbf{x}, \mathbf{y}) = x_1 y_2$, but the matrix

$$B(\mathbf{e}_i, \mathbf{e}_j) = \begin{pmatrix} 0 & 1 \\ 0 & 0 \end{pmatrix}$$

cannot be symmetrized by adding an alternating matrix with entries in $\mathbb{F}_2 = \{0, 1\}$. ∎

Remark. We shall not be concerned with characteristic 2, or non-symmetric B, except at the end of this book when considering the relation between the exterior algebra and the Clifford algebras. The role of non-symmetric B will be described in Chevalley's construction $\mathcal{C}\ell(Q) \subset \mathrm{End}(\bigwedge V)$. ∎

A non-zero vector $\mathbf{x}$ is *null* or *isotropic* if $Q(\mathbf{x}) = 0$. A quadratic form is *anisotropic* if $Q(\mathbf{x}) = 0$ implies $\mathbf{x} = 0$, and *isotropic* if $Q(\mathbf{x}) = 0$ for some $\mathbf{x} \neq 0$. A bilinear form is *non-degenerate* if $B(\mathbf{x}, \mathbf{y}) = 0$ for all $\mathbf{y} \in V$ implies $\mathbf{x} = 0$. An anisotropic quadratic form is always non-degenerate.

A 2-dimensional isotropic but non-degenerate quadratic space is known as the *hyperbolic plane*. A hyperbolic plane has a quadratic form $y_1 y_2$ or equivalently $x_1^2 - x_2^2$ (choose $y_1 = x_1 + x_2$ and $y_2 = x_1 - x_2$ in characteristic $\neq 2$).

A subspace with a vanishing quadratic form is *totally isotropic*. In a non-degenerate quadratic space with a totally isotropic subspace S, there is another totally isotropic subspace S' such that $S \cap S' = \{0\}$ and $\dim S = \dim S'$. A non-degenerate quadratic space is *neutral* or *hyperbolic* if it is a direct sum of two totally isotropic subspaces (necessarily of the same dimension). A neutral quadratic space is even-dimensional.

Examples. 1. A Euclidean space $\mathbb{R}^n$ has an anisotropic (and positive definite) quadratic form on itself, i.e., $\mathbf{x} \neq 0$ implies $\mathbf{x} \cdot \mathbf{x} > 0$. This enables us to introduce the *norm* or *length* $|\mathbf{x}| = \sqrt{\mathbf{x} \cdot \mathbf{x}}$ of $\mathbf{x} \in \mathbb{R}^n$.
2. The real quadratic space $\mathbb{R}^{p,q}$ is non-degenerate and for non-zero p, q also isotropic (indefinite). The dimension of its maximal totally isotropic subspace is p or q according as $p \leq q$ or $p \geq q$, respectively. This number is called the *isotropy index* (or Witt index) of $\mathbb{R}^{p,q}$.
3. The quadratic form $x_1^2 + x_2^2$ on $\mathbb{F}_5^2$ is non-degenerate but isotropic, since $1^2 + 2^2 = 0 \bmod 5$. It is also neutral.
4. The quadratic forms $x_1^2 + 2x_2^2$ and $x_1^2 + 3x_2^2$ are anisotropic on $\mathbb{F}_5^2$. ∎

Two quadratic spaces (V,Q) and (V',Q') are *isometric* if there is a linear isomorphism $L : V \to V'$ such that $Q'(L\mathbf{x}) = Q(\mathbf{x})$, or equivalently, in characteristic $\neq 2$,

$$B'(L\mathbf{x}, L\mathbf{y}) = B(\mathbf{x},\mathbf{y}) \quad \text{for all} \quad \mathbf{x},\mathbf{y} \in V.$$

We will express the isometry as $(V,Q) \simeq (V',Q')$ or simply $Q \simeq Q'$. A self-isometry (or automorphism) of Q on V is a linear isomorphism $L : V \to V$ such that $Q(L(\mathbf{x})) = Q(\mathbf{x})$ for all $\mathbf{x} \in V$; these self-isometries of Q form the *orthogonal group* $O(V,Q)$.

Two vectors $\mathbf{x},\mathbf{y}$ such that $B(\mathbf{x},\mathbf{y}) = 0$ are said to be *orthogonal* (in the case of a symmetric B). If a quadratic space is a direct sum of two subspaces, $(V_1,Q_1) \oplus (V_2,Q_2)$, such that $B(\mathbf{x}_1,\mathbf{x}_2) = 0$ for all $\mathbf{x}_1 \in V_1$ and $\mathbf{x}_2 \in V_2$, it is an *orthogonal sum* denoted by $V_1 \perp V_2$ or $Q_1 \perp Q_2$.

There is also, for two quadratic spaces (V_1,Q_1) and (V_2,Q_2), the *tensor product* $V_1 \otimes V_2$ of dimension $(\dim V_1)(\dim V_2)$, with a quadratic form satisfying

$$Q(\mathbf{x}_1 \otimes \mathbf{x}_2) = Q_1(\mathbf{x}_1)Q_2(\mathbf{x}_2)$$

for decomposable elements $\mathbf{x}_1 \otimes \mathbf{x}_2$ with $\mathbf{x}_1 \in V_1$ and $\mathbf{x}_2 \in V_2$.

The symmetric matrix $B(\mathbf{e}_i,\mathbf{e}_j)$ can be diagonalized (in characteristic $\neq 2$); as a consequence any quadratic form is isometric to a diagonal form $d_1x_1^2 + d_2x_2^2 + \ldots + d_nx_n^2$ for some $d_1, d_2, \ldots, d_n \in \mathbb{F}$. We shall write

$$\langle d_1, d_2, \ldots, d_n\rangle \quad \text{to denote} \quad d_1x_1^2 + d_2x_2^2 + \ldots + d_nx_n^2.$$

The orthogonal sum and the tensor product appear in diagonal form as

$$\langle a_1, \ldots, a_m\rangle \perp \langle b_1, \ldots, b_n\rangle = \langle a_1, \ldots, a_m, b_1, \ldots, b_n\rangle,$$
$$\langle a_1, \ldots, a_m\rangle \otimes \langle b_1, \ldots, b_n\rangle = \langle a_1b_1, \ldots, a_ib_j, \ldots, a_mb_n\rangle.$$

Examples. 1. The real quadratic space $\mathbb{R}^{p,p}$ is neutral and an orthogonal sum of p copies of hyperbolic planes $\mathbb{R}^{1,1}$ each with a quadratic form $\langle 1,-1\rangle$.
2. The hyperbolic plane over $\mathbb{F}$, $\operatorname{char}\mathbb{F} \neq 2$, is isometric with x_1x_2 and $x_1^2 - x_2^2 \simeq \langle 1,-1\rangle$.
3. In $\mathbb{F}_5$, $\langle 1\rangle \not\simeq \langle 2\rangle$ since $2 \notin \mathbb{F}_5^{\square}$, the set of non-zero squares, but $\langle 2\rangle \simeq \langle 3\rangle$ since $2x_1^2 = 3(2x_1)^2$.
4. The quadratic forms $\langle 1,2\rangle$ and $\langle 1,3\rangle$ are isometric on $\mathbb{F}_5^2$ as can be seen by the identity $x_1^2 + 2x_2^2 = x_1^2 + 3(2x_2)^2 \bmod 5$ and the linear isomorphism $(x_1,x_2) \to (x_1,2x_2)$.
5. The quadratic forms $\langle 1,1\rangle$ and $\langle 1,-1\rangle$ are isometric on $\mathbb{F}_5^2$ since $x_1^2 - x_2^2 = x_1 + (2x_2)^2$. ∎

15.2 Witt rings

Two quadratic spaces V, Q and V', Q' are said to be in the same *Witt class* if $Q \perp (-Q')$ is neutral. A non-degenerate quadratic space is an orthogonal sum $V = V_a \perp V_h$ of an anisotropic subspace V_a and a neutral subspace V_h. The anisotropic part V_a is unique up to isometry. It follows that V and V' are in the same Witt class iff $Q_a \perp (-Q'_a)$ is neutral, or equivalently iff $Q_a \simeq Q'_a$. This results in a correspondence

$$\boxed{\text{Witt class of } V = V_a \oplus V_h \longleftrightarrow \text{isometry class of } V_a}$$

exactly one anisotropic isometry class being included in each Witt class.

Examples. Let the ground field be $\mathbb{F}_5$.
1. From the orthogonal sum $\langle 1\rangle \perp \langle 1,2\rangle \simeq \langle 1,1,2\rangle$ we can cancel the hyperbolic plane $\langle 1,1\rangle$ to extract the anisotropic part $\langle 2\rangle$. So the sum $\langle 1\rangle \perp \langle 1,2\rangle$ is in the Witt class of $\langle 2\rangle$.
2. From the tensor product $\langle 1,2\rangle \otimes \langle 3\rangle \simeq \langle 3, 2\cdot 3 = 1\rangle$ there is nothing to cancel since $\langle 3,1\rangle \simeq \langle 1,2\rangle$ is anisotropic. So $\langle 1,2\rangle \otimes \langle 3\rangle$ is in the Witt class of $\langle 1,2\rangle$. ∎

The orthogonal sum $\perp$ and the tensor product $\otimes$ provide the set of all the Witt classes over $\mathbb{F}$ with a ring structure yielding the *Witt ring* $W(\mathbb{F})$.

The opposite of Q is represented by $-Q$ in $W(\mathbb{F})$. The neutral quadratic forms, in particular $\langle 0\rangle$, represent the zero of $W(\mathbb{F})$. The 1-dimensional form $\langle 1\rangle$ corresponds to the multiplicative unity of $W(\mathbb{F})$. The zero $\langle 0\rangle$ and the unity $\langle 1\rangle$ are the only idempotents in $W(\mathbb{F})$. The even-dimensional quadratic forms induce an ideal of $W(\mathbb{F})$.

Examples. 1. The Witt ring $W(\mathbb{F}_5)$ contains four anisotropic isometry classes $\langle 0\rangle$, $\langle 1\rangle$, $\langle 2\rangle$, $\langle 1,2\rangle$. The addition and multiplication tables of $W(\mathbb{F}_5)$ are

$\perp$	$\langle 1\rangle$	$\langle 2\rangle$	$\langle 1,2\rangle$
$\langle 1\rangle$	$\langle 0\rangle$	$\langle 1,2\rangle$	$\langle 2\rangle$
$\langle 2\rangle$	$\langle 1,2\rangle$	$\langle 0\rangle$	$\langle 1\rangle$
$\langle 1,2\rangle$	$\langle 2\rangle$	$\langle 1\rangle$	$\langle 0\rangle$

$\otimes$	$\langle 1\rangle$	$\langle 2\rangle$	$\langle 1,2\rangle$
$\langle 1\rangle$	$\langle 1\rangle$	$\langle 2\rangle$	$\langle 1,2\rangle$
$\langle 2\rangle$	$\langle 2\rangle$	$\langle 1\rangle$	$\langle 1,2\rangle$
$\langle 1,2\rangle$	$\langle 1,2\rangle$	$\langle 1,2\rangle$	$\langle 0\rangle$

The Witt ring $W(\mathbb{F}_5)$ is isomorphic to the group algebra $\mathbb{Z}_2[\mathbb{F}_5^\times/\mathbb{F}_5^\square]$; and the additive group of $W(\mathbb{F}_5)$ is $\mathbb{Z}_2 \times \mathbb{Z}_2$, but as a ring $W(\mathbb{F}_5) \not\simeq {}^2\mathbb{Z}_2 = \mathbb{Z}_2 \times \mathbb{Z}_2$.
2. The 1-dimensional line $\mathbb{F}_5$ with quadratic form $\langle 2\rangle$ has the quadratic field extension $\mathbb{F}_5(\sqrt{2}) \simeq \mathbb{F}_{25}$ as its Clifford algebra. The Clifford algebras of both $\langle 1\rangle$ and $\langle -1\rangle$ split as the double-ring ${}^2\mathbb{F}_5 = \mathbb{F}_5 \times \mathbb{F}_5$. Therefore, as algebras $\mathcal{C}\ell(\langle 2\rangle, \mathbb{F}_5) \not\simeq \mathcal{C}\ell(\langle 1\rangle, \mathbb{F}_5) \simeq \mathcal{C}\ell(\langle -1\rangle, \mathbb{F}_5)$. [1]

1 Denoting the Clifford algebra of the n-dimensional quadratic space $p\langle 1\rangle \perp q\langle -1\rangle$ by

3. The Witt ring $W(\mathbb{F}_7)$ contains four anisotropic isometry classes $\langle 0\rangle$, $\langle 1\rangle$, $\langle 1,1\rangle$, $\langle 1,1,1\rangle$, and $W(\mathbb{F}_7) \simeq \mathbb{Z}_4$. ∎

The finite fields $\mathbb{F}_q$. The Witt ring of $\mathbb{F}_q$, $q = p^m$, $\operatorname{char} p \neq 2$, contains four anisotropic isometry classes

$$\begin{array}{lll} \langle 0\rangle,\ \langle 1\rangle,\ \langle s\rangle,\ \langle 1,s\rangle \quad \text{where} \quad s \notin \mathbb{F}_q^{\square} & \text{for} & p = 1 \bmod 4, \\ \langle 0\rangle,\ \langle 1\rangle,\ \langle 1,1\rangle,\ \langle 1,1,1\rangle & \text{for} & p = 3 \bmod 4. \end{array}$$

The corresponding Witt rings are $W(\mathbb{F}_q) \simeq \mathbb{Z}_2[\mathbb{F}_q^{\times}/\mathbb{F}_q^{\square}]$, $q = 1 \bmod 4$, and $W(\mathbb{F}_q) \simeq \mathbb{Z}_4$, $q = 3 \bmod 4$. All quadratic forms over the finite fields are isotropic in dimensions $n > 3$.

The real field $\mathbb{R}$. A field $\mathbb{F}$ is *ordered* if there is a subset $P \subset \mathbb{F}$ (of positive numbers) such that for all $a, b \in P$ also $a+b, ab \in P$, and, for all $a \in \mathbb{F}$ exactly one of $a \in P$, $a = 0$, and $-a \in P$ holds. The statement $a - b \in P$ is also written $a > b$. An ordered field $\mathbb{F}$ has an *absolute value* $\mathbb{F} \to P$, $x \to |x|$ defined by setting $|0| = 0$, $|x| = x$ for $x > 0$, and $|x| = -x$ for $-x > 0$.

In an ordered field, $\mathbb{F}^{\square} \subset P$. If all the positive numbers have square roots, then there is a unique ordering with $P = \mathbb{F}^{\square}$. The following holds for any ordered field $\mathbb{F}$ such that $P = \mathbb{F}^{\square}$, but we shall only consider the real field $\mathbb{R}$.

There exist exactly two anisotropic forms on $\mathbb{R}^n$, namely the positive definite $\langle 1,1,\ldots,1\rangle$ and the negative definite $\langle -1,-1,\ldots,-1\rangle$. [2] A non-degenerate quadratic form on $\mathbb{R}^n$ is isometric to

$$x_1^2 + \ldots + x_p^2 - x_{p+1}^2 - \ldots - x_{p+q}^2, \quad p + q = n,$$

which we abbreviate as $p\langle 1\rangle \perp q\langle -1\rangle$. The real quadratic space with this quadratic form is denoted by $\mathbb{R}^{p,q}$. The integer $p - q$ is called the *signature* of $\mathbb{R}^{p,q}$.

The signature map sending $\mathbb{R}^{p,q}$ to $p-q$ gives a ring isomorphism $W(\mathbb{R}) \simeq \mathbb{Z}$. As a consequence, the Clifford algebras of non-degenerate real quadratic spaces can be listed by the symbols $C\ell_{p,q}$, denoted more fully by $C\ell_{p,q}(\mathbb{R}^n) = C\ell(\mathbb{R}^{p,q})$ or $C\ell(p\langle 1\rangle \perp q\langle -1\rangle, \mathbb{R}^n)$.

The complex field $\mathbb{C}$. The Witt ring of $\mathbb{C}$ contains only two anisotropic isometry classes, namely $\langle 0\rangle$ and $\langle 1\rangle$, and $W(\mathbb{C}) \simeq \mathbb{Z}_2$. We only have to distinguish between even- and odd-dimensional spaces over $\mathbb{C}$.

Exercises 1,2

$C\ell_{p,q}(\mathbb{F}^n)$, this example shows that the notion $C\ell_{p,q}$ does not reach all the Clifford algebras over arbitrary fields $\mathbb{F}$.

2 The real linear space $\mathbb{R}^n$ with the positive definite quadratic form $n\langle 1\rangle = \langle 1,1,\ldots,1\rangle$ is the *Euclidean space* $\mathbb{R}^n$.

15.3 Algebras

An algebra A over a field $\mathbb{F}$ is a linear (that is a vector) space A over $\mathbb{F}$ together with a bilinear map $A \times A \to A$, $(a, b) \to ab$, the algebra product. Bilinearity means distributivity $(a + b)c = ac + bc$, $a(b + c) = ab + ac$ and $(\lambda a)b = a(\lambda b) = \lambda(ab)$ for all $a, b, c \in A$ and $\lambda \in \mathbb{F}$.

Examples. 1. The 2-dimensional real linear space $\mathbb{R}^2$ together with the product $(x_1, y_1)(x_2, y_2) = (x_1x_2 - y_1y_2, x_1y_2 + x_2y_1)$ results in the real algebra of complex numbers $\mathbb{C}$.
2. The double-ring ${}^2\mathbb{F}$ of a field $\mathbb{F}$ has a product $(a_1, b_1)(a_2, b_2) = (a_1a_2, b_1b_2)$ making it a 2-dimensional algebra over the subfield denoted by $\mathbb{F}(1, 1) = \{(\lambda, \lambda) \mid \lambda \in \mathbb{F}\}$.
3. The matrix algebra of real 2×2-matrices $\text{Mat}(2, \mathbb{R})$ is a 4-dimensional real associative algebra with unity I.
4. The real linear space $\mathbb{R}^3$ together with the cross product $\mathbf{a} \times \mathbf{b}$ is a (non-associative) Lie algebra. ∎

An algebra is without *zero-divisors* if $ab = 0$ implies $a = 0$ or $b = 0$. In a *division algebra* $\mathbb{D}$ the equations $ax = b$ and $ya = b$ have unique solutions x, y for all non-zero $a, b \in \mathbb{D}$. A division algebra is without zero-divisors, and conversely, every finite-dimensional algebra without zero-divisors is a division algebra. If a division algebra is associative, then it has a multiplicative unity and each non-zero element admits a unique inverse (on both sides).

An algebra with a multiplicative unity is said to *admit inverses* if each non-zero element admits an inverse.

Examples. 1. The quaternions $\mathbb{H}$ form a real associative but non-commutative division algebra with unity 1.
2. Define the following product for pairs of quaternions:

$$(x_1, y_1) \circ (x_2, y_2) = (x_1x_2 - \bar{y}_2y_2, y_2x_1 + y_1\bar{x}_2).$$

This makes the real linear space $\mathbb{H} \times \mathbb{H}$ a real algebra, the Cayley algebra of octonions $\mathbb{O}$. The Cayley algebra is non-associative, $a \circ (b \circ c) \neq (a \circ b) \circ c$, but alternative, $(a \circ a) \circ b = a \circ (a \circ b)$, $a \circ (b \circ b) = (a \circ b) \circ b$. It is a division algebra with unity 1.
3. Consider a 3-dimensional real algebra with basis $\{1, i, j\}$ such that 1 is the unity and $i^2 = j^2 = -1$ but $ij = ji = 0$. The algebra is commutative, non-associative and non-alternative. It admits inverses, but the inverses of the elements of the form $xi + yj$ are not unique, $(xi + yj)^{-1} = \lambda(yi - xj) - \frac{ix+jy}{x^2+y^2}$. It has by definition zero-divisors, and cannot be a division algebra. ∎

An isomorphism or anti-isomorphism of algebras A and B is a linear isomorphism $f : A \to B$ such that

$$f(xy) = f(x)f(y) \quad \text{or} \quad f(xy) = f(y)f(x),$$

respectively. An *automorphism* or *anti-automorphism* of an algebra A is an isomorphism or anti-isomorphism $A \to A$, respectively. An automorphism or anti-automorphism f of A such that $f(f(x)) = x$ for all $x \in A$ is an *involution* or an *anti-involution*, respectively.

The only algebra automorphisms of $\mathbb{C}$, regarded as a real algebra, are the identity and the complex conjugation $z \to \bar{z}$.

The only automorphisms of the real algebra $^2\mathbb{R}$ are the identity and the *swap*

$$^2\mathbb{R} \to {}^2\mathbb{R}, \quad (\lambda, \mu) \to \text{swap}(\lambda, \mu) = (\mu, \lambda).$$

The swap acts like the complex conjugation of $\mathbb{C}$, since

$$\text{swap}[\lambda(1,1) + \mu(1,-1)] = \lambda(1,1) - \mu(1,-1).$$

Two automorphisms or anti-automorphisms α, β of an algebra A are said to be *similar* if there is an automorphism γ of A such that $\alpha\gamma = \gamma\beta$. If no such γ exists then α and β are said to be *dissimilar*.

The identity automorphism is similar only to itself. Consequently, the two involutions of the real algebra $\mathbb{C}$ are dissimilar, and the two involutions of the real algebra $^2\mathbb{R}$ are dissimilar.

Exercises 3,4

15.4 Tensor products of algebras, Brauer groups

The tensor product of two algebras A and B over a field $\mathbb{F}$ is the linear space $A \otimes B$ made into an algebra with the product satisfying

$$(a \otimes b)(a' \otimes b') = (aa') \otimes (bb')$$

for $a, a' \in A$ and $b, b' \in B$. This algebra is also denoted by $A \otimes B$ or, to emphasize the ground field, by $A \otimes_{\mathbb{F}} B$.

In the special case of finite-dimensional associative algebras with multiplicative unity, the statement $C = A \otimes B$ can be tested by the following conditions posed on the subalgebras A and B of C:

(i) $ab = ba$ for any $a \in A$ and $b \in B$,
(ii) C is generated as an algebra by A and B,
(iii) $\dim C = (\dim A)(\dim B)$.

Examples. 1. $\mathbb{C} \otimes_{\mathbb{R}} \mathbb{H} \simeq \text{Mat}(2, \mathbb{C})$ the real matrix algebra of 2×2-matrices with complex numbers as entries.

2. $\mathrm{Mat}(p,\mathbb{R})\otimes\mathrm{Mat}(q,\mathbb{R})\simeq\mathrm{Mat}(pq,\mathbb{R})$.
3. $\mathrm{Mat}(p,\mathbb{R})\otimes\mathrm{Mat}(q,\mathbb{C})\simeq\mathrm{Mat}(pq,\mathbb{C})$. ∎

Exercises 5,6

A two-sided ideal T of an algebra A is a subalgebra such that $ta\in T$ and $at\in T$ for all $t\in T$ and $a\in A$. An algebra A is called *simple* if it has no two-sided ideals other than 0 and A. The *center* $\mathrm{Cen}(A)$ of an algebra A consists of the elements commuting with all the elements of A:

$$\mathrm{Cen}(A)=\{c\in A\mid ac=ca \text{ for all } a\in A\}.$$

An algebra with multiplicative unity 1 is called *central* if $\mathrm{Cen}(A)=\mathbb{F}\cdot 1\simeq\mathbb{F}$. A finite-dimensional central simple associative $\mathbb{F}$-algebra A with multiplicative unity is isomorphic to $\mathrm{Mat}(d,\mathbb{D})$ for some suitable division ring $\mathbb{D}$ (and division algebra over $\mathbb{F}$).

The *opposite* A^{opp} of an algebra A is the linear space A with a new product $\mathrm{opp}[ab]$ of $a,b\in A$ given by $\mathrm{opp}[ab]=ba$. Two central simple associative $\mathbb{F}$-algebras A and B are in the same *Brauer class* if $A\otimes B^{\mathrm{opp}}\simeq\mathrm{Mat}(d,\mathbb{F})$ for some integer d. The tensor product of algebras induces a product for Brauer classes, making the set of Brauer classes a group, called the *Brauer group* $Br(\mathbb{F})$ of the field $\mathbb{F}$.

Examples. $Br(\mathbb{R})\simeq\{\mathbb{R},\mathbb{H}\}$, $Br(\mathbb{C})\simeq\{\mathbb{C}\}$, $Br(\mathbb{F}_5)\simeq\{\mathbb{F}_5\}$. ∎

An algebra A is graded over $\mathbb{Z}_2=\{0,1\}$ if it is a direct sum of two subalgebras $A=A_0\oplus A_1$ so that $A_iA_j\subset A_{i+j}$ [the indices are added modulo 2]. For two graded algebras $A=A_0\oplus A_1$ and $B=B_0\oplus B_1$ the *graded tensor product* $A\hat{\otimes}B$ is the linear space $A\otimes B$ provided with the product determined by the formula

$$(a\otimes b)(a'\otimes b')=(-1)^{ij}(aa')\otimes(bb')$$

for homogeneous $a'\in A_i$ and $b\in B_j$. The graded opposite A^{opp} of a graded algebra $A=A_0\oplus A_1$ is the linear space A with a new product $\mathrm{opp}[ab]$ of $a,b\in A$ given by $\mathrm{opp}[ab]=(-1)^{ij}ba$ for homogeneous elements $a\in A_i$ and $b\in B_j$.

Exercises

1. Determine the addition and the multiplication tables of the anisotropic isometry classes $\langle 0\rangle$, $\langle 1\rangle$, $\langle 1,1\rangle$, $\langle 1,1,1\rangle$ of $W(\mathbb{F}_7)$.
2. Identify as matrix algebras all the Clifford algebras of non-degenerate quadratic forms over $\mathbb{F}_5$.

3. Show that the two involutions $\alpha(\lambda,\mu)=(\mu,\lambda)$ and $\beta(\lambda,\mu)=(\bar{\mu},\bar{\lambda})$ are similar involutions of the real or complex algebra ${}^2\mathbb{C}$.
4. Consider the four anti-involutions of $\mathrm{Mat}(2,\mathbb{R})$ sending

$$\begin{pmatrix} a & b \\ c & d \end{pmatrix} \text{ to } \begin{pmatrix} a & c \\ b & d \end{pmatrix}, \begin{pmatrix} a & -c \\ -b & d \end{pmatrix}, \begin{pmatrix} d & b \\ c & a \end{pmatrix}, \begin{pmatrix} d & -b \\ -c & a \end{pmatrix}.$$

Determine which ones of these four anti-involutions are similar or dissimilar to each other. Hint: keep track of what happens to the matrices

$$\begin{pmatrix} 1 & 0 \\ 0 & -1 \end{pmatrix}, \begin{pmatrix} 0 & 1 \\ 1 & 0 \end{pmatrix}, \begin{pmatrix} 0 & -1 \\ 1 & 0 \end{pmatrix}$$

with squares I, I, and $-I$.
5. Show that $\mathbb{C}\otimes_{\mathbb{R}}\mathbb{C}\simeq\mathbb{C}\oplus\mathbb{C}$.
6. Show that $\mathbb{H}\otimes_{\mathbb{R}}\mathbb{H}\simeq\mathrm{Mat}(4,\mathbb{R})$.

Solutions

1.

$\perp$	$\langle 1\rangle$	$\langle 1,1\rangle$	$\langle 1,1,1\rangle$
$\langle 1\rangle$	$\langle 1,1\rangle$	$\langle 1,1,1\rangle$	$\langle 0\rangle$
$\langle 1,1\rangle$	$\langle 1,1,1\rangle$	$\langle 0\rangle$	$\langle 1\rangle$
$\langle 1,1,1\rangle$	$\langle 0\rangle$	$\langle 1\rangle$	$\langle 1,1\rangle$

$\otimes$	$\langle 1\rangle$	$\langle 1,1\rangle$	$\langle 1,1,1\rangle$
$\langle 1\rangle$	$\langle 1\rangle$	$\langle 1,1\rangle$	$\langle 1,1,1\rangle$
$\langle 1,1\rangle$	$\langle 1,1\rangle$	$\langle 0\rangle$	$\langle 1,1\rangle$
$\langle 1,1,1\rangle$	$\langle 1,1,1\rangle$	$\langle 1,1\rangle$	$\langle 1\rangle$

2. A non-degenerate quadratic form $\langle a_1,a_2,\ldots,a_n\rangle$ over $\mathbb{F}_5$ is isometric to $p\langle 1\rangle\perp q\langle 2\rangle$ where the numbers p and q mean, respectively, occurrences of $1,4$ and $2,3$ in $a_1,a_2,\ldots,a_n$. The Clifford algebra $C\ell(p\langle 1\rangle\perp q\langle 2\rangle,\mathbb{F}_5^n)$ is isomorphic, as an associative algebra, to the matrix algebra

$\mathrm{Mat}(2^{n/2},\mathbb{F}_5)$	p and q even
$\mathrm{Mat}(2^{n/2},\mathbb{F}_5)$	p and q odd
${}^2\mathrm{Mat}(2^{(n-1)/2},\mathbb{F}_5)$	p odd, q even
$\mathrm{Mat}(2^{(n-1)/2},\mathbb{F}_5(\sqrt{2}))$	p even, q odd.

For instance, $C\ell(\langle 1,2\rangle,\mathbb{F}_5^2)\simeq\mathrm{Mat}(2,\mathbb{F}_5)$ by the correspondences

$$\mathbf{e}_1\simeq\begin{pmatrix} 1 & 0 \\ 0 & -1 \end{pmatrix},\qquad \mathbf{e}_2\simeq\begin{pmatrix} 0 & 2 \\ 1 & 0 \end{pmatrix}$$

of an orthogonal basis $\{e_1, e_2\}$ of $\langle 1, 2\rangle$ on $\mathbb{F}_5^2$.

3. Choose $\gamma(\lambda, \mu) = (\bar{\lambda}, \mu)$ or $\gamma(\lambda, \mu) = (\lambda, \bar{\mu})$ to find $\alpha\gamma = \gamma\beta$.
4. Only two of the anti-involutions are similar,

$$\alpha\begin{pmatrix} a & b \\ c & d \end{pmatrix} = \begin{pmatrix} a & -c \\ -b & d \end{pmatrix}, \quad \beta\begin{pmatrix} a & b \\ c & d \end{pmatrix} = \begin{pmatrix} d & b \\ c & a \end{pmatrix},$$

as can be seen by choosing the intertwining automorphism

$$\gamma\begin{pmatrix} a & b \\ c & d \end{pmatrix} = \frac{1}{\sqrt{2}}\begin{pmatrix} 1 & -1 \\ 1 & 1 \end{pmatrix}\begin{pmatrix} a & b \\ c & d \end{pmatrix}\frac{1}{\sqrt{2}}\begin{pmatrix} 1 & 1 \\ -1 & 1 \end{pmatrix}$$

for which $\alpha\gamma = \gamma\beta$.

5. We must have $1 \otimes 1 \simeq (1, 1)$ and may choose $i \otimes i \simeq (1, -1)$ or $i \otimes i \simeq (-1, 1)$. If we choose the latter, we may still choose $1 \otimes i \simeq (i, i)$, $i \otimes 1 \simeq (i, -i)$ or $1 \otimes i \simeq (i, -i)$, $i \otimes 1 \simeq (i, i)$ or opposites of both.
6. Choose for $a = a_0 + ia_1 + ja_2 + ka_3$, $b = b_0 + ib_1 + jb_2 + kb_3$ in $\mathbb{H}$ the matrix representations

$$a \simeq \begin{pmatrix} a_0 & -a_1 & -a_2 & -a_3 \\ a_1 & a_0 & -a_3 & a_2 \\ a_2 & a_3 & a_0 & -a_1 \\ a_3 & -a_2 & a_1 & a_0 \end{pmatrix}, \quad b \simeq \begin{pmatrix} b_0 & b_1 & b_2 & b_3 \\ -b_1 & b_0 & -b_3 & b_2 \\ -b_2 & b_3 & b_0 & -b_1 \\ -b_3 & -b_2 & b_1 & b_0 \end{pmatrix},$$

and check that the matrices commute and form two isomorphic images of the ring $\mathbb{H}$.

Bibliography

M.-A. Knus: *Quadratic Forms, Clifford Algebras and Spinors.* Univ. Estadual de Campinas, SP, 1988.

T.Y. Lam: *The Algebraic Theory of Quadratic Forms.* Benjamin, Reading, MA, 1973, 1980.

E. Witt: Theorie der quadratischen Formen in beliebigen Körpern. *J. Reine Angew. Math.* **176** (1937), 31-44.

16

Matrix Representations and Periodicity of 8

The Clifford algebra $\mathcal{C}\ell(Q)$ of a quadratic form Q on a linear space V over a field $\mathbb{F}$ contains an isometric copy of the vector space V. In this chapter we will temporarily forget this special feature of the Clifford algebra $\mathcal{C}\ell(Q)$. Then the Clifford algebra of a non-degenerate quadratic form is nothing but a matrix algebra or a direct sum of two matrix algebras. We have already identified the following Clifford algebras:

$$\begin{aligned}
&\mathcal{C}\ell_2 \simeq \mathrm{Mat}(2,\mathbb{R}), \quad \mathcal{C}\ell_{0,2} \simeq \mathbb{H},\\
&\mathcal{C}\ell_3 \simeq \mathrm{Mat}(3,\mathbb{C}), \quad \mathcal{C}\ell_{0,3} \simeq \mathbb{H}\oplus\mathbb{H},\\
&\mathcal{C}\ell_4 \simeq \mathrm{Mat}(2,\mathbb{H}), \quad \mathcal{C}\ell_{3,1} \simeq \mathrm{Mat}(4,\mathbb{R}), \quad \mathcal{C}\ell_{1,3} \simeq \mathrm{Mat}(2,\mathbb{H}).
\end{aligned}$$

We will find a general pattern for matrix images of Clifford algebras $\mathcal{C}\ell_{p,q}$ of non-degenerate quadratic spaces $\mathbb{R}^{p,q}$. We will see that $\mathcal{C}\ell_{p,q}$ are isomorphic to real matrix algebras with entries in $\mathbb{R},\mathbb{C},\mathbb{H}$ or in ${}^2\mathbb{R} = \mathbb{R}\oplus\mathbb{R}$, ${}^2\mathbb{H} = \mathbb{H}\oplus\mathbb{H}$, that is, their matrix images are

$$\begin{aligned}
&\mathrm{Mat}(d,\mathbb{R}),\ \mathrm{Mat}(d,\mathbb{C}),\ \mathrm{Mat}(d,\mathbb{H}) \quad \text{or}\\
&{}^2\mathrm{Mat}(d,\mathbb{R}) = \mathrm{Mat}(d,{}^2\mathbb{R}), \quad {}^2\mathrm{Mat}(d,\mathbb{H}) = \mathrm{Mat}(d,{}^2\mathbb{H}).
\end{aligned}$$

Review of Matrix Images of $\mathcal{C}\ell_{p,q}$, $p+q<5$

The quadratic space $\mathbb{R}^{p,q}$ is an n-dimensional real vector space $\mathbb{R}^n$, $n=p+q$, with a non-degenerate symmetric *scalar product*

$$\mathbf{x}\cdot\mathbf{y} = x_1y_1 + \ldots + x_py_p - x_{p+1}y_{p+1} - \ldots - x_{p+q}y_{p+q}.$$

The scalar product induces the *quadratic form*

$$\mathbf{x}\cdot\mathbf{x} = x_1^2 + \ldots + x_p^2 - x_{p+1}^2 - \ldots - x_{p+q}^2.$$

A real associative algebra with unity 1 is the *Clifford algebra* $\mathcal{C}\ell_{p,q}$ on $\mathbb{R}^{p,q}$

if it contains $\mathbb{R}^{p,q}$ and $\mathbb{R} = \mathbb{R} \cdot 1 \not\subset \mathbb{R}^{p,q}$ as subspaces so that $\mathbb{R}^{p,q}$ generates $\mathcal{C}\ell_{p,q}$ as a real algebra and

$$\mathbf{x}^2 = \mathbf{x} \cdot \mathbf{x}$$

for all $\mathbf{x} \in \mathbb{R}^{p,q}$. Furthermore, we require that $\mathcal{C}\ell_{p,q}$ is not generated by any proper subspace of $\mathbb{R}^{p,q}$.

The identity $\mathbf{x}^2 = \mathbf{x} \cdot \mathbf{x}$ has a polarized form $\mathbf{xy} + \mathbf{yx} = 2\mathbf{x} \cdot \mathbf{y}$. In an orthonormal basis $\mathbf{e}_1, \mathbf{e}_2, \ldots, \mathbf{e}_n$ of $\mathbb{R}^{p,q}$ this means

$$\mathbf{e}_i\mathbf{e}_j + \mathbf{e}_j\mathbf{e}_i = 2g_{ij}$$

where $g_{ij} = \mathbf{e}_i \cdot \mathbf{e}_j$ or $g_{ii} = 1$, $i \leq p$, $g_{ii} = -1$, $i > p$, and $g_{ij} = 0$, $i \neq j$. The above identity is a condensed form of the relations

$$\mathbf{e}_i^2 = 1,\ 1 \leq i \leq p, \quad \mathbf{e}_i^2 = -1,\ p < i \leq n, \quad \mathbf{e}_i\mathbf{e}_j = -\mathbf{e}_j\mathbf{e}_i,\ i < j.$$

The requirement that no proper subspace of $\mathbb{R}^{p,q}$ generates $\mathcal{C}\ell_{p,q}$ results in the constraint $\mathbf{e}_1\mathbf{e}_2 \ldots \mathbf{e}_n \neq \pm 1$, needed only in the case $p - q = 1 \bmod 4$.

The Clifford algebra $\mathcal{C}\ell_{p,q}$, $p + q = n$, is of dimension 2^n. If the constraint $\mathbf{e}_1\mathbf{e}_2 \ldots \mathbf{e}_n \neq \pm 1$ is omitted, then the resulting algebra could be of dimension 2^n or 2^{n-1}, the lower value being possible only if $p-q = 1 \bmod 4$. In the lower-dimensional case we have $\mathbf{e}_1\mathbf{e}_2 \ldots \mathbf{e}_n = \pm 1$, the algebra itself being isomorphic to the two-sided ideal $\frac{1}{2}(1 \pm \mathbf{e}_{12\ldots n})\mathcal{C}\ell_{p,q}$. For instance, the negative definite quadratic space $\mathbb{R}^{0,3}$ has an 8-dimensional Clifford algebra $\mathcal{C}\ell_{0,3} \simeq \mathbb{H} \oplus \mathbb{H}$, which is a direct sum of two ideals $\frac{1}{2}(1 \pm \mathbf{e}_{123})\mathcal{C}\ell_{0,3}$, both isomorphic to the 4-dimensional quaternion algebra $\mathbb{H}$.

16.1 The Euclidean spaces $\mathbb{R}^n$

In the positive definite case, $p = n$, $q = 0$, of the Euclidean space we abbreviate $\mathbb{R}^{n,0}$ to $\mathbb{R}^n$ and its Clifford algebra $\mathcal{C}\ell_{n,0}$ to $\mathcal{C}\ell_n$. In the Euclidean case we can speak of the *length* $|\mathbf{x}|$ of a vector $\mathbf{x} \in \mathbb{R}^n$ given by $|\mathbf{x}|^2 = \mathbf{x} \cdot \mathbf{x}$. [1]

The Euclidean plane $\mathbb{R}^2$. Consider the Euclidean plane $\mathbb{R}^2$. The Clifford algebra $\mathcal{C}\ell_2$ of $\mathbb{R}^2$ is generated by an orthonormal basis $\mathbf{e}_1, \mathbf{e}_2$ of $\mathbb{R}^2$. We have the multiplication rules

$$\begin{array}{lcl} \mathbf{e}_1^2 = 1,\ \mathbf{e}_2^2 = 1 & & |\mathbf{e}_1| = 1,\ |\mathbf{e}_2| = 1 \\ \mathbf{e}_1\mathbf{e}_2 = -\mathbf{e}_2\mathbf{e}_1 & \text{corresponding to} & \mathbf{e}_1 \perp \mathbf{e}_2. \end{array}$$

Using $\mathbf{e}_1\mathbf{e}_2 = -\mathbf{e}_2\mathbf{e}_1$ and associativity we find $(\mathbf{e}_1\mathbf{e}_2)^2 = -\mathbf{e}_1^2\mathbf{e}_2^2$ which implies $(\mathbf{e}_1\mathbf{e}_2)^2 = -1$. This indicates that $\mathbf{e}_1\mathbf{e}_2$ is neither a scalar nor a vector, but a

1 In the negative definite case we can also speak of the length $|\mathbf{x}|$ of $\mathbf{x} \in \mathbb{R}^{0,n}$ given by $|\mathbf{x}|^2 = -\mathbf{x} \cdot \mathbf{x}$.

new kind of unit, called a *bivector*. The Clifford algebra $\mathcal{C}\ell_2$ is 4-dimensional with a basis consisting of

1	a scalar
$\mathbf{e}_1$, $\mathbf{e}_2$	vectors
$\mathbf{e}_1\mathbf{e}_2$	a bivector.

Write for short $\mathbf{e}_{12} = \mathbf{e}_1\mathbf{e}_2$. The Clifford algebra $\mathcal{C}\ell_2$ has the following multiplication table:

	$\mathbf{e}_1$	$\mathbf{e}_2$	$\mathbf{e}_{12}$
$\mathbf{e}_1$	1	$\mathbf{e}_{12}$	$\mathbf{e}_2$
$\mathbf{e}_2$	$-\mathbf{e}_{12}$	1	$-\mathbf{e}_1$
$\mathbf{e}_{12}$	$-\mathbf{e}_2$	$\mathbf{e}_1$	-1

The Clifford algebra $\mathcal{C}\ell_2$ of the Euclidean plane $\mathbb{R}^2$ is isomorphic, as an associative algebra, to the matrix algebra of real 2×2-matrices $\mathrm{Mat}(2, \mathbb{R})$. This is seen by the correspondences

$\mathcal{C}\ell_2$	$\mathrm{Mat}(2, \mathbb{R})$
1	$\begin{pmatrix} 1 & 0 \\ 0 & 1 \end{pmatrix}$
$\mathbf{e}_1$, $\mathbf{e}_2$	$\begin{pmatrix} 1 & 0 \\ 0 & -1 \end{pmatrix}, \begin{pmatrix} 0 & 1 \\ 1 & 0 \end{pmatrix}$
$\mathbf{e}_{12}$	$\begin{pmatrix} 0 & 1 \\ -1 & 0 \end{pmatrix}$

It should be emphasized that the Clifford algebra $\mathcal{C}\ell_2$ has more structure than the matrix algebra $\mathrm{Mat}(2, \mathbb{R})$. The Clifford algebra $\mathcal{C}\ell_2$ is the matrix algebra $\mathrm{Mat}(2, \mathbb{R})$ with a specific subspace singled out (and a quadratic form on that subspace making it isometric to the Euclidean plane $\mathbb{R}^2$). ∎

The 3-dimensional Euclidean space $\mathbb{R}^3$. Consider the 3-dimensional Euclidean space $\mathbb{R}^3$. The Clifford algebra $\mathcal{C}\ell_3$ is generated by an orthonormal basis $\{\mathbf{e}_1, \mathbf{e}_2, \mathbf{e}_3\}$ of $\mathbb{R}^3$. This time there are three linearly independent bivectors $\mathbf{e}_{12}, \mathbf{e}_{13}, \mathbf{e}_{23}$, each being a square root of -1. In addition, there is the volume element $\mathbf{e}_{123} = \mathbf{e}_1\mathbf{e}_2\mathbf{e}_3$ which squares to -1 and commutes with all the vectors $\mathbf{e}_1, \mathbf{e}_2, \mathbf{e}_3$ and thereby also with all the elements of the algebra $\mathcal{C}\ell_3$.

The Clifford algebra $\mathcal{C}\ell_3$ is 8-dimensional over $\mathbb{R}$ and has a basis consisting of

1	a scalar
$\mathbf{e}_1, \mathbf{e}_2, \mathbf{e}_3$	vectors
$\mathbf{e}_{12}, \mathbf{e}_{13}, \mathbf{e}_{23}$	bivectors
$\mathbf{e}_{123}$	a volume element.

The Clifford algebra $\mathcal{C}\ell_3$ is isomorphic, as a real associative algebra, to the matrix algebra $\text{Mat}(2, \mathbb{C})$ of 2×2-matrices with entries in $\mathbb{C}$. The isomorphism $\mathcal{C}\ell_3 \simeq \text{Mat}(2, \mathbb{C})$ of real associative algebras is fixed by the correspondences

$$\mathbf{e}_1 \simeq \begin{pmatrix} 0 & 1 \\ 1 & 0 \end{pmatrix}, \quad \mathbf{e}_2 \simeq \begin{pmatrix} 0 & -i \\ i & 0 \end{pmatrix}, \quad \mathbf{e}_3 \simeq \begin{pmatrix} 1 & 0 \\ 0 & -1 \end{pmatrix}.$$

The matrices above are known as *Pauli spin matrices.* The multiplication of the unit vectors, $\mathbf{e}_1\mathbf{e}_2\mathbf{e}_3 = \mathbf{e}_{123}$, results in the correspondence

$$\mathbf{e}_{123} \simeq \begin{pmatrix} i & 0 \\ 0 & i \end{pmatrix}.$$

As noted above, the volume element $\mathbf{e}_{123}$, such that $\mathbf{e}_{123}^2 = -1$, commutes with all the elements of the algebra $\mathcal{C}\ell_3$; that is, it belongs to the center of $\mathcal{C}\ell_3$. This enables us to view $\mathcal{C}\ell_3$ as a complex algebra isomorphic, as an associative algebra, to the matrix algebra of complex 2×2-matrices $\text{Mat}(2, \mathbb{C})$. ∎

The 4-dimensional Euclidean space $\mathbb{R}^4$. The Clifford algebra $\mathcal{C}\ell_4$ of the Euclidean space $\mathbb{R}^4$ is isomorphic, as an associative algebra, to the real algebra $\text{Mat}(2, \mathbb{H})$ of 2×2-matrices with entries in the division ring of quaternions $\mathbb{H}$. Using an orthonormal basis $\{\mathbf{e}_1, \mathbf{e}_2, \mathbf{e}_3, \mathbf{e}_4\}$ of $\mathbb{R}^4$ we can find the correspondences

$$\mathbf{e}_1 = \begin{pmatrix} 0 & -i \\ i & 0 \end{pmatrix}, \quad \mathbf{e}_2 = \begin{pmatrix} 0 & -j \\ j & 0 \end{pmatrix}, \quad \mathbf{e}_3 = \begin{pmatrix} 0 & -k \\ k & 0 \end{pmatrix}$$

$$\mathbf{e}_4 = \begin{pmatrix} 1 & 0 \\ 0 & -1 \end{pmatrix}.$$

The Clifford algebra $\mathcal{C}\ell_4$ is of dimension 16 and has a basis consisting of

1	a scalar
$\mathbf{e}_1, \mathbf{e}_2, \mathbf{e}_3, \mathbf{e}_4$	vectors
$\mathbf{e}_{12}, \mathbf{e}_{13}, \mathbf{e}_{14}, \mathbf{e}_{23}, \mathbf{e}_{24}, \mathbf{e}_{34}$	bivectors
$\mathbf{e}_{123}, \mathbf{e}_{124}, \mathbf{e}_{134}, \mathbf{e}_{234}$	3-vectors
$\mathbf{e}_{1234}$	a 4-volume element.

An arbitrary element $u = \langle u\rangle_0 + \langle u\rangle_1 + \langle u\rangle_2 + \langle u\rangle_3 + \langle u\rangle_4$ in $\mathcal{C}\ell_4$ is a sum of a scalar $\langle u\rangle_0$, a vector $\langle u\rangle_1$, a bivector $\langle u\rangle_2$, a 3-vector $\langle u\rangle_3$ and a volume element $\langle u\rangle_4$.

Split complex numbers $\mathbb{R}\oplus\mathbb{R}$. The Clifford algebra $\mathcal{C}\ell_1$ of the Euclidean line $\mathbb{R}^1=\mathbb{R}$ is spanned by $1, \mathbf{e}_1$ where $\mathbf{e}_1^2=1$. Its multiplication table is

	1	$\mathbf{e}_1$
1	1	$\mathbf{e}_1$
$\mathbf{e}_1$	$\mathbf{e}_1$	1

The Clifford algebra $\mathcal{C}\ell_1$ is isomorphic, as an associative algebra, to the double-field $\mathbb{R}\oplus\mathbb{R}$ of *split complex numbers*. The product of two elements (α_1,α_2) and (β_1,β_2) in $\mathbb{R}\oplus\mathbb{R}$ is defined component-wise:

$$(\alpha_1,\alpha_2)(\beta_1,\beta_2)=(\alpha_1\beta_1,\alpha_2\beta_2).$$

The isomorphism $\mathcal{C}\ell_1\simeq\mathbb{R}\oplus\mathbb{R}$ can be seen by the correspondences

$\mathcal{C}\ell_1$	$\mathbb{R}\oplus\mathbb{R}$
1	$(1,1)$
$\mathbf{e}_1$	$(1,-1)$

The Clifford algebra $\mathcal{C}\ell_1\simeq\mathbb{R}\oplus\mathbb{R}$ is a direct sum of two ideals spanned by the idempotents $\frac{1}{2}(1+\mathbf{e}_1)\simeq(1,0)$ and $\frac{1}{2}(1-\mathbf{e}_1)\simeq(0,1)$. ∎

The 5-dimensional Euclidean space $\mathbb{R}^5$. The Clifford algebra $\mathcal{C}\ell_5$ of $\mathbb{R}^5$ is isomorphic to ${}^2\mathrm{Mat}(2,\mathbb{H})=\mathrm{Mat}(2,{}^2\mathbb{H})$, as can be seen by the correspondences

$$\mathbf{e}_1=\begin{pmatrix}(0,0) & (\text{-}i,i)\\ (i,\text{-}i) & (0,0)\end{pmatrix},\quad \mathbf{e}_2=\begin{pmatrix}(0,0) & (\text{-}j,j)\\ (j,\text{-}j) & (0,0)\end{pmatrix},\quad \mathbf{e}_3=\begin{pmatrix}(0,0) & (\text{-}k,k)\\ (k,\text{-}k) & (0,0)\end{pmatrix}$$

$$\mathbf{e}_4=\begin{pmatrix}(1,\text{-}1) & (0,0)\\ (0,0) & (\text{-}1,1)\end{pmatrix},\quad \mathbf{e}_5=\begin{pmatrix}(0,0) & (1,\text{-}1)\\ (1,\text{-}1) & (0,0)\end{pmatrix}.$$

The Clifford algebra $\mathcal{C}\ell_5$ has two central idempotents

$$\frac{1}{2}(1+\mathbf{e}_{12345})=\begin{pmatrix}(1,0) & (0,0)\\ (0,0) & (1,0)\end{pmatrix},\quad \frac{1}{2}(1-\mathbf{e}_{12345})=\begin{pmatrix}(0,1) & (0,0)\\ (0,0) & (0,1)\end{pmatrix}$$

which both project out of $\mathcal{C}\ell_5$ an isomorphic copy of $\mathrm{Mat}(2,\mathbb{H})$, that is, $\frac{1}{2}(1\pm\mathbf{e}_{12345})\mathcal{C}\ell_5\simeq\mathrm{Mat}(2,\mathbb{H})$. An isomorphic copy of $\frac{1}{2}(1\pm\mathbf{e}_{12345})\mathcal{C}\ell_5$ is constructed within another subspace of $\mathcal{C}\ell_5$ in the following counter-example.

Counter-example. Consider the subspace of scalars, vectors and bivectors $\mathbb{R}\oplus\mathbb{R}^5\oplus\bigwedge^2\mathbb{R}^5$ of dimension $1+5+\frac{1}{2}5(5-1)=\frac{1}{2}2^5$. Introduce in this subspace a new product $u\circ v$ defined by (one of the following)

$$u\circ v=\langle uv(1\pm\mathbf{e}_{12345})\rangle_{0,1,2}$$

where $\langle w\rangle_{0,1,2} = \langle w\rangle_0 + \langle w\rangle_1 + \langle w\rangle_2$. This new product is associative and satisfies

$$\mathbf{x}\circ\mathbf{x} = |\mathbf{x}|^2 \quad \text{for} \quad \mathbf{x}\in\mathbb{R}^5.$$

However, $\mathbf{e}_1\circ\mathbf{e}_2\circ\mathbf{e}_3\circ\mathbf{e}_4\circ\mathbf{e}_5 = \pm 1$. As a sample, this new product satisfies

$$\mathbf{e}_1\circ\mathbf{e}_2 = \mathbf{e}_{12},\ \mathbf{e}_1\circ\mathbf{e}_{12} = \mathbf{e}_2,\ \mathbf{e}_{12}\circ\mathbf{e}_{12} = -1,\ \mathbf{e}_{12}\circ\mathbf{e}_{23} = \mathbf{e}_{13}$$
$$\mathbf{e}_1\circ\mathbf{e}_{23} = \mp\mathbf{e}_{45},\ \mathbf{e}_{12}\circ\mathbf{e}_{34} = \pm\mathbf{e}_5.$$

The multiplication table of this new product is given by the following matrices

$$\mathbf{e}_1 = \pm\begin{pmatrix} 0 & -i \\ i & 0 \end{pmatrix},\quad \mathbf{e}_2 = \pm\begin{pmatrix} 0 & -j \\ j & 0 \end{pmatrix},\quad \mathbf{e}_3 = \pm\begin{pmatrix} 0 & -k \\ k & 0 \end{pmatrix},$$
$$\mathbf{e}_4 = \pm\begin{pmatrix} 1 & 0 \\ 0 & -1 \end{pmatrix},\quad \mathbf{e}_5 = \pm\begin{pmatrix} 0 & 1 \\ 1 & 0 \end{pmatrix}.$$

This serves as a counter-example to a belief that the Clifford algebra would be uniquely generated by its subspaces $\mathbb{R}$ and $\mathbb{R}^n$. ∎

The 3-dimensional anti-Euclidean space $\mathbb{R}^{0,3}$

The anti-Euclidean space $\mathbb{R}^{0,3}$ has a negative definite quadratic form sending a vector $\mathbf{x} = x_1\mathbf{e}_1 + x_2\mathbf{e}_2 + x_3\mathbf{e}_3$ to the scalar

$$\mathbf{x}\cdot\mathbf{x} = -(x_1^2 + x_2^2 + x_3^2).$$

An orthonormal basis $\{\mathbf{e}_1, \mathbf{e}_2, \mathbf{e}_3\}$ of $\mathbb{R}^{0,3}$ obeys the multiplication rules

$$\mathbf{e}_1^2 = \mathbf{e}_2^2 = \mathbf{e}_3^2 = -1 \quad \text{and}$$
$$\mathbf{e}_1\mathbf{e}_2 = -\mathbf{e}_2\mathbf{e}_1,\ \mathbf{e}_1\mathbf{e}_3 = -\mathbf{e}_3\mathbf{e}_1,\ \mathbf{e}_2\mathbf{e}_3 = -\mathbf{e}_3\mathbf{e}_2.$$

These relations are satisfied by the unit quaternions

$$i = \mathbf{e}_1,\quad j = \mathbf{e}_2,\quad k = \mathbf{e}_3$$

in $\mathbb{H}$. The rule $ijk = -1$, or $\mathbf{e}_1\mathbf{e}_2\mathbf{e}_3 = -1$, means that the real algebra $\mathbb{H} = \mathbb{R}\oplus\mathbb{R}^{0,3}$ is generated by a proper subspace $\mathbb{R}^{0,2}$ of $\mathbb{R}^{0,3}$. In other words, each quaternion can be expressed in the form $x = x_0 + x_1\mathbf{e}_1 + x_2\mathbf{e}_2 + x_3\mathbf{e}_1\mathbf{e}_2$ where $\mathbf{e}_3 = \mathbf{e}_1\mathbf{e}_2$. This matter is expressed by saying that $\mathbb{H}$ is an *algebra of the quadratic form*

$$x_1\mathbf{e}_1 + x_2\mathbf{e}_2 + x_3\mathbf{e}_3 \to -(x_1^2 + x_2^2 + x_3^2)$$

although it is not the Clifford algebra $\mathcal{C}\ell_{0,3}$. The 8-dimensional Clifford algebra $\mathcal{C}\ell_{0,3}$ is isomorphic, as an associative algebra, to the direct sum $\mathbb{H}\oplus\mathbb{H}$. This can be seen by the correspondences

$\mathcal{C}\ell_{0,3}$	$\mathbb{H} \oplus \mathbb{H}$
1	$(1, 1)$
$\mathbf{e}_1, \mathbf{e}_2, \mathbf{e}_3$	$(i, -i)$, $(j, -j)$, $(k, -k)$
$\mathbf{e}_{23}, \mathbf{e}_{31}, \mathbf{e}_{12}$	(i, i), (j, j), (k, k)
$\mathbf{e}_{123}$	$(-1, 1)$

The Clifford algebra $\mathcal{C}\ell_{0,3}$ of $\mathbb{R}^{0,3}$ is the **universal object** in the category of *algebras of the quadratic form*

$$x_1\mathbf{e}_1 + x_2\mathbf{e}_2 + x_3\mathbf{e}_3 \to -(x_1^2 + x_2^2 + x_3^2)$$

or for short in the *category of quadratic algebras.* [2] If there are other objects in this category, they are quotients of the universal object with respect to a two-sided ideal. This gives us two other algebras of dimension 4; in one of them we have the relation $\mathbf{e}_1\mathbf{e}_2\mathbf{e}_3 = 1$ and in the other $\mathbf{e}_1\mathbf{e}_2\mathbf{e}_3 = -1$. These two algebras of dimension 4 are linearly isomorphic to $\mathbb{R} \oplus \mathbb{R}^{0,3}$. In the category of quadratic algebras these two algebras of dimension 4 are *not isomorphic* with each other, which means that the relations $\mathbf{e}_1\mathbf{e}_2\mathbf{e}_3 = 1$ and $\mathbf{e}_1\mathbf{e}_2\mathbf{e}_3 = -1$ prevent the identity mapping on $\mathbb{R}^{0,3}$ being extended to an isomorphism in this category. However, in the category of all associative algebras these two algebras of dimension 4 are isomorphic with each other (and with the quaternion algebra $\mathbb{H} = \mathbb{R} \oplus \mathbb{R}^{0,3}$). The isomorphism can be seen by the mappings

$$\mathbf{e}_1 \to \mathbf{e}_1, \quad \mathbf{e}_2 \to \mathbf{e}_2 \quad \text{and} \quad \mathbf{e}_3 \to -\mathbf{e}_3.$$

■

16.2 Indefinite metrics $\mathbb{R}^{p,q}$

The hyperbolic plane $\mathbb{R}^{1,1}$. The hyperbolic plane is the linear space $\mathbb{R}^2$ endowed with a quadratic form

$$(u, v) \to uv$$

which by change of variables $u = x_1 + x_2$, $v = x_1 - x_2$ is seen to be

$$(x_1, x_2) \to x_1^2 - x_2^2.$$

Thus the hyperbolic plane is indefinite, neutral and has the Lorentz signature $\mathbb{R}^{1,1}$. The Clifford algebra $\mathcal{C}\ell_{1,1}$ of $\mathbb{R}^{1,1}$ is isomorphic, as an associative algebra,

---

2 The term quadratic algebra is customarily used for something else: in a quadratic algebra x^2 is linearly dependent on x and 1.

to the matrix algebra $\mathrm{Mat}(2,\mathbb{R})$ by the correspondences

$C\ell_{1,1}$	$\mathrm{Mat}(2,\mathbb{R})$
1	$\begin{pmatrix}1&0\\0&1\end{pmatrix}$
$\mathbf{e}_1$, $\mathbf{e}_2$	$\begin{pmatrix}1&0\\0&-1\end{pmatrix}$, $\begin{pmatrix}0&1\\-1&0\end{pmatrix}$
$\mathbf{e}_{12}$	$\begin{pmatrix}0&1\\1&0\end{pmatrix}$

Note that the Clifford algebras $C\ell_{1,1}$ and $C\ell_2 \simeq \mathrm{Mat}(2,\mathbb{R})$ are isomorphic *as associative algebras* but non-isomorphic *as quadratic algebras*. ∎

The Minkowski space-time $\mathbb{R}^{3,1}$. The elements of an orthonormal basis $\{\mathbf{e}_1,\mathbf{e}_2,\mathbf{e}_3,\mathbf{e}_4\}$ of $\mathbb{R}^{3,1}$ anticommute, $\mathbf{e}_\mu\mathbf{e}_\nu = -\mathbf{e}_\nu\mathbf{e}_\mu$, and have unit squares, $\mathbf{e}_1^2 = \mathbf{e}_2^2 = \mathbf{e}_3^2 = 1$, $\mathbf{e}_4^2 = -1$. The basis vectors are often given the following representation by complex 4×4-matrices:

$$\mathbf{e}_k = \mathbf{e}^k = \begin{pmatrix}0&\sigma_k\\\sigma_k&0\end{pmatrix} \text{ for } k=1,2,3 \text{ and } \mathbf{e}_4 = -\mathbf{e}^4 = \begin{pmatrix}i&0\\0&-i\end{pmatrix}$$

where we recognize the 2×2 Pauli spin matrices $\sigma_1,\sigma_2,\sigma_3$. It is possible to represent $C\ell_{3,1}$ by real matrices as follows:

$$\mathbf{e}_1 = \begin{pmatrix}\sigma_3&0\\0&-\sigma_3\end{pmatrix},\quad \mathbf{e}_2 = \begin{pmatrix}\sigma_1&0\\0&\sigma_1\end{pmatrix},\quad \mathbf{e}_3 = \begin{pmatrix}0&\sigma_3\\\sigma_3&0\end{pmatrix},$$

$$\mathbf{e}_4 = \begin{pmatrix}-i\sigma_2&0\\0&-i\sigma_2\end{pmatrix}.$$

This implies $C\ell_{3,1} \simeq \mathrm{Mat}(4,\mathbb{R})$. ∎

The Minkowski time-space $\mathbb{R}^{1,3}$. In the signature $\mathbb{R}^{1,3}$ one usually gives the following representation by complex 4×4-matrices:

$$\gamma_0 = \gamma^0 = \begin{pmatrix}1&0\\0&-1\end{pmatrix},\quad \text{and}$$

$$\gamma_k = -\gamma^k = \begin{pmatrix}0&-\sigma_k\\\sigma_k&0\end{pmatrix},\quad \text{for } k=1,2,3.$$

In addition to the above matrix representation one can represent the Clifford algebra $C\ell_{1,3}$ by the following 2×2-matrices with quaternion entries:

$$\gamma_0 = \begin{pmatrix} 1 & 0 \\ 0 & -1 \end{pmatrix},$$

$$\gamma_1 = \begin{pmatrix} 0 & i \\ i & 0 \end{pmatrix}, \quad \gamma_2 = \begin{pmatrix} 0 & j \\ j & 0 \end{pmatrix}, \quad \gamma_3 = \begin{pmatrix} 0 & k \\ k & 0 \end{pmatrix}.$$

Since the Clifford algebra $C\ell_{1,3}$ and the matrix algebra $\mathrm{Mat}(2,\mathbb{H})$ of 2×2-matrices with entries in $\mathbb{H}$ are both real algebras of dimension 16, the above correspondences establish an isomorphism of associative algebras, that is, $C\ell_{1,3} \simeq \mathrm{Mat}(2,\mathbb{H})$.

A short look at physics: A vector $u = u_0\gamma^0 + u_1\gamma^1 + u_2\gamma^2 + u_3\gamma^3$ with square $u^2 = u_0^2 - u_1^2 - u_2^2 - u_3^2$ can be time-like $u^2 > 0$, null $u^2 = 0$, or space-like $u^2 < 0$. A time-like vector or non-zero null vector can be future oriented $u_0 > 0$ or past oriented $u_0 < 0$. A time-like future oriented unit vector u, $u^2 = 1$, gives the velocity $v < c$ of a real particle by

$$u_0 = \frac{1}{\sqrt{1 - \dfrac{v^2}{c^2}}}.$$

∎

Physicists might want to observe that the Clifford algebras $C\ell_{3,1} \simeq \mathrm{Mat}(4,\mathbb{R})$ and $C\ell_{1,3} \simeq \mathrm{Mat}(2,\mathbb{H})$ are *not isomorphic* as associative algebras, even though both of them have the same complexification $\mathrm{Mat}(4,\mathbb{C})$ with the same complex structure but with different real structures (= different real subalgebras). The complexified Clifford algebras $\mathbb{C} \otimes C\ell_{1,3} \simeq \mathbb{C} \otimes C\ell_{3,1}$ have a 4-dimensional irreducible left ideal (8-dimensional real subspace). As a graded left ideal this ideal is also irreducible. The real algebra $C\ell_{1,3}$ has an 8-dimensional irreducible left ideal, which is also graded. However, the real algebra $C\ell_{3,1}$ has a 4-dimensional irreducible ideal, which is not graded (that is $C\ell_{3,1}$ does not have primitive idempotents sitting in $C\ell_{3,1}^+$), and an 8-dimensional irreducible graded ideal.

The Table of Clifford Algebras

The Clifford algebra $C\ell_{p,q}$, where $p - q \neq 1 \bmod 4$, is a simple algebra of dimension 2^n, where $n = p + q$, and therefore isomorphic with a full matrix algebra with entries in $\mathbb{R}$, $\mathbb{C}$, or $\mathbb{H}$. The Clifford algebra $C\ell_{p,q}$, where $p - q = 1 \bmod 4$, is a semi-simple algebra of dimension 2^n so that the two central idempotents $\frac{1}{2}(1 \pm \mathbf{e}_1\mathbf{e}_2 \ldots \mathbf{e}_n)$ project out two copies of a full matrix algebra with entries in $\mathbb{R}$ or $\mathbb{H}$. To put it slightly differently, the Clifford algebra $C\ell_{p,q}$

has a faithful representation as a matrix algebra with entries in $\mathbb{R}$, $\mathbb{C}$, $\mathbb{H}$ or $\mathbb{R}\oplus\mathbb{R}$, $\mathbb{H}\oplus\mathbb{H}$. In the rings ${}^2\mathbb{R}=\mathbb{R}\oplus\mathbb{R}$ and ${}^2\mathbb{H}=\mathbb{H}\oplus\mathbb{H}$ the multiplication is defined component-wise:

$$(a_1,a_2)(b_1,b_2)=(a_1b_1,a_2b_2).$$

16.3 Matrix representation $\mathcal{C}\ell_{p+1,q+1}\simeq \mathrm{Mat}(2,\mathcal{C}\ell_{p,q})$

Let $\{\mathbf{e}_1,\mathbf{e}_2,\ldots,\mathbf{e}_n\}$ be an orthonormal basis of $\mathbb{R}^{p,q}$, $n=p+q$, generating the Clifford algebra $\mathcal{C}\ell_{p,q}$. The 2×2-matrices

$$\begin{pmatrix}\mathbf{e}_i & 0\\ 0 & -\mathbf{e}_i\end{pmatrix}\quad \text{for } i=1,2,\ldots,n,\qquad \begin{pmatrix}0 & 1\\ 1 & 0\end{pmatrix},\quad \begin{pmatrix}0 & -1\\ 1 & 0\end{pmatrix}$$

anticommute and generate the Clifford algebra $\mathcal{C}\ell_{p+1,q+1}$. In other words, the Clifford algebra $\mathcal{C}\ell_{p+1,q+1}$ is isomorphic, as an associative algebra, to the algebra of 2×2-matrices with entries in the Clifford algebra $\mathcal{C}\ell_{p,q}$. This can be condensed by writing $\mathcal{C}\ell_{p+1,q+1}\simeq \mathrm{Mat}(2,\mathcal{C}\ell_{p,q})$.

Examples. Recall that $\mathcal{C}\ell_1\simeq {}^2\mathbb{R}=\mathbb{R}\oplus\mathbb{R}$ by setting $\mathbf{e}_1\simeq(1,-1)$. This implies the isomorphism $\mathcal{C}\ell_{2,1}\simeq {}^2\mathrm{Mat}(2,\mathbb{R})$. Recall that $\mathcal{C}\ell_{0,3}\simeq {}^2\mathbb{H}=\mathbb{H}\oplus\mathbb{H}$, which implies $\mathcal{C}\ell_{1,4}\simeq {}^2\mathrm{Mat}(2,\mathbb{H})$. Recall that $\mathcal{C}\ell_{1,3}\simeq \mathrm{Mat}(2,\mathbb{H})$ which implies $\mathcal{C}\ell_{2,4}\simeq \mathrm{Mat}(4,\mathbb{H})$. ∎

Supplement an orthonormal basis $\{\mathbf{e}_1,\mathbf{e}_2,\ldots,\mathbf{e}_n\}$ of $\mathbb{R}^{p,q}$ with two more anticommuting basis vectors $\mathbf{e}_+$ and $\mathbf{e}_-$ such that $\mathbf{e}_+^2=1$ and $\mathbf{e}_-^2=-1$ to form an orthonormal basis of $\mathbb{R}^{p+1,q+1}$. The generators $\mathbf{e}_1,\mathbf{e}_2,\ldots,\mathbf{e}_n,\mathbf{e}_+,\mathbf{e}_-$ of $\mathcal{C}\ell_{p+1,q+1}$ correspond to the generators

$$\mathbf{e}_i\simeq\begin{pmatrix}\mathbf{e}_i & 0\\ 0 & -\mathbf{e}_i\end{pmatrix}\quad\text{for } i=1,2,\ldots,n,$$

$$\mathbf{e}_+\simeq\begin{pmatrix}0 & 1\\ 1 & 0\end{pmatrix},\quad \mathbf{e}_-\simeq\begin{pmatrix}0 & -1\\ 1 & 0\end{pmatrix}$$

of $\mathrm{Mat}(2,\mathcal{C}\ell_{p,q})$, so that the element $a\in\mathcal{C}\ell_{p,q}$ is represented by a matrix

$$a\simeq\begin{pmatrix}a & 0\\ 0 & \hat{a}\end{pmatrix}$$

where the hat means the grade involution $\hat{a}=a_0-a_1$ with $a_0=\mathrm{even}(a)$ and $a_1=\mathrm{odd}(a)$. There is another possibility to embed $\mathcal{C}\ell_{p,q}$ into $\mathrm{Mat}(2,\mathcal{C}\ell_{p,q})$, so that $a\in\mathcal{C}\ell_{p,q}$ is represented by

$$a'=a_0+a_1\mathbf{e}_+\mathbf{e}_-\simeq\begin{pmatrix}a & 0\\ 0 & a\end{pmatrix}$$

which is just a multiple of the identity matrix. Since $a' = a_0 + a_1\mathbf{e}_+\mathbf{e}_-$ commutes with

$$\tfrac{1}{2}(1+\mathbf{e}_+\mathbf{e}_-) \simeq \begin{pmatrix} 1 & 0 \\ 0 & 0 \end{pmatrix}, \quad \tfrac{1}{2}(\mathbf{e}_+ - \mathbf{e}_-) \simeq \begin{pmatrix} 0 & 1 \\ 0 & 0 \end{pmatrix}$$

$$\tfrac{1}{2}(\mathbf{e}_+ + \mathbf{e}_-) \simeq \begin{pmatrix} 0 & 0 \\ 1 & 0 \end{pmatrix}, \quad \tfrac{1}{2}(1-\mathbf{e}_+\mathbf{e}_-) \simeq \begin{pmatrix} 0 & 0 \\ 0 & 1 \end{pmatrix}$$

we have the correspondence $\text{Mat}(2, C\ell_{p,q}) \simeq C\ell_{p+1,q+1}$, given by

$$\begin{aligned}\begin{pmatrix} a & b \\ c & d \end{pmatrix} &= a\begin{pmatrix} 1 & 0 \\ 0 & 0 \end{pmatrix} + b\begin{pmatrix} 0 & 1 \\ 0 & 0 \end{pmatrix} + c\begin{pmatrix} 0 & 0 \\ 1 & 0 \end{pmatrix} + d\begin{pmatrix} 0 & 0 \\ 0 & 1 \end{pmatrix} \\ &\simeq a'\tfrac{1}{2}(1+\mathbf{e}_+\mathbf{e}_-) + b'\tfrac{1}{2}(\mathbf{e}_+ - \mathbf{e}_-) + c'\tfrac{1}{2}(\mathbf{e}_+ + \mathbf{e}_-) + d'\tfrac{1}{2}(1-\mathbf{e}_+\mathbf{e}_-) \\ &= a\tfrac{1}{2}(1+\mathbf{e}_+\mathbf{e}_-) + b\tfrac{1}{2}(\mathbf{e}_+ - \mathbf{e}_-) + \hat{c}\tfrac{1}{2}(\mathbf{e}_+ + \mathbf{e}_-) + \hat{d}\tfrac{1}{2}(1-\mathbf{e}_+\mathbf{e}_-) \\ &= \tfrac{1}{2}(1+\mathbf{e}_+\mathbf{e}_-)a + \tfrac{1}{2}(\mathbf{e}_+ - \mathbf{e}_-)\hat{b} + \tfrac{1}{2}(\mathbf{e}_+ + \mathbf{e}_-)c + \tfrac{1}{2}(1-\mathbf{e}_+\mathbf{e}_-)\hat{d}.\end{aligned}$$

To put all this in another way: The Clifford algebra $C\ell_{p+1,q+1}$ contains an isomorphic copy of $C\ell_{p,q}$ generated by the elements $\mathbf{e}'_i = \mathbf{e}_i\mathbf{e}_+\mathbf{e}_-$, where $i = 1, 2, \ldots, n = p+q$, in such a way that each element of $C\ell_{p,q}$ commutes with every element of a copy of $C\ell_{1,1}$ generated by $\mathbf{e}_+$ and $\mathbf{e}_-$, and further that $C\ell_{p,q}$ and $C\ell_{1,1}$ together generate all of $C\ell_{p+1,q+1}$. These considerations can be condensed by writing

$$\boxed{C\ell_{p,q} \otimes C\ell_{1,1} \simeq C\ell_{p+1,q+1}}$$

where $C\ell_{1,1} \simeq \text{Mat}(2, \mathbb{R})$.

Symmetry $C\ell_{p,q} \simeq C\ell_{q+1,p-1}$. Take an orthonormal basis $\{\mathbf{e}_1, \mathbf{e}_2, \ldots, \mathbf{e}_n\}$ of $\mathbb{R}^{p,q}$ where $p \geq 1$ and set

$$\mathbf{e}'_1 = \mathbf{e}_1 \quad \text{and} \quad \mathbf{e}'_i = \mathbf{e}_i\mathbf{e}_1 \quad \text{for} \quad i > 1.$$

The elements $\mathbf{e}'_i$ where $i = 1, 2, \ldots, n$ anticommute with each other so that $\mathbf{e}'^2_1 = \mathbf{e}^2_1$ and $\mathbf{e}'^2_i = -\mathbf{e}^2_i$ for $i > 1$. Therefore, the subset $\{\mathbf{e}'_1, \mathbf{e}'_2, \ldots, \mathbf{e}'_n\}$ of $C\ell_{p,q}$ is a generating set for $C\ell_{q+1,p-1}$. This proves the isomorphism

$$\boxed{C\ell_{p,q} \simeq C\ell_{q+1,p-1}}$$

when $p \geq 1$.

Examples. Recall that $C\ell_3 \simeq \text{Mat}(2, \mathbb{C})$, which by symmetry implies $C\ell_{1,2} \simeq \text{Mat}(2, \mathbb{C})$. Recall that $C\ell_{3,1} \simeq \text{Mat}(4, \mathbb{R})$, which implies $C\ell_{2,2} \simeq \text{Mat}(4, \mathbb{R})$. From $C\ell_{0,4} \simeq \text{Mat}(2, \mathbb{H})$ we can first deduce $C\ell_{1,5} \simeq \text{Mat}(4, \mathbb{H})$ (by adding a hyperbolic plane) which implies $C\ell_6 \simeq \text{Mat}(4, \mathbb{H})$. ∎

16.4 Periodicity of 8

Table 1, of Clifford algebras, contains or continues with two kinds of periodicities of 8, namely for algebras of the same dimension $\mathcal{C}\ell_{p,q} \simeq \mathcal{C}\ell_{p-4,q+4}$ where $p \geq 4$, and for algebras of different dimension $\mathcal{C}\ell_{p+8,q} \simeq \mathrm{Mat}(16, \mathcal{C}\ell_{p,q})$. Let us first prove $\mathcal{C}\ell_{p,q} \simeq \mathcal{C}\ell_{p-4,q+4}$ where $p \geq 4$. Take an orthonormal basis $\{\mathbf{e}_1, \mathbf{e}_2, \ldots, \mathbf{e}_n\}$ of $\mathbb{R}^{p,q}$ and set

$$\begin{aligned} \mathbf{e}'_i &= \mathbf{e}_i \mathbf{h} \qquad \text{for} \quad i = 1, 2, 3, 4, \\ \mathbf{e}'_i &= \mathbf{e}_i \qquad \text{for} \quad i > 4, \end{aligned}$$

where $\mathbf{h} = \mathbf{e}_1\mathbf{e}_2\mathbf{e}_3\mathbf{e}_4$. Then the subset $\{\mathbf{e}'_1, \mathbf{e}'_2, \ldots, \mathbf{e}'_n\}$ of $\mathcal{C}\ell_{p,q}$ is a generating set for $\mathcal{C}\ell_{p-4,q+4}$, which implies the isomorphism

$$\mathcal{C}\ell_{p,q} \simeq \mathcal{C}\ell_{p-4,q+4}$$

where $p \geq 4$. These isomorphisms are due to Cartan 1908 p. 464.

Examples. Recall that $\mathcal{C}\ell_6 \simeq \mathrm{Mat}(4, \mathbb{H})$, which implies $\mathcal{C}\ell_{2,4} \simeq \mathrm{Mat}(4, \mathbb{H})$. From $\mathcal{C}\ell_3 \simeq \mathrm{Mat}(2, \mathbb{C})$ deduce first $\mathcal{C}\ell_{4,1} \simeq \mathrm{Mat}(4, \mathbb{C})$, which implies $\mathcal{C}\ell_{0,5} \simeq \mathrm{Mat}(4, \mathbb{C})$. From $\mathcal{C}\ell_{2,2} \simeq \mathrm{Mat}(4, \mathbb{R})$ we first deduce $\mathcal{C}\ell_{3,3} \simeq \mathrm{Mat}(8, \mathbb{R})$; then by $\mathcal{C}\ell_{3,3} \simeq \mathcal{C}\ell_{4,2}$ and $\mathcal{C}\ell_{4,2} \simeq \mathcal{C}\ell_{0,6}$ we find $\mathcal{C}\ell_{0,6} \simeq \mathrm{Mat}(8, \mathbb{R})$. From $\mathcal{C}\ell_{3,3} \simeq \mathrm{Mat}(8, \mathbb{R})$ we find $\mathcal{C}\ell_{4,4} \simeq \mathrm{Mat}(16, \mathbb{R})$ and also $\mathcal{C}\ell_8 \simeq \mathrm{Mat}(16, \mathbb{R})$ and $\mathcal{C}\ell_{0,8} \simeq \mathrm{Mat}(16, \mathbb{R})$. ∎

Next, prove $\mathcal{C}\ell_{p+8,q} \simeq \mathrm{Mat}(16, \mathcal{C}\ell_{p,q})$ by showing that $\mathcal{C}\ell_{p,q+8} \simeq \mathrm{Mat}(16, \mathcal{C}\ell_{p,q})$. Take an orthonormal basis $\{\mathbf{e}_1, \mathbf{e}_2, \ldots, \mathbf{e}_n, \mathbf{e}_{n+1}, \ldots, \mathbf{e}_{n+8}\}$ of $\mathbb{R}^{p,q+8}$ where $n = p + q$ and set

$$\mathbf{e}'_i = \mathbf{e}_i \mathbf{e}_{n+1} \ldots \mathbf{e}_{n+8} \quad \text{for} \quad i = 1, 2, \ldots, n = p + q.$$

Then the subset $\{\mathbf{e}'_1, \mathbf{e}'_2, \ldots, \mathbf{e}'_n\}$ of $\mathcal{C}\ell_{p,q+8}$ generates a subalgebra isomorphic to $\mathcal{C}\ell_{p,q}$. The subalgebra generated by $\mathbf{e}_{n+1}, \ldots, \mathbf{e}_{n+8}$ is isomorphic to $\mathcal{C}\ell_{0,8} \simeq \mathrm{Mat}(16, \mathbb{R})$. These two subalgebras commute with each other element-wise and generate all of $\mathcal{C}\ell_{p,q+8}$, which shows that

$$\boxed{\mathcal{C}\ell_{p,q+8} \simeq \mathcal{C}\ell_{p,q} \otimes \mathrm{Mat}(16, \mathbb{R}) \simeq \mathrm{Mat}(16, \mathcal{C}\ell_{p,q})}$$

Similarly, $\mathcal{C}\ell_{p+8,q} \simeq \mathcal{C}\ell_{p,q} \otimes \mathrm{Mat}(16, \mathbb{R}) \simeq \mathrm{Mat}(16, \mathcal{C}\ell_{p,q})$. These isomorphisms are due to Cartan 1908.

Example. Note that $\mathcal{C}\ell_{0,1} \simeq \mathbb{C}$ which implies $\mathcal{C}\ell_{8,1} \simeq \mathrm{Mat}(16, \mathbb{C})$. Recall that $\mathcal{C}\ell_{1,1} \simeq \mathrm{Mat}(2, \mathbb{R})$ and so $\mathcal{C}\ell_{1,9} \simeq \mathrm{Mat}(32, \mathbb{R})$. ∎

Table 1. Clifford Algebras $C\ell_{p,q}$, $p+q<8$.

$p+q$ \ $p-q$	-7	-6	-5	-4	-3	-2	-1	0	1	2	3	4	5	6	7
0								$\mathbb{R}$							
1							$\mathbb{C}$		${}^2\mathbb{R}$						
2						$\mathbb{H}$		$\mathbb{R}(2)$		$\mathbb{R}(2)$					
3					${}^2\mathbb{H}$		$\mathbb{C}(2)$		${}^2\mathbb{R}(2)$		$\mathbb{C}(2)$				
4				$\mathbb{H}(2)$		$\mathbb{H}(2)$		$\mathbb{R}(4)$		$\mathbb{R}(4)$		$\mathbb{H}(2)$			
5			$\mathbb{C}(4)$		${}^2\mathbb{H}(2)$		$\mathbb{C}(4)$		${}^2\mathbb{R}(4)$		$\mathbb{C}(4)$		${}^2\mathbb{H}(2)$		
6		$\mathbb{R}(8)$		$\mathbb{H}(4)$		$\mathbb{H}(4)$		$\mathbb{R}(8)$		$\mathbb{R}(8)$		$\mathbb{H}(4)$		$\mathbb{H}(4)$	
7	${}^2\mathbb{R}(8)$		$\mathbb{C}(8)$		${}^2\mathbb{H}(4)$		$\mathbb{C}(8)$		${}^2\mathbb{R}(8)$		$\mathbb{C}(8)$		${}^2\mathbb{H}(4)$		$\mathbb{C}(8)$

$\mathbb{A}(d)$ means the real algebra of $d \times d$-matrices $\mathrm{Mat}(d, \mathbb{A})$ with entries in the ring $\mathbb{A} = \mathbb{R}$, $\mathbb{C}$, $\mathbb{H}$, ${}^2\mathbb{R}$, ${}^2\mathbb{H}$.

16.5 Complex Clifford algebras and their periodicity of 2

Complex quadratic spaces $\mathbb{C}^n$ have quadratic forms

$$z_1^2 + z_2^2 + \ldots + z_n^2.$$

The type of their Clifford algebras $C\ell(\mathbb{C}^n)$ depends only on the parity of n. Denote $\ell = \lfloor n \rfloor$. In even dimensions $C\ell(\mathbb{C}^n) \simeq \mathrm{Mat}(2^\ell, \mathbb{C})$ and in odd dimensions $C\ell(\mathbb{C}^n) \simeq {}^2\mathrm{Mat}(2^\ell, \mathbb{C})$.

Table 2. Complex Clifford Algebras $C\ell(\mathbb{C}^n)$, $n < 8$.

n		
0	$\mathbb{C}$	
1		${}^2\mathbb{C}$
2	$\mathbb{C}(2)$	
3		${}^2\mathbb{C}(2)$
4	$\mathbb{C}(4)$	
5		${}^2\mathbb{C}(4)$
6	$\mathbb{C}(8)$	
7		${}^2\mathbb{C}(8)$

Exercises

1. Show that $\mathcal{C}\ell^+_{p,q} \simeq \mathcal{C}\ell_{p,q-1}$ and $\mathcal{C}\ell^+_n = \mathcal{C}\ell^+_{n,0} \simeq \mathcal{C}\ell_{0,n-1}$.
2. Show that all the algebra isomorphisms presented in this chapter are special cases of the following:
$$\mathcal{C}\ell(V_1 \oplus V_2, Q_1 \perp Q_2) \simeq \mathcal{C}\ell(V_1, \lambda Q_1) \otimes \mathcal{C}\ell(V_2, Q_2),$$
$$(\mathbf{x}, \mathbf{y}) \to \mathbf{x} \otimes \omega + 1 \otimes \mathbf{y},$$
where Q_2 is non-degenerate and V_2 is even-dimensional, $\dim V_2 = 2k$, $\omega \in \bigwedge^{2k} V_2$, $\omega^2 = \lambda \in \mathbb{R} \setminus \{0\}$.

Bibliography

M.F. Atiyah, R. Bott, A. Shapiro: Clifford modules. *Topology* **3**, suppl. 1 (1964), 3-38. Reprinted in R. Bott: *Lectures on* $K(X)$. Benjamin, New York, 1969, pp. 143-178. Reprinted in *Michael Atiyah: Collected Works*, Vol. 2. Clarendon Press, Oxford, 1988, pp. 301-336.

E. Cartan (exposé d'après l'article allemand de E. Study): Nombres complexes; pp. 329-468 in J. Molk (red.): *Encyclopédie des sciences mathématiques*, Tome **I**, vol. 1, Fasc. **4**, art. **I5** (1908). Reprinted in E. Cartan: *Œuvres complètes*, Partie II. Gauthier-Villars, Paris, 1953, pp. 107-246.

W.K. Clifford: On the classification of geometric algebras, pp. 397-401 in R. Tucker (ed.): *Mathematical Papers by William Kingdon Clifford*, Macmillan, London, 1882. Reprinted by Chelsea, New York, 1968. Title of talk announced already in *Proc. London Math. Soc.* **7** (1876), p. 135.

F.R. Harvey: *Spinors and Calibrations*. Academic Press, San Diego, 1990.

T.Y. Lam: *The Algebraic Theory of Quadratic Forms*. Benjamin, Reading, MA, 1973, 1980.

I.R. Porteous: *Topological Geometry*. Van Nostrand Reinhold, London, 1969. Cambridge University Press, Cambridge, 1981.

I.R. Porteous: *Clifford Algebras and the Classical Groups*. Cambridge University Press, Cambridge, 1995.

17

Spin Groups and Spinor Spaces

We have already met in some lower-dimensional special cases the spinor spaces, minimal left ideals of Clifford algebras, and the spin groups, which operate on spinor spaces. In this chapter we shall study the general case of $\mathbb{R}^{p,q}$.

Spin Groups and the Two Exponentials

Review first the special case of the 3-dimensional Euclidean space $\mathbb{R}^3$.

17.1 Spin group Spin(3) and $SU(2)$

The traceless Hermitian matrices $x\sigma_1 + y\sigma_2 + z\sigma_3$, with $x, y, z \in \mathbb{R}$, represent vectors $\mathbf{r} = x\mathbf{e}_1 + y\mathbf{e}_2 + z\mathbf{e}_3 \in \mathbb{R}^3$. The group of unitary and unimodular matrices

$$SU(2) = \{U \in \mathrm{Mat}(2, \mathbb{C}) \mid U^\dagger U = I, \ \det U = I\}$$

represents the *spin group* $\mathbf{Spin}(3) = \{u \in C\ell_3 \mid u\tilde{u} = 1, \ u\bar{u} = 1\}$ or

$$\mathbf{Spin}(3) = \{u \in C\ell_3^+ \mid u\tilde{u} = 1\}.$$

Both these groups are isomorphic with the group of unit quaternions $S^3 = \{q \in \mathbb{H} \mid q\bar{q} = 1\}$. For an element $u \in \mathbf{Spin}(3)$ the mapping $\mathbf{r} \to u\mathbf{r}\tilde{u}$ is a rotation of $\mathbb{R}^3$. Every element of $SO(3)$ can be represented in this way by an element in $\mathbf{Spin}(3)$. In fact, there are two elements u and $-u$ in $\mathbf{Spin}(3)$ representing the same rotation of $\mathbb{R}^3$. This can be written as $\mathbf{Spin}(3)/\{\pm 1\} \simeq SO(3)$ and one can say that $\mathbf{Spin}(3)$ is a double covering of $SO(3)$.

17.2 The Lipschitz groups and the spin groups

The Lipschitz group $\mathbf{\Gamma}_{p,q}$, also called the Clifford group although invented by Lipschitz 1880/86, could be defined as the subgroup in $\mathcal{C}\ell_{p,q}$ generated by invertible vectors $\mathbf{x} \in \mathbb{R}^{p,q}$, or equivalently in either of the following ways:

$$\mathbf{\Gamma}_{p,q} = \{s \in \mathcal{C}\ell_{p,q} \mid \forall \mathbf{x} \in \mathbb{R}^{p,q},\ s\mathbf{x}\hat{s}^{-1} \in \mathbb{R}^{p,q}\}$$
$$\mathbf{\Gamma}_{p,q} = \{s \in \mathcal{C}\ell^+_{p,q} \cup \mathcal{C}\ell^-_{p,q} \mid \forall \mathbf{x} \in \mathbb{R}^{p,q},\ s\mathbf{x}s^{-1} \in \mathbb{R}^{p,q}\}.$$

Note the presence of the grade involution $s \to \hat{s}$, and/or the restriction to the even/odd parts $\mathcal{C}\ell^\pm_{p,q}$. For $s \in \mathbf{\Gamma}_{p,q}$, $s\tilde{s} \in \mathbb{R}$. The Lipschitz group has a normalized subgroup

$$\mathbf{Pin}(p,q) = \{s \in \mathbf{\Gamma}_{p,q} \mid s\tilde{s} = \pm 1\}.$$

The group $\mathbf{Pin}(p,q)$ has an even subgroup

$$\mathbf{Spin}(p,q) = \mathbf{Pin}(p,q) \cap \mathcal{C}\ell^+_{p,q}.$$

The spin group $\mathbf{Spin}(p,q)$ has a subgroup

$$\mathbf{Spin}_+(p,q) = \{s \in \mathbf{Spin}(p,q) \mid s\tilde{s} = 1\}.$$

Write $\mathbf{Spin}(n) = \mathbf{Spin}(n,0)$, and note that $\mathbf{Spin}_+(n) = \mathbf{Spin}(n)$. Because of the algebra isomorphisms $\mathcal{C}\ell^+_{p,q} \simeq \mathcal{C}\ell^+_{q,p}$ we have the group isomorphisms $\mathbf{Spin}(p,q) \simeq \mathbf{Spin}(q,p)$. However, in general $\mathbf{Pin}(p,q) \not\simeq \mathbf{Pin}(q,p)$. In particular, $\mathbf{Pin}(1) \simeq \mathbb{Z}_2 \times \mathbb{Z}_2$ and $\mathbf{Pin}(0,1) \simeq \mathbb{Z}_4$.

The groups $\mathbf{Pin}(p,q)$, $\mathbf{Spin}(p,q)$, $\mathbf{Spin}_+(p,q)$ are two-fold covering groups of $O(p,q)$, $SO(p,q)$, $SO_+(p,q)$. Although $SO_+(p,q)$ is connected, its two-fold cover $\mathbf{Spin}_+(p,q)$ need not be connected. However, the groups $\mathbf{Spin}_+(p,q)$, $p+q \geq 2$, are connected with the exception of

$$\mathbf{Spin}_+(1,1) = \{x + y\mathbf{e}_{12} \mid x, y \in \mathbb{R};\ x^2 - y^2 = 1\},$$

which has two components, two branches of a hyperbola. The group

$$\mathbf{Spin}(1,1) = \{x + y\mathbf{e}_{12} \mid x, y \in \mathbb{R};\ x^2 - y^2 = \pm 1\}$$

has four components.

The groups $\mathbf{Spin}(n)$, $n \geq 3$, and $\mathbf{Spin}_+(n-1,1) \simeq \mathbf{Spin}_+(1,n-1)$, $n \geq 4$, are simply connected and therefore universal covering groups of $SO(n)$ and $SO_+(n-1,1) \simeq SO_+(1,n-1)$. However, the maximal compact subgroup of $SO_+(3,3)$ is $SO(3) \times SO(3)$ which has a four-fold universal cover $\mathbf{Spin}(3) \times \mathbf{Spin}(3)$. Consequently, $\mathbf{Spin}_+(3,3)$ is not simply connected, but rather doubly connected, and therefore not a universal cover of $SO_+(3,3)$.

17.3 The two exponentials of bivectors

The Lie algebra of $\mathbf{Spin}_+(p,q)$ is the space of bivectors $\bigwedge^2 \mathbb{R}^{p,q}$. For two bivectors $\mathbf{A},\mathbf{B} \in \bigwedge^2 \mathbb{R}^{p,q}$ the commutator is again a bivector,

$$\mathbf{AB} - \mathbf{BA} \in \bigwedge^2 \mathbb{R}^{p,q}.$$

This can be seen by considering the reverse of

$$\mathbf{AB} = \langle \mathbf{AB} \rangle_0 + \langle \mathbf{AB} \rangle_2 + \langle \mathbf{AB} \rangle_4 \in \mathbb{R} \oplus \bigwedge^2 \mathbb{R}^{p,q} \oplus \bigwedge^4 \mathbb{R}^{p,q}$$

for which

$$(\mathbf{AB})^\sim = \langle \mathbf{AB} \rangle_0 - \langle \mathbf{AB} \rangle_2 + \langle \mathbf{AB} \rangle_4$$

and on the other hand

$$(\mathbf{AB})^\sim = \tilde{\mathbf{B}}\tilde{\mathbf{A}} = (-\mathbf{B})(-\mathbf{A}) = \mathbf{BA}.$$

The exponentials of bivectors generate the group $\mathbf{Spin}_+(p,q)$.

In this section we consider two different exponentials of bivectors, the ordinary or Clifford exponential

$$e^{\mathbf{B}} = 1 + \mathbf{B} + \frac{1}{2}\mathbf{B}^2 + \frac{1}{6}\mathbf{B}^3 + \ldots,$$

where $\mathbf{B}^2 = \mathbf{BB}$, and the exterior exponential

$$e^{\wedge \mathbf{B}} = 1 + \mathbf{B} + \frac{1}{2}\mathbf{B}^{\wedge 2} + \frac{1}{6}\mathbf{B}^{\wedge 3} + \ldots,$$

where $\mathbf{B}^{\wedge 2} = \mathbf{B} \wedge \mathbf{B}$. The series of the exterior exponential is finite. The ordinary exponential is always in the spin group, that is,

$$e^{\mathbf{B}} \in \mathbf{Spin}_+(p,q) \quad \text{for} \quad \mathbf{B} \in \bigwedge^2 \mathbb{R}^{p,q}.$$

The exterior exponential is in the Lipschitz group, if it is invertible in the Clifford algebra,

$$e^{\wedge \mathbf{B}} \in \mathbf{\Gamma}_{p,q} \quad \text{for} \quad \mathbf{B} \in \bigwedge^2 \mathbb{R}^{p,q} \quad \text{such that} \quad e^{\wedge \mathbf{B}} e^{\wedge(-\mathbf{B})} \neq 0.$$

Note that $e^{\wedge \mathbf{B}} \wedge e^{\wedge(-\mathbf{B})} = 1$, and so the exterior inverse of $e^{\wedge \mathbf{B}}$ is $e^{\wedge(-\mathbf{B})}$. The reverse of $s = e^{\wedge \mathbf{B}}$ is $\tilde{s} = e^{\wedge(-\mathbf{B})}$, and so the exterior inverse $s^{\wedge(-1)}$ of s equals $\tilde{s}$. The ordinary inverse of s, in the Clifford algebra $\mathcal{C}\ell_3$, is given by

$$s^{-1} = \frac{\tilde{s}}{s\tilde{s}}$$

where $s\tilde{s} \in \mathbb{R}$.

The Euclidean spaces $\mathbb{R}^n$. The bivector $\mathbf{B}$ can be written as a sum

$$\mathbf{B} = \mathbf{B}_1 + \mathbf{B}_2 + \ldots + \mathbf{B}_\ell$$

of at most $\ell = \lfloor n/2 \rfloor$ simple bivectors $\mathbf{B}_i$, $\mathbf{B}_i \wedge \mathbf{B}_i = 0$, which are mutually completely orthogonal so that their planes have only one point in common, $\mathbf{B}_i \wedge \mathbf{B}_j = \mathbf{B}_i\mathbf{B}_j$, $i \neq j$. This decomposition is unique unless $\mathbf{B}_i^2 = \mathbf{B}_j^2$. Notwithstanding, the product

$$(1+\mathbf{B}_1) \wedge (1+\mathbf{B}_2) \wedge \ldots \wedge (1+\mathbf{B}_\ell) = (1+\mathbf{B}_1)(1+\mathbf{B}_2)\ldots(1+\mathbf{B}_\ell)$$

depends only on $\mathbf{B}$ and equals the exterior exponential $e^{\wedge \mathbf{B}}$. The square norm of $e^{\wedge \mathbf{B}}$ is seen to be $|e^{\wedge \mathbf{B}}|^2 = (1-\mathbf{B}_1^2)(1-\mathbf{B}_2^2)\ldots(1-\mathbf{B}_\ell^2)$.

The Cayley transform. An antisymmetric $n \times n$-matrix A is sent by the Cayley transform to the rotation matrix

$$U = (I+A)(I-A)^{-1} \in SO(n).$$

There corresponds to A a bivector $\mathbf{B} \in \bigwedge^2 \mathbb{R}^n$ such that $A(\mathbf{x}) = \mathbf{B} \llcorner \mathbf{x}$ for all $\mathbf{x} \in \mathbb{R}^{p,q}$. If $\mathbf{y} = U\mathbf{x}$, then $\mathbf{y} - A\mathbf{x} = \mathbf{x} + A\mathbf{x}$, or equivalently

$$\mathbf{y} + \mathbf{y} \lrcorner \mathbf{B} = \mathbf{x} + \mathbf{B} \llcorner \mathbf{x}. \tag{1}$$

Next, compute $s \wedge (\mathbf{x} + \mathbf{B} \llcorner \mathbf{x}) = s \wedge \mathbf{x} + s \wedge (\mathbf{B} \llcorner \mathbf{x})$ for $s = e^{\wedge \mathbf{B}}$. Sum up

$$\frac{1}{k!}(\underbrace{\mathbf{B} \wedge \mathbf{B} \wedge \ldots \wedge \mathbf{B}}_{k}) \wedge (\mathbf{B} \llcorner \mathbf{x}) = \frac{1}{(k+1)!}(\underbrace{\mathbf{B} \wedge \mathbf{B} \wedge \ldots \wedge \mathbf{B}}_{k+1}) \llcorner \mathbf{x}$$

for $k = 0, 1, 2, \ldots, \ell$ to obtain $s \wedge (\mathbf{B} \llcorner \mathbf{x}) = s \llcorner \mathbf{x}$. Since $s \wedge \mathbf{x} + s \llcorner \mathbf{x} = s\mathbf{x}$, it follows that $s \wedge (\mathbf{x} + \mathbf{B} \llcorner \mathbf{x}) = s\mathbf{x}$. Similarly, $s \wedge (\mathbf{y} + \mathbf{y} \lrcorner \mathbf{B}) = s \wedge \mathbf{y} - s \llcorner \mathbf{y} = \mathbf{y} \wedge s + \mathbf{y} \lrcorner s = \mathbf{y}s$. Therefore, the equation (1) is equivalent to $s\mathbf{x} = \mathbf{y}s$ or

$$U(\mathbf{x}) = s\mathbf{x}s^{-1}.$$

This representation of rotations was first discovered by R. Lipschitz 1880/86.

Thus we have the following result: An antisymmetric $n \times n$-matrix A and the rotation matrix $U = (I+A)(I-A)^{-1} \in SO(n)$ correspond, respectively, to the bivector $\mathbf{B} \in \bigwedge^2 \mathbb{R}^n$, $A(\mathbf{x}) = \mathbf{B} \llcorner \mathbf{x}$, and its exterior exponential $s = e^{\wedge \mathbf{B}} \in \mathbf{\Gamma}_n^+$, which is the unique element of $\mathbf{\Gamma}_n^+$, with scalar part 1, inducing the rotation U, $U(\mathbf{x}) = s\mathbf{x}s^{-1}$. For every rotation $U \in SO(n)$, which does not rotate any plane by a half-turn (all eigenvalues are different from -1), there is a unique element $s \in \mathbf{\Gamma}_n^+$, $\langle s \rangle_0 = 1$, such that $U(\mathbf{x}) = s\mathbf{x}s^{-1}$.

For an element $s \in \mathbf{\Gamma}_n^+$, $s\tilde{s} \in \mathbb{R}$, $s\tilde{s} > 0$. Therefore $|s| = \sqrt{s\tilde{s}}$, and

$$\frac{s}{|s|} \in \mathbf{Spin}(n) \quad \text{for} \quad s = e^{\wedge \mathbf{B}}.$$

Every element $u \in \mathbf{Spin}(n)$, $\langle u \rangle_0 \neq 0$, can be written in the form

$$u = \pm \frac{e^{\wedge \mathbf{B}}}{|e^{\wedge \mathbf{B}}|},$$

which corresponds to the rotation $U = (I+A)(I-A)^{-1} \in SO(n)$. This should be contrasted with the fact that every element in $\mathbf{Spin}(n)$ can be written in the form $e^{\mathbf{B}/2}$, which corresponds to the rotation e^A in $SO(n)$.

Lorentz signatures. In the Lorentz signatures the decomposition

$$\mathbf{B} = \mathbf{B}_1 + \mathbf{B}_2 + \ldots + \mathbf{B}_\ell$$

still exists and can be used to test invertibility of $\mathbf{e}^{\wedge \mathbf{B}}$. The exterior exponential

$$e^{\wedge \mathbf{B}} = (1+\mathbf{B}_1)(1+\mathbf{B}_2)\ldots(1+\mathbf{B}_\ell)$$

is invertible in the Clifford algebra if $\mathbf{B}_i^2 \neq 1$ for all i. In other words, $e^{\wedge \mathbf{B}} \in \mathbf{\Gamma}^+_{n-1,1}$ if all $\mathbf{B}_i^2 \neq 1$.

Indefinite metrics. Every isometry U of $\mathbb{R}^{p,q}$, connected to the identity of $SO_+(p,q)$, is an exponential of an antisymmetric transformation A of $\mathbb{R}^{p,q}$, $U = e^A$, if and only if

$$\mathbb{R}^{p,q} \quad \text{is} \quad \mathbb{R}^{n,0},\ \mathbb{R}^{0,n},\ \mathbb{R}^{n-1,1} \quad \text{or} \quad \mathbb{R}^{1,n-1},$$

see M. Riesz 1958/93 pp. 150-152. In these Euclidean and Lorentz signatures there is always a bivector $\mathbf{B}$, $\mathbf{B} \llcorner \mathbf{x} = A(\mathbf{x})$ such that $U(\mathbf{x}) = e^{\mathbf{B}}\mathbf{x}e^{-\mathbf{B}}$, see M. Riesz 1958/93 p. 160.

Given a bivector $\mathbf{B}$ one can, in general, find other bivectors $\mathbf{F}$ such that $e^{\mathbf{B}} = -e^{\mathbf{F}}$ and hence $e^{\mathbf{B}}\mathbf{x}e^{-\mathbf{B}} = e^{\mathbf{F}}\mathbf{x}e^{-\mathbf{F}}$. The only exceptions concern the following cases:

$$\begin{array}{lllll} \mathbb{R}^{1,1} & \text{for all} & \mathbf{B} & & \\ \mathbb{R}^{2,1} & \text{and} \ \mathbb{R}^{1,2} & \text{for all} & \mathbf{B} \neq 0 & \text{such that} \ \mathbf{B}^2 \geq 0 \\ \mathbb{R}^{3,1} & \text{and} \ \mathbb{R}^{1,3} & \text{for all} & \mathbf{B} \neq 0 & \text{such that} \ \mathbf{B}^2 = 0, \end{array}$$

see M. Riesz 1958/93 p. 172.

To summarize with special cases: All the elements of the compact spin groups $\mathbf{Spin}(n)$ are exponentials of bivectors [when $n \geq 2$]. Among the other spin groups the same holds only for $\mathbf{Spin}_+(n-1,1) \simeq \mathbf{Spin}_+(1,n-1)$, $n \geq 5$. In particular, the two-fold cover $\mathbf{Spin}_+(1,3) \simeq SL(2,\mathbb{C})$ of the Lorentz group $SO_+(1,3)$ contains elements which are not exponentials of bivectors: take $(\gamma_0 + \gamma_1)\gamma_2 \in \bigwedge^2 \mathbb{R}^{1,3}$, $[(\gamma_0+\gamma_1)\gamma_2]^2 = 0$, then $-e^{(\gamma_0+\gamma_1)\gamma_2} = -1-(\gamma_0+\gamma_1)\gamma_2 \neq e^{\mathbf{B}}$ for any $\mathbf{B} \in \bigwedge^2 \mathbb{R}^{1,3}$. [1] However, all the elements of $\mathbf{Spin}_+(1,3)$ are of the

1 In contrast, $-e^{(\mathbf{e}_1+\mathbf{e}_5)\mathbf{e}_2} = -1-(\mathbf{e}_1+\mathbf{e}_5)\mathbf{e}_2 = e^{(\mathbf{e}_1+\mathbf{e}_5)\mathbf{e}_2+\pi\mathbf{e}_{34}}$ in $\mathbf{Spin}_+(4,1) \simeq Sp(2,2)$.

form $\pm e^{\mathbf{B}}$, $\mathbf{B} \in \bigwedge^2 \mathbb{R}^{1,3}$. Therefore, the exponentials of bivectors do not form a group.

Every element L of the Lorentz group $SO_+(1,3)$ is an exponential of an antisymmetric matrix, $L = e^A$, $gA^\top g^{-1} = -A$; a similar property is not shared by $SO_+(2,2)$. There are elements in $\mathbf{Spin}_+(2,2)$ which cannot be written in the form $\pm e^{\mathbf{B}}$, $\mathbf{B} \in \bigwedge^2 \mathbb{R}^{2,2}$; for instance $\pm \mathbf{e}_{1234} e^{\beta \mathbf{B}}$, $\mathbf{B} = \mathbf{e}_{12} + 2\mathbf{e}_{14} + \mathbf{e}_{34}$, $\beta > 0$, see M. Riesz 1958/93 p. 168-171. [2]

Lower-dimensional spin groups. The dimension of the Lie group $\mathbf{Spin}(n)$ is $\frac{1}{2}n(n-1)$. The groups $\mathbf{Spin}(n)$, $n \leq 6$, and $\mathbf{Spin}_+(p,q)$, $p+q \leq 6$, are identified in Table 1.

Table 1. Spin Groups $\mathbf{Spin}_+(p,q)$, $p+q \leq 6$.

$q \backslash p$	0	1	2	3	4	5	6
0	$\{\pm 1\}$	$O(1)$	$U(1)$	$Sp(2)$	${}^2Sp(2)$	$Sp(4)$	$SU(4)$
1	$O(1)$	$GL(1,\mathbb{R})$	$Sp(2,\mathbb{R})$	$Sp(2,\mathbb{C})$	$Sp(2,2)$	$SU^*(4)$	
2	$U(1)$	$Sp(2,\mathbb{R})$	${}^2Sp(2,\mathbb{R})$	$Sp(4,\mathbb{R})$	$SU(2,2)$		
3	$Sp(2)$	$Sp(2,\mathbb{C})$	$Sp(4,\mathbb{R})$	$SL(4,\mathbb{R})$			
4	${}^2Sp(2)$	$Sp(2,2)$	$SU(2,2)$				
5	$Sp(4)$	$SU^*(4)$					
6	$SU(4)$						

Note that $\mathbf{Spin}_+(p,q) = \{s \in \mathcal{C}\ell^+_{p,q} \mid s\tilde{s} = 1\}$ for $p+q \leq 5$. In dimension 6 the group $\{s \in \mathcal{C}\ell^+_6 \mid s\tilde{s} = 1\} \simeq U(4)$ has a proper subgroup $\mathbf{Spin}(6) \simeq SU(4)$. The groups $\mathbf{Spin}(7)$ and $\mathbf{Spin}(8)$ are not directly related to classical matrix groups; their study will be postponed till the discussion on triality.

In the case of the complex quadratic spaces $\mathbb{C}^n$ we define the complex pin group slightly differently: [3]

$$\mathbf{Pin}(n,\mathbb{C}) = \{s \in \mathcal{C}\ell(\mathbb{C}^n) \mid s\tilde{s} = 1;\ \forall \mathbf{x} \in \mathbb{C}^n,\ s\mathbf{x}\hat{s}^{-1} \in \mathbb{C}^n\}.$$

2 Riesz also showed, by the same construction on pp. 170-171, that there are bivectors which cannot be written as sums of simple and completely orthogonal bivectors; for instance $\mathbf{B} = \mathbf{e}_{12} + 2\mathbf{e}_{14} + \mathbf{e}_{34} \in \bigwedge^2 \mathbb{R}^{2,2}$.

3 The structures of square classes are different for $\mathbb{R}$ and $\mathbb{C}$. In $\mathbb{R}^\times = \mathbb{R} \setminus \{0\}$, $\mathbb{R}^\times = \pm\mathbb{R}^\square$, $\mathbb{R}^\square = \{\lambda^2 \mid \lambda \in \mathbb{R}^\times\}$; so to pick up one representative out of each square class we set $s\tilde{s} = \pm 1$. In contrast, in $\mathbb{C}^\times = \mathbb{C} \setminus \{0\}$, $\mathbb{C}^\times = \mathbb{C}^\square$; so to pick up one representative out of each square class we set $s\tilde{s} = 1$.

The complex spin groups $\mathbf{Spin}(n,\mathbb{C})$, $n \leq 6$, are seen to be as follows:

0	1	2	3	4	5	6
$\{\pm 1\}$	$O(1,\mathbb{C})$	$GL(1,\mathbb{C})$	$Sp(2,\mathbb{C})$	${}^2Sp(2,\mathbb{C})$	$Sp(4,\mathbb{C})$	$SL(4,\mathbb{C})$

We also define the Lipschitz group $\$\varGamma_{q+1,p}$ for paravectors in $\mathbb{R}\oplus\mathbb{R}^{p,q}$ as the group containing the products of invertible paravectors, or equivalently,

$$\$\varGamma_{q+1,p} = \{s \in \mathcal{C}\ell_{p,q} \mid \forall x \in \mathbb{R}\oplus\mathbb{R}^{p,q},\ sx\hat{s}^{-1} \in \mathbb{R}\oplus\mathbb{R}^{p,q}\}.$$

For any non-null paravector $a \in \mathbb{R}\oplus\mathbb{R}^{p,q}$, the mapping $x \to ax\hat{a}^{-1}$ is a special orthogonal transformation of $\mathbb{R}\oplus\mathbb{R}^{p,q}$ with metric $x \to x\bar{x}$. Therefore $\$\varGamma_{q+1,p} \simeq \mathbf{\Gamma}^{+}_{q+1,p}$. Note that $\mathbf{\Gamma}_{p,q} \subset \$\varGamma_{q+1,p}$ and $\$\varGamma^{\pm}_{q+1,p} = \mathbf{\Gamma}^{\pm}_{p,q}$. The normalized subgroup $\$pin(q+1,p) = \{s \in \$\varGamma_{q+1,p} \mid s\bar{s} = \pm 1\}$ is isomorphic to $\mathbf{Spin}(q+1,p)$.

Idempotents, Left Ideals and Spinors

Review first the Clifford algebra $\mathcal{C}\ell_3$ of the Euclidean space $\mathbb{R}^3$.

17.4 Pauli spinors

In the non-relativistic theory of the electron, spinors are regarded as columns

$$\begin{pmatrix}\psi_1\\ \psi_2\end{pmatrix} \quad \text{where} \quad \psi_1, \psi_2 \in \mathbb{C}.$$

We shall instead introduce spinors as square matrices

$$\psi \simeq \begin{pmatrix}\psi_1 & 0\\ \psi_2 & 0\end{pmatrix}.$$

If we multiply ψ on the left by an arbitrary element u in $\mathcal{C}\ell_3$ we obtain another element $u\psi = \varphi$ in $\mathcal{C}\ell_3$ whose matrix is also of spinor type:

$$\begin{pmatrix}u_{11} & u_{12}\\ u_{21} & u_{22}\end{pmatrix}\begin{pmatrix}\psi_1 & 0\\ \psi_2 & 0\end{pmatrix} = \begin{pmatrix}\varphi_1 & 0\\ \varphi_2 & 0\end{pmatrix}.$$

The spinors make up a *left ideal* S of $\mathcal{C}\ell_3$, that is,

$$\text{for all} \quad u \in \mathcal{C}\ell_3 \quad \text{and} \quad \psi \in S \quad \text{we also have} \quad u\psi \in S.$$

The left ideal S contains no left ideals of $\mathcal{C}\ell_3$ other than the zero ideal $\{0\}$ and S itself. Such a left ideal is called *minimal* in $\mathcal{C}\ell_3$.

The element $f = \frac{1}{2}(1+\mathbf{e}_3)$ is an *idempotent*, that is, $f^2 = f$, which is *primitive* in $\mathcal{C}\ell_3$, that is, it is not a sum of two annihilating idempotents, $f \neq f_1 + f_2$, $f_1 f_2 = f_2 f_1 = 0$. The left ideal $S = \mathcal{C}\ell_3 f$ can be provided with a right linear structure over the division ring $\mathbb{D} = f C\ell_3 f$ as follows: $S \times \mathbb{D} \to S$, $(\psi, \lambda) \to \psi\lambda$. With this right linear structure over $\mathbb{D} \simeq \mathbb{C}$ the left ideal S becomes a *spinor space*.

17.5 Primitive idempotents and minimal left ideals

An orthonormal basis of $\mathbb{R}^{p,q}$ induces a basis of $C\ell_{p,q}$, called the standard basis. Take a non-scalar element $\mathbf{e}_T$, $\mathbf{e}_T^2 = 1$, from the standard basis of $C\ell_{p,q}$. Set $e = \frac{1}{2}(1 + \mathbf{e}_T)$ and $f = \frac{1}{2}(1 - \mathbf{e}_T)$, then $e + f = 1$ and $ef = fe = 0$. So $C\ell_{p,q}$ decomposes into a sum of two left ideals $C\ell_{p,q} = C\ell_{p,q}e \oplus C\ell_{p,q}f$, where $\dim C\ell_{p,q}e = \dim C\ell_{p,q}f = \frac{1}{2}C\ell_{p,q} = 2^{n-1}$. Furthermore, if $\{\mathbf{e}_{T_1}, \mathbf{e}_{T_2}, \ldots, \mathbf{e}_{T_k}\}$ is a set of non-scalar basis elements such that

$$\mathbf{e}_{T_i}^2 = 1 \quad \text{and} \quad \mathbf{e}_{T_i}\mathbf{e}_{T_j} = \mathbf{e}_{T_j}\mathbf{e}_{T_i},$$

then letting the signs vary independently in the product $\frac{1}{2}(1 \pm \mathbf{e}_{T_1})\frac{1}{2}(1 \pm \mathbf{e}_{T_2}) \ldots \frac{1}{2}(1 \pm \mathbf{e}_{T_k})$, one obtains 2^k idempotents which are mutually annihilating and sum up to 1. The Clifford algebra $C\ell_{p,q}$ is thus decomposed into a direct sum of 2^k left ideals, and by construction, each left ideal has dimension 2^{n-k}. In this way one obtains a minimal left ideal by forming a maximal product of non-annihilating and commuting idempotents.

The Radon-Hurwitz number r_i for $i \in \mathbb{Z}$ is given by

i	0	1	2	3	4	5	6	7
r_i	0	1	2	2	3	3	3	3

and the recursion formula

$$r_{i+8} = r_i + 4.$$

For the negative values of i one may observe that $r_{-1} = -1$ and $r_{-i} = 1 - i + r_{i-2}$ for $i > 1$.

Theorem. In the standard basis of $C\ell_{p,q}$ there are always $k = q - r_{q-p}$ non-scalar elements $\mathbf{e}_{T_i}$, $\mathbf{e}_{T_i}^2 = 1$, which commute, $\mathbf{e}_{T_i}\mathbf{e}_{T_j} = \mathbf{e}_{T_j}\mathbf{e}_{T_i}$, and generate a group of order 2^k. The product of the corresponding mutually non-annihilating idempotents,

$$f = \frac{1}{2}(1 + \mathbf{e}_{T_1})\frac{1}{2}(1 + \mathbf{e}_{T_2}) \ldots \frac{1}{2}(1 + \mathbf{e}_{T_k}),$$

is primitive in $C\ell_{p,q}$. Thus, the left ideal $S = C\ell_{p,q}f$ is minimal in $C\ell_{p,q}$. ∎

Examples. 1. In the case of $\mathbb{R}^{0,7}$ we have $k = 7 - r_7 = 4$. Therefore the idempotent $f = \frac{1}{2}(1 + \mathbf{e}_{124})\frac{1}{2}(1 + \mathbf{e}_{235})\frac{1}{2}(1 + \mathbf{e}_{346})\frac{1}{2}(1 + \mathbf{e}_{457})$ is primitive in $C\ell_{0,7} \simeq {}^2\mathrm{Mat}(8, \mathbb{R})$.

2. In the case of $\mathbb{R}^{2,1}$ we have $k = 1 - r_{-1} = 1 - (r_7 - 4) = 2$. Therefore the idempotent $f = \frac{1}{2}(1 + \mathbf{e}_1)\frac{1}{2}(1 + \mathbf{e}_{23}) = \frac{1}{4}(1 + \mathbf{e}_1 + \mathbf{e}_{23} + \mathbf{e}_{123})$ is primitive in $\mathcal{C}\ell_{2,1} \simeq {}^2\mathrm{Mat}(2, \mathbb{R})$. ∎

If e and f are commuting idempotents of a ring R, then ef and $e + f - ef$ are also idempotents of R. The idempotents ef and $e+f-ef$ are a greatest lower bound and a least upper bound relative to the partial ordering given by

$$e \leq f \quad \text{if and only if} \quad ef = fe = e.$$

A set of commuting idempotents induces a lattice of idempotents.

Example. In the Clifford algebra $\mathcal{C}\ell_{3,1}$, $k = 1 - r_{-2} = 1 - (r_6 - 4) = 2$. Since $2^k = 4$, $\mathcal{C}\ell_{3,1} \simeq \mathrm{Mat}(4, \mathbb{R})$ and there are $2^{2^k} = 16$ commuting idempotents in the lattice generated by the following four mutually annihilating primitive idempotents:

$$f_1 = \tfrac{1}{2}(1 + \mathbf{e}_1)\tfrac{1}{2}(1 + \mathbf{e}_{24}), \quad f_2 = \tfrac{1}{2}(1 - \mathbf{e}_1)\tfrac{1}{2}(1 + \mathbf{e}_{24}),$$
$$f_3 = \tfrac{1}{2}(1 + \mathbf{e}_1)\tfrac{1}{2}(1 - \mathbf{e}_{24}), \quad f_4 = \tfrac{1}{2}(1 - \mathbf{e}_1)\tfrac{1}{2}(1 - \mathbf{e}_{24}).$$

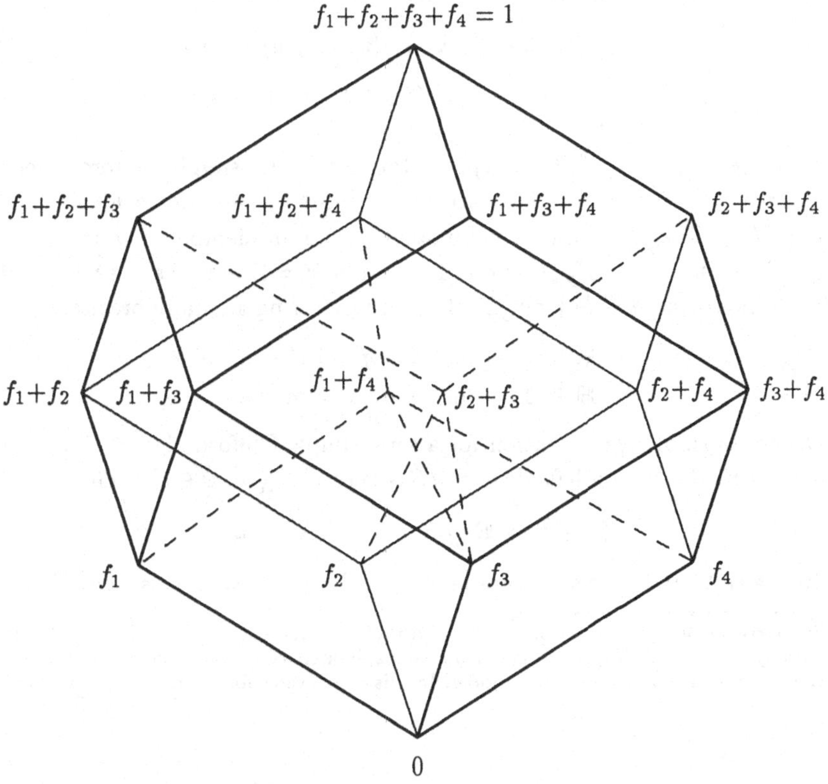

The lattice induced by the primitive idempotents f_1, f_2, f_3, f_4 looks like a rhombidodecahedron, see diagram. ∎

17.6 Spinor spaces

For a primitive idempotent $f \in C\ell_{p,q}$ the division ring $\mathbb{D} = fC\ell_{p,q}f$ is isomorphic to

$$\begin{array}{lll} \mathbb{R} & \text{for} & p-q = 0,1,2 \bmod 8 \\ \mathbb{C} & \text{for} & p-q = 3 \bmod 4 \\ \mathbb{H} & \text{for} & p-q = 4,5,6 \bmod 8 \end{array}$$

and the map

$$S \times \mathbb{D} \to S, \ (\psi, \lambda) \to \psi\lambda$$

defines a *right* $\mathbb{D}$-linear structure on the minimal left ideal $S = C\ell_{p,q}f$. Provided with this right $\mathbb{D}$-linear structure the minimal left ideal S becomes a *spinor space.* [4]

The spinor space provides an *irreducible representation*

$$C\ell_{p,q} \to \mathrm{End}_{\mathbb{D}}(S), \ u \to \gamma(u), \quad \gamma(u)\psi = u\psi,$$

of $C\ell_{p,q}$. This representation is also faithful for all simple Clifford algebras $C\ell_{p,q}$, $p-q \neq 1 \bmod 4$.

Next, we construct a faithful representation for semi-simple Clifford algebras $C\ell_{p,q}$, $p-q = 1 \bmod 4$, which are direct sums of two simple ideals $\frac{1}{2}(1 \pm \mathbf{e}_{12\ldots n})C\ell_{p,q}$. Take a primitive idempotent f and an idempotent $e = f + \hat{f}$ in $C\ell_{p,q}$. The ring $\mathbb{E} = eC\ell_{p,q}e$ is the direct sum $\mathbb{E} = \mathbb{D} \oplus \hat{\mathbb{D}}$, $\hat{\mathbb{D}} = \{\hat{\lambda} \mid \lambda \in \mathbb{D}\}$, isomorphic to the *double ring* ${}^2\mathbb{D}$ of the division ring $\mathbb{D}$, more precisely,

$$\begin{array}{lll} \mathbb{R} \oplus \mathbb{R} & \text{for} & p-q = 1 \bmod 8 \\ \mathbb{H} \oplus \mathbb{H} & \text{for} & p-q = 5 \bmod 8. \end{array}$$

To find a faithful representation for a semi-simple Clifford algebra $C\ell_{p,q}$ with $p-q = 1 \bmod 4$ take a left ideal $S \oplus \hat{S}$ where $\hat{S} = \{\hat{\psi} \mid \psi \in S\}$. The map

$$(S \oplus \hat{S}) \times \mathbb{E} \to S \oplus \hat{S}, \ (\psi, \lambda) \to \psi\lambda$$

defines a right $\mathbb{E}$-linear structure on $S \oplus \hat{S}$. Provided with this right $\mathbb{E}$-linear

4 Similarly, beginning with a minimal left ideal of the even subalgebra $C\ell^+_{p,q}$ we obtain an *even spinor space*. The dimension of the even spinor space is lower than the dimension of the spinor space, when $p-q = 0 \bmod 4$. In this case, even spinors are called semi-spinors.

structure the left ideal $S \oplus \hat{S}$ of $C\ell_{p,q}$ becomes a *double spinor space.* [5] The double spinor space provides a faithful but reducible representation

$$C\ell_{p,q} \to \mathrm{End}_{\mathbb{E}}(S \oplus \hat{S}), \ u \to \gamma(u), \quad \gamma(u)\psi = u\psi,$$

for a semi-simple $C\ell_{p,q}$, $p - q = 1 \bmod 4$.

In order to be able to consider faithful representations of simple and semi-simple Clifford algebras at the same time, we adopt the following notation:

$$\begin{array}{ccccc} \check{\mathbb{D}} & \text{is} & \mathbb{D} & \text{or} & \mathbb{D} \oplus \hat{\mathbb{D}} \\ \check{S} & \text{is} & S & \text{or} & S \oplus \hat{S} \end{array}$$

according as $C\ell_{p,q}$ is simple or semi-simple, respectively. Thus, the ring $\check{\mathbb{D}}$ is isomorphic to $\mathbb{R}$, $\mathbb{C}$, $\mathbb{H}$, ${}^2\mathbb{R}$ or ${}^2\mathbb{H}$. In this way we have a faithful representation

$$C\ell_{p,q} \to \mathrm{End}_{\check{\mathbb{D}}} \check{S}, \ u \to \gamma(u), \quad \gamma(u)\psi = u\psi,$$

for all $C\ell_{p,q}$. However, this representation is reducible in the cases $p - q = 1 \bmod 4$.

Questions

1. Do the exponentials of bivectors form a group?
2. Do the exterior exponentials of bivectors form a group?
3. Are $\mathbf{Spin}_+(p, q)$, $p + q \geq 3$, $p, q \neq 2$, universal covers of $SO_+(p, q)$?
4. Are double spinor spaces needed to construct a faithful representation for $C\ell_{2,5}$?

Answers

1. No. 2. No. 3. No. 4. Yes.

Bibliography

I.M. Benn, R.W. Tucker: *An Introduction to Spinors and Geometry with Applications in Physics*. Adam Hilger, Bristol, 1987.

P. Budinich, A. Trautman: *The Spinorial Chessboard*. Springer, Berlin, 1988.

C. Chevalley: *The Algebraic Theory of Spinors*. Columbia University Press, New York, 1954.

J.S.R. Chisholm, A.K. Common (eds.): *Proceedings of the NATO and SERC Workshop on 'Clifford Algebras and their Applications in Mathematical Physics' (Canterbury, 1985)*. Reidel, Dordrecht, The Netherlands, 1986.

5 Similarly, by doubling of a minimal left ideal of the even subalgebra $C\ell^+_{p,q}$ we obtain a *double even spinor space*.

A. Crumeyrolle: *Orthogonal and Symplectic Clifford Algebras, Spinor Structures.* Kluwer, Dordrecht, The Netherlands, 1990.

R. Deheuvels: *Formes quadratiques et groupes classiques.* Presses Universitaires de France, Paris, 1981.

F.R. Harvey: *Spinors and Calibrations.* Academic Press, San Diego, 1990.

R. Lipschitz: Principes d'un calcul algébrique qui contient comme espèces particulières le calcul des quantités imaginaires et des quaternions. *C.R. Acad. Sci. Paris* **91** (1880), 619-621, 660-664. Reprinted in *Bull. Soc. Math.* (2) **11** (1887), 115-120.

R. Lipschitz: *Untersuchungen über die Summen von Quadraten.* Max Cohen und Sohn, Bonn, 1886, pp. 1-147. The first chapter of pp. 5-57 translated into French by J. Molk: Recherches sur la transformation, par des substitutions réelles, d'une somme de deux ou troix carrés en elle-même. *J. Math. Pures Appl.* (4) **2** (1886), 373-439. French résumé of all three chapters in *Bull. Sci. Math.* (2) **10** (1886), 163-183.

R. Lipschitz (signed): Correspondence. *Ann. of Math.* **69** (1959), 247-251.

P. Lounesto: Cayley transform, outer exponential and spinor norm, in Proc. Winter School of Geometry and Physics, Srní. *Suppl. Rend. Circ. Mat. Palermo*, Ser. II (1987), 191-198.

P. Lounesto, G.P. Wene: Idempotent structure of Clifford algebras. *Acta Applic. Math.* **9** (1987), 165-173.

I.R. Porteous: *Topological Geometry.* Van Nostrand Reinhold, London, 1969. Cambridge University Press, Cambridge, 1981.

I.R. Porteous: *Clifford Algebras and the Classical Groups.* Cambridge University Press, Cambridge, 1995.

M. Riesz: Sur certaines notions fondamentales en théorie quantique relativiste. *C.R. 10ᵉ Congrès Math. Scandinaves, Copenhagen, 1946.* Jul. Gjellerups Forlag, Copenhagen, 1947, pp. 123-148. Collected Papers, pp. 545-570.

M. Riesz: *Clifford Numbers and Spinors.* The Institute for Fluid Dynamics and Applied Mathematics, Lecture Series No. **38**, University of Maryland, 1958. Reprinted as facsimile (eds.: E.F. Bolinder, P. Lounesto) by Kluwer, The Netherlands, 1993.

18

Scalar Products of Spinors and the Chessboard

The Euclidean space $\mathbb{R}^3$ has a scalar product $\mathbf{x} \cdot \mathbf{y} = x_1y_1 + x_2y_2 + x_3y_3$ with the automorphism group $O(3)$. Pauli spinors of $\mathbb{R}^3$ are of the form

$$\psi = \begin{pmatrix} \psi_1 \\ \psi_2 \end{pmatrix} \quad \text{where} \quad \psi_1, \psi_2 \in \mathbb{C}$$

and belong to a complex linear space $\mathbb{C}^2$. There are two kinds of scalar products for Pauli spinors $\psi, \varphi \in \mathbb{C}^2$,

$$\psi^{*\mathsf{T}}\varphi = \psi_1^*\varphi_1 + \psi_2^*\varphi_2 \quad \text{and}$$
$$\psi^{\mathsf{T}} i\sigma_2\varphi = \psi_1\varphi_2 - \psi_2\varphi_1,$$

which have automorphism groups $U(2)$ and $Sp(2, \mathbb{C}) = SL(2, \mathbb{C})$, respectively. The Minkowski space $\mathbb{R}^{1,3}$ has a scalar product

$$\mathbf{x} \cdot \mathbf{y} = x_0y_0 - x_1y_1 - x_2y_2 - x_3y_3$$

with the automorphism group $O(1, 3)$. Dirac spinors of $\mathbb{R}^{1,3}$ belong to a complex linear space $\mathbb{C}^4$. There is a scalar product of Dirac spinors $\psi, \varphi \in \mathbb{C}^4$,

$$\psi^{*\mathsf{T}}\gamma_0\varphi = \psi_1^*\varphi_1 + \psi_2^*\varphi_2 - \psi_3^*\varphi_3 - \psi_4^*\varphi_4,$$

with the automorphism group $U(2, 2)$.

One might wonder about the following things:

(i) Why do spinors with complex entries arise in conjunction with the real quadratic spaces $\mathbb{R}^3$ and $\mathbb{R}^{1,3}$?

(ii) If we consider generalizations to arbitrary $\mathbb{R}^{p,q}$, are the scalar products of spinors still Hermitian or antisymmetric?

(iii) Are the scalar products of spinors definite or neutral for all $\mathbb{R}^{p,q}$?

(iv) Is there a general pattern in higher dimensions for the changes from $\mathbb{R}^3$ to $\bar{\mathbb{C}}^2$ or from $\mathbb{R}^{1,3}$ to $\bar{\mathbb{C}}^{2,2}$?

We will answer these questions in the following general form: What is the automorphism group of the scalar product of spinors in the case of the quadratic space $\mathbb{R}^{p,q}$? The scalar products of spinors can be collected into two equivalence classes when p and q are kept fixed in $\mathbb{R}^{p,q}$. There are altogether

$$32 = \frac{8 \times 8}{2}$$

different kinds of scalar products of spinors when we let p and q vary in $\mathbb{R}^{p,q}$.

The situation is much simplified if we consider instead of the real quadratic spaces $\mathbb{R}^{p,q}$ their complexifications $\mathbb{C} \otimes \mathbb{R}^{p,q}$. Then there remain only four different types of scalar products of spinors to be considered.

The reader will notice that the unitary group $U(2,2)$ can be adjoined to the Minkowski space-times $\mathbb{R}^{1,3}$ and $\mathbb{R}^{3,1}$ in two different ways by

- complexifying, or
- adding one extra dimension (of positive signature),

which respectively result in

- $\mathbb{C} \otimes \mathbb{R}^{1,3}$ and $\mathbb{C} \otimes \mathbb{R}^{3,1}$, or
- $\mathbb{R}^{2,3}$ and $\mathbb{R}^{4,1}$.

In both cases $U(2,2)$ is the automorphism group of the scalar product of spinors. The latter case gives a hint of a relation to the conformal group of the Minkowski space. [1]

18.1 Scalar products on spinor spaces

We start with spinors ψ, φ in spinor spaces $S = \mathcal{C}\ell_{p,q} f$ which are linear spaces over division rings $\mathbb{D} = f\mathcal{C}\ell_{p,q} f$. We will consider two cases:

(i) The minimal left ideals $S = \mathcal{C}\ell_{p,q} f$ providing irreducible representations for all $\mathcal{C}\ell_{p,q}$; these representations are also faithful for simple $\mathcal{C}\ell_{p,q}$.

(ii) The left ideals $S \oplus \hat{S} = \mathcal{C}\ell_{p,q} e$, $e = f + \hat{f}$, providing faithful representations for semi-simple $\mathcal{C}\ell_{p,q}$.

---

1 The Vahlen matrices of the Minkowski space are such that $\mathrm{Mat}(2, \mathcal{C}\ell_{1,3}) \simeq \mathcal{C}\ell_{2,4}$ and $\mathrm{Mat}(2, \mathcal{C}\ell_{3,1}) \simeq \mathcal{C}\ell_{4,2}$, where the even subalgebras are isomorphic: $\mathcal{C}\ell_{2,4}^+ \simeq \mathcal{C}\ell_{4,2}^+ \simeq \mathrm{Mat}(4, \mathbb{C})$ or $\mathcal{C}\ell_{2,3} \simeq \mathcal{C}\ell_{4,1} \simeq \mathrm{Mat}(4, \mathbb{C})$. The (connected components of the) conformal groups of $\mathbb{R}^{1,3}$ and $\mathbb{R}^{3,1}$ are isomorphic to

$$\frac{SO_+(2,4)}{\{I,-I\}} \simeq \frac{SO_+(4,2)}{\{I,-I\}} \simeq \frac{SU(2,2)}{\{\pm I, \pm iI\}}.$$

The automorphism group $U(2,2)$ of the scalar product of Dirac spinors contains as a subgroup the universal cover $SU(2,2)$ of the conformal group of the Minkowski space.

As before, let

$$\begin{array}{llll} \check{\mathbb{D}} & \text{be either} & \mathbb{D} & \text{or} \quad \mathbb{D}\oplus\hat{\mathbb{D}}, \\ \check{S} & \text{be either} & S & \text{or} \quad S\oplus\hat{S} \end{array}$$

according as $\mathcal{C}\ell_{p,q}$ is simple or semi-simple, respectively.

Let β be either of the anti-automorphisms $u \to \tilde{u}$ and $u \to \bar{u}$ of $\mathcal{C}\ell_{p,q}$. The real linear spaces

$$P_+ = \{\psi \in S \mid \beta(\psi) = +\psi\},$$
$$P_- = \{\psi \in S \mid \beta(\psi) = -\psi\}$$

have real dimensions 0, 1, 2 or 3 and

$$P = P_+ \oplus P_- = \{\psi \in S \mid \beta(\psi) \in S\}$$

has real dimension 0, 1, 2 or 4 no matter how large the dimension of S is. To prove this we may use periodicity, $\mathcal{C}\ell_{p,q} \otimes \mathcal{C}\ell_{0,8} \simeq \mathcal{C}\ell_{p,q+8}$, and the fact that for $\mathcal{C}\ell_{0,8}$ the dimension of $P = P_+$ is 1 (over $\mathbb{R}$).

Define the real linear space

$$\check{P} = \{\psi \in \check{S} \mid \beta(\psi) \in \check{S}\}$$

which has real dimension 1, 2, 3 or 4. For all ψ, φ in S or $\check{S}$ we have $\beta(\psi)\varphi$ in P or $\check{P}$. There is an invertible element s in $\mathcal{C}\ell_{p,q}$ with the property $P \subset s^{-1}\mathbb{D}$ and which is, in the case $\dim P \neq 0$, such that for all λ in $\mathbb{D}$ also $\lambda^\sigma = s\beta(\lambda)s^{-1}$ is in $\mathbb{D}$. [2] To prove that such an element s exists in every $\mathcal{C}\ell_{p,q}$ we may first consider the lower-dimensional cases and then proceed by making use of the fact that $\beta(f) = f$ for

$$f = \frac{1}{2}(1+\mathbf{e}_{1248})\frac{1}{2}(1+\mathbf{e}_{2358})\frac{1}{2}(1+\mathbf{e}_{3468})\frac{1}{2}(1+\mathbf{e}_{4578})$$

in $\mathcal{C}\ell_{0,8}$, and therefore $s = 1$ is such an element in $\mathcal{C}\ell_{0,8}$.

In the same way, there is an invertible element s in $\mathcal{C}\ell_{p,q}$ with the property $\check{P} = s^{-1}\check{\mathbb{D}}$ and which is moreover such that for all λ in $\check{\mathbb{D}}$ also $\lambda^\sigma = s\beta(\lambda)s^{-1}$ is in $\check{\mathbb{D}}$. Both the maps

$$\check{S} \times \check{S} \to \check{\mathbb{D}}, \quad (\psi, \varphi) \to \begin{cases} s\tilde{\psi}\varphi \\ s\bar{\psi}\varphi \end{cases}$$

are scalar products on $\check{S}$. Similarly, we may construct a scalar product on S. The element s can be chosen from the standard basis of $\mathcal{C}\ell_{p,q}$ [when f is constructed by the standard basis of $\mathcal{C}\ell_{p,q}$]. In particular, $\beta(s) = \pm s$, and so the scalar product is symmetric or antisymmetric [on both S and $\check{S}$]. [3] The

2 The mapping $\lambda \to \lambda^\sigma$ is an (anti-)automorphism of the division ring $\mathbb{D}$.

3 More precisely, the scalar product on S is $\mathbb{D}^\sigma$-symmetric or $\mathbb{D}^\sigma$-skew, and the scalar product on $\check{S}$ is $\check{\mathbb{D}}^\sigma$-symmetric or $\check{\mathbb{D}}^\sigma$-skew.

scalar product on $\check{S}$ is more interesting; it is

symmetric or antisymmetric

non-degenerate

positive definite (for the choice $s = 1$) on $\begin{cases} \mathcal{C}\ell_{n,0} & \text{with} \quad s\tilde{\psi}\varphi \\ \mathcal{C}\ell_{0,n} & \text{with} \quad s\bar{\psi}\varphi \end{cases}$

neutral except on $\begin{cases} \mathcal{C}\ell_{n,0},\ \mathcal{C}\ell_{0,1},\ \mathcal{C}\ell_{0,2},\ \mathcal{C}\ell_{0,3} & \text{with} \quad s\tilde{\psi}\varphi \\ \mathcal{C}\ell_{0,n},\ \mathcal{C}\ell_{1,0} \quad \text{with} \quad s\bar{\psi}\varphi. \end{cases}$

The scalar product is definite or neutral except for $\mathcal{C}\ell_{0,1}$, $\mathcal{C}\ell_{0,2}$, $\mathcal{C}\ell_{0,3}$ or $\mathcal{C}\ell_{1,0}$. In these lower-dimensional exceptional cases neutrality is not possible, because the spinor space $\check{S}$ is 1-dimensional over $\check{\mathbb{D}} = \mathbb{C}$, $\mathbb{H}$, ${}^2\mathbb{H}$ or ${}^2\mathbb{R}$, respectively.

For a fixed $\mathcal{C}\ell_{p,q}$, the neutral scalar products on $\check{S}$, induced by arbitrary anti-automorphisms of $\mathcal{C}\ell_{p,q}$, can be collected into *two* equivalence classes, the equivalence relation being

$$\langle\psi,\varphi\rangle_1 \simeq \langle\psi,\varphi\rangle_2 \iff \exists\, U \in \mathrm{End}_{\check{\mathbb{D}}}\,\check{S},\ \langle U\psi, U\varphi\rangle_1 = \langle\psi,\varphi\rangle_2$$

for all $\psi,\varphi \in \check{S}$. In each class there is a scalar product induced by such an anti-automorphism of $\mathcal{C}\ell_{p,q}$ (extending an orthogonal transformation of $\mathbb{R}^{p,q}$) that does not single out any distinguished direction in $\mathbb{R}^{p,q}$, namely, the reversion $u \to \tilde{u}$ or the Clifford-conjugation $u \to \bar{u}$ of $\mathcal{C}\ell_{p,q}$.

18.2 Automorphism groups of scalar products of spinors

Examples. 1. The Clifford algebra $\mathcal{C}\ell_{2,1}$ is isomorphic to $\mathrm{Mat}(2, {}^2\mathbb{R})$. The idempotent $f = \frac{1}{2}(1+\mathbf{e}_1)\frac{1}{2}(1+\mathbf{e}_{23})$ is primitive in $\mathcal{C}\ell_{2,1}$. The subalgebra $\mathbb{D} = f\mathcal{C}\ell_{2,1}f$ is just the line $\{\lambda f \mid \lambda \in \mathbb{R}\}$; with unity f it is isomorphic to the division ring $\mathbb{R}$. The basis elements

$$f_1 = \tfrac{1}{4}(1+\mathbf{e}_1+\mathbf{e}_{23}+\mathbf{e}_{123})$$
$$f_2 = \tfrac{1}{4}(\mathbf{e}_2-\mathbf{e}_{12}+\mathbf{e}_3-\mathbf{e}_{13})$$

of $S = \mathcal{C}\ell_{2,1}f$ are such that

$$\begin{matrix} \tilde{f}_1 f_1 = 0, & \tilde{f}_1 f_2 = 0 \\ \tilde{f}_2 f_1 = 0, & \tilde{f}_2 f_2 = 0 \end{matrix} \quad \text{and} \quad \begin{matrix} \bar{f}_1 f_1 = 0, & \bar{f}_1 f_2 = f_2 \\ \bar{f}_2 f_1 = -f_2, & \bar{f}_2 f_2 = 0. \end{matrix}$$

The products $s\tilde{\psi}\varphi$, $s = 1$, and $s\bar{\psi}\varphi$, $s = \mathbf{e}_2$, have values in $\mathbb{D}$; they are scalar products on S. The scalar product $\tilde{\psi}\varphi$ vanishes identically; its automorphism group is the full linear group $GL(2,\mathbb{R})$. The scalar product $\mathbf{e}_2\bar{\psi}\varphi$ is antisymmetric; its automorphims group is $Sp(2,\mathbb{R})$. If we consider $\check{S} = S \oplus \hat{S}$ instead of S, then the automorphism group of the scalar product $s\tilde{\psi}\varphi$ becomes non-degenerate (because of the swap) and the automorphism group of the scalar

product $s\bar{\psi}\varphi$ splits: ${}^2Sp(2,\mathbb{R}) = Sp(2,\mathbb{R}) \times Sp(2,\mathbb{R})$.
2. The Clifford algebra $\mathcal{C}\ell_{1,3}$ of the Minkowski space $\mathbb{R}^{1,3}$ is isomorphic to the real matrix algebra $\mathrm{Mat}(2,\mathbb{H})$. Take an orthonormal basis $\{\gamma_0,\gamma_1,\gamma_2,\gamma_3\}$ for $\mathbb{R}^{1,3}$. The idempotent $f = \frac{1}{2}(1+\gamma_0)$ is primitive in $\mathcal{C}\ell_{1,3}$. As a real linear space the minimal left ideal $S = \mathcal{C}\ell_{1,3}f$ is 8-dimensional and the elements

$$\begin{aligned} h_1 &= \tfrac{1}{2}(1+\gamma_0), & h_2 &= \tfrac{1}{2}(-\gamma_{123}+\gamma_{0123}) \\ i_1 &= \tfrac{1}{2}(\gamma_{23}+\gamma_{023}), & i_2 &= \tfrac{1}{2}(\gamma_1-\gamma_{01}) \\ j_1 &= \tfrac{1}{2}(\gamma_{31}+\gamma_{031}), & j_2 &= \tfrac{1}{2}(\gamma_2-\gamma_{02}) \\ k_1 &= \tfrac{1}{2}(\gamma_{12}+\gamma_{012}), & k_2 &= \tfrac{1}{2}(\gamma_3-\gamma_{03}) \end{aligned}$$

form a basis for $S_{\mathbb{R}}$. The set $\{h_1,i_1,j_1,k_1\}$ is a basis for the real linear space $\mathbb{D} = f\mathcal{C}\ell_{1,3}f$. As a ring $\mathbb{D}$ is isomorphic to the quaternion ring $\mathbb{H}$, and the right $\mathbb{D}$-linear module $S_{\mathbb{D}}$ is two-dimensional with basis $\{h_1,h_2\}$. In the basis $\{h_1,h_2\}$ left multiplication by $\gamma_0,\gamma_1,\gamma_2,\gamma_3$ is represented by the following 2×2-matrices with quaternion entries:

$$\gamma_0 \simeq \begin{pmatrix} 1 & 0 \\ 0 & -1 \end{pmatrix},$$

$$\gamma_1 \simeq \begin{pmatrix} 0 & i \\ i & 0 \end{pmatrix}, \quad \gamma_2 \simeq \begin{pmatrix} 0 & j \\ j & 0 \end{pmatrix}, \quad \gamma_3 \simeq \begin{pmatrix} 0 & k \\ k & 0 \end{pmatrix}.$$

The real linear spaces P_+ and P_- have bases

$$\begin{array}{ccc} & P_+ & P_- \\ \tilde{u}: & \{h_1\} & \{i_1,j_1,k_1\} \\ \bar{u}: & \{h_2\} & \{i_2,j_2,k_2\}. \end{array}$$

In the scalar products $S\times S\to\mathbb{D}$, $(\psi,\varphi)\to s\beta(\psi)\varphi$ one can take $s=1$ for $s\tilde{\psi}\varphi$ and $s=\gamma_{123}$ for $s\bar{\psi}\varphi$. Direct computation shows that

$$\begin{matrix} \tilde{h}_1h_1 = h_1, & \tilde{h}_1h_2 = 0 \\ \tilde{h}_2h_1 = 0, & \tilde{h}_2h_2 = -h_1 \end{matrix} \quad \text{and} \quad \begin{matrix} \bar{h}_1h_1 = 0, & \bar{h}_1h_2 = h_2 \\ \bar{h}_2h_1 = h_2, & \bar{h}_2h_2 = 0. \end{matrix}$$

Both the scalar products have the automorphism group $Sp(2,2)$. ∎

The Tables 1 and 2 list automorphism groups of the scalar products on $\check{S}$; they are nothing but the groups

$$\{s\in\mathcal{C}\ell_{p,q} \mid s\tilde{s}=1\} \quad\text{and}\quad \{s\in\mathcal{C}\ell_{p,q} \mid s\bar{s}=1\}.$$

If the Clifford algebra $\mathcal{C}\ell_{p,q}$ is semi-simple and if the automorphism group on $\check{S}$ is a direct product ${}^2G = G\times G$, then the automorphism group on S is G.

Table 1. Automorphism Groups of $s\tilde{\psi}\varphi$ on $\check{S}$ in $\mathcal{C}\ell_{p,q}$.

$p+q$ \ $p-q$	-7	-6	-5	-4	-3	-2	-1	0	1	2	3	4	5	6	7
0								$O(1)$							
1							$O(1,\mathbb{C})$		${}^2O(1)$						
2						$SO^*(2)$		$O(1,1)$		$O(2)$					
3					$GL(1,\mathbb{H})$		$U(1,1)$		$GL(2,\mathbb{R})$		$U(2)$				
4				$Sp(2,2)$		$Sp(2,2)$		$Sp(4,\mathbb{R})$		$Sp(4,\mathbb{R})$		$Sp(4)$			
5			$Sp(4,\mathbb{C})$		${}^2Sp(2,2)$		$Sp(4,\mathbb{C})$		${}^2Sp(4,\mathbb{R})$		$Sp(4,\mathbb{C})$		${}^2Sp(4)$		
6		$Sp(8,\mathbb{R})$		$Sp(4,4)$		$Sp(4,4)$		$Sp(8,\mathbb{R})$		$Sp(8,\mathbb{R})$		$Sp(4,4)$		$Sp(8)$	
7	$GL(8,\mathbb{R})$		$U(4,4)$		$GL(4,\mathbb{H})$		$U(4,4)$		$GL(8,\mathbb{R})$		$U(4,4)$		$GL(4,\mathbb{H})$		$U(8)$

Table 2. Automorphism Groups of $s\bar{\psi}\varphi$ on $\check{S}$ in $\mathcal{C}\ell_{p,q}$.

$p+q$ \ $p-q$	-7	-6	-5	-4	-3	-2	-1	0	1	2	3	4	5	6	7
0								$O(1)$							
1							$U(1)$		$GL(1,\mathbb{R})$						
2						$Sp(2)$		$Sp(2,\mathbb{R})$		$Sp(2,\mathbb{R})$					
3					${}^2Sp(2)$		$Sp(2,\mathbb{C})$		${}^2Sp(2,\mathbb{R})$		$Sp(2,\mathbb{C})$				
4				$Sp(4)$		$Sp(2,2)$		$Sp(4,\mathbb{R})$		$Sp(4,\mathbb{R})$		$Sp(2,2)$			
5			$U(4)$		$GL(2,\mathbb{H})$		$U(2,2)$		$GL(4,\mathbb{R})$		$U(2,2)$		$GL(2,\mathbb{H})$		
6		$O(8)$		$SO^*(8)$		$SO^*(8)$		$O(4,4)$		$O(4,4)$		$SO^*(8)$		$SO^*(8)$	
7	${}^2O(8)$		$O(8,\mathbb{C})$		${}^2SO^*(8)$		$O(8,\mathbb{C})$		${}^2O(4,4)$		$O(8,\mathbb{C})$		${}^2SO^*(8)$		$O(8,\mathbb{C})$

Examples. 1. $\mathcal{C}\ell_{0,2}$, $s\tilde{\psi}\varphi$: $SO^*(2) = \{U \in SO(2,\mathbb{C}) \mid U^*J = JU\} \simeq SO(2)$.
2. $\mathcal{C}\ell_2$, $s\bar{\psi}\varphi$: $Sp(2,\mathbb{R}) = \{U \in \mathrm{Mat}(2,\mathbb{R}) \mid U^\top JU = U\} \simeq SL(2,\mathbb{R})$.
3. $\mathcal{C}\ell_5$, $s\bar{\psi}\varphi$: ${}^2Sp(4) = Sp(4) \times Sp(4)$, $Sp(4)/\{\pm I\} \simeq SO(5)$.
4. $\mathcal{C}\ell_{1,3}$, $Sp(2,2) = U(2,2) \cap Sp(4,\mathbb{C})$, $Sp(2,2)/\{\pm I\} \simeq SO_+(4,1)$. ∎

Note that the group $U(2,2)$ appears as an automorphism group of the scalar product $s\bar{\psi}\varphi$ for $\mathcal{C}\ell_{2,3}$ and $\mathcal{C}\ell_{4,1}$. To explain the presence of $U(2,2)$ in the Dirac theory by the real Clifford algebras $\mathcal{C}\ell_{p,q}$, we must add one dimension of positive square to the Minkowski spaces $\mathbb{R}^{1,3}$ and $\mathbb{R}^{3,1}$.

There is another explanation: use complexifications $\mathbb{C} \otimes \mathcal{C}\ell_{p,q}$. For a fixed $n = p+q$ we have the isomorphisms of algebras $\mathbb{C} \otimes \mathcal{C}\ell_{p,q} \simeq \mathcal{C}\ell(\mathbb{C}^n)$. Although the complex linear space $\mathbb{C}^n$ has a symmetric (= not sesquilinear) bilinear form on itself, we may equip the spinor spaces of $\mathbb{C} \otimes \mathcal{C}\ell_{p,q}$ with sesquilinear forms

$s\tilde{\psi}^*\varphi$ and $s\bar{\psi}^*\varphi$. These sesquilinear products have automorphism groups

$$\{s \in \mathbb{C} \otimes C\ell_{p,q} \mid s\tilde{s}^* = 1\} \quad \text{and} \quad \{s \in \mathbb{C} \otimes C\ell_{p,q} \mid s\bar{s}^* = 1\}.$$

For a fixed $n = p + q$ these groups depend on the values of p and q [although the algebra $\mathbb{C} \otimes C\ell_{p,q}$ is independent of p and q].

Table 3. Automorphism Groups of $s\tilde{\psi}^*\varphi$ in $\mathbb{C}^* \otimes C\ell_{p,q}$.

$p+q$ \ $p-q$	-7	-6	-5	-4	-3	-2	-1	0	1	2	3	4	5	6	7
0								$U(1)$							
1							$GL(1,\mathbb{C})$		$^2U(1)$						
2						$U(1,1)$		$U(1,1)$		$U(2)$					
3					$GL(2,\mathbb{C})$		$^2U(1,1)$		$GL(2,\mathbb{C})$		$^2U(2)$				
4				$U(2,2)$		$U(2,2)$		$U(2,2)$		$U(2,2)$		$U(4)$			
5			$GL(4,\mathbb{C})$		$^2U(2,2)$		$GL(4,\mathbb{C})$		$^2U(2,2)$		$GL(4,\mathbb{C})$		$^2U(4)$		
6		$U(4,4)$		$U(4,4)$		$U(4,4)$		$U(4,4)$		$U(4,4)$		$U(4,4)$		$U(8)$	
7	$GL(8,\mathbb{C})$		$^2U(4,4)$		$GL(8,\mathbb{C})$		$^2U(4,4)$		$GL(8,\mathbb{C})$		$^2U(4,4)$		$GL(8,\mathbb{C})$		$^2U(8)$

Table 4. Automorphism Groups of $s\bar{\psi}^*\varphi$ in $\mathbb{C}^* \otimes C\ell_{p,q}$.

$p+q$ \ $p-q$	-7	-6	-5	-4	-3	-2	-1	0	1	2	3	4	5	6	7
0								$U(1)$							
1							$^2U(1)$		$GL(1,\mathbb{C})$						
2						$U(2)$		$U(1,1)$		$U(1,1)$					
3					$^2U(2)$		$GL(2,\mathbb{C})$		$^2U(1,1)$		$GL(2,\mathbb{C})$				
4				$U(4)$		$U(2,2)$		$U(2,2)$		$U(2,2)$		$U(2,2)$			
5			$^2U(4)$		$GL(4,\mathbb{C})$		$^2U(2,2)$		$GL(4,\mathbb{C})$		$^2U(2,2)$		$GL(4,\mathbb{C})$		
6		$U(8)$		$U(4,4)$		$U(4,4)$		$U(4,4)$		$U(4,4)$		$U(4,4)$		$U(4,4)$	
7	$^2U(8)$		$GL(8,\mathbb{C})$		$^2U(4,4)$		$GL(8,\mathbb{C})$		$^2U(4,4)$		$GL(8,\mathbb{C})$		$^2U(4,4)$		$GL(8,\mathbb{C})$

See Porteous 1969 p. 271 ll. 1-8. Note that complexification explains the occurrence of $U(2,2)$ in conjunction with the Minkowski spaces.

In complexifications of real algebras we replaced the ground field $\mathbb{R}$ by $\mathbb{C}$, a field extension with an involution, the complex conjugation [to emphasize that $\mathbb{C}$ comes with a complex conjugation we denote $\bar{\mathbb{C}}$ or $\mathbb{C}^*$].

We could also tensor $C\ell_{p,q}$ by the real algebra $^2\mathbb{R}$, a commutative ring with an irreducible involution, the swap. See Porteous 1969 pp. 193, 251. This leads to the automorphism groups shown in Table 5 [isomorphic to the subgroup of invertible elements in $C\ell_{p,q}$].

Table 5. Automorphism Groups for ${}^2\mathbb{R} \otimes C\ell_{p,q}$.

$p+q$ \ $p-q$	-7	-6	-5	-4	-3	-2	-1	0	1	2	3	4	5	6	7
0								$GL(1,\mathbb{R})$							
1							$GL(1,\mathbb{C})$		${}^2GL(1,\mathbb{R})$						
2						$GL(1,\mathbb{H})$		$GL(2,\mathbb{R})$		$GL(2,\mathbb{R})$					
3					${}^2GL(1,\mathbb{H})$		$GL(2,\mathbb{C})$		${}^2GL(2,\mathbb{R})$		$GL(2,\mathbb{C})$				
4				$GL(2,\mathbb{H})$		$GL(2,\mathbb{H})$		$GL(4,\mathbb{R})$		$GL(4,\mathbb{R})$		$GL(2,\mathbb{H})$			
5			$GL(4,\mathbb{C})$		${}^2GL(2,\mathbb{H})$		$GL(4,\mathbb{C})$		${}^2GL(4,\mathbb{R})$		$GL(4,\mathbb{C})$		${}^2GL(2,\mathbb{H})$		
6		$GL(8,\mathbb{R})$		$GL(4,\mathbb{H})$		$GL(4,\mathbb{H})$		$GL(8,\mathbb{R})$		$GL(8,\mathbb{H})$		$GL(4,\mathbb{H})$		$GL(4,\mathbb{H})$	
7	${}^2GL(8,\mathbb{R})$		$GL(8,\mathbb{C})$		${}^2GL(4,\mathbb{H})$		$GL(8,\mathbb{C})$		${}^2GL(8,\mathbb{R})$		$GL(8,\mathbb{C})$		${}^2GL(4,\mathbb{H})$		$GL(8,\mathbb{C})$

See Porteous 1969 p. 271 ll. 11-18.

In the case of the complex Clifford algebras $C\ell(\mathbb{C}^n)$ we may further equip the spinor space with a symmetric (= not sesquilinear) form on itself, sending (ψ,φ) to $s\tilde{\psi}\varphi$ or $s\bar{\psi}\varphi$, see Table 6.

Table 6. Automorphism Groups for $\mathbb{C}^n$.

n	$s\tilde{\psi}\varphi$	n	$s\bar{\psi}\varphi$
0	$O(1,\mathbb{C})$	0	$O(1,\mathbb{C})$
1	${}^2O(1,\mathbb{C})$	1	$GL(1,\mathbb{C})$
2	$O(2,\mathbb{C})$	2	$Sp(2,\mathbb{C})$
3	$GL(2,\mathbb{C})$	3	${}^2Sp(2,\mathbb{C})$
4	$Sp(4,\mathbb{C})$	4	$Sp(4,\mathbb{C})$
5	${}^2Sp(4,\mathbb{C})$	5	$GL(4,\mathbb{C})$
6	$Sp(8,\mathbb{C})$	6	$O(8,\mathbb{C})$
7	$GL(8,\mathbb{C})$	7	${}^2O(8,\mathbb{C})$

See Porteous 1969 p. 271 l. 9.

As the last extension we consider the tensor product ${}^2\mathbb{C} \otimes_{\mathbb{C}} \mathbb{C}^n$. The scalar products of spinors are formed by reversion or Clifford-conjugation composed with swap (no complex conjugation), see Table 7.

Table 7. Automorphism Groups for ${}^2\mathbb{C} \otimes \mathbb{C}^n$.

n	$s\tilde{\psi}\varphi$ or $s\bar{\psi}\varphi$
0	$GL(1,\mathbb{C})$
1	${}^2GL(1,\mathbb{C})$
2	$GL(2,\mathbb{C})$
3	${}^2GL(2,\mathbb{C})$
4	$GL(4,\mathbb{C})$
5	${}^2GL(4,\mathbb{C})$
6	$GL(8,\mathbb{C})$
7	${}^2GL(8,\mathbb{C})$

See Porteous 1969 p. 271 l. 10.

18.3 Brauer-Wall-Porteous groups

As before, we consider only finite-dimensional associative algebras.

Central simple algebras over $\mathbb{R}$ are isomorphic to the real matrix algebras $\mathrm{Mat}(d,\mathbb{R})$ and $\mathrm{Mat}(d,\mathbb{H})$. A tensor product of two matrix algebras with entries in $\mathbb{H}$ is a matrix algebra with entries in $\mathbb{R}$. This can be expressed by saying that the Brauer group $Br(\mathbb{R})$ of $\mathbb{R}$ is a two-element group $\{\mathbb{R},\mathbb{H}\}$.

Tensor products of graded central simple algebras over $\mathbb{R}$ lead to the Brauer-Wall group $BW(\mathbb{R})$ of $\mathbb{R}$; this is a cyclic group of eight elements,

$$\left\{\frac{\mathbb{R}(2\nu)}{{}^2\mathbb{R}(\nu)},\frac{{}^2\mathbb{R}(\nu)}{\mathbb{R}(\nu)},\frac{\mathbb{R}(2\nu)}{\mathbb{C}(\nu)},\frac{\mathbb{C}(2\nu)}{\mathbb{H}(\nu)},\frac{\mathbb{H}(2\nu)}{{}^2\mathbb{H}(\nu)},\frac{{}^2\mathbb{H}(\nu)}{\mathbb{H}(\nu)},\frac{\mathbb{H}(\nu)}{\mathbb{C}(\nu)},\frac{\mathbb{C}(\nu)}{\mathbb{R}(\nu)}\right\}.$$

Here we use the abbreviation $\mathbb{A}(\nu)=\mathrm{Mat}(\nu,\mathbb{A})$; the notation

$$\frac{A}{B}$$

means that B is the even subalgebra of A. The elements of $BW(\mathbb{R})$ can be represented by the graded algebras

$$\frac{C\ell_{0,n}}{C\ell^{+}_{0,n}}$$

where n is taken modulo 8. This is just another way of expressing Cartan's periodicity of 8.

Graded algebras are algebras with an involution (= involutory automorphism). We could further consider tensor products in graded central simple algebras with an anti-involution (= involutory anti-automorphism). When the involution and the anti-involution commute, this leads to the Brauer-Wall-Porteous group $BWP(\mathbb{R})$ of $\mathbb{R}$; its elements are graded subgroups (of a graded algebra A)

$$\frac{G}{H}$$

where G is the subgroup determinded by the anti-involution β, $G = \{s \in A \mid \beta(s)s = 1\}$, and H is its even subgroup, $H = B \cap G$ (B is the even part of A).

Table 8. Scalar Product $s\tilde{\psi}\varphi$ in $\mathcal{C}\ell_{p,q}$ and $BWP(\mathbb{R})$.

	$p-q$	0	1	2	3	4	5	6	7
$q=0$	$p+q$								
$O(2\nu)$	0	$O(\nu,\nu)$		$O(\nu,\nu)$		$SO^*(2\nu)$		$SO^*(2\nu)$	
${}^2O(2\nu)$	1		${}^2O(\nu,\nu)$		$O(2\nu,\mathbb{C})$		${}^2SO^*(2\nu)$		$O(2\nu,\mathbb{C})$
$O(2\nu)$	2	$O(\nu,\nu)$		$O(\nu,\nu)$		$SO^*(2\nu)$		$SO^*(2\nu)$	
$U(2\nu)$	3		$GL(2\nu,\mathbb{R})$		$U(\nu,\nu)$		$GL(\nu,\mathbb{H})$		$U(\nu,\nu)$
$Sp(2\nu)$	4	$Sp(2\nu,\mathbb{R})$		$Sp(2\nu,\mathbb{R})$		$Sp(\nu,\nu)$		$Sp(\nu,\nu)$	
${}^2Sp(2\nu)$	5		${}^2Sp(2\nu,\mathbb{R})$		$Sp(2\nu,\mathbb{C})$		${}^2Sp(\nu,\nu)$		$Sp(2\nu,\mathbb{C})$
$Sp(2\nu)$	6	$Sp(2\nu,\mathbb{R})$		$Sp(2\nu,\mathbb{R})$		$Sp(\nu,\nu)$		$Sp(\nu,\nu)$	
$U(2\nu)$	7		$GL(2\nu,\mathbb{R})$		$U(\nu,\nu)$		$GL(\nu,\mathbb{H})$		$U(\nu,\nu)$

Table 9. Scalar Product $s\bar{\psi}\varphi$ in $\mathcal{C}\ell_{p,q}$ and $BWP(\mathbb{R})$.

	$p-q$	0	1	2	3	4	5	6	7
$p=0$	$p+q$								
$O(2\nu)$	0	$O(\nu,\nu)$		$O(\nu,\nu)$		$SO^*(2\nu)$		$SO^*(2\nu)$	
$U(2\nu)$	1		$GL(2\nu,\mathbb{R})$		$U(\nu,\nu)$		$GL(\nu,\mathbb{H})$		$U(\nu,\nu)$
$Sp(2\nu)$	2	$Sp(2\nu,\mathbb{R})$		$Sp(2\nu,\mathbb{R})$		$Sp(\nu,\nu)$		$Sp(\nu,\nu)$	
${}^2Sp(2\nu)$	3		${}^2Sp(2\nu,\mathbb{R})$		$Sp(2\nu,\mathbb{C})$		${}^2Sp(\nu,\nu)$		$Sp(2\nu,\mathbb{C})$
$Sp(2\nu)$	4	$Sp(2\nu,\mathbb{R})$		$Sp(2\nu,\mathbb{R})$		$Sp(\nu,\nu)$		$Sp(\nu,\nu)$	
$U(2\nu)$	5		$GL(2\nu,\mathbb{R})$		$U(\nu,\nu)$		$GL(\nu,\mathbb{H})$		$U(\nu,\nu)$
$O(2\nu)$	6	$O(\nu,\nu)$		$O(\nu,\nu)$		$SO^*(2\nu)$		$SO^*(2\nu)$	
${}^2O(2\nu)$	7		${}^2O(\nu,\nu)$		$O(2\nu,\mathbb{C})$		${}^2SO^*(2\nu)$		$O(2\nu,\mathbb{C})$

The Brauer-Wall-Porteous group $BWP(\mathbb{R})$ is a commutative group of 32 elements,

$$BWP(\mathbb{R}) \simeq \{(x,y) \in \mathbb{Z}_8 \times \mathbb{Z}_8 \mid x,y \in 2\mathbb{Z}\}.$$

We see that the elements of $BWP(\mathbb{R})$ are (graded) automorphism groups of scalar products of spinors for $C\ell_{p,q}$,

$$\frac{\{s \in C\ell_{p,q} \mid s\beta(s) = 1\}}{\{s \in C\ell^+_{p,q} \mid s\beta(s) = 1\}}.$$

The even subgroup $\{s \in C\ell^+_{p,q} \mid s\beta(s) = 1\}$ is isomorphic to $\{s \in C\ell_{p,q-1} \mid s\beta(s) = 1\}$, obtained by taking a step to the North-East. Tensor products of real graded central simple algebras with an anti-involution correspond to movements of a bishop on the chessboard.

Recall that the Brauer group $Br(\mathbb{C})$ of $\mathbb{C}$ is a one-element group $\{\mathbb{C}\}$. The Brauer-Wall group $BW(\mathbb{C})$ of $\mathbb{C}$ is a group of two elements

$$\left\{\frac{\mathrm{Mat}(2,\mathbb{C})}{{}^2\mathbb{C}}, \frac{{}^2\mathbb{C}}{\mathbb{C}}\right\}.$$

Thus, complex Clifford algebras have a periodicity of 2. The Brauer-Wall-Porteous group $BWP(\mathbb{C})$ of $\mathbb{C}$ is a cyclic group of eight elements; in other words complex Clifford algebras with an anti-involution have a periodicity of 8, see Table 10.

Table 10. $C\ell(\mathbb{C}^n)$ and $BWP(\mathbb{C})$.

n	$s\tilde{\psi}\varphi$
0	$O(2\nu,\mathbb{C})$
1	${}^2O(2\nu,\mathbb{C})$
2	$O(2\nu,\mathbb{C})$
3	$GL(2\nu,\mathbb{C})$
4	$Sp(2\nu,\mathbb{C})$
5	${}^2Sp(2\nu,\mathbb{C})$
6	$Sp(2\nu,\mathbb{C})$
7	$GL(2\nu,\mathbb{C})$

n	$s\bar{\psi}\varphi$
0	$O(2\nu,\mathbb{C})$
1	$GL(2\nu,\mathbb{C})$
2	$Sp(2\nu,\mathbb{C})$
3	${}^2Sp(2\nu,\mathbb{C})$
4	$Sp(2\nu,\mathbb{C})$
5	$GL(2\nu,\mathbb{C})$
6	$O(2\nu,\mathbb{C})$
7	${}^2O(2\nu,\mathbb{C})$

The Brauer-Wall-Porteous group $BWP({}^2\mathbb{R})$ of the double ring ${}^2\mathbb{R}$ with swap is also a cyclic group of eight elements, see Table 11.

Table 11. ${}^2\mathbb{R} \otimes C\ell_{p,q}$ and $BWP({}^2\mathbb{R})$.

$p-q$	0	1	2	3	4	5	6	7
	$GL(2\nu,\mathbb{R})$	${}^2GL(2\nu,\mathbb{R})$	$GL(2\nu,\mathbb{R})$	$GL(2\nu,\mathbb{C})$	$GL(\nu,\mathbb{H})$	${}^2GL(\nu,\mathbb{H})$	$GL(\nu,\mathbb{H})$	$GL(2\nu,\mathbb{C})$

Tensoring $\mathcal{C}\ell_{p,q}$ by $\mathbb{C}^*$, the complex field with complex conjugation, results in a Brauer-Wall-Porteous group isomorphic to $\mathbb{Z}_2 \times \mathbb{Z}_2$, see Table 12.

Table 12. $\mathbb{C}^* \otimes \mathcal{C}\ell_{p,q}$ and $BWP(\mathbb{C}^*)$.

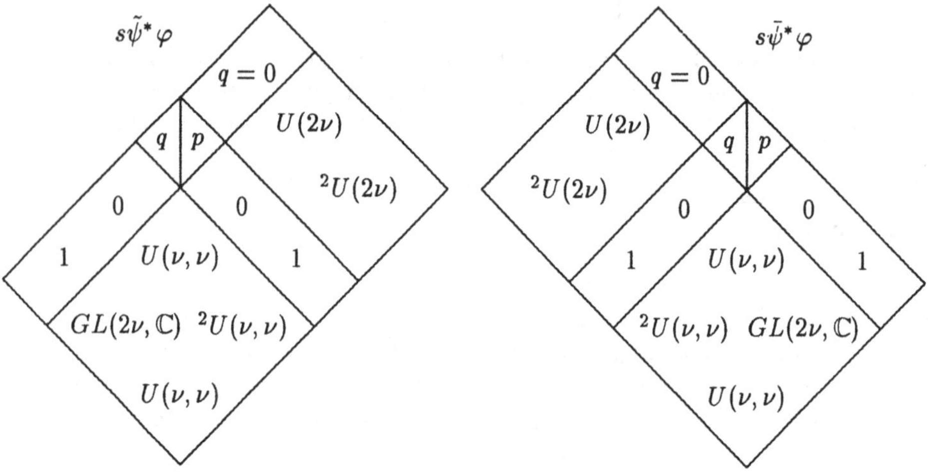

As our last extension we tensor $\mathcal{C}\ell(\mathbb{C}^n)$ by $^2\mathbb{C}$ (Table 13).

Table 13. $^2\mathbb{C} \otimes_{\mathbb{C}} \mathbb{C}^n$ and $BWP(^2\mathbb{C})$.

n	$s\tilde{\psi}\varphi$ and $\bar{\psi}\varphi$	
0	$GL(2\nu,\mathbb{C})$	
1		$^2GL(2\nu,\mathbb{C})$

In total, we have the following Brauer-Wall-Porteous groups (of $\mathbb{R}$ and $\mathbb{C}$ and their extensions with an irreducible involution).

$\mathbb{R}^{p,q}$	$BWP(\mathbb{R}) \simeq (\mathbb{Z}_8 \times \mathbb{Z}_8)/\mathbb{Z}_2$
$^2\mathbb{R} \otimes_{\mathbb{R}} \mathbb{R}^{p,q}$	$BWP(^2\mathbb{R}) \simeq \mathbb{Z}_8$
$\mathbb{C}^* \otimes \mathbb{R}^{p,q}$	$BWP(\mathbb{C}^*) \simeq \mathbb{Z}_2 \times \mathbb{Z}_2$
$\mathbb{C}^n$	$BWP(\mathbb{C}) \simeq \mathbb{Z}_8$
$^2\mathbb{C} \otimes_{\mathbb{C}} \mathbb{C}^n$	$BWP(^2\mathbb{C}) \simeq \mathbb{Z}_2$.

It is convenient to be able to characterize the automorphism groups of scalar products on spinor spaces S directly by making use of real dimensions of the subspaces $P_\pm = \{\psi \in S \mid \beta(\psi) = \pm\psi\}$, see Table 14.

Table 14. Scalar products on S.

$\dim P_+$	0	1	2	3
$\dim P_-$				
0	$GL(\nu,\mathbb{R})$	$O(\nu,\nu)$	$O(2\nu,\mathbb{C})$	
1	$Sp(2\nu,\mathbb{R})$	$U(\nu,\nu)$		$SO^*(4\nu)$
2	$Sp(2\nu,\mathbb{C})$			
3		$Sp(2\nu,2\nu)$		

Bibliography

I.M. Benn, R.W. Tucker: *An Introduction to Spinors and Geometry with Applications in Physics*. Adam Hilger, Bristol, 1987.

P. Budinich, A. Trautman: *The Spinorial Chessboard*. Springer, Berlin, 1988.

J.S.R. Chisholm, A.K. Common (eds.): *Proceedings of the NATO and SERC Workshop on 'Clifford Algebras and their Applications in Mathematical Physics' (Canterbury, 1985)*. Reidel, Dordrecht, The Netherlands, 1986.

F.R. Harvey: *Spinors and Calibrations*. Academic Press, San Diego, 1990.

P. Lounesto: Scalar products of spinors and an extension of Brauer-Wall groups. *Found. Phys.* **11** (1981), 721-740.

I.R. Porteous: *Topological Geometry*. Van Nostrand Reinhold, London, 1969. Cambridge University Press, Cambridge, 1981.

I.R. Porteous: *Clifford Algebras and the Classical Groups*. Cambridge University Press, Cambridge, 1995.

19

Möbius Transformations and Vahlen Matrices

Classical complex analysis can be generalized from the complex plane to higher dimensions in three different ways: function theory of several complex variables (commutative), higher-dimensional one-variable hypercomplex analysis (anti-commutative), and conformal transformations (geometric). In this chapter we study the third possibility: conformal transformations in n dimensions, $n \geq 3$.

A function f sending a region in $\mathbb{R}^2 = \mathbb{C}$ into $\mathbb{C}$ is conformal at z, if it is complex analytic and has a non-zero derivative, $f'(z) \neq 0$ (we consider only sense-preserving conformal mappings). The only conformal transformations of the whole plane $\mathbb{C}$ are affine linear transformations: compositions of rotations, dilations and translations. The *Möbius mapping*

$$f(z) = \frac{az+b}{cz+d}, \quad a,b,c,d \in \mathbb{C}, \quad ad - bc \neq 0,$$

is affine linear when $c = 0$; otherwise it is conformal at each $z \in \mathbb{C}$ except when $z = -\frac{d}{c}$. The Möbius mapping f sends $\mathbb{C} \setminus \{-\frac{d}{c}\}$ onto $\mathbb{C} \setminus \{\frac{a}{c}\}$. If we agree that $f(-\frac{d}{c}) = \infty$ and $f(\infty) = \frac{a}{c}$, then f becomes a (one-to-one) transformation of $\mathbb{C} \cup \{\infty\}$, the complex plane compactified by the point at infinity. [1] These transformations are called *Möbius transformations* of $\mathbb{C} \cup \{\infty\}$. Möbius transformations are compositions of rotations, translations, dilations and transversions. [2] Möbius transformations send circles (and affine lines) to circles (or affine lines). The derivative of a Möbius transformation is a composition of a rotation and a dilation.

By definition, a conformal mapping preserves angles between intersecting curves. Formally, let D be a region in a Euclidean space $\mathbb{R}^n$. A continuously

1 Möbius mappings $f(z) = \frac{az+b}{cz+d}$ are defined almost everywhere in $\mathbb{C}$. The set of Möbius mappings can be used to compactify $\mathbb{C}$, the compactification being $\mathbb{C} \cup \{\infty\}$.

2 A transversion is a composition of an inversion in the unit circle, a translation and another inversion. Thus, transversions are conjugate (by the inversion) to translations.

differentiable function $g : D \to \mathbb{R}^n$ is *conformal* in D if there is a continuous function $\lambda : D \to \mathbb{R}^\times = \mathbb{R} \setminus \{0\}$ such that

$$<f'(\mathbf{x})\mathbf{a}, f'(\mathbf{x})\mathbf{b}> = <\mathbf{a}\lambda(\mathbf{x}), \mathbf{b}\lambda(\mathbf{x})>$$

for all $\mathbf{x} \in D$ and $\mathbf{a}, \mathbf{b} \in \mathbb{R}^n$. In higher-dimensional Euclidean spaces $\mathbb{R}^n$, $n \geq 3$, the only conformal mappings [sending a region in $\mathbb{R}^n$ into $\mathbb{R}^n$] are restrictions of Möbius transformations of $\mathbb{R}^n \cup \{\infty\}$. [3] The case $n = 3$ was proved by Liouville 1850. The analogous statement for indefinite quadratic spaces is also true by a theorem of Haantjes 1937.

19.1 Quaternion representation of conformal transformations of $\mathbb{R}^4$

Conformal transformations of $\mathbb{R}^4$ can be represented by quaternion computation:

$$\mathbb{R}^4 = \mathbb{H} \to \mathbb{H}, \quad q \to (aq + b)(cq + d)^{-1}, \quad a, b, c, d \in \mathbb{H}.$$

In order to exclude constant functions we require the matrix

$$\begin{pmatrix} a & b \\ c & d \end{pmatrix}$$

to be invertible, that is, $|a|^2|d|^2 + |b|^2|c|^2 - 2\,\mathrm{Re}(a\,\bar{c}\,d\,\bar{b}) \neq 0$. This matrix representation renders composition of non-linear conformal transformations into multiplication of matrices.

19.2 Möbius transformations of $\mathbb{R}^n$

Möbius transformations might be sense-preserving with $\det f'(\mathbf{x}) > 0$ or sense-reversing with $\det f'(\mathbf{x}) < 0$. The Möbius transformations form a group, the full Möbius group, which has two components, the identity component being the sense-preserving Möbius group. The full Möbius group of $\mathbb{R}^n$ is generated by translations, reflections and the inversion

$$\mathbf{x} \to \mathbf{x}^{-1} = \frac{\mathbf{x}}{\mathbf{x}^2},$$

or equivalently, by reflections in affine hyperplanes and inversions in spheres (not necessarily centered at the origin). The sense-preserving Möbius group is generated by the following four types of transformations:

3 We shall often refer to Möbius transformations of the Euclidean space $\mathbb{R}^n$ whereby we tacitly mean transformations of the compactification $\mathbb{R}^n \cup \{\infty\}$.

rotations	$a\mathbf{x}a^{-1}$	$a \in \mathbf{Spin}(n)$
translations	$\mathbf{x}+\mathbf{b}$	$\mathbf{b} \in \mathbb{R}^n$
dilations	$\mathbf{x}\delta$	$\delta > 0$
transversions	$\dfrac{\mathbf{x}+\mathbf{x}^2\mathbf{c}}{1+2\mathbf{x}\cdot\mathbf{c}+\mathbf{x}^2\mathbf{c}^2}$	$\mathbf{c} \in \mathbb{R}^n$.

Rewriting the transversion into the form $(\mathbf{x}^{-1}+\mathbf{c})^{-1}$, one sees that it is a composition of the inversion, a translation and the inversion. Using the multiplicative notation of the Clifford algebra $C\ell_n$, the transversion can further be written in the form

$$\mathbf{x} \to \mathbf{x}(\mathbf{c}\mathbf{x}+1)^{-1}.$$

This might suggest the following: Let a, b, c, d be in $C\ell_n$. If $(a\mathbf{x}+\mathbf{x})(c\mathbf{x}+d)^{-1}$ is in $\mathbb{R}^n$ for almost all $\mathbf{x} \in \mathbb{R}^n$ and if the range of

$$g(\mathbf{x}) = (a\mathbf{x}+b)(c\mathbf{x}+d)^{-1}$$

is dense in $\mathbb{R}^n$, then g is a Möbius transformation of $\mathbb{R}^n$. Although this is true, the group so obtained is too large to be a practical covering group of the full Möbius group. [4] Therefore, we introduce:

Definition (Maass 1949, Ahlfors 1984). [5] The matrix $\begin{pmatrix} a & b \\ c & d \end{pmatrix} \in \mathrm{Mat}(2, C\ell_n)$ fulfilling the conditions

(i) $a, b, c, d \in \mathbf{\Gamma}_n \cup \{0\}$
(ii) $a\tilde{b},\ \tilde{b}d,\ d\tilde{c},\ \tilde{c}a \in \mathbb{R}^n$
(iii) $a\tilde{d} - b\tilde{c} \in \mathbb{R}\setminus\{0\}$

is called a *Vahlen matrix* of the Möbius transformation g of $\mathbb{R}^n$ given by $g(\mathbf{x}) = (a\mathbf{x}+b)(c\mathbf{x}+d)^{-1}$. ∎

By condition (i) the diagonal entries of a Vahlen matrix are either even or odd. Conditions (i) and (ii) imply that if the diagonal entries are even then the off-diagonal entries must be odd, and if the diagonal entries are odd then

4 This group is the Vahlen group multiplied by the group generated by invertible matrices of the form

$$\begin{pmatrix} \alpha+\beta\mathbf{e}_{12\ldots n} & 0 \\ 0 & \alpha-\beta\hat{\mathbf{e}}_{12\ldots n} \end{pmatrix}, \quad \alpha, \beta \in \mathbb{R}.$$

5 Vahlen 1902 originally wrote the second condition in the form

$$\text{(ii)}\quad \bar{a}b,\ b\bar{d},\ \bar{d}c,\ c\bar{a} \in \mathbb{R}^n,$$

which gives an equivalent characterization of the Vahlen group.

the off-diagonal entries must be even. Condition (iii) tells us that the pseudo-determinant $a\tilde{d}-b\tilde{c}$ is real and non-zero, in particular, that the Vahlen matrix is invertible.

The Vahlen matrices form a group under matrix multiplication, the Vahlen group. The Vahlen group has a normalized subgroup where condition (iii) is replaced by

(iii′) $a\tilde{d}-b\tilde{c}=\pm 1$.

The normalized Vahlen group is a four-fold, or rather double two-fold, covering group of the full Möbius group of $\mathbb{R}^n$; the identity Möbius transformation is represented by the following four matrices:

$$\pm\begin{pmatrix}1 & 0\\ 0 & 1\end{pmatrix},\quad \pm\begin{pmatrix}\mathbf{e}_{12\ldots n} & 0\\ 0 & -\hat{\mathbf{e}}_{12\ldots n}\end{pmatrix}.$$

The sense-preserving Möbius group has a non-trivial two-fold covering group formed by normalized Vahlen matrices with even diagonal (and odd off-diagonal) and pseudo-determinant equal to 1. The full Möbius group has a non-trivial two-fold covering group with two components, the non-identity component consisting of normalized Vahlen matrices with odd diagonal (and even off-diagonal) and pseudo-determinant equal to -1.

19.3 Opposite of a Euclidean space

Consider the $(n-1)$-dimensional real quadratic space $\mathbb{R}^{0,n-1}$ having a negative definite quadratic form

$$\mathbf{x}\to\mathbf{x}^2=-x_1^2-\ldots-x_{n-1}^2.$$

The sums of scalars and vectors are called *paravectors*. Paravectors span the linear space $\mathbb{R}\oplus\mathbb{R}^{0,n-1}$, which we denote by

$$\$R^n=\mathbb{R}\oplus\mathbb{R}^{0,n-1}.$$

The linear space of paravectors, $\$R^n$, can be made isometric to the Euclidean space $\mathbb{R}^n$ by introducing for $x=x_0+\mathbf{x}\in\mathbb{R}\oplus\mathbb{R}^{0,n-1}$, where $x_0\in\mathbb{R}$ and $\mathbf{x}\in\mathbb{R}^{0,n-1}$, a quadratic form

$$x=x_0+\mathbf{x}\to x\bar{x}=x_0^2-\mathbf{x}^2=x_0^2+x_1^2+\ldots+x_{n-1}^2.$$

As an extension of the Lipschitz group Porteous 1969 pp. 254-259 introduced the group of products of invertible paravectors, defined equivalently by

$$\$\Gamma_n=\{s\in C\ell_{0,n-1}\mid \forall x\in \$R^n,\ sx\hat{s}^{-1}\in \$R^n\}.$$

For a non-zero paravector $a\in \$R^n$ the mapping $x\to ax\bar{a}^{-1}$ is a rotation of

$\$R^n$. Thus, we have a group isomorphism $\$\Gamma_n \simeq \mathbf{\Gamma}_n^+$. Note that $\mathbf{\Gamma}_{0,n-1} \subset \$\Gamma_n$ and $\mathbf{\Gamma}_{0,n-1}^{\pm} = \$\Gamma_n^{\pm}$.

Vahlen originally considered the sense-preserving Möbius group of the paravector space $\$R^n$.

Definition (Vahlen 1902). [6] The matrix $\begin{pmatrix} a & b \\ c & d \end{pmatrix} \in \mathrm{Mat}(2, C\ell_{0,n-1})$ fulfilling the conditions

(i) $a, b, c, d \in \$\Gamma_n \cup \{0\}$
(ii) $\bar{a}b,\ b\bar{d},\ \bar{d}c,\ c\bar{a} \in \$R^n$
(iii) $a\tilde{d} - b\tilde{c} = 1$

is a *Vahlen matrix*, with pseudo-determinant or norm 1, of the sense-preserving Möbius transformation g of $\$R^n$ given by $g(x) = (ax + b)(cx + d)^{-1}$. ∎

These Vahlen matrices with norm 1 form a group, which is a non-trivial two-fold cover of the sense-preserving Möbius group of $\$R^n$.

19.4 Indefinite quadratic spaces

The full Möbius group of $\mathbb{R}^{p,q}$ contains two components (if either p or q is even) or four components (if both p and q are odd).

The identity component of the Möbius group of $\mathbb{R}^{p,q}$ is generated by rotations, translations, dilations and transversions which are represented, respectively, as follows:

$a\mathbf{x}a^{-1}$	$a \in \mathbf{Spin}_+(p,q)$	$\begin{pmatrix} a & 0 \\ 0 & a \end{pmatrix}$
$\mathbf{x} + \mathbf{b}$	$\mathbf{b} \in \mathbb{R}^{p,q}$	$\begin{pmatrix} 1 & \mathbf{b} \\ 0 & 1 \end{pmatrix}$
$\mathbf{x}\delta$	$\delta > 0$	$\begin{pmatrix} \sqrt{\delta} & 0 \\ 0 & 1/\sqrt{\delta} \end{pmatrix}$
$\dfrac{\mathbf{x} + \mathbf{x}^2\mathbf{c}}{1 + 2\mathbf{x}\cdot\mathbf{c} + \mathbf{x}^2\mathbf{c}^2}$	$\mathbf{c} \in \mathbb{R}^{p,q}$	$\begin{pmatrix} 1 & 0 \\ \mathbf{c} & 1 \end{pmatrix}$.

On the right we have the Vahlen matrices of the respective Möbius transformations.

6 Maass 1949 and Ahlfors 1984 presented an equivalent characterization of Vahlen matrices where the second condition was replaced by

$$\text{(ii)}\quad a\tilde{b},\ \tilde{b}d,\ d\tilde{c},\ \tilde{c}a \in \$R^n.$$

Theorem (J. Maks 1989). Consider four Vahlen matrices which represent a rotation, a translation, a dilation and a transversion. A product of these four matrices, in any order, always has an invertible entry in its diagonal (there are $4! = 24$ such products).

Proof. To complete the proof of the fact that a product of a rotation, a translation, a dilation and a transversion, in any order, is such that its Vahlen matrix always has an invertible entry in its diagonal, one can (or rather must) check the claim for all the 24 orderings. For instance, in the product

$$\begin{pmatrix} a & 0 \\ 0 & a \end{pmatrix}\begin{pmatrix} 1 & \mathbf{b} \\ 0 & 1 \end{pmatrix}\begin{pmatrix} \sqrt{\delta} & 0 \\ 0 & 1/\sqrt{\delta} \end{pmatrix}\begin{pmatrix} 1 & 0 \\ \mathbf{c} & 1 \end{pmatrix}$$
$$= \begin{pmatrix} a\sqrt{\delta} + a\mathbf{bc}/\sqrt{\delta} & a\mathbf{b}/\sqrt{\delta} \\ a\mathbf{c}/\sqrt{\delta} & a/\sqrt{\delta} \end{pmatrix}$$

the lower right-hand diagonal element $a/\sqrt{\delta}$ is invertible. We leave the verification of the remaining 23 orderings to the reader. ∎

Counter-example (Maks 1989). In the general case ($p \neq 0,\ q \neq 0$) J. Maks 1989 p. 41 gave an example of a Vahlen matrix where none of the entries is invertible (and all are non-zero).

Consider the Minkowski space-time $\mathbb{R}^{3,1}$ and its Clifford algebra $\mathcal{C}\ell_{3,1} \simeq \mathrm{Mat}(4,\mathbb{R})$ generated by $\mathbf{e}_1, \mathbf{e}_2, \mathbf{e}_3, \mathbf{e}_4$ satisfying $\mathbf{e}_1^2 = \mathbf{e}_2^2 = \mathbf{e}_3^2 = 1,\ \mathbf{e}_4^2 = -1$. Take a Vahlen matrix

$$M = \frac{1}{2}\begin{pmatrix} 1 + \mathbf{e}_{14} & \mathbf{e}_1 + \mathbf{e}_4 \\ -\mathbf{e}_1 + \mathbf{e}_4 & 1 - \mathbf{e}_{14} \end{pmatrix}.$$

By the theorem of Maks the matrix M cannot be a product of just one rotation, one translation, one dilation and one transversion (in any order). However, the matrix M is in the identity component of the normalized Vahlen group, the four-fold covering group of the Möbius group of the Minkowski space-time. This can be concluded while M has pseudo-determinant equal to 1 and even diagonal. This can also be deduced by factoring M into a product of a transversion, a translation and a transversion as follows:

$$M = \begin{pmatrix} 1 & 0 \\ \frac{1}{2}(-\mathbf{e}_1 + \mathbf{e}_4) & 1 \end{pmatrix}\begin{pmatrix} 1 & \frac{1}{2}(\mathbf{e}_1 + \mathbf{e}_4) \\ 0 & 1 \end{pmatrix}\begin{pmatrix} 1 & 0 \\ \frac{1}{2}(-\mathbf{e}_1 + \mathbf{e}_4) & 1 \end{pmatrix}.$$

Topologically, we can see this by connecting M to the identity matrix by the following path (here β grows from 0 to $\pi/4$):

$$M = M_{\pi/4}, \qquad M_\beta = \exp\left\{\beta\begin{pmatrix} 0 & \mathbf{e}_1 + \mathbf{e}_4 \\ -\mathbf{e}_1 + \mathbf{e}_4 & 0 \end{pmatrix}\right\}.$$

Maks' counter-example proves that condition (i) has to be modified in the definition of a Vahlen matrix. ∎

Recall that the Lipschitz group $\boldsymbol{\Gamma}_{p,q}$ consists of products of non-isotropic vectors of $\mathbb{R}^{p,q}$. In the sequel we need the set $\boldsymbol{\Pi}_{p,q}$ of products of vectors, possibly isotropic, of $\mathbb{R}^{p,q}$. The set $\boldsymbol{\Pi}_{p,q}$ is the closure of $\boldsymbol{\Gamma}_{p,q}$. [7]

Definition (Fillmore & Springer 1990). The matrix $\begin{pmatrix} a & b \\ c & d \end{pmatrix} \in \mathrm{Mat}(2, \mathcal{C}\ell_{p,q})$ fulfilling the conditions

(i) $a, b, c, d \in \boldsymbol{\Pi}_{p,q}$
(ii) $\bar{a}b,\ b\bar{d},\ \bar{d}c,\ c\bar{a} \in \mathbb{R}^{p,q}$
(iii) $a\tilde{d} - b\tilde{c} \in \mathbb{R} \setminus \{0\}$

is a *Vahlen matrix* of the Möbius transformation g of $\mathbb{R}^{p,q}$ given by $g(\mathbf{x}) = (a\mathbf{x} + b)(c\mathbf{x} + d)^{-1}$. ∎

The Vahlen matrices form a group under matrix multiplication, the *Vahlen group*. The normalized Vahlen matrices, with pseudo-determinant satisfying $a\tilde{d} - b\tilde{c} = \pm 1$, form a four-fold, possibly trivial, covering group of the full Möbius group of $\mathbb{R}^{p,q}$. When both p and q are odd, the normalized Vahlen group is a non-trivial four-fold covering group of the full Möbius group of $\mathbb{R}^{p,q}$. When either p or q is even, we may find a non-trivial two-fold covering group of the full Möbius group of $\mathbb{R}^{p,q}$. It consists of the identity component of the normalized Vahlen group, that is, normalized Vahlen matrices with even diagonal and pseudo-determinant equal to 1, and another component not containing the (non-trivial) pre-images of the identity:

$$\pm \begin{pmatrix} \mathbf{e}_{12\ldots n} & 0 \\ 0 & -\hat{\mathbf{e}}_{12\ldots n} \end{pmatrix}.$$

The identity component of the normalized Vahlen group is a two-fold (either p or q is even) or four-fold (both p and q are odd) covering group of the sense-preserving Möbius group.

Conditions (i), (iii) and $\bar{a}b,\ b\bar{d},\ \bar{d}c,\ c\bar{a} \in \mathbb{R}^{p,q}$ imply $a\tilde{b},\ \tilde{b}d,\ d\tilde{c},\ \tilde{c}a \in \mathbb{R}^{p,q}$. In contrast to the Euclidean case, conditions (i), (iii) and $a\tilde{b},\ \tilde{b}d,\ d\tilde{c},\ \tilde{c}a \in \mathbb{R}^{p,q}$ do not imply $\bar{a}b,\ b\bar{d},\ \bar{d}c,\ c\bar{a} \in \mathbb{R}^{p,q}$.

Counter-example (Cnops 1996). Consider the Minkowski space-time $\mathbb{R}^{3,1}$ and its Clifford algebra $\mathcal{C}\ell_{3,1} \simeq \mathrm{Mat}(4, \mathbb{R})$ generated by $\mathbf{e}_1$, $\mathbf{e}_2$, $\mathbf{e}_3$, $\mathbf{e}_4$ satisfying $\mathbf{e}_1^2 = \mathbf{e}_2^2 = \mathbf{e}_3^2 = 1$, $\mathbf{e}_4^2 = -1$. The Vahlen matrix

$$C = \frac{1}{2}\begin{pmatrix} 1 + \mathbf{e}_{14} & (\mathbf{e}_1 + \mathbf{e}_4)\mathbf{e}_{23} \\ (-\mathbf{e}_1 + \mathbf{e}_4)\mathbf{e}_{23} & 1 - \mathbf{e}_{14} \end{pmatrix}$$

7 The set $\boldsymbol{\Pi}_{p,q} \subset \mathcal{C}\ell_{p,q} \simeq \bigwedge \mathbb{R}^n$, considered as a subset of the exterior algebra $\bigwedge \mathbb{R}^n$, is independent of p, q for a fixed $n = p + q$.

satisfies $a, b, c, d \in \mathbf{\Pi}_{3,1}$, $a\tilde{d} - b\tilde{c} = 1$ and $a\tilde{b}$, $\tilde{b}d$, $d\tilde{c}$, $\tilde{c}a = 0 \in \mathbb{R}^{3,1}$, but even then $\bar{a}b$, $b\bar{d}$, $\bar{d}c$, $c\bar{a} \notin \mathbb{R}^{3,1}$. The mapping $g_C(\mathbf{x}) = (a\mathbf{x} + b)(c\mathbf{x} + d)^{-1}$ is conformal. If the matrix C is multiplied on either side by

$$D = \frac{1}{\sqrt{2}} \begin{pmatrix} 1 + \mathbf{e}_{1234} & 0 \\ 0 & 1 - \mathbf{e}_{1234} \end{pmatrix},$$

then $B = CD = DC$ is such that $g_B(\mathbf{x}) = g_C(\mathbf{x})$ for almost all $\mathbf{x} \in \mathbb{R}^{3,1}$. Furthermore, B does satisfy $\bar{a}b$, $b\bar{d}$, $\bar{d}c$, $c\bar{a} \in \mathbb{R}^{3,1}$.

The matrices satisfying $a, b, c, d \in \mathbf{\Pi}_{3,1}$, $a\tilde{d} - b\tilde{c} = 1$ and $a\tilde{b}$, $\tilde{b}d$, $d\tilde{c}$, $\tilde{c}a \in \mathbb{R}^{3,1}$ do not form a group, but only a set which is not closed under multiplication. This set generates a group which is the Vahlen group with norm 1 multiplied by the group consisting of the matrices

$$\begin{pmatrix} \cos\varphi + \mathbf{e}_{1234} \sin\varphi & 0 \\ 0 & \cos\varphi - \mathbf{e}_{1234} \sin\varphi \end{pmatrix}.$$

All these matrices are pre-images of the identity Möbius transformation. ∎

19.5 Indefinite paravectors

Let $\$\Pi_{q+1,p}$ be the set of products of paravectors in $\$R^{q+1,p} = \mathbb{R} \oplus \mathbb{R}^{p,q}$.

Definition. The matrix $\begin{pmatrix} a & b \\ c & d \end{pmatrix} \in \mathrm{Mat}(2, \mathcal{C}\ell_{p,q})$ fulfilling the conditions

(i) $a, b, c, d \in \$\Pi_{q+1,p}$
(ii) $\bar{a}b$, $b\bar{d}$, $\bar{d}c$, $c\bar{a} \in \$R^{q+1,p}$
(iii) $a\tilde{d} - b\tilde{c} = 1$

is a *Vahlen matrix* with norm 1 of the sense-preserving Möbius transformation g of $\$R^{q+1,p}$ given by $g(x) = (ax + b)(cx + d)^{-1}$. ∎

The Vahlen matrices with norm 1 form a two-fold or four-fold covering group of the sense-preserving Möbius group of $\$R^{q+1,p}$. Conditions (i), (ii), (iii) imply $a\tilde{b}$, $\tilde{b}d$, $d\tilde{c}$, $\tilde{c}a \in \$R^{q+1,p}$ [although (i), (iii) and $a\tilde{b}$, $\tilde{b}d$, $d\tilde{c}$, $\tilde{c}a \in \$R^{q+1,p}$ do not imply $\bar{a}b$, $b\bar{d}$, $\bar{d}c$, $c\bar{a} \in \$R^{q+1,p}$].

19.6 The derivative of a Möbius transformation

The difference of the Möbius transformations of x, y in $\$R^{q+1,p}$ is given by

$$g(x) - g(y) = (cy + d)^{\sim -1}(x - y)(cx + d)^{-1}.$$

Letting x approach y we may compute the derivative of a Möbius transformation. Denoting $z = cx + d$ and using $N(z) = z\bar{z} \in \mathbb{R}$, we see that in the case $N(z) \neq 0$ the derivative of $x \to g(x)$ is the composition of the rotation

$$x \to \hat{z}xz^{-1}$$

and the dilation

$$x \to \frac{x}{N(z)}.$$

19.7 The Lie algebra of the Vahlen group

If the matrix $\begin{pmatrix} A & B \\ C & D \end{pmatrix} \in \mathrm{Mat}(2, C\ell_n)$ is in the Lie algebra of the Vahlen group of $\mathbb{R}^n$, then $A\mathbf{x} + B - \mathbf{x}C\mathbf{x} - \mathbf{x}D \in \mathbb{R}^n$ for all $\mathbf{x} \in \mathbb{R}^n$. It follows that $B, C \in \mathbb{R}^n$ and $A, D \in \mathbb{R} \oplus \bigwedge^2 \mathbb{R}^2 \oplus \bigwedge^n \mathbb{R}^n$ so that $\langle A \rangle_2 = \langle D \rangle_2$ and $\langle A \rangle_n = \pm \langle D \rangle_n$. Actually, for the Lie algebra of the Vahlen group $\langle A \rangle_n$, $\langle D \rangle_n$ vanish and for the Lie algebra of the normalized Vahlen group $\langle A \rangle_0 = -\langle D \rangle_0$. In fact, matrices in the Lie algebra of the normalized Vahlen group can be characterized by

(i) $A, D \in \mathbb{R} \oplus \bigwedge^2 \mathbb{R}^n$
(ii) $B, C \in \mathbb{R}^n$
(iii) $A + \tilde{D} = 0$.

The Lie algebra is spanned by the matrices

$$L_{\mu\nu} = \begin{pmatrix} -\frac{1}{2}\mathbf{e}_{\mu\nu} & 0 \\ 0 & -\frac{1}{2}\mathbf{e}_{\mu\nu} \end{pmatrix}, \quad P_\mu = \begin{pmatrix} 0 & \mathbf{e}_\mu \\ 0 & 0 \end{pmatrix},$$

$$D = \begin{pmatrix} \frac{1}{2} & 0 \\ 0 & \frac{1}{2} \end{pmatrix}, \quad K_\mu = \begin{pmatrix} 0 & 0 \\ \mathbf{e}_\mu & 0 \end{pmatrix}.$$

These matrices represent rotations, translations, dilations and transversions.

19.8 Compactification and the isotropic cone at infinity

The set of Möbius mappings on $\mathbb{R}^{p,q}$ can be used to compactify $\mathbb{R}^{p,q}$. The compactification is homeomorphic to

$$\frac{S^p \times S^q}{\mathbb{Z}_2}.$$

In particular, the compactification of a Euclidean space $\mathbb{R}^n$ is the sphere S^n, and the compactification of the hyperbolic plane $\mathbb{R}^{1,1}$ is the torus $S^1 \times S^1$. The conformal compactification adjoins an isotropic cone at infinity to the quadratic space.

Questions

1. Is an element in the identity component of the conformal group necessarily a product of a rotation, a translation, a dilation and a transversion?
2. The group $SU(2,2)$ is a covering group of the identity component of the conformal group of $\mathbb{R}^{1,3}$. Is it a two-fold or a four-fold covering group?

Answers

1. No, as the counter-example of Maks shows.
2. As the identity component of the normalized Vahlen group it is a four-fold covering group.

Exercises

1. The counter-example M of Maks can be factored into a product of two 'diversions':
$$M = \frac{1}{\sqrt{2}}\begin{pmatrix} 1 & \mathbf{e}_1 \\ -\mathbf{e}_1 & 1 \end{pmatrix} \frac{1}{\sqrt{2}}\begin{pmatrix} 1 & \mathbf{e}_4 \\ \mathbf{e}_4 & 1 \end{pmatrix}.$$
Show that a 'diversion' is a product of just one transversion, one dilation, one translation and one rotation.
2. Show that in the case of a Euclidean space $\mathbb{R}^n$ the conditions $a\tilde{b}, \tilde{b}d, d\tilde{c}, \tilde{c}a \in \mathbb{R}^n$ and $\bar{a}b, b\bar{d}, \bar{d}c, c\bar{a} \in \mathbb{R}^n$ are equivalent.
3. Show that the conformal compactification of the Minkowksi space $\mathbb{R}^{1,3}$ is homeomorphic to $U(2)$.

Solutions

1. The first factor is a product of just one transversion, one dilation and one translation as follows:
$$\frac{1}{\sqrt{2}}\begin{pmatrix} 1 & \mathbf{e}_1 \\ -\mathbf{e}_1 & 1 \end{pmatrix} = \begin{pmatrix} 1 & 0 \\ -\mathbf{e}_1 & 1 \end{pmatrix}\begin{pmatrix} 1/\sqrt{2} & 0 \\ 0 & \sqrt{2} \end{pmatrix}\begin{pmatrix} 1 & \mathbf{e}_1 \\ 0 & 1 \end{pmatrix}.$$
One can insert the identity rotation as the last factor.
2. For $a \in \mathbf{\Gamma}_n$, $\bar{a} = \tilde{a}$. If $\tilde{a}b \in \mathbb{R}^n$, then we have two cases to consider: either a is zero, and so $a\tilde{b}$ is a vector, or a is in the Lipschitz group $\mathbf{\Gamma}_n$, but then $a(\tilde{a}b)\tilde{a}$ is a vector and $a\tilde{a} \in \mathbb{R} \setminus \{0\}$, and so $b\tilde{a}$ is a vector, which implies $\tilde{a}b = (b\tilde{a})^{\sim} \in \mathbb{R}^n$.
3. This follows as a special case from the matrix isomorphism
$$U(n) \simeq \frac{SU(n) \times U(1)}{\mathbb{Z}_n}.$$

Bibliography

L.V. Ahlfors: Old and new in Möbius groups. *Ann. Acad. Sci. Fenn.* Ser. A I Math. **9** (1984), 93-105.

L.V. Ahlfors: Möbius transformations and Clifford numbers; pp. 65-73 in I. Chavel, H.M. Farkas (eds.): *Differential Geometry and Complex Analysis.* Springer, Berlin, 1985.

L.V. Ahlfors: Möbius transformations in $\mathbb{R}^n$ expressed through 2×2 matrices of Clifford numbers. *Complex Variables Theory Appl.* **5** (1986), 215-224.

P. Anglès: Construction de revêtements du groupe conforme d'un espace vectoriel muni d'une 'métrique' de type (p, q). *Ann. Inst. H. Poincaré* Sect. A**33** (1985), 33-51.

P. Anglès: Real conformal spin structures on manifolds. *Studia Scientiarum Mathematicarum Hungarica* **23** (1988), 11-139.

J. Cnops: Vahlen matrices for non-definite metrics; pp. 155-164 in R. Abłamowicz, P. Lounesto, J.M. Parra (eds.): *Clifford Algebras with Numeric and Symbolic Computations*, Birkhäuser, Boston, MA, 1996.

A. Crumeyrolle: *Orthogonal and Symplectic Clifford Algebras, Spinor Structures.* Kluwer, Dordrecht, The Netherlands, 1990. [Conformal group on pages 149-164.]

R. Deheuvels: Groupes conformes et algèbres de Clifford. *Rend. Sem. Mat. Univ. Politec. Torino* **43** (1985), 205-226.

J. Elstrodt, F. Grunewald, J. Mennicke: Vahlen's groups of Clifford matrices and spin-groups. *Math. Z.* **196** (1987), 369-390.

J.P. Fillmore, A. Springer: Möbius groups over general fields using Clifford algebras associated with spheres. *Internat. J. Theoret. Phys.* **29** (1990), 225-246.

J. Gilbert, M. Murray: *Clifford Algebras and Dirac Operators in Harmonic Analysis.* Cambridge Studies in Advanced Mathematics, Cambridge University Press, Cambridge, 1991. [Conformal transformations on pp. 37-38, 278-296.]

J. Haantjes: Conformal representations of an n-dimensional euclidean space with a non-definite fundamental form on itself. *Nederl. Akad. Wetensch. Proc.* **40** (1937), 700-705.

J. Liouville: Extension au cas de trois dimensions de la question du trace géographique; pp. 609-616 in G. Monge (ed.): *Application de l'analyse à la géometrie*, Paris, 1850.

P. Lounesto, A. Springer: Möbius transformations and Clifford algebras of Euclidean and anti-Euclidean spaces; pp. 79-90 in J. Lawrynowicz (ed.): *Deformations of Mathematical Structures.* Kluwer, Dordrecht, The Netherlands, 1989.

H. Maass: Automorphe Funktionen von mehreren Veränderlichen und Dirichletsche Reihen. *Abh. Math. Sem. Univ. Hamburg* **16** (1949), 72-100.

J. Maks: *Modulo (1,1) periodicity of Clifford algebras and the generalized (anti-)Möbius transformations.* Thesis, Technische Universiteit Delft, 1989.

J. Maks: Clifford algebras and Möbius transformations; pp. 57-63 in A. Micali et al. (eds.): *Clifford Algebras and their Applications in Mathematical Physics.* Kluwer, Dordrecht, The Netherlands, 1992.

J. Ryan: Conformal Clifford manifolds arising in Clifford analysis. *Proc. Roy. Irish Acad.* Sect. A**85** (1985), 1-23.

J. Ryan: Clifford matrices, Cauchy-Kowalewski extension and analytic functionals. *Proc. Centre Math. Anal. Austral. Nat. Univ.* **16** (1988), 284-299.

K. Th. Vahlen: Über Bewegungen und complexe Zahlen. *Math. Ann.* **55** (1902), 585-593.

20
Hypercomplex Analysis

Complex analysis has applications in the theory of heat, fluid dynamics and electrostatics. Such versatility gives occasion to explore whether function theory of complex variables can be generalized from the plane to higher dimensions. Are there hypercomplex number systems which could provide a higher-dimensional analog for complex analytic functions?

Function theory can be generalized to higher dimensions in several different ways, for instance, to quasiconformal mappings, several complex variables or to hypercomplex analysis. Clearly, these generalizations cannot maintain all the features of complex analysis.

In the theory of quasiconformal mappings one retains some geometric features, related to similar appearance of images, and renounces some algebraic features, like multiplication of complex numbers. In the theory of quasiconformal mappings one does not multiply vectors in $\mathbb{R}^n$.

The starting point of hypercomplex analysis is the introduction of a suitable multiplication of vectors in $\mathbb{R}^n$. In contrast to the theory of several complex variables, which commute, hypercomplex analysis is a one-variable theory – the argument being in higher dimensions, where orthogonal vectors anticommute.

20.1 Formulation of complex analysis in $\mathcal{C}\ell_2$

For a complex valued function $u+iv = f(x+iy)$ of the complex variable $x+iy$ the *Cauchy-Riemann equations* are

$$\frac{\partial u}{\partial x} = \frac{\partial v}{\partial y}, \qquad \frac{\partial u}{\partial y} = -\frac{\partial v}{\partial x}.$$

The second equation tells us that the vector $(u, -v)$ is the gradient of a function $\phi : \mathbb{R}^2 \to \mathbb{R}$:

$$u = \frac{\partial \phi}{\partial x}, \qquad -v = \frac{\partial \phi}{\partial y}. \tag{1}$$

Using the first relation of the Cauchy-Riemann equations we obtain

$$\frac{\partial^2 \phi}{\partial x^2} + \frac{\partial^2 \phi}{\partial y^2} = \frac{\partial u}{\partial x} - \frac{\partial v}{\partial y} = 0,$$

that is, ϕ is a harmonic function, $\nabla^2 \phi = 0$. Conversely, if ϕ is harmonic, then u and v defined by the relation (1) satisfy the Cauchy-Riemann equations.

The Cauchy-Riemann equations can be condensed into a single equation as follows:

$$\Big(\frac{\partial}{\partial x} + i\frac{\partial}{\partial y}\Big)(u + iv) = 0.$$

Recall that $i = \mathbf{e}_1\mathbf{e}_2$ and multiply this equation on the left and on the right by $\mathbf{e}_1$, then use associativity and anticommutativity to get

$$\Big(\mathbf{e}_1\frac{\partial}{\partial x} + \mathbf{e}_2\frac{\partial}{\partial y}\Big)(\mathbf{e}_1 u - \mathbf{e}_2 v) = 0.$$

As we know, this relation holds if and only if the vector $(u, -v)$ is the gradient of a harmonic function. It follows that

$$\Big(\mathbf{e}_1\frac{\partial}{\partial x} + \mathbf{e}_2\frac{\partial}{\partial y}\Big)(\mathbf{e}_1 u + \mathbf{e}_2 v) = 0 \tag{2}$$

if and only if (u, v) is the gradient of a harmonic function.

There are three possible ways to formulate the Cauchy-Riemann equations employing the Clifford algebra $\mathcal{C}\ell_2$ (these possibilities will be generalized to higher dimensions in three different ways).

1) Firstly, we may consider the Cauchy-Riemann equations to be a condition on *vector fields*, sending a vector $x\mathbf{e}_1 + y\mathbf{e}_2$ in $\mathbb{R}^2$ to a vector $u\mathbf{e}_1 + v\mathbf{e}_2$ in $\mathbb{R}^2$. The above condition (2),

$$(i) \qquad \Big(\mathbf{e}_1\frac{\partial}{\partial x} + \mathbf{e}_2\frac{\partial}{\partial y}\Big)(u\mathbf{e}_1 + v\mathbf{e}_2) = 0,$$

gives us Cauchy-Riemann equations up to sign and results in those conformal maps which reverse the orientation of $\mathbb{R}^2$. In higher dimensions this alternative means the study of those vector fields, that is, mappings from $\mathbb{R}^n$ to $\mathbb{R}^n$, which are gradients of harmonic functions, mappings from $\mathbb{R}^n$ to $\mathbb{R}$.

2) Secondly, we may reformulate the Cauchy-Riemann equations as a condition

on the *even fields* sending a vector $x\mathbf{e}_1 + y\mathbf{e}_2$ in $\mathbb{R}^2$ to an even element in $C\ell_2^+ = \{u + v\mathbf{e}_1\mathbf{e}_2 \mid u, v \in \mathbb{R}\} \simeq \mathbb{C}$ of the Clifford algebra $C\ell_2$. The condition

$$(ii) \qquad \left(\mathbf{e}_1\frac{\partial}{\partial x} + \mathbf{e}_2\frac{\partial}{\partial y}\right)(u + v\mathbf{e}_{12}) = 0$$

gives us the Cauchy-Riemann equations. This alternative has non-trivial generalizations in higher dimensions sending the vector space $\mathbb{R}^n$ to the even subalgebra $C\ell_n^+$ of the Clifford algebra $C\ell_n$.

3) Thirdly, we may focus our attention on *spinor fields* sending the vector plane $\mathbb{R}^2$ to a minimal left ideal of $C\ell_2$. Before studying this alternative closer, let us recall that the Clifford algebra $C\ell_2$ is isomorphic to the matrix algebra of real 2×2-matrices $\mathrm{Mat}(2, \mathbb{R})$. The isomorphism is seen by the correspondences

$$1 \simeq \begin{pmatrix} 1 & 0 \\ 0 & 1 \end{pmatrix}$$

$$\mathbf{e}_1 \simeq \begin{pmatrix} 1 & 0 \\ 0 & -1 \end{pmatrix}, \qquad \mathbf{e}_2 \simeq \begin{pmatrix} 0 & 1 \\ 1 & 0 \end{pmatrix}$$

$$\mathbf{e}_{12} \simeq \begin{pmatrix} 0 & 1 \\ -1 & 0 \end{pmatrix}.$$

In this case one sends a vector $x\mathbf{e}_1 + y\mathbf{e}_2$ in $\mathbb{R}^2$ to a spinor

$$uf_1 + vf_2 \simeq \begin{pmatrix} u & 0 \\ v & 0 \end{pmatrix},$$

where

$$f_1 = \frac{1}{2}(1 + \mathbf{e}_1) \simeq \begin{pmatrix} 1 & 0 \\ 0 & 0 \end{pmatrix} \quad \text{and} \quad f_2 = \frac{1}{2}(\mathbf{e}_2 - \mathbf{e}_{12}) \simeq \begin{pmatrix} 0 & 0 \\ 1 & 0 \end{pmatrix}.$$

Here $f_1^2 = f_1$, so f_1 is an *idempotent*, and the spinor space $S = C\ell_2 f_1 = \{af_1 \mid a \in C\ell_2\}$ is a *left ideal* of $C\ell_2$, for which $a\psi \in S$ for all $a \in C\ell_2$ and $\psi \in S$. Since

$$\mathbf{e}_1 f_1 = f_1, \quad \mathbf{e}_1 f_2 = -f_2$$
$$\mathbf{e}_2 f_1 = f_2, \quad \mathbf{e}_2 f_2 = f_1$$

one verifies that

$$(iii) \qquad \left(\mathbf{e}_1\frac{\partial}{\partial x} + \mathbf{e}_2\frac{\partial}{\partial y}\right)(uf_1 + vf_2) = 0$$

is equivalent to (i).

To summarize, there are three alternatives for the 2-dimensional target:

(i) the Euclidean vector space itself $\mathbb{R}^2$,
(ii) the even subalgebra $C\ell_2^+$ of the Clifford algebra $C\ell_2$,
(iii) the spinor space $S = C\ell_2 f_1$.

In the next section, we will generalize, as a preliminary construction, the first alternative to higher dimensions.

20.2 Vector fields

The Dirac operator. It is possible to extract a certain kind of square root of the n-dimensional Laplace-operator

$$\nabla^2 = \frac{\partial^2}{\partial x_1^2} + \frac{\partial^2}{\partial x_2^2} + \cdots + \frac{\partial^2}{\partial x_n^2}$$

and consider instead a first-order differential operator

$$\nabla = \mathbf{e}_1 \frac{\partial}{\partial x_1} + \mathbf{e}_2 \frac{\partial}{\partial x_2} + \cdots + \mathbf{e}_n \frac{\partial}{\partial x_n}$$

called the *Dirac operator.* Since the Dirac operator applied twice equals the Laplace operator, the elements $\mathbf{e}_1, \mathbf{e}_2, \ldots, \mathbf{e}_n$ are subject to the relations

$$\begin{aligned} \mathbf{e}_i^2 &= 1, \qquad & i &= 1, 2, \ldots, n, \\ \mathbf{e}_i \mathbf{e}_j &= -\mathbf{e}_j \mathbf{e}_i, & i &< j. \end{aligned}$$

The linear combinations $\mathbf{x} = x_1 \mathbf{e}_1 + x_2 \mathbf{e}_2 + \cdots + x_n \mathbf{e}_n$ can be considered as vectors building up an n-dimensional vector space $\mathbb{R}^n$ with quadratic form $\mathbf{x}^2 = x_1^2 + x_2^2 + \cdots + x_n^2$. The above relations generate an associative algebra of dimension 2^n, the Clifford algebra $C\ell_n$ of $\mathbb{R}^n$ [or of dimension $\frac{1}{2}2^n$, isomorphic to an ideal $\frac{1}{2}(1 \pm \mathbf{e}_{12\ldots n})C\ell_n$ of $C\ell_n$].

Operating on a vector field $\mathbf{f}$ with ∇ gives

$$\nabla \mathbf{f} = \nabla \cdot \mathbf{f} + \nabla \wedge \mathbf{f}$$

where $\nabla \cdot \mathbf{f}$ is the *divergence* of $\mathbf{f}$ and $\nabla \wedge \mathbf{f}$ is the *curl*, which in this approach is bivector valued.

Sourceless and irrotational vector fields. Consider a steady motion of incompressible fluid in an n-dimensional Euclidean space $\mathbb{R}^n$. Represent the velocity of the flow by the vector field $\mathbf{f}$. The integral

$$\Psi = \int_S d\mathbf{S} \wedge \mathbf{f}$$

over an orientable hypersurface S, $\dim S = n - 1$, is the *stream* across S. We regard $d\mathbf{S}$ as a tangent $(n-1)$-vector measure, rather than the normal vector measure; this makes the stream n-vector valued. If a vector field $\mathbf{f}$ is *sourceless*, $\nabla \cdot \mathbf{f} = 0$, its stream across S depends in a contractible domain only on the boundary ∂S of S. In particular, no stream emerges through a closed hypersurface. If $n \geq 2$, a sourceless vector field $\mathbf{f}$ has a bivector valued potential $\mathbf{v}$ such that $\mathbf{f} = \nabla \rfloor \mathbf{v}$. If $n \geq 3$, the bivector potential can be subjected to a supplementary condition $\nabla \wedge \mathbf{v} = 0$, in which case $\mathbf{f} = \nabla \mathbf{v}$.

The *circulation* of the vector field $\mathbf{f}$ around a closed path C is given by the line integral

$$\int_C d\mathbf{x} \cdot \mathbf{f}.$$

If a vector field $\mathbf{f}$ is *irrotational*, $\nabla \wedge \mathbf{f} = 0$, the circulation vanishes in a simply connected domain, and the line integral

$$u(\mathbf{x}) = -\int_{\mathbf{x}_0}^{\mathbf{x}} \mathbf{f} \cdot d\mathbf{x}$$

is independent of path. The function u is called the scalar potential of $\mathbf{f}$. The irrotational vector field $\mathbf{f}$ is the gradient of its scalar potential u, $\mathbf{f} = -\nabla u$.

If a vector field $\mathbf{f}$ is sourceless and irrotational, that is $\nabla \cdot \mathbf{f} = 0$ and $\nabla \wedge \mathbf{f} = 0$, its scalar potential is harmonic, $\nabla^2 u = 0$. A vector field $\mathbf{f}$ is called *monogenic*, if $\nabla \mathbf{f} = 0$. For a monogenic vector field $\nabla \mathbf{f} = \nabla \cdot \mathbf{f} + \nabla \wedge \mathbf{f} = 0$, and so it is sourceless and irrotational. A monogenic vector field has a potential which is a sum or complex of the scalar and bivector potentials: $w = u + \mathbf{v}$. The complex of potentials is also monogenic, $\nabla w = 0$.

Example. A monogenic vector field $\mathbf{f}$, homogeneous of degree ℓ, has a scalar potential

$$u = -\frac{\mathbf{x} \cdot \mathbf{f}}{\ell + 1}, \quad \ell \neq -1,$$

and a bivector potential

$$\mathbf{v} = \frac{\mathbf{x} \wedge \mathbf{f}}{\ell + n - 1}, \quad \ell \neq -(n-1).$$

In the singular case, a monogenic vector field $\mathbf{f}$, homogeneous of degree $\ell = -(n-1)$, might still have a bivector/complex potential. For instance, the Cauchy kernel $\mathbf{q}(\mathbf{x}) = \mathbf{x}/r^n$, $r = |\mathbf{x}|$, has a complex potential

$$-\log r + \mathbf{i}\theta \quad \text{for} \quad n = 2,$$

$$\frac{1}{r}(1 + \mathbf{i} \tan \frac{\theta}{2}) \quad \text{for} \quad n = 3,$$

where θ is the angle between $\mathbf{x}$ and a fixed direction $\mathbf{a}$, and $\mathbf{i}$ is the imaginary unit of the plane $\mathbf{x} \wedge \mathbf{a}$, given by $\mathbf{i} = \mathbf{x} \wedge \mathbf{a}/|\mathbf{x} \wedge \mathbf{a}|$. ∎

The plane case. In the plane the stream is a bivector valued line integral. If the plane vector field is sourceless, its stream across any line in a simply connected domain depends only on the two end points of the line. Integrating from a fixed point $\mathbf{x}_0$ to a variable point $\mathbf{x}$ the stream becomes a bivector valued function, called the stream function,

$$\psi = \int_{\mathbf{x}_0}^{\mathbf{x}} d\mathbf{x} \wedge \mathbf{f},$$

which can serve as the bivector potential of $\mathbf{f} = \nabla \rfloor \psi$ $[\mathbf{v} = \psi]$.

If $\mathbf{f}$ is monogenic, $\nabla \mathbf{f} = 0$, then there is an even valued function

$$w = u + \psi = -\int_{\mathbf{x}_0}^{\mathbf{x}} d\mathbf{x} \cdot \mathbf{f} + \int_{\mathbf{x}_0}^{\mathbf{x}} d\mathbf{x} \wedge \mathbf{f} = -\int_{\mathbf{x}_0}^{\mathbf{x}} \mathbf{f}\, d\mathbf{x},$$

which serves as the complex potential of $\mathbf{f}$. This complex potential is also monogenic, $\nabla w = 0$. (In higher dimensions there is no correspondence for such line integrals representing complex potentials – unless one is confined to axially symmetric vector fields.) ∎

Even fields. Instead of vector fields, we could instead examine the even fields

$$\mathbb{R}^n \to C\ell_n^+, \quad \mathbf{x} \to f(\mathbf{x}).$$

Here we replace the target $\mathbb{R}^n$ of dimension n by a wider target $C\ell_n^+$ of dimension $\frac{1}{2}2^n$. The even subalgebra $C\ell_n^+$ is a direct sum of the k-vector spaces $\bigwedge^k \mathbb{R}^n$ with even k. If we require the even functions to be monogenic, $\nabla f(\mathbf{x}) = 0$, then we have a system of coupled equations:

$$\nabla \wedge f_{k-2} + \nabla \rfloor f_k = 0, \quad \nabla \wedge f_k + \nabla \rfloor f_{k+2} = 0,$$

where f_k is the homogeneous part of degree k of $f = f(\mathbf{x})$. These equations are invariant under the rotation group $SO(n)$.

Irreducible fields. Instead of vector fields or even fields, we could examine functions with values in an irreducible representation of $SO(n)$ or $\mathbf{Spin}(n)$. In the Clifford algebra realm this would mean studying k-vector fields or spinor fields. Important physical fields fall into this category: the Maxwell equations are of the form $\nabla \wedge \mathbf{F} = 0$, $\nabla \rfloor \mathbf{F} = \mathbf{J}$, where $\mathbf{F} \in \bigwedge^2 \mathbb{R}^{3,1}$, and the Dirac field has as its target the spinor space, a minimal left ideal of $\mathbb{C} \otimes C\ell_{3,1}$. It should be emphasized though that in modern treatment of the Dirac theory the spinor space is replaced by the even subalgebra $C\ell_{3,1}^+$. ∎

20.3 Tangential integration

Here we consider integration over surfaces, that is, smooth manifolds embedded in a linear space $\mathbb{R}^n$. Our surface S is compact, connected, orientable and contractible. The surface S is k-dimensional if there are k linearly independent vectors tangent to S at each point $\mathbf{x}$ of S. The tangent vectors span a tangent space $T_{\mathbf{x}}$.

In a Euclidean space $\mathbb{R}^n$ the tangent space $T_{\mathbf{x}}$ generates a tangent algebra $C\ell(T_{\mathbf{x}})$ isomorphic to $C\ell_k$. A multivector field on S is a smooth function $f : S \to C\ell_n$; it is *tangential* if $f(\mathbf{x}) \in C\ell(T_{\mathbf{x}})$ for each $\mathbf{x} \in S$.

There are exactly two continuous tangential unit k-vector fields on an orientable k-dimensional surface S, each corresponding to one of the two orientations attached to S. So tangent to each point $\mathbf{x}$ of an oriented k-dimensional surface S there is a unique unit k-vector $\tau(\mathbf{x})$ characterizing the orientation of S at $\mathbf{x} \in S$. The value of the map τ at $\mathbf{x}$ is called the *tangent* of S at $\mathbf{x}$.

Consider a multivector field $f : S \to C\ell_n$ on a k-dimensional surface $S \subset \mathbb{R}^n$, $1 \leq k \leq n$. Define the *tangential integral* of f over S by

$$\int_S d\mathbf{S}\, f(\mathbf{x}) = \int_S \tau(\mathbf{x}) f(\mathbf{x})\, dV,$$

where dV is the usual scalar measure of the k-dimensional volume element of S and $d\mathbf{S}$ is the k-vector valued *tangential measure*,

$$d\mathbf{S} = \tau(\mathbf{x})\, dV.$$

So the tangential integral of $f(\mathbf{x})$ is equivalent to the usual (Riemann) integral of $\tau(\mathbf{x})f(\mathbf{x})$ over S.

Since multiplication of multivectors is not commutative, the above equation is not the most general form for a tangential integral. The appropriate generalization is the following:

$$\int_S g(\mathbf{x})\, d\mathbf{S}\, f(\mathbf{x}) = \int_S g(\mathbf{x})\tau(\mathbf{x})f(\mathbf{x})\, dV.$$

Consider an oriented k-dimensional surface S with boundary ∂S of dimension $k-1$. Set at the point $\mathbf{x} \in \partial S$ a tangent $\tau_{\partial S}(\mathbf{x})$ of the boundary ∂S and a tangent $\tau_S(\mathbf{x})$ of the surface S. Then the expression $(\tau_S(\mathbf{x}))^{-1}\tau_{\partial S}(\mathbf{x})$ is a vector normal to the boundary ∂S. There are two alternatives: the normal vector points inwards, in which case it is tangent to the surface S, or outwards, in which case it is opposite to the inward tangent of S. The orientations of S and ∂S are *compatible*, when the vector $(\tau_S(\mathbf{x}))^{-1}\tau_{\partial S}(\mathbf{x})$ points outwards. Define the *normal integral* of f over the boundary ∂S by

$$\int_{\partial S} \partial \mathrm{s}\, f(\mathbf{x}) = \int_{\partial S} (\tau_S(\mathbf{x}))^{-1}\, d\mathrm{s}\, f(\mathbf{x})$$

where $d\mathrm{s}$ is the $(k-1)$-vector valued tangential measure on the boundary ∂S, and $\partial \mathrm{s}$ is an outward pointing vector normal to ∂S,

$$\partial \mathrm{s} = (\tau_S(\mathbf{x}))^{-1}\, d\mathrm{s} \quad \text{for} \quad \mathbf{x} \in \partial S.$$

The Dirac operator without coordinates. Take a k-dimensional surface S which is contractible to a point $\mathbf{x} \in S$ in such a way that the tangent of S at $\mathbf{x}$ remains a fixed k-vector τ. Define a differential operator

$$\nabla_\tau f(\mathbf{x}) = \lim_{\mathrm{d}(S)\to 0} \frac{1}{\mathrm{vol}(S)} \int_{\partial S} \partial \mathrm{s}\, f(\mathbf{x}),$$

where $\mathrm{d}(S)$ is the diameter of S and $\mathrm{vol}(S)$ is the scalar volume of S.

For instance, when τ is a 1-vector, the partial derivative ∂_τ in the direction of τ could be expressed as $\partial_\tau = \tau \nabla_\tau$.

The case when τ is an oriented volume element is important. The same result is obtained for $+\mathbf{e}_{12\ldots n}$ and $-\mathbf{e}_{12\ldots n}$. So it is convenient to drop the subscript and write ∇. The differential operator ∇ is called the *Dirac operator*. Applying an orthonormal basis $\{\mathbf{e}_1, \mathbf{e}_2, \ldots, \mathbf{e}_n\}$ of $\mathbb{R}^n$ the Dirac operator is seen to be

$$\nabla = \mathbf{e}_1 \frac{\partial}{\partial x_1} + \mathbf{e}_2 \frac{\partial}{\partial x_2} + \cdots + \mathbf{e}_n \frac{\partial}{\partial x_n},$$

where $\dfrac{\partial}{\partial x_i} = \partial_{\mathbf{e}_i}$.

The relation of ∇_τ to the Dirac operator ∇ is obtained by computing

$$\nabla = \tau^{-1} \tau \nabla = \tau^{-1}(\tau \,\llcorner\, \nabla + \tau \wedge \nabla)$$

where

$$\nabla_\tau = \tau^{-1}(\tau \,\llcorner\, \nabla).$$

20.4 Stokes' theorem

Consider a compact, contractible and oriented surface $S \subset \mathbb{R}^n$ with boundary ∂S and a real differentiable function $f : S \to \mathcal{C}\ell_n$. Stokes' theorem relates tangential integrals over S and ∂S, with compatible orientations,

$$\int_S d\mathrm{S}\, \nabla_\tau f = \int_{\partial S} d\mathrm{s}\, f,$$

where the left hand side becomes, using $\nabla_\tau = \tau^{-1}(\tau \llcorner \nabla)$,

$$\int_S (d\mathbf{S} \llcorner \nabla) f = \int_{\partial S} ds\, f.$$

Here $d\mathbf{S}$ and ds are of dimension degree k and $k-1$, respectively.

Examples. 1. Consider a 2-dimensional surface S in $\mathbb{R}^3$, and a vector field $\mathbf{f}$ on S. Then the scalar part of Stokes' theorem says

$$\int_S (d\mathbf{S} \llcorner \nabla) \cdot \mathbf{f} = \int_{\partial S} d\mathbf{s} \cdot \mathbf{f}.$$

Use a vector measure $d\mathbf{A} = \mathbf{e}_{123}^{-1} d\mathbf{S}$, normal to the surface S, to write the left hand side as

$$\int_S (d\mathbf{S} \llcorner \nabla) \cdot \mathbf{f} = \int_S (d\mathbf{A} \times \nabla) \cdot \mathbf{f},$$

then use the interchange rule (of dot and cross) to get the usual *Stokes' theorem*

$$\int_S d\mathbf{A} \cdot (\nabla \times \mathbf{f}) = \int_{\partial S} d\mathbf{s} \cdot \mathbf{f},$$

where $d\mathbf{s}$ and $d\mathbf{A}$ form a right-hand system.
2. Consider a 2-dimensional surface $S \in \mathbb{R}^n$, with bounding line $C = \partial S$, and a circulation of vector field $\mathbf{f}$ around C. First, convert the line integral to a surface integral by Stokes' theorem,

$$\int_C d\mathbf{x} \cdot \mathbf{f} = \int_S (d\mathbf{S} \llcorner \nabla) \cdot \mathbf{f},$$

and then compare the homogeneous components of degree 0 to obtain

$$\int_S (d\mathbf{S} \llcorner \nabla) \cdot \mathbf{f} = \int_S d\mathbf{S} \llcorner (\nabla \wedge \mathbf{f}).$$

This shows that in a simply connected domain the circulation vanishes if the divergence vanishes. ∎

By convention ∇ differentiates only quantities to its right, unless otherwise indicated. Because of non-commutativity of multiplication, it is good to have a notation indicating differentiation both to the right and to the left, when desired. Accordingly, we have, for instance, the Leibniz rule,

$$\dot{g}\dot{\nabla}\dot{f} = g\dot{\nabla}\dot{f} + \dot{g}\dot{\nabla} f,$$

where the dots indicate where the differentiation is applied. Stokes' theorem is now generalized to the form

$$\int_{\partial S} g\, ds\, f = \int_S g(d\mathbf{S} \llcorner \dot{\nabla})\dot{f} - (-1)^k \int_S \dot{g}(\dot{\nabla} \lrcorner d\mathbf{S}) f.$$

Here $\dim S = k \in \{1, 2, ..., n\}$. The minus sign on the last term comes from $\mathbf{a} \rfloor \mathbf{b} = -(-1)^k \mathbf{b} \llcorner \mathbf{a}$ for $\mathbf{a} \in \mathbb{R}^n$ and $\mathbf{b} \in \bigwedge^k \mathbb{R}^n$.

20.5 Positive and negative definite metrics

The monogenic homogeneous polynomials play a part in hypercomplex analysis similar to that of powers of a complex variable in the classical function theory. In constructing an explicit basis for the function space of monogenic polynomials it is customary to single out a special direction, say $\mathbf{e}_n$, in an orthonormal basis $\{\mathbf{e}_1, \mathbf{e}_2, \ldots, \mathbf{e}_{n-1}, \mathbf{e}_n\}$ of the Euclidean space $\mathbb{R}^n$. The unit bivectors

$$\mathbf{i}_k = \mathbf{e}_n \mathbf{e}_k, \quad k = 1, 2, \ldots, n-1,$$

generate the even subalgebra $\mathcal{C}\ell_n^+$; they anticommute and square up to -1. Thus, they form an orthonormal basis $\{\mathbf{i}_1, \mathbf{i}_2, \ldots, \mathbf{i}_{n-1}\}$ of a negative definite quadratic space $\mathbb{R}^{0,n-1}$ generating a Clifford algebra $\mathcal{C}\ell_{0,n-1} \simeq \mathcal{C}\ell_n^+$. A closer contact with the classical function theory is obtained if a vector

$$\mathbf{x} = x_1 \mathbf{e}_1 + x_2 \mathbf{e}_2 + \ldots + x_{n-1} \mathbf{e}_{n-1} + x_n \mathbf{e}_n$$

in the Euclidean space $\mathbb{R}^n$ is replaced by a sum of a vector and a scalar, a paravector,

$$z = x_1 \mathbf{i}_1 + x_2 \mathbf{i}_2 + \ldots + x_{n-1} \mathbf{i}_{n-1} + y, \quad y = x_n,$$

in $\mathbb{R}^{0,n-1} \oplus \mathbb{R}$ [the special direction is the scalar/real part y, also denoted by $x_0 = x_n$]. By the above correspondence $z \leftrightarrow \mathbf{e}_n \mathbf{x}$, we have established a correspondence between the following two mappings:

$$\mathbb{R}^n \to \mathcal{C}\ell_n^+, \quad \mathbf{x} \to f(\mathbf{x}),$$
$$\mathbb{R} \oplus \mathbb{R}^{0,n-1} \to \mathcal{C}\ell_{0,n-1}, \quad z \to f(z);$$

both are denoted for convenience by f.

In the case of a Euclidean space $\mathbb{R}^n$, the Dirac operator is homogeneous,

$$\nabla = \mathbf{e}_1 \frac{\partial}{\partial x_1} + \mathbf{e}_2 \frac{\partial}{\partial x_2} + \ldots + \mathbf{e}_{n-1} \frac{\partial}{\partial x_{n-1}} + \mathbf{e}_n \frac{\partial}{\partial x_n},$$

but it is replaced by a differential operator (inhomogeneous in the dimension degrees)

$$D = \frac{\partial}{\partial x_0} + \mathbf{i}_1 \frac{\partial}{\partial x_1} + \mathbf{i}_2 \frac{\partial}{\partial x_2} + \ldots + \mathbf{i}_{n-1} \frac{\partial}{\partial x_{n-1}},$$

$D = \frac{\partial}{\partial x_0} + \nabla$, in the paravector space $\mathbb{R} \oplus \mathbb{R}^{0,n-1}$.

20.6 Cauchy's integral formula

Consider a region $S \subset \mathbb{R}^n$ of dimension n with boundary ∂S of dimension $n-1$ and multivector function $f : S \to \mathcal{C}\ell_n$. In this case Stokes' theorem is of the form

$$\int_S d\mathbf{S}\, \nabla f = \int_{\partial S} d\mathbf{s}\, f.$$

If a multivector function f is monogenic, $\nabla f = 0$, then

$$\int_{\partial S} d\mathbf{s}\, f = 0 \quad \text{or equivalently} \quad \int_{\partial S} \partial s\, f = 0,$$

which means that the 'stream' of a monogenic function across any closed hypersurface vanishes. This is *Cauchy's theorem*.

In the following we need the *Cauchy kernel*

$$\mathbf{q}(\mathbf{x}) = \frac{\mathbf{x}}{|\mathbf{x}|^n},$$

which is both left monogenic, $\nabla \mathbf{q} = 0$, and right monogenic, $\dot{\mathbf{q}}\dot{\nabla} = 0$, at $\mathbf{x} \neq 0$. Substitute the Cauchy kernel $g(\mathbf{x}) = \mathbf{q}(\mathbf{x} - \mathbf{a})$ and a left monogenic function $f(\mathbf{x})$, $\nabla f(\mathbf{x}) = 0$, into Stokes' theorem. On the right hand side the first term vanishes, and the second term can be evaluated by a limiting process. One obtains *Cauchy's integral formula*

$$\int_{\partial S} \mathbf{q}(\mathbf{x} - \mathbf{a})\, d\mathbf{s}\, f(\mathbf{x}) = -(-1)^n \mathbf{e}_{12\ldots n} n\omega_n f(\mathbf{a}),$$

where $\omega_n = \pi^{n/2}/(n/2)!$ and the sign in $-(-1)^n \mathbf{e}_{12\ldots n}$ comes from the choice of orientation $\tau_S = \mathbf{e}_{12\ldots n}$ for S.

Example. In the special case $n = 2$ the above formula is

$$\int_{\partial S} \frac{\mathbf{x} - \mathbf{a}}{|\mathbf{x} - \mathbf{a}|^2}\, d\mathbf{s}\, f(\mathbf{x}) = -\mathbf{e}_{12} 2\pi f(\mathbf{a}),$$

since according to our convention $\mathbf{e}_{12}$ is compatible with the clockwise orientation. The classical formula corresponds to both the special cases $f : \mathbb{R}^2 \to \mathbb{R}^2$ and $f : \mathbb{R}^2 \to \mathcal{C}\ell_2^+$. [A better matching with the classical case would be obtained by mappings $f(x + y\mathbf{e}_1) = u + v\mathbf{e}_1$, where $\mathbf{e}_1^2 = -1$, in the Clifford algebra $\mathcal{C}\ell_{0,1} \simeq \mathbb{C}$ of $\mathbb{R}^{0,1}$.] ∎

Cauchy's integral formula can also be written in the form

$$\int_{\partial S} \mathbf{q}(\mathbf{x} - \mathbf{a})\, \partial s\, f(\mathbf{x}) = n\omega_n f(\mathbf{a}) \quad \text{for} \quad f : \mathbb{R}^n \to \mathcal{C}\ell_n^+,$$

and in the form

$$\int_{\partial S} q(z-a)\,\partial s\, f(z) = n\omega_n f(a) \quad \text{for} \quad f:\mathbb{R}\oplus\mathbb{R}^{0,n-1}\to C\ell_{0,n-1},$$

where the Cauchy kernel is

$$q(z) = \frac{\bar{z}}{|z|^n} = \frac{z^{-1}|z|^2}{|z|^n}.$$

The paravector $z\in\mathbb{R}\oplus\mathbb{R}^{0,n-1}$ has a norm $|z|=\sqrt{z\bar{z}}$, and an arbitrary element $u\in C\ell_{0,n-1}$ has a norm given by $|u|^2=\langle u\bar{u}\rangle_0$.

As in the classical case, we conclude that in a simply connected domain the values of a left monogenic function are determined by its values on the boundary.

20.7 Monogenic homogeneous functions

A function $f:\mathbb{R}^n\to C\ell_n$ homogeneous of degree ℓ satisfies

$$f(\lambda\mathbf{x}) = \lambda^\ell f(\mathbf{x}) \quad \text{for} \quad \lambda\in\mathbb{R},$$

which implies Euler's formula

$$r\frac{\partial}{\partial r}f(\mathbf{x}) = \ell f(\mathbf{x}), \quad \text{where} \quad r=|\mathbf{x}|.$$

If a multivector function $f(\mathbf{x})$ is monogenic, $\nabla f(\mathbf{x})=0$, then also $\mathbf{x}\nabla f(\mathbf{x}) = (\mathbf{x}\cdot\nabla+\mathbf{x}\wedge\nabla)f(\mathbf{x})=0$, and so

$$(r\frac{\partial}{\partial r}+\mathbf{L})f(\mathbf{x}) = 0$$

where

$$r\frac{\partial}{\partial r}f(\mathbf{x}) = \mathbf{x}\cdot\nabla = \sum_{i=0}^{n} x_i\frac{\partial}{\partial x_i}, \quad r=|\mathbf{x}|,$$

and

$$\mathbf{L} = \mathbf{x}\wedge\nabla = \sum_{i<j}\mathbf{e}_{ij}\Big(x_i\frac{\partial}{\partial x_j}-x_j\frac{\partial}{\partial x_i}\Big).$$

If a multivector function f is monogenic, $\nabla f(\mathbf{x})=0$, and homogeneous of degree ℓ, that is $r\frac{\partial}{\partial r}f=\ell f$, then $\mathbf{L}f=\kappa f$, where $\kappa=-\ell$. If a multivector function $f:\mathbb{R}^n\to C\ell_n$ is harmonic, $\nabla^2 f=0$, and homogeneous of degree ℓ, that is $r\frac{\partial}{\partial r}f = \ell f$, then $\mathbf{L}f = \kappa f$, where

$$\kappa = \ell + n - 2 \quad \text{spin up}$$
$$\kappa = -\ell \qquad\qquad \text{spin down.}$$

This can be seen by writing the Laplace operator ∇^2 as the square of the Dirac operator $\nabla = \mathbf{x}^{-1}(r\frac{\partial}{\partial r} + \mathbf{L})$ and factoring it by the relations

$$r\frac{\partial}{\partial r}\mathbf{x} - \mathbf{x}\,r\frac{\partial}{\partial r} = \mathbf{x}$$
$$\mathbf{L}\mathbf{x} + \mathbf{x}\mathbf{L} = (n-1)\mathbf{x}$$

as follows:

$$\nabla^2 = \mathbf{x}^{-2}\Big(r\frac{\partial}{\partial r} + n - 2 - \mathbf{L}\Big)\Big(r\frac{\partial}{\partial r} + \mathbf{L}\Big).$$

If a multivector function $f(\mathbf{x})$ is monogenic, $\nabla f(\mathbf{x}) = 0$, then $h(\mathbf{x}) = \mathbf{x}f(\mathbf{x})$ is harmonic, $\nabla^2 h(\mathbf{x}) = 0$.

If a multivector function $f : \mathbb{R}^n \to C\ell_n$ is monogenic at $\mathbf{x} \neq 0$, that is $(r\frac{\partial}{\partial r} + \mathbf{L})f(\mathbf{x}) = 0$, then $f(\mathbf{x}^{-1})$ satisfies $(-r\frac{\partial}{\partial r} + \mathbf{L})f(\mathbf{x}) = 0$, and the function

$$g(\mathbf{x}) = \mathbf{q}(\mathbf{x})f(\mathbf{x}^{-1}), \quad \mathbf{q}(\mathbf{x}) = \frac{\mathbf{x}}{|\mathbf{x}|^n},$$

is monogenic,

$$\Big(r\frac{\partial}{\partial r} + \mathbf{L}\Big)g(\mathbf{x}) = 0.$$

The plane case. Consider conformal mappings sending $\mathbf{x} \in \mathbb{R}^2$ to $f(\mathbf{x}) \in \mathbb{R}^2$. For sense-preserving conformal mappings $(-r\frac{\partial}{\partial r} + \mathbf{L})f = 0$, and for sense-reversing conformal mappings $(r\frac{\partial}{\partial r} + \mathbf{L})f = 0$. ∎

Example. Using the vector identity $\mathbf{a} \rfloor (\mathbf{b} \wedge \mathbf{c}) = (\mathbf{a}\cdot\mathbf{b})\mathbf{c} - \mathbf{b}(\mathbf{a}\cdot\mathbf{c})$ we find that

$$\nabla \rfloor (\mathbf{x} \wedge \mathbf{f}) = (\nabla\cdot\mathbf{x})\mathbf{f} - \mathbf{x}(\nabla\cdot\mathbf{f}) + (\mathbf{x}\cdot\dot{\nabla})\dot{\mathbf{f}} - \dot{\mathbf{x}}(\dot{\nabla}\cdot\mathbf{f})$$
$$= n\mathbf{f} - \mathbf{x}(\nabla\cdot\mathbf{f}) + r\frac{\partial}{\partial r}\mathbf{f} - \mathbf{f}$$

for a vector field $f : \mathbb{R}^n \to \mathbb{R}^n$.

If the vector field $\mathbf{f}$ is sourceless, $\nabla\cdot\mathbf{f} = 0$, and homogeneous of degree ℓ, that is $r\frac{\partial}{\partial r}\mathbf{f} = \ell\mathbf{f}$, then

$$\nabla \rfloor (\mathbf{x} \wedge \mathbf{f}) = (\ell + n - 1)\mathbf{f},$$

which shows that

$$\mathbf{v} = \frac{1}{\ell+n-1}\mathbf{x}\wedge\mathbf{f}, \qquad \ell \neq -(n-1),$$

is a bivector potential for $\mathbf{f}$.

If the vector field $\mathbf{f}$ is irrotational, $\nabla\wedge\mathbf{f} = 0$, and homogeneous of degree ℓ, that is $r\frac{\partial}{\partial r}\mathbf{f} = \ell\mathbf{f}$, then

$$\nabla(\mathbf{x}\cdot\mathbf{f}) = (\ell+1)\mathbf{f},$$

which shows that

$$u = -\frac{1}{\ell+1}\mathbf{x}\wedge\mathbf{f}, \qquad \ell \neq -1,$$

is a scalar potential for $\mathbf{f}$. ∎

20.8 A basis for monogenic homogeneous polynomials

A monogenic function is real analytic, that is, a power series of the components of the argument. The homogeneous part of degree 1 of the Taylor series expansion

$$f'(a)z = x_1\frac{\partial f}{\partial x_1}\bigg|_a + x_2\frac{\partial f}{\partial x_2}\bigg|_a + \ldots + x_{n-1}\frac{\partial f}{\partial x_{n-1}}\bigg|_a + y\frac{\partial f}{\partial y}\bigg|_a$$

can be written, in virtue of a monogenic f, as follows:

$$f'(a)z = (x_1 - y\mathbf{i}_1)\frac{\partial f}{\partial x_1}\bigg|_a + (x_2 - y\mathbf{i}_2)\frac{\partial f}{\partial x_2}\bigg|_a + \ldots + (x_{n-1} - y\mathbf{i}_{n-1})\frac{\partial f}{\partial x_{n-1}}\bigg|_a.$$

The functions

$$z_1 = x_1 - y\mathbf{i}_1,\ z_2 = x_2 - y\mathbf{i}_2,\ \ldots,\ z_{n-1} = x_{n-1} - y\mathbf{i}_{n-1}$$

are monogenic; note that $z_k = x_k + y\mathbf{e}_k\mathbf{e}_n$, $k = 1, 2, \ldots, n-1$. Write $\underline{l} = (l_1, l_2, \ldots, l_{n-1})$, and define the symmetrized polynomials

$$p_{\underline{l}}(z) = \frac{1}{l!}\sum_{\pi\in S_l} z_{\pi(1)}z_{\pi(2)}\cdots z_{\pi(l)},$$

each term being homogeneous of degree l_k with respect to z_k, and homogeneous of degree $l = l_1 + l_2 + \ldots + l_{n-1}$ with respect to z. The symmetric polynomials are monogenic, $Dp_{\underline{l}}(z) = 0$; they appear as multipliers of partial derivatives

$$\partial_{\underline{l}} f\,|_a = \frac{\partial^l f}{\partial x_1^{l_1}\partial x_2^{l_2}\cdots\partial x_{n-1}^{l_{n-1}}}\bigg|_a$$

in the Taylor series expansion of $f(a+z)$.

Examples. 1. For $l = (2,1,0,0,\ldots,0)$ we have

$$\begin{aligned} p_{\underline{l}}(z) &= \tfrac{1}{3}(z_1^2 z_2 + z_1 z_2 z_1 + z_2 z_1^2) \\ &= (x_1^2 - y^2)x_2 - 2x_1x_2y\mathbf{i}_1 - (x_1^2 - \tfrac{y^2}{3})y\mathbf{i}_2 \\ &= (x_1^2 - y^2)x_2 + 2x_1x_2y\mathbf{e}_1\mathbf{e}_n + (x_1^2 - \tfrac{y^2}{3})y\mathbf{e}_2\mathbf{e}_n. \end{aligned}$$

2. The functions $q_{\underline{l}}(z) = \partial_{\underline{l}} q(z)$ are homogeneous of degree $-(l+n-1)$ and monogenic when $z \neq 0$. ∎

For all $z \in \mathbb{R} \oplus \mathbb{R}^{0,n-1}$, we also have $p_{\underline{l}}(z) \in \mathbb{R} \oplus \mathbb{R}^{0,n-1}$.

The monogenic polynomials homogeneous of degree l span a right module over the ring $C\ell_{0,n-1}$; the polynomials $p_{\underline{l}}(z)$ form a basis of this module of dimension $\binom{l+n-2}{l}$. Harmonic polynomials homogeneous of degree l form a module over the rotation group $SO(n)$ of $\mathbb{R} \oplus \mathbb{R}^{0,n-1}$, namely the irreducible module of traceless symmetric tensors of degree l, the dimension of this module being

$$\binom{l+n-1}{l} - \binom{l+n-3}{l-2} = \binom{l+n-2}{l} + \binom{l+n-3}{l-1}.$$

The Laurent series expansion is formulated as follows:

$$f(z) = \sum_{l=0}^{\infty} \sum_{\underline{l}} \left[p_{\underline{l}}(z-a) b_{\underline{l}} + q_{\underline{l}}(z-a) c_{\underline{l}} \right],$$

where

$$b_{\underline{l}} = \frac{1}{n\omega_n} \int_{\partial S} q_{\underline{l}}(z-a)\, \partial s\, f(z),$$

$$c_{\underline{l}} = \frac{1}{n\omega_n} \int_{\partial S} p_{\underline{l}}(z-a)\, \partial s\, f(z),$$

when $f(z)$ is monogenic in a region $S \subset \mathbb{R} \oplus \mathbb{R}^{0,n-1}$ except at $a \in S$.

Axial Vector Fields

Single out a distinguished direction or *axis* $\mathbf{a}$ in $\mathbb{R}^n$, say

$$\mathbf{a} = \mathbf{e}_n.$$

Write $y = x_n$ and $\mathbf{x} = x_1\mathbf{e}_1 + x_2\mathbf{e}_2 + \cdots + x_{n-1}\mathbf{e}_{n-1}$ so that

$$\mathbf{r} = \mathbf{x} + y\mathbf{a} = x_1\mathbf{e}_1 + x_2\mathbf{e}_2 + \cdots + x_{n-1}\mathbf{e}_{n-1} + x_n\mathbf{e}_n.$$

Write $x = |\mathbf{x}|$ and let

$$\mathbf{i} = \frac{\mathbf{r} \wedge \mathbf{a}}{|\mathbf{r} \wedge \mathbf{a}|} = \frac{\mathbf{x}\mathbf{a}}{x}$$

be the unit bivector of the plane determined by $\mathbf{a}$ and $\mathbf{r}$. Since $\nabla(\mathbf{ra}) = (\nabla\mathbf{r})\mathbf{a} = n\mathbf{a}$ and $\nabla(\mathbf{ar}) = -(n-2)\mathbf{a}$, it is evident that the polynomial

$$w = \frac{1}{2}[n\mathbf{ar} + (n-2)\mathbf{ra}] = (n-1)\mathbf{a}\cdot\mathbf{r} + \mathbf{a}\wedge\mathbf{r} = (n-1)y - x\mathbf{i}$$

satisfies the equation $\nabla w = 0$; such a w is called *monogenic.* The polynomial w is a complex of its scalar part $u = (n-1)\mathbf{a}\cdot\mathbf{r} = (n-1)y$ and its bivector part $\mathbf{v} = \mathbf{a}\wedge\mathbf{r} = -x\mathbf{i}$. The vector field

$$\mathbf{f} = -(n-1)\mathbf{a}$$

is irrotational, that is $\nabla\wedge\mathbf{f} = 0$, with a (scalar) potential $u = (n-1)y$, such that $\mathbf{f} = -\nabla u$, and also sourceless, that is $\nabla\cdot\mathbf{f} = 0$, with a *stream function* $\psi = -x^{n-1}$ and a *bivector potential* $\mathbf{v} = -x\mathbf{i}$, such that $\mathbf{f} = \nabla\lrcorner\mathbf{v}$ and $\nabla\wedge\mathbf{v} = 0$. In fact, $\mathbf{f}$ is monogenic, $\nabla\mathbf{f} = \nabla\cdot\mathbf{f} + \nabla\wedge\mathbf{f} = 0$, with a complex potential $w = u + v\mathbf{i} = (n-1)y - x\mathbf{i}$, which is also monogenic, $\nabla w = 0$.

The vector function $\mathbf{q}(\mathbf{r}) = \mathbf{r}/r^n$, where $r = |\mathbf{r}|$, is called the *Cauchy kernel,* and

$$\mathbf{q}_1(\mathbf{r}) = \nabla[\mathbf{a}\cdot\mathbf{q}(\mathbf{r})] = \frac{1}{r^n}(\mathbf{a} - n(\mathbf{a}\cdot\mathbf{r})\mathbf{r}^{-1})$$

is the field of an n-dimensional *dipole.* These vector fields can be used to reproduce the complex potential

$$w = -\mathbf{q}(\mathbf{r})\mathbf{q}_1(\mathbf{r}^{-1})$$

of the vector field $\mathbf{f} = -(n-1)\mathbf{a}$.

Axial monogenic polynomials of degree 2. In the above we just found that

$$\nabla(\mathbf{ar}) = -(n-2)\mathbf{a}$$
$$\nabla\mathbf{ra} = n\mathbf{a}.$$

Next, we differentiate the second powers of $\mathbf{r}$. The product $\mathbf{ara} = (\mathbf{ar}+\mathbf{ra})\mathbf{a} - \mathbf{a}^2\mathbf{r} = 2(\mathbf{a}\cdot\mathbf{r})\mathbf{a} - a^2\mathbf{r}$ is a vector in the plane determined by $\mathbf{a}$ and $\mathbf{r}$. So

$$(\mathbf{ar})^2 = 2(\mathbf{a}\cdot\mathbf{r})\mathbf{ar} - a^2r^2$$
$$(\mathbf{ra})^2 = 2(\mathbf{a}\cdot\mathbf{r})\mathbf{ra} - a^2r^2$$

and $(\mathbf{ar})(\mathbf{ra}) = a^2r^2$. Since

$$\nabla(\mathbf{ar})^2 = -2(n-2)(\mathbf{a}\cdot\mathbf{r})\mathbf{a}$$
$$\nabla(\mathbf{ar})(\mathbf{ra}) = 2a^2\mathbf{r}$$
$$\nabla(\mathbf{ra})^2 = 2(n+2)(\mathbf{a}\cdot\mathbf{r})\mathbf{a} - 4a^2\mathbf{r}$$

it is evident that the polynomial

$$\begin{aligned} p_2(\mathbf{r}) &= \tfrac{n}{4}[(n+2)(\mathbf{ar})^2 + 2(n-2)(\mathbf{ar})(\mathbf{ra}) + (n-2)(\mathbf{ra})^2] \\ &= \tfrac{n}{2}[n(\mathbf{a}\cdot\mathbf{r})^2 - a^2r^2 + 2(\mathbf{a}\cdot\mathbf{r})(\mathbf{a}\wedge\mathbf{r})] \end{aligned}$$

is monogenic, $\nabla p_2(\mathbf{r}) = 0$. In cylindrical coordinates $y, \mathbf{x}$, the polynomial

$$p_2(\mathbf{r}) = \frac{n}{2}[(n-1)y^2 - x^2 - 2yx\mathbf{i}]$$

is a complex potential of the monogenic vector field $\mathbf{f} = n(\mathbf{x} - (n-1)y\mathbf{a})$ which has a stream function $\psi = -nyx^{n-1}$.

20.9 Axial monogenic polynomials of homogeneous degree

In the following, monogenic polynomials, homogeneous of degree l in the factors $\mathbf{ar}$ and $\mathbf{ra}$ (and then also in $\mathbf{r}$ or in y and $\mathbf{x}$), will be introduced. First,

$$\begin{aligned} \nabla(\mathbf{ar})^l = -(n-2)\mathbf{a}[(\mathbf{ar})^{l-1} + (\mathbf{ar})^{l-2}(\mathbf{ra}) + \\ \cdots + (\mathbf{ar})(\mathbf{ra})^{l-2} + (\mathbf{ra})^{l-1}], \end{aligned} \tag{3}$$

$$\begin{aligned} \nabla(\mathbf{ra})^l = (n-2)\mathbf{a}[(\mathbf{ar})^{l-1} + (\mathbf{ar})^{l-2}(\mathbf{ra}) + \\ \cdots + (\mathbf{ar})(\mathbf{ra})^{l-2} + (\mathbf{ra})^{l-1}] + 2l\mathbf{a}(\mathbf{ra})^{l-1}. \end{aligned} \tag{4}$$

To verify these, observe that $\mathbf{e}_1\mathbf{f}\mathbf{e}_1 + \cdots + \mathbf{e}_n\mathbf{f}\mathbf{e}_n = -(n-2)\mathbf{f}$ for any vector $\mathbf{f}$. Then

$$\begin{aligned} \sum_{i=1}^{n} \mathbf{e}_i(\mathbf{ar})^{j-1}\mathbf{a}\mathbf{e}_i(\mathbf{ar})^{l-j} &= -(n-2)(\mathbf{ar})^{j-1}\mathbf{a}(\mathbf{ar})^{l-j} \\ &= -(n-2)\mathbf{a}(\mathbf{ra})^{j-1}(\mathbf{ar})^{l-j} \\ &= -(n-2)\mathbf{a}(\mathbf{ar})^{l-j}(\mathbf{ra})^{j-1} \end{aligned}$$

because $(\mathbf{ar})^{j-1}\mathbf{a}$ is a vector in the plane determined by $\mathbf{a}$ and $\mathbf{r}$, as can be seen by inspection on $(\mathbf{ra})^j\mathbf{r} = \mathbf{r}((\mathbf{ar})^{j-1}\mathbf{a})\mathbf{r}$ and induction on j. The equation (3) is now proved. Similarly, observe that $\sum_{i=1}^{n} \mathbf{e}_i\mathbf{u}\mathbf{e}_i = (-1)^l(n-2l)\mathbf{u}$ for any element $\mathbf{u}$, homogeneous of grade l. Then

$$\begin{aligned} \sum_{i=1}^{n} \mathbf{e}_i(\mathbf{ra})^j\mathbf{e}_i &= \sum_{i=1}^{n} \tfrac{1}{2}\mathbf{e}_i[((\mathbf{ra})^j + (\mathbf{ar})^j) + ((\mathbf{ra})^j - (\mathbf{ar})^j)]\mathbf{e}_i \\ &= \tfrac{n}{2}[(\mathbf{ra})^j + (\mathbf{ar})^j] + \tfrac{n-4}{2}[(\mathbf{ra})^j - (\mathbf{ar})^j] \\ &= (n-2)(\mathbf{ra})^j + 2(\mathbf{ar}^j), \end{aligned}$$

because $(\mathbf{ra})^j + (\mathbf{ar})^j$ is a scalar of grade 0 and $(\mathbf{ra})^j - (\mathbf{ar})^j$ is a bivector of grade 2. It follows that

$$\sum_{i=1}^{n} \mathbf{e}_i(\mathbf{ra})^j \mathbf{e}_i \mathbf{a} = (n-2)\mathbf{a}(\mathbf{ar})^j + 2\mathbf{a}(\mathbf{ra})^j,$$

which multiplied by $(\mathbf{ra})^{l-j-1}$ on the right gives the equation (4) after summing up the terms $j = 0, 1, \cdots, l-1$.

To calculate $\nabla(\mathbf{ar})^{l-j}(\mathbf{ra})^j$ note that $(\mathbf{ar})^{l-j}(\mathbf{ra})^j = r^{2j}a^{2j}(\mathbf{ar})^{l-2j}$ when $l \geq 2j$ and $(\mathbf{ar})^{l-j}(\mathbf{ra})^j = r^{2(l-j)}(\mathbf{ra})^{2j-1}$ when $l \leq 2j$. Use results (3), (4) and

$$\nabla \mathbf{r}^l = \begin{cases} l\mathbf{r}^{l-1} & \text{for } l \text{ even} \\ (l+n-1)r^{l-1} & \text{for } l \text{ odd.} \end{cases}$$

Then

$$\nabla(\mathbf{ar})^{l-j}(\mathbf{ra})^j = \sum_{k=1}^{l} m_{jk}\mathbf{a}(\mathbf{ar})^{l-k}(\mathbf{ra})^k$$

where the $(l+1) \times l$-matrix m_{jk} is

$$\begin{pmatrix} -(n-2) & -(n-2) & -(n-2) & \cdots & -(n-2) & -(n-2) & -(n-2) \\ 2 & -(n-2) & -(n-2) & \cdots & -(n-2) & -(n-2) & 0 \\ 0 & 4 & -(n-2) & \cdots & -(n-2) & 0 & 0 \\ 0 & 0 & 6 & & 0 & 0 & 0 \\ \vdots & \vdots & \vdots & \vdots & \vdots & \vdots & \vdots \\ 0 & 0 & (n-2) & \cdots & (n+2(l-3)) & 0 & 0 \\ 0 & (n-2) & (n-2) & \cdots & (n-2) & (n+2(l-2)) & 0 \\ (n-2) & (n-2) & (n-2) & \cdots & (n-2) & (n-2) & (n+2(l-1)) \end{pmatrix}$$

To get the coefficients $p_{l,j}$ in

$$p_l(\mathbf{r}) = \sum_{j=0}^{l} p_{l,j}(\mathbf{ar})^{l-j}(\mathbf{ra})^j$$

such that $\nabla p_l(\mathbf{r}) = 0$, multiply the rows of m_{jk} by the corresponding coefficients $p_{l,j}$ and determine $p_{l,j}$ so that the sum of the resulting elements in each column is zero. To calculate the coefficients $p_{l,j}$ one has useful algorithms such as [1]

$$p_{l,j} = \frac{1}{l}\left(\frac{n-2}{2} + l - j\right)(p_{l-1,j} + p_{l-1,l-j})$$

1 It is worth noting that the formula (5) for the coefficients $p_{l,j}$ is valid for all signatures and not only for positive definite quadratic forms.

which gives

$$p_{l,j} = \binom{\frac{n-2}{2}+l-j}{l-j}\binom{\frac{n-4}{2}+j}{j}. \tag{5}$$

Example. Let $n = 3$ and $l = 2$. Then

$$\begin{array}{r|l} \nabla(\mathbf{ar})^2 = -\mathbf{a}(\mathbf{ar}) - \mathbf{a}(\mathbf{ra}) & 5 \\ \nabla(\mathbf{arra}) = 2\mathbf{a}(\mathbf{ar}) & 2 \\ \nabla(\mathbf{ra})^2 = \mathbf{a}(\mathbf{ar}) + 5\mathbf{a}(\mathbf{ra}) & 1 \end{array}$$

where the right-hand sides of the identities give the matrix

$$\begin{pmatrix} -1 & -1 \\ 2 & 0 \\ 1 & 5 \end{pmatrix}$$

and the right column gives

$$p_2(\mathbf{r}) = 5(\mathbf{ar})^2 + 2(\mathbf{arra}) + (\mathbf{ra})^2$$

with coefficients $p_{2,0} = 5$, $p_{2,1} = 2$ and $p_{2,2} = 1$. ∎

In some cases it is worth knowing the smallest integer coefficients, which in a few lower-dimensional cases are as follows:

	$n=3$					$n=4$					$n=5$				
$l\backslash j$	0	1	2	3	4	0	1	2	3	4	0	1	2	3	4
1	3	1				2	1				5	3			
2	5	2	1			3	2	1			7	6	3		
3	35	15	9	5		4	3	2	1		21	21	15	7	
4	63	28	18	12	7	5	4	3	2	1	33	36	30	20	9

20.10 Differential equations in cylindrical coordinates

Consider an axially symmetric vector field $\mathbf{f}$ in the cylindrical coordinates y, $\mathbf{x}$. Write $\mathbf{r} = \mathbf{x} + y\mathbf{a}$, $\nabla = \nabla_{\mathbf{x}} + \mathbf{a}(\partial/\partial y)$ and $\mathbf{f} = \mathbf{g} + h\mathbf{a}$ with $\mathbf{g} = (\mathbf{x}/x)g$. Then

$$\nabla \cdot \mathbf{f} = \nabla_{\mathbf{x}} \cdot \mathbf{g} + \frac{\partial h}{\partial y} = \frac{\partial g}{\partial x} + \frac{\partial h}{\partial y} + (n-2)\frac{g}{x}$$

$$\nabla \wedge \mathbf{f} = \frac{\mathbf{x}}{x}\frac{\partial h}{\partial x}\mathbf{a} + \mathbf{a}\frac{\partial \mathbf{g}}{\partial y} = \mathbf{i}\Big(\frac{\partial h}{\partial x} - \frac{\partial g}{\partial y}\Big).$$

If now $\mathbf{f}$ is monogenic, $\nabla\mathbf{f} = 0$, then there is a complex potential $w = u + v\mathbf{i} = u + \mathbf{v}$ such that $\mathbf{f} = -\nabla u = \nabla \rfloor \mathbf{v}$, $\nabla \wedge \mathbf{v} = 0$. The condition to be monogenic,

$\nabla w = 0$, means that

$$\left(\nabla_{\mathbf{x}} + \mathbf{a}\frac{\partial}{\partial y}\right)\left(u + v\frac{\mathbf{xa}}{x}\right)$$
$$= \frac{\mathbf{x}}{x}\frac{\partial u}{\partial x} + \frac{\mathbf{x}}{x}\frac{\partial v}{\partial x}\frac{\mathbf{xa}}{x} + v(n-2)\frac{\mathbf{a}}{x} + \mathbf{a}\frac{\partial u}{\partial y} + \mathbf{a}\frac{\partial v}{\partial y}\frac{\mathbf{xa}}{x} = 0,$$

which decomposed gives an n-dimensional analog of the Cauchy-Riemann equations

$$\begin{cases} \dfrac{\partial u}{\partial x} - \dfrac{\partial v}{\partial y} = 0 \\ \dfrac{\partial v}{\partial x} + \dfrac{\partial u}{\partial y} + (n-2)\dfrac{v}{x} = 0. \end{cases} \tag{6}$$

The components of $\mathbf{f} = \mathbf{g} + h\mathbf{a}$ are then expressed as

$$\begin{cases} g = -\dfrac{\partial u}{\partial x} = -\dfrac{\partial v}{\partial y} \\ h = -\dfrac{\partial u}{\partial y} = \dfrac{\partial v}{\partial x} + (n-2)\dfrac{v}{x}. \end{cases}$$

Of course, u is harmonic,

$$\nabla^2 u = \left(\frac{\partial^2}{\partial x^2} + \frac{\partial^2}{\partial y^2} + \frac{n-2}{x}\frac{\partial}{\partial x}\right)u = 0,$$

and also $\mathbf{v}$ is harmonic,

$$\nabla^2 \mathbf{v} = \nabla^2(v\mathbf{i}) = \mathbf{i}\left\{\frac{\partial^2 v}{\partial x^2} + \frac{\partial^2 v}{\partial y^2} + \frac{n-2}{x}\left(\frac{\partial v}{\partial x} - \frac{v}{x}\right)\right\} = 0,$$

and so

$$\nabla^2 v = \frac{n-2}{x^2}v.$$

As an axially symmetric and sourceless vector field $\mathbf{f}$ has a *stream function* $\psi = x^{n-2}v$, satisfying

$$\left(\frac{\partial^2}{\partial x^2} + \frac{\partial^2}{\partial y^2} - \frac{n-2}{x}\frac{\partial}{\partial x}\right)\psi = 0,$$

such that

$$\frac{\partial \psi}{\partial x} = x^{n-2}h, \quad \frac{\partial \psi}{\partial y} = -x^{n-2}g.$$

The stream function can be expressed by a line integral

$$\psi - \psi_0 = \int_{P_0}^{P} x^{n-2}(g\,dy - h\,dx)$$

independent of path in a fixed plane containing the symmetry axis $\mathbf{a}$. Let the path of integration sweep around the symmetry axis $\mathbf{a}$ and form an axially symmetric hypersurface S. The stream

$$\Psi = \int_S d\mathbf{S} \wedge \mathbf{f}$$

across S is $\Psi = \mathbf{e}_{12\ldots n}(n-1)\omega_{n-1}(\psi - \psi_0)$ where $(n-1)\omega_{n-1}$ is the measure of the unit sphere S^{n-2} in the $(n-1)$-dimensional space (orthogonal to the axis $\mathbf{a} \in \mathbb{R}^n$).

The n-dimensional Cauchy-Riemann equations (6), by a change of variables $x = r\sin\theta$, $y = r\cos\theta$, become

$$\begin{cases} \dfrac{\partial u}{\partial r} + \dfrac{1}{r}\dfrac{\partial v}{\partial \theta} + (n-2)\cot\theta \dfrac{v}{r} = 0 \\ \dfrac{\partial v}{\partial r} - \dfrac{1}{r}\dfrac{\partial u}{\partial \theta} + (n-2)\dfrac{v}{r} = 0. \end{cases}$$

If the complex potential $w = u + v\mathbf{i}$ is homogeneous of degree l, then

$$\begin{cases} \dfrac{\partial v}{\partial \theta} + (n-2)\cot\theta v + lu = 0 \\ \dfrac{\partial u}{\partial \theta} = (l+n-2)v. \end{cases}$$

Differentiation with respect to θ and a further change of variable $\mu = \cos\theta$ then result in

$$(1-\mu^2)\frac{d^2v}{d\mu^2} - (n-1)\mu\frac{dv}{d\mu} + \left\{l(l+n-2) - \frac{n-2}{1-\mu^2}\right\}v = 0$$

and

$$(1-\mu^2)\frac{d^2u}{d\mu^2} - (n-1)\mu\frac{du}{d\mu} + l(l+n-2)u = 0.$$

The solutions u/r^l of this last equation can be expressed as hypergeometric series or ultraspherical (Gegenbauer) functions

$${}_2F_1(-l, l+n-2, \frac{n-1}{2}; \frac{1-\mu}{2}) = \frac{l!(n-3)!}{(l+n-3)!}C_l^{((n-2)/2)}(\mu)$$

and $v = [\sqrt{1-\mu^2}/(l+n-2)](du/d\mu)$. The previously introduced monogenic polynomials $p_l(\mathbf{r})$, homogeneous of degree l, are now $p_l(\mathbf{r}) = u + v\mathbf{i}$.

In some lower-dimensional cases the scalar part u of $p_l(\mathbf{r})$ divided by r^l is

n	u/r^l	
2	$\cos l\theta = T_l(\cos\theta)$	Chebyshev
3	$P_l(\cos\theta)$	Legendre
4	$(\frac{l}{2}+1)(\cos l\theta + \cot\theta \sin l\theta) = (\frac{l}{2}+1)\dfrac{\sin(l+1)\theta}{\sin\theta}$	

20.11 Inversion of multipoles in unit sphere

If a function $f(\mathbf{r})$ with values in the Clifford algebra is monogenic, then the function $\mathbf{q}(\mathbf{r})f(\mathbf{r}^{-1})$, obtained by inversion, is also monogenic for $\mathbf{r} \neq 0$. The vector field

$$\mathbf{q}_l(\mathbf{r}) = \frac{\partial^l}{\partial y^l}\mathbf{q}(\mathbf{r})$$

is axially symmetric, monogenic and homogeneous of degree $-(l+n-1)$, and it describes an axial multipole of order 2^l. The previously introduced monogenic complex polynomials are obtained by inversion in the unit sphere,

$$p_l(\mathbf{r}) = \frac{(-1)^l}{l!}\mathbf{q}(\mathbf{r})\mathbf{q}_l(\mathbf{r}^{-1}).$$

These axially symmetric monogenic complex polynomials $p_l(\mathbf{r})$ should be distinguished from the monogenic complex polynomials introduced by Haefeli, who defined symmetrized products of the functions $z_i = x_i + y\mathbf{e}_i\mathbf{e}_n$, where $i = 1, 2, \ldots, n-1$.

Example. The polynomial $\frac{1}{3}(z_1^2 z_2 + z_1 z_2 z_1 + z_2 z_1^2) = x_2(x_1^2 - y^2) + 2x_1 x_2 y\mathbf{e}_1\mathbf{e}_n + y(x_1^2 - (y^2/3))\mathbf{e}_2\mathbf{e}_n$ is such a monogenic symmetrized product. ∎

Haefeli's monogenic symmetrized products form a basis of the right module (over the Clifford algebra) which consists of monogenic polynomials.

Example. The axially symmetric monogenic polynomials, homogeneous of degree 1 and 2, can be expressed in this basis in the forms

$$p_1(\mathbf{r}) = -\sum_{i=1}^{n-1} z_i\mathbf{e}_i\mathbf{e}_n \quad \text{and} \quad p_2(\mathbf{r}) = -\frac{n}{2}\sum_{i=1}^{n-1} z_i^2$$

respectively. ∎

Finally, the complex polynomial $p_l(\mathbf{r})$ is such that its bivector part determines the plane spanned by $\mathbf{a}$ and $\mathbf{r} = \mathbf{x} + y\mathbf{a}$. So the function $\mathbf{r}p_l(\mathbf{r})$ is vector valued in this same plane. Since the complex function $p_l(\mathbf{r})$ is monogenic,

$\nabla p_l(\mathbf{r}) = 0$, the vector function $\mathbf{r}p_l(\mathbf{r})$ is harmonic, $\nabla^2 \mathbf{r}p_l(\mathbf{r}) = 0$. This can be seen from

$$\mathbf{r}^2\nabla^2 = \left(r\frac{\partial}{\partial r} + \frac{n-2}{2}\right)^2 - \left(\mathbf{L} - \frac{n-2}{2}\right)^2,$$

where $\mathbf{r}\nabla = \mathbf{r}\cdot\nabla + \mathbf{r}\wedge\nabla$, $r(\partial/\partial r) = \mathbf{r}\cdot\nabla$ and $\mathbf{L} = \mathbf{r}\wedge\nabla$, which has axially symmetric eigenfunctions $\mathbf{L}p_l(\mathbf{r}) = -lp_l(\mathbf{r})$ and $\mathbf{L}[\mathbf{r}p_l(\mathbf{r})] = (l+n-1)\mathbf{r}p_l(\mathbf{r}.)$

History and survey of research

Hypercomplex analysis attempts to generalize one-variable complex analysis to a higher-dimensional one-variable theory using Clifford algebras of Euclidean spaces. It was first examined by Moisil (in terms of integrals), and rediscovered in quaternion form by Fueter, who introduced the symmetrized polynomials. In quaternion analysis the central result was Cauchy's integral formula in dimension 4. The notion of monogenic functions with values in a Clifford algebra is due to Iftimie and Bosshard. Habetha showed that if an algebra gives rise to Cauchy's integral formula, then it is sufficient that it contains a linear subspace where all non-zero vectors are invertible in the algebra; that is, the algebra is almost a Clifford algebra.

Lounesto & Bergh 1983 initiated a study of axially symmetric functions with values in a Clifford algebra. The research was later taken over by Sommen.

Presently, there are several schools studying hypercomplex analysis with different emphasis: harmonic analysis (J. Ryan, J. Gilbert), functional analysis (R. Delanghe, F. Brackx), and function theory (K. Habetha, R. Gilbert).

Exercises

1. Show that $\nabla\mathbf{x} = n$, $\nabla\mathbf{x}^2 = 2\mathbf{x}$, $\nabla\mathbf{x}^3 = (n+2)\mathbf{x}^2$, $\nabla\mathbf{x}^4 = 4\mathbf{x}^3$.
2. Show that $\nabla\mathbf{x}^k = k\mathbf{x}^{k-1}$ for k even, and $\nabla\mathbf{x}^k = (n+k-1)\mathbf{x}^{k-1}$ for k odd.
3. Show that $\nabla\cos\mathbf{x} = -\sin\mathbf{x}$, $\nabla\sin\mathbf{x} = \cos\mathbf{x} + (n-1)\frac{1}{\mathbf{x}}\sin\mathbf{x}$.
4. Show that $\nabla\exp\mathbf{x} = \exp\mathbf{x} + (n-1)\frac{1}{\mathbf{x}}\sinh\mathbf{x}$,

$$\nabla\log(1+\mathbf{x}) = \frac{1}{1+\mathbf{x}} + (n-1)\frac{1}{\mathbf{x}}\arctan\mathbf{x}.$$

Bibliography

P. Bosshard: Die Cliffordschen Zahlen, ihre Algebra und ihre Funktionentheorie. Ph.D. Thesis, Univ. Zürich, 1940.

D. Constales: The relative position of L^2 domains in complex and in Clifford analysis. Ph.D. Thesis, Univ. Gent, 1989.

R. Delanghe, F. Brackx: Hypercomplex function theory and Hilbert modules with reproducing kernel. *Proc. London Math. Soc.* **37** (1978), 545-576.

R. Delanghe, F. Sommen, V. Souček: *Clifford Algebra and Spinor Valued Functions: A Function Theory for the Dirac Operator.* Kluwer, Dordrecht, The Netherlands, 1992.

J. Gilbert, M. Murray: *Clifford Algebras and Dirac Operators in Harmonic Analysis.* Cambridge Studies in Advanced Mathematics, Cambridge University Press, Cambridge, 1991.

R.P. Gilbert, G.N. Hile: Hypercomplex function theory in the sense of L. Bers. *Math. Nachr.* **72** (1976), 187-200.

K. Gürlebeck, W. Sprössig: *Quaternionic Analysis and Elliptic Boundary Value Problems.* Akademie-Verlag, Berlin, 1989. Birkhäuser, Basel, 1990.

K. Gürlebeck, W. Sprössig: *Quaternionic and Clifford Calculus for Physicists and Engineers.* Wiley, Chichester, 1997.

K. Habetha: Function theory in algebras; pp. 225-237 in E. Lanckau, W. Tutschke (eds.): *Complex Analysis: Methods, Trends and Applications*, Akademie-Verlag, Berlin, 1983.

G. Haefeli: Hyperkomplexe Differentiale. *Comm. Math. Helv.* **20** (1947), 382-420.

D. Hestenes: Multivector calculus. *J. Math. Anal. Appl.* **24** (1968), 313-325.

V. Iftimie: Fonctions hypercomplexes. *Bull. Math. Soc. Sci. Math. R.S. Roumanie* **9**,57 (1965), 279-332.

P. Lounesto, P. Bergh: Axially symmetric vector fields and their complex potentials. *Complex Variables, Theory and Application* **2** (1983), 139-150.

G.C. Moisil: *Les fonctions monogènes dans espaces à plusieurs dimensions.* C.R. Congrès des Soc. Savantes de Clermont-Ferrand, 1931.

J. Ryan (ed.): *Clifford Algebras in Analysis and Related Topics.* CRC Press, Boca Raton, FL, 1996.

F. Sommen: Special functions in Clifford analysis and axial symmetry. *J. Math. Anal. Appl.* **130** (1988), 110-133.

A. Sudbery: Quaternionic analysis. *Math. Proc. Camb. Phil. Soc.* **85** (1979), 199-225.

21

Binary Index Sets and Walsh Functions

The present chapter scrutinizes how the sign of the product of two elements in the basis for the Clifford algebra of dimension 2^n can be computed by the Walsh functions of degree less than 2^n. In the multiplication formula the basis elements are labelled by binary n-tuples, which form an abelian group Ω, which in turn gives rise to the maximal grading of the Clifford algebra. The group of binary n-tuples is also employed in the Cayley-Dickson process.

WALSH FUNCTIONS

Consider n-tuples $\underline{a} = a_1a_2\ldots a_n$ of binary digits $a_i = 0, 1$. For two such n-tuples $\underline{a}$ and $\underline{b}$ the sum $\underline{a} \oplus \underline{b} = \underline{c}$ is defined by termwise addition modulo 2, that is,

$$c_i = a_i + b_i \mod 2.$$

These n-tuples form a group so that the group characters are *Walsh functions*

$$w_{\underline{a}}(\underline{b}) = (-1)^{\sum_{i=1}^n a_i b_i}.$$

The Walsh functions have only two values, ± 1, and they satisfy $w_{\underline{k}}(\underline{a} \oplus \underline{b}) = w_{\underline{k}}(\underline{a}) w_{\underline{k}}(\underline{b})$, as group characters, and $w_{\underline{a}}(\underline{b}) = w_{\underline{b}}(\underline{a})$. The Walsh functions $w_{\underline{k}}$, labelled by binary n-tuples $\underline{k} = k_1k_2\ldots k_n$, can be ordered by integers $k = \sum_{i=1}^n k_i 2^{n-i}$.

21.1 Sequency order

In applications one often uses the *sequency order* of the Walsh functions,

$$\tilde{w}_{\underline{k}}(\underline{x}) = (-1)^{k_1 x_1 + \sum_{i=2}^n (k_{i-1}+k_i) x_i},$$

for instance, in special analysis of time series, signal processing, communications and filtering, Harmuth 1977 and Maqusi 1981. In the sequency order the index $\underline{k}$ is often replaced by an integer $k = \sum_{i=1}^{n} k_i 2^{n-i}$ and the argument $\underline{x}$ by a real number on the unit interval $x = 2^{-n} \sum_{i=1}^{n} x_i 2^{i-1}$ (Fig. 1 and Fig. 2).

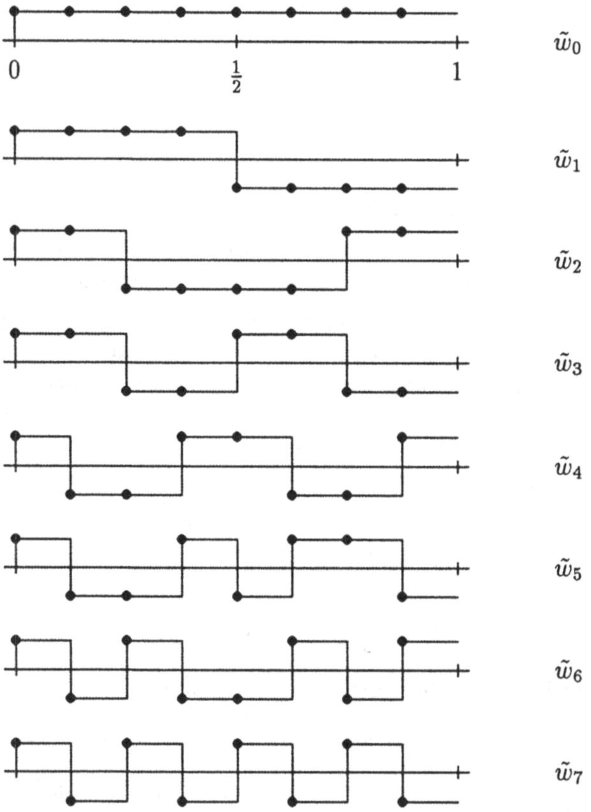

Figure 1. The first eight Walsh functions $\tilde{w}_{\underline{k}}(x)$, $k = 0, 1, \ldots, 7$.

In Figure 1 the first eight Walsh functions are given:

$$\tilde{w}_{\underline{k}}(x) = (-1)^{k_1 x_1 + (k_1 + k_2) x_2 + (k_2 + k_3) x_3}$$

with $k = 4k_1 + 2k_2 + k_3$ and $x = \frac{1}{8}(x_1 + 2x_2 + 4x_3)$. Observe that the number of zero crossings per unit interval equals k.

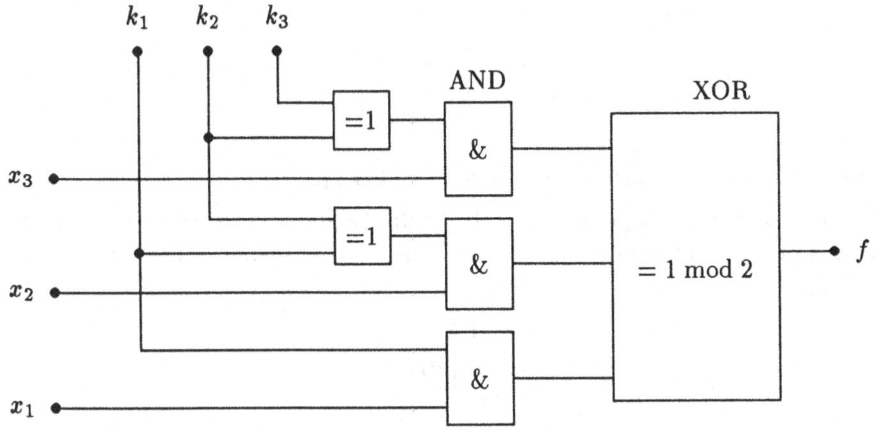

Figure 2. The first eight Walsh functions in hardware, $\tilde{w}_k(x) = (-1)^f$.

21.2 Gray code

The passage to the sequency order is related to the *Gray code* g defined by

$$g(\underline{k})_1 = k_1, \quad g(\underline{k})_i = k_{i-1} + k_i \bmod 2, \quad i = 2, \dots, n.$$

The formula $\tilde{w}_{\underline{a}}(\underline{x}) = w_{g(\underline{a})}(\underline{x})$ reorders the Walsh functions. The Gray code is a single digit change code, that is, the codes of two consecutive integers differ only in one bit (Table 1).

Table 1. The Gray code for $k < 8$.

k	$\underline{k}$	$g(\underline{k})$
0	000	000
1	001	001
2	010	011
3	011	010
4	100	110
5	101	111
6	110	101
7	111	100

The Gray code is a group isomorphism among the binary n-tuples, that is,

$g(\underline{a} \oplus \underline{b}) = g(\underline{a}) \oplus g(\underline{b})$. The *inverse* h of the Gray code is obtained by

$$h(\underline{a})_i = \sum_{j=1}^{i} a_j \bmod 2.$$

BINARY REPRESENTATIONS OF CLIFFORD ALGEBRAS

As a preliminary example, consider the Clifford algebra $\mathcal{C}\ell_{0,2}$, isomorphic to the division ring of quaternions $\mathbb{H}$. Relabel the basis elements of $\mathcal{C}\ell_{0,2}$ by binary 2-tuples

$$\begin{array}{r|l} 1 & \mathbf{e}_{00} \\ \mathbf{e}_1, \mathbf{e}_2 & \mathbf{e}_{10}, \mathbf{e}_{01} \\ \mathbf{e}_{12} & \mathbf{e}_{11} \end{array}$$

and verify the multiplication rule

$$\mathbf{e}_{\underline{a}}\mathbf{e}_{\underline{b}} = w_{\underline{a}}(h(\underline{b}))\mathbf{e}_{\underline{a}\oplus\underline{b}}.$$

For an alternative representation reorder the basis elements by the formula

$$\tilde{\mathbf{e}}_{\underline{a}} = \mathbf{e}_{g(\underline{a})} \quad \text{or} \quad \mathbf{e}_{\underline{a}} = \tilde{\mathbf{e}}_{h(\underline{a})}$$

to get the correspondences

$$\begin{array}{r|l} 1 & \tilde{\mathbf{e}}_{00} \\ \mathbf{e}_1, \mathbf{e}_2 & \tilde{\mathbf{e}}_{11}, \tilde{\mathbf{e}}_{01} \\ \mathbf{e}_{12} & \tilde{\mathbf{e}}_{10}. \end{array}$$

This yields the multiplication rule

$$\tilde{\mathbf{e}}_{\underline{a}}\tilde{\mathbf{e}}_{\underline{b}} = \tilde{w}_{\underline{a}}(\underline{b})\tilde{\mathbf{e}}_{\underline{a}\oplus\underline{b}}.$$

21.3 Clifford multiplication

In general, consider the Clifford algebra $\mathcal{C}\ell_{0,n}$ with n generators $\mathbf{e}_1, \mathbf{e}_2, \ldots, \mathbf{e}_n$ such that

$$\begin{aligned} \mathbf{e}_i^2 &= -1 \quad \text{for} \quad i = 1, 2, \ldots, n, \\ \mathbf{e}_i\mathbf{e}_j &= -\mathbf{e}_j\mathbf{e}_i \quad \text{for} \quad i \neq j. \end{aligned}$$

Theorem 1. If a real 2^n-dimensional algebra A has the multiplication rule

$$\mathbf{e}_{\underline{a}}\mathbf{e}_{\underline{b}} = w_{\underline{a}}(h(b))\mathbf{e}_{\underline{a}\oplus\underline{b}}$$

between the basis elements labelled by the binary n-tuples, then A is isomorphic to the Clifford algebra $\mathcal{C}\ell_{0,n}$.

Proof. It is sufficient to show that A is associative, has a unit element and is generated by n anticommuting elements with square -1.

The element $\mathbf{e}_{\underline{0}} = \mathbf{e}_{000\ldots00}$ is the unit, since $\mathbf{e}_{\underline{a}}\mathbf{e}_{\underline{0}} = w_{\underline{a}}(h(\underline{0}))\mathbf{e}_{\underline{a}\oplus\underline{0}} = w_{\underline{0}}(\underline{0})\mathbf{e}_{\underline{a}} = +\mathbf{e}_{\underline{a}}$ and similarly $\mathbf{e}_{\underline{0}}\mathbf{e}_{\underline{a}} = +\mathbf{e}_{\underline{a}}$. The n basis elements

$$\mathbf{e}_{100\ldots00},\ \mathbf{e}_{010\ldots00},\ \ldots,\ \mathbf{e}_{000\ldots01}$$

generate by definition all of A. Each generator has square $-\mathbf{e}_{\underline{0}}$; in particular for the i:th generator $\mathbf{e}_{\underline{a}}$

$$\underline{a} = \underset{1}{0}0\ldots0\underset{i}{1}0\ldots0\underset{n}{0}, \quad h(\underline{a}) = \underset{1}{0}0\ldots0\underset{i}{1}1\ldots1\underset{n}{1}$$

and so $w_{\underline{a}}(h(\underline{a})) = -1$, from which one concludes that $\mathbf{e}_{\underline{a}}\mathbf{e}_{\underline{a}} = w_{\underline{a}}(h(\underline{a}))\mathbf{e}_{\underline{a}\oplus\underline{a}} = -\mathbf{e}_{\underline{0}}$. In a similar manner one finds that generators anticommute with each other.

Finally, A is associative, since for three arbitrary basis elements $\mathbf{e}_{\underline{a}}, \mathbf{e}_{\underline{b}}, \mathbf{e}_{\underline{c}}$ the condition $(\mathbf{e}_{\underline{a}}\mathbf{e}_{\underline{b}})\mathbf{e}_{\underline{c}} = \mathbf{e}_{\underline{a}}(\mathbf{e}_{\underline{b}}\mathbf{e}_{\underline{c}})$ is equivalent to

$$w_{\underline{a}}(h(\underline{b}))w_{\underline{a}\oplus\underline{b}}(h(\underline{c})) = w_{\underline{a}}(h(\underline{b}\oplus\underline{c}))w_{\underline{b}}(h(\underline{c})),$$

which is a consequence of $w_{\underline{a}\oplus\underline{b}}(\underline{x}) = w_{\underline{a}}(\underline{x})w_{\underline{b}}(\underline{x})$ and $w_{\underline{a}}(\underline{x}\oplus\underline{y}) = w_{\underline{a}}(\underline{x})w_{\underline{a}}(\underline{y})$ and h being a group isomorphism. ∎

It is convenient to assume the correspondences

$$\mathbf{e}_i = \mathbf{e}_{\underset{1}{0}0\ldots0\underset{i}{1}0\ldots0\underset{n}{0}} \quad \text{for} \quad i = 1, 2, \ldots, n$$

between the ordinary and binary representations of the generators of the Clifford algebra $\mathcal{C}\ell_{0,n}$. Then the basis elements of $\mathcal{C}\ell_{0,n}$ are labelled by the binary n-tuples $\underline{a} = a_1a_2\ldots a_n$ as follows:

$$\mathbf{e}_{\underline{a}} = \mathbf{e}_1^{a_1}\mathbf{e}_2^{a_2}\ldots\mathbf{e}_n^{a_n}, \quad a_i = 0, 1.$$

Since the Gray code is a group isomorphism among the binary n-tuples, we can reorder the basis of the Clifford algebra $\mathcal{C}\ell_{0,n}$ by

$$\tilde{\mathbf{e}}_{\underline{a}} = \mathbf{e}_{g(\underline{a})}.$$

This reordering results in a simple multiplication formula:

Corollary. The product of the basis elements of the Clifford algebra $\mathcal{C}\ell_{0,n}$ is given by

$$\tilde{\mathbf{e}}_{\underline{a}}\tilde{\mathbf{e}}_{\underline{b}} = \tilde{w}_{\underline{a}}(\underline{b})\tilde{\mathbf{e}}_{\underline{a}\oplus\underline{b}}.$$

Proof.

$$\tilde{\mathbf{e}}_{\underline{a}}\tilde{\mathbf{e}}_{\underline{b}} = \mathbf{e}_{g(\underline{a})}\mathbf{e}_{g(\underline{b})} = w_{g(\underline{a})}(h(g(\underline{b})))\mathbf{e}_{g(\underline{a})\oplus g(\underline{b})}$$

$$= w_{g(\underline{a})}(\underline{b})\mathbf{e}_{g(\underline{a}\oplus\underline{b})} = \tilde{w}_{\underline{a}}(\underline{b})\tilde{\mathbf{e}}_{\underline{a}\oplus\underline{b}}. \qquad \blacksquare$$

If you choose the signs in $\mathbf{e}_{\underline{a}}\mathbf{e}_{\underline{b}} = \pm\mathbf{e}_{\underline{a}\oplus\underline{b}}$ in some other way, you get other algebras than $\mathcal{C}\ell_{0,n}$. For instance, the Clifford algebra $\mathcal{C}\ell_{p,q}$ over the quadratic form $x_1^2 + \ldots + x_p^2 - x_{p+1}^2 - \ldots - x_{p+q}^2$ has the multiplication formula

$$\mathbf{e}_{\underline{a}}\mathbf{e}_{\underline{b}} = (-1)^{\sum_{i=1}^{p} a_i b_i} w_{\underline{a}}(h(\underline{b}))\mathbf{e}_{\underline{a}\oplus\underline{b}}.$$

Of course, this might also be written without Walsh functions:

$$\mathbf{e}_{\underline{a}}\mathbf{e}_{\underline{b}} = (-1)^{\sum_{i=p+1}^{n} a_i b_i}(-1)^{\sum_{i>j} a_i b_j}\mathbf{e}_{\underline{a}\oplus\underline{b}},$$

a formula essentially obtained by Brauer & Weyl 1935. See also Artin 1957 and Delanghe & Brackx 1978 for a related definition of the product on the Clifford algebras (based on sums of multi-indices).

21.4 An iterative process to form Clifford algebras

Clifford algebras can be obtained by a method analogous to the Cayley-Dickson process. Consider pairs (u, v) of elements u and v in the Clifford algebra $\mathcal{C}\ell_{p,q}$. Define a product for two such pairs,

$$(u_1, v_1)(u_2, v_2) = (u_1 u_2 \pm v_1 \hat{v}_2, u_1 v_2 + v_1 \hat{u}_2),$$

where $u \to \hat{u}$ is the grade involution of $\mathcal{C}\ell_{p,q}$. This results in an algebra isomorphic to the Clifford algebra

$$\mathcal{C}\ell_{p+1,q}$$

or

$$\mathcal{C}\ell_{p,q+1}$$

according to the $\pm$ sign. This iterative process could be repeated by noting that $(u, v)\hat{} = (\hat{u}, -\hat{v})$.

For more details on the Clifford algebras see Micali & Revoy 1977 and Porteous 1969, 1981.

Some Clifford-like Algebras

All the above algebras are special cases of the following. Let A be a real linear space of dimension 2^n. Label a basis for A by binary n-tuples $\underline{a}$ to get the basis elements $\mathbf{e}_{\underline{a}}$. Then define a multiplication between the basis elements $\mathbf{e}_{\underline{a}}$ and extend it to all of A by linearity. The definition is of the form

$$\mathbf{e}_{\underline{a}}\mathbf{e}_{\underline{b}} = \pm\mathbf{e}_{\underline{a}\oplus\underline{b}}$$

for a certain choice of signs. Then the algebra A is a direct sum of the 1-dimensional subspaces $U_{\underline{a}}$, spanned by $\mathbf{e}_{\underline{a}}$, satisfying

$$U_{\underline{a}} U_{\underline{b}} \subset U_{\underline{a} \oplus \underline{b}}.$$

In other words A is an algebra graded by the abelian group of binary n-tuples Ω. This grading is maximal (Kwasniewski 1985), and these algebras will be called Clifford-like algebras. Next we shall study some Clifford-like algebras.

21.5 Cayley-Dickson process

Consider a generalized quaternion ring Q with $i^2 = \gamma_1$, $j^2 = \gamma_2$ and $k^2 = \gamma_1 \gamma_2$, where $\gamma_1, \gamma_2 = \pm 1$. The conjugation-involution $u \to u^L$ of Q is given by

$$i^L = -i, \quad j^L = -j, \quad k^L = -k.$$

Introduce a multiplication in the 8-dimensional real linear space $Q \times Q$ by the formula

$$(u_1, v_1) \circ (u_2, v_2) = (u_1 u_2 + \gamma_3 v_2^L v_1, v_2 u_1 + v_1 u_2^L)$$

where $\gamma_3 = \pm 1$. Inducing an anti-involution $(u, v)^L = (u^L, -v)$ of $Q \times Q = CD(\gamma_1, \gamma_2, \gamma_3)$ makes it possible to repeat this *Cayley-Dickson process* to get an algebra $CD(\gamma_1, \gamma_2, \ldots, \gamma_n)$, where $\gamma_i = \pm 1$. In fact, the Cayley-Dickson process could be started with $\mathbb{R}$ to give $CD(\gamma_1)$ and $Q = CD(\gamma_1, \gamma_2)$.

Example. $CD(-1) \simeq \mathbb{C}$, $CD(-1, -1) \simeq \mathbb{H}$, and $CD(-1, -1, -1) \simeq \mathbb{O}$, the real 8-dimensional alternative division algebra of octonions (Porteous 1969, 1981). ∎

The algebras $CD(\gamma_1, \gamma_2, \ldots, \gamma_n)$ obtained by the Cayley-Dickson process are simple flexible algebras of dimension 2^n (Schafer 1954). Every element of such an algebra satisfies a quadratic equation with real coefficients.

21.6 Binary representation of the Cayley-Dickson process

The algebras formed by the Cayley-Dickson process are Clifford-like algebras. For instance, choose a basis of $CD(\gamma_1) = \mathbb{R} \times \mathbb{R}$,

$$\mathbf{e}_0 = (1, 0), \quad \mathbf{e}_1 = (0, 1),$$

and introduce the multiplication rule

$$\mathbf{e}_{\underline{a}} \mathbf{e}_{\underline{b}} = \gamma_1^{a_1 b_1} \mathbf{e}_{\underline{a} \oplus \underline{b}} \quad (\underline{a} = a_1, \underline{b} = b_1).$$

The involution is given by

$$\mathbf{e}_0^L = (1,0) = \mathbf{e}_0, \quad \mathbf{e}_1^L = (0,-1) = -\mathbf{e}_1$$

or in a condensed form $\mathbf{e}_{\underline{a}}^L = (-1)^{a_1}\mathbf{e}_{\underline{a}}$.

Theorem 2. A Clifford-like algebra A, $\dim A = 2^n$, with multiplication rule

$$\mathbf{e}_{\underline{a}}\mathbf{e}_{\underline{b}} = f(\underline{a},\underline{b})\mathbf{e}_{\underline{a}\oplus\underline{b}}$$
$$f(\underline{a},\underline{b}) = (-1)^{\sum_{i=1}^{n-1}((S_i(\underline{a})+S_i(\underline{b})+S_i(\underline{a}\oplus\underline{b}))b_{i+1}+S_i(\underline{b})a_{i+1})} \times \prod_{i=1}^{n} \gamma_i^{a_i b_i},$$

where $S_i(\underline{a})$ is the maximum of a_j for $1 \le j \le i$, is isomorphic to the Cayley-Dickson algebra $CD(\gamma_1, \gamma_2, \ldots, \gamma_n)$. The anti-involution is

$$\mathbf{e}_{\underline{a}}^L = (-1)^{S_n(\underline{a})}\mathbf{e}_{\underline{a}}.$$

Proof. The first case of the mathematical induction is proved in the example above.

Assume that the statement holds up to the n:th step, and apply the Cayley-Dickson process. If the new basis elements are denoted by

$$\mathbf{e}_{a_1a_2\ldots a_n a_{n+1}} = \begin{cases} (\mathbf{e}_{\underline{a}}, 0), & a_{n+1} = 0 \\ (0, \mathbf{e}_{\underline{a}}), & a_{n+1} = 1 \end{cases}$$

or $\mathbf{e}_{\underline{a}a_{n+1}} = \mathbf{e}_{a_1a_2\ldots a_n a_{n+1}}$ for short, then

$$\mathbf{e}_{\underline{a}0}\mathbf{e}_{\underline{b}0} = (\mathbf{e}_{\underline{a}},0)(\mathbf{e}_{\underline{b}},0) = (\mathbf{e}_{\underline{a}}\mathbf{e}_{\underline{b}},0) = f(\underline{a},\underline{b})(\mathbf{e}_{\underline{a}\oplus\underline{b}},0) = f(\underline{a},\underline{b})\mathbf{e}_{\underline{a}\oplus\underline{b}0}$$
$$\mathbf{e}_{\underline{a}1}\mathbf{e}_{\underline{b}0} = (0,\mathbf{e}_{\underline{a}})(\mathbf{e}_{\underline{b}},0) = (0,\mathbf{e}_{\underline{a}}\mathbf{e}_{\underline{b}}^L) = (-1)^{S_n(\underline{b})}f(\underline{a},\underline{b})\mathbf{e}_{\underline{a}\oplus\underline{b}1}$$
$$\mathbf{e}_{\underline{a}0}\mathbf{e}_{\underline{b}1} = (\mathbf{e}_{\underline{a}},0)(0,\mathbf{e}_{\underline{b}}) = (0,\mathbf{e}_{\underline{b}}\mathbf{e}_{\underline{a}}) = f(\underline{b},\underline{a})\mathbf{e}_{\underline{a}\oplus\underline{b}1}$$
$$\mathbf{e}_{\underline{a}1}\mathbf{e}_{\underline{b}1} = (0,\mathbf{e}_{\underline{a}})(0,\mathbf{e}_{\underline{b}}) = (\gamma_{n+1}\mathbf{e}_{\underline{b}}^L\mathbf{e}_{\underline{a}},0) = \gamma_{n+1}(-1)^{S_n(\underline{b})}f(\underline{b},\underline{a})\mathbf{e}_{\underline{a}\oplus\underline{b}0}.$$

These four equations can be condensed into one equation

$$\mathbf{e}_{\underline{a}a_{n+1}}\mathbf{e}_{\underline{b}b_{n+1}} = f(\underline{a},\underline{b})^{1-b_{n+1}}f(\underline{b},\underline{a})^{b_{n+1}} \times \gamma_{n+1}^{a_{n+1}b_{n+1}}(-1)^{a_{n+1}S_n(\underline{b})}\mathbf{e}_{\underline{a}\oplus\underline{b}(a_{n+1}\oplus b_{n+1})},$$

where

$$f(\underline{b},\underline{a}) = f(\underline{a},\underline{b})(-1)^{S_n(\underline{a})+S_n(\underline{b})+S_n(\underline{a}\oplus\underline{b})},$$

which is a consequence of $(\mathbf{e}_{\underline{a}}\mathbf{e}_{\underline{b}})^L = \mathbf{e}_{\underline{b}}^L\mathbf{e}_{\underline{a}}^L$. Thus we have proved the desired multiplication rule in the case $n+1$. The induced anti-involution is also of the assumed type:

$$\mathbf{e}_{\underline{a}0}^L = (\mathbf{e}_{\underline{a}}^L,0) = (-1)^{S_n(\underline{a})}\mathbf{e}_{\underline{a}0}$$
$$\mathbf{e}_{\underline{a}1}^L = (0,-\mathbf{e}_{\underline{a}}) = -\mathbf{e}_{\underline{a}1}$$

or in a condensed form

$$e^L_{\underline{a}a_{n+1}} = (-1)^{\max(S_n(\underline{a}),a_{n+1})} e_{\underline{a}a_{n+1}}.$$ ∎

The algebra $CD(\gamma_1, \gamma_2, \ldots, \gamma_n)$ is generated by an n-dimensional vector space, whose elements

$$x_1 e_{100\ldots00} + x_2 e_{010\ldots00} + \ldots + x_n e_{000\ldots01}$$

have squares $(\gamma_1 x_1^2 + \gamma_2 x_2^2 + \ldots + \gamma_n x_n^2)e_{\underline{0}}$. In contrast to the Clifford algebras, different orderings of the parameters γ_i in $CD(\gamma_1, \gamma_2, \ldots, \gamma_n)$ may result in non-isomorphic algebras in the case where $n > 3$.

Another construction relating Clifford algebras and Cayley-Dickson algebras is found in Wene 1984.

For more details of the algebraic extensions of the group of binary n-tuples Ω see Hagmark 1980.

Bibliography

E. Artin: *Geometric Algebra*. Interscience, New York, 1957.

R. Brauer, H. Weyl: Spinors in n dimensions. *Amer. J. Math.* **57** (1935), 425-449.

R. Delanghe, F. Brackx: Hypercomplex function theory and Hilbert modules with reproducing kernel. *Proc. London Math. Soc.* (3) **37** (1978),545-576.

N.J. Fine: On the Walsh functions. *Trans. Amer. Math. Soc.* **65** (1949), 372-414.

P.-E. Hagmark: Construction of some 2^n-dimensional algebras. *Helsinki UT, Math. Report* A**177**, 1980.

H.F. Harmuth: *Sequency Theory, Foundations and Applications*. Academic Press, New York, 1977.

A.K. Kwasniewski: Clifford- and Grassmann-like algebras – old and new. *J. Math. Phys.* **26** (1985), 2234-2238.

M. Maqusi: *Walsh Analysis and Applications*. Heyden, London, 1981.

A. Micali, Ph. Revoy: *Modules quadratiques*. Cahiers Mathématiques **10**, Montpellier, 1977. *Bull. Soc. Math. France* **63**, suppl. (1979), 5-144.

I.R. Porteous: *Topological Geometry*. VNR, London, 1969. Cambridge University Press, Cambridge, 1981.

R.D. Schafer: On the algebras formed by the Cayley-Dickson process. *Amer. J. Math.* **76** (1954), 435-446.

K. Th. Vahlen: Über höhere komplexe Zahlen. *Schriften der phys.-ökon. Gesellschaft zu Königsberg* **38** (1897), 72-78.

G.P. Wene: A construction relating Clifford algebras and Cayley-Dickson algebras. *J. Math. Phys.* **25** (1984), 2351-2353.

22

Chevalley's Construction and Characteristic 2

Consider an n-dimensional linear space V over a field $\mathbb{F}$, $\operatorname{char}\mathbb{F} \neq 2$, and the symmetric bilinear form

$$<\mathbf{x},\mathbf{y}> = \frac{1}{2}[Q(\mathbf{x}+\mathbf{y}) - Q(\mathbf{x}) - Q(\mathbf{y})]$$

associated with the quadratic form Q. Later in this chapter we will discuss the case $\operatorname{char}\mathbb{F} = 2$. As before, we denote the exterior algebra of V by $\bigwedge V$ and the Clifford algebra of Q, with $\mathbf{x}^2 = Q(\mathbf{x})$, by $\mathcal{C}\ell(Q)$ or $\mathcal{C}\ell_{p,q}$ when $V = \mathbb{R}^{p,q}$ and

$$Q(\mathbf{x}) = x_1^2 + x_2^2 + \ldots + x_p^2 - x_{p+1}^2 - \ldots - x_{p+q}^2, \qquad n = p + q.$$

We shall construct a natural linear isomorphism $\bigwedge V \to \mathcal{C}\ell(Q)$, review how Riesz goes backwards and derives $\bigwedge V$ from $\mathcal{C}\ell(Q)$ and compare Riesz's method to an alternative construction due to Chevalley but known to some theoretical physicists in the disguise of the Kähler-Atiyah isomorphism.

22.1 Construction of the linear isomorphism $\bigwedge V \to \mathcal{C}\ell(Q)$

Here we start from the exterior algebra $\bigwedge V$. Recall that a k-vector $\mathbf{a} \in \bigwedge^k V$ has the grade involute $\hat{\mathbf{a}} = (-1)^k\mathbf{a}$ and the reverse $\tilde{\mathbf{a}} = (-1)^{\frac{1}{2}k(k-1)}\mathbf{a}$. The symmetric bilinear form associated with Q on V can be extended to simple k-vectors in $\bigwedge^k V$ by way of

$$<\mathbf{x}_1 \wedge \mathbf{x}_2 \wedge \ldots \wedge \mathbf{x}_k, \mathbf{y}_1 \wedge \mathbf{y}_2 \wedge \ldots \wedge \mathbf{y}_k> = \det<\mathbf{x}_i, \mathbf{y}_j>,$$

where

$$\det<\mathbf{x}_i,\mathbf{y}_j> = \begin{vmatrix} <\mathbf{x}_1,\mathbf{y}_1> & <\mathbf{x}_1,\mathbf{y}_2> & \cdots & <\mathbf{x}_1,\mathbf{y}_k> \\ <\mathbf{x}_2,\mathbf{y}_1> & <\mathbf{x}_2,\mathbf{y}_2> & \cdots & <\mathbf{x}_2,\mathbf{y}_k> \\ \vdots & \vdots & \ddots & \vdots \\ <\mathbf{x}_k,\mathbf{y}_1> & <\mathbf{x}_k,\mathbf{y}_2> & \cdots & <\mathbf{x}_k,\mathbf{y}_k> \end{vmatrix},$$

and further by linearity to all of $\bigwedge^k V$ and by orthogonality to all of $\bigwedge V$.

Example. Let $Q(x_1\mathbf{e}_1 + x_2\mathbf{e}_2) = ax_1^2 + bx_2^2$. Then $<\mathbf{x},\mathbf{y}> = ax_1y_1 + bx_2y_2$ and $\mathbf{x}\wedge\mathbf{y} = (x_1y_2 - x_2y_1)\mathbf{e}_1\wedge\mathbf{e}_2$. The identity $(ax_1^2+bx_2^2)(ay_1^2+by_2^2) = (ax_1y_1+bx_2y_2)^2 + ab(x_1y_2 - x_2y_1)^2$ can be written as $Q(\mathbf{x})Q(\mathbf{y}) = <\mathbf{x},\mathbf{y}>^2 + Q(\mathbf{x}\wedge\mathbf{y})$, where $Q(\mathbf{x}\wedge\mathbf{y}) = ab(x_1y_2 - x_2y_1)^2$. ∎

In the case of a non-degenerate Q on V we can introduce the dual of the exterior product called the left contraction $u \rfloor v$ of $v \in \bigwedge V$ by $u \in \bigwedge V$ through the requirement

$$<u \rfloor v, w> = <v, \tilde{u} \wedge w> \quad \text{for all} \quad w \in \bigwedge V.$$

Examples. 1. Let $\mathbf{x},\mathbf{y} \in V$, $w \in \mathbb{F}$. Then $<\mathbf{x} \rfloor \mathbf{y}, w> = <\mathbf{y}, \mathbf{x}\wedge w> = <\mathbf{y}, \mathbf{x}w> = <\mathbf{y},\mathbf{x}>w$ and since $<\mathbf{x} \rfloor \mathbf{y}, w> = <\mathbf{x} \rfloor \mathbf{y}, 1>w$ we have the rule $\mathbf{x} \rfloor \mathbf{y} = <\mathbf{x},\mathbf{y}>$.

2. Let $\mathbf{x}$, $\mathbf{y}$, $\mathbf{z}$, $\mathbf{w} \in V$. Then $<\mathbf{x} \rfloor (\mathbf{y}\wedge\mathbf{z}), \mathbf{w}> = <\mathbf{y}\wedge\mathbf{z}, \mathbf{x}\wedge\mathbf{w}>$

$$= \begin{vmatrix} <\mathbf{y},\mathbf{x}> & <\mathbf{y},\mathbf{w}> \\ <\mathbf{z},\mathbf{x}> & <\mathbf{z},\mathbf{w}> \end{vmatrix}$$

$= (\mathbf{x} \rfloor \mathbf{y})<\mathbf{z},\mathbf{w}> - (\mathbf{x} \rfloor \mathbf{z})<\mathbf{y},\mathbf{w}> = <(\mathbf{x} \rfloor \mathbf{y})\mathbf{z} - (\mathbf{x} \rfloor \mathbf{z})\mathbf{y}, \mathbf{w}>$ and so we have the rule $\mathbf{x} \rfloor (\mathbf{y}\wedge\mathbf{z}) = (\mathbf{x} \rfloor \mathbf{y})\mathbf{z} - (\mathbf{x} \rfloor \mathbf{z})\mathbf{y}$.

3. Let $\mathbf{x} \in V$, $\mathbf{x}_i \in V$ and $w = \mathbf{w}_1 \wedge \mathbf{w}_2 \wedge \ldots \wedge \mathbf{w}_{k-1} \in \bigwedge^{k-1} V$. Then $<\mathbf{x} \rfloor (\mathbf{x}_1\wedge\mathbf{x}_2\wedge\ldots\wedge\mathbf{x}_k), w> = <\mathbf{x}_1\wedge\mathbf{x}_2\wedge\ldots\wedge\mathbf{x}_k, \mathbf{x}\wedge\mathbf{w}_1\wedge\mathbf{w}_2\wedge\ldots\wedge\mathbf{w}_{k-1}>$

$$= \begin{vmatrix} <\mathbf{x}_1,\mathbf{x}> & <\mathbf{x}_1,\mathbf{w}_1> & <\mathbf{x}_1,\mathbf{w}_2> & \cdots & <\mathbf{x}_1,\mathbf{w}_{k-1}> \\ <\mathbf{x}_2,\mathbf{x}> & <\mathbf{x}_2,\mathbf{w}_1> & <\mathbf{x}_2,\mathbf{w}_2> & \cdots & <\mathbf{x}_2,\mathbf{w}_{k-1}> \\ \vdots & \vdots & \vdots & \ddots & \vdots \\ <\mathbf{x}_k,\mathbf{x}> & <\mathbf{x}_k,\mathbf{w}_1> & <\mathbf{x}_k,\mathbf{w}_2> & \cdots & <\mathbf{x}_k,\mathbf{w}_{k-1}> \end{vmatrix}$$

$$= \sum_{i=1}^{k}(-1)^{i-1}<\mathbf{x},\mathbf{x}_i><\mathbf{x}_1\wedge\mathbf{x}_2\wedge\ldots\wedge\mathbf{x}_{i-1}\wedge\mathbf{x}_{i+1}\wedge\ldots\wedge\mathbf{x}_k, w>$$

and so

$$\mathbf{x} \rfloor (\mathbf{x}_1\wedge\mathbf{x}_2\wedge\ldots\wedge\mathbf{x}_k) = \sum_{i=1}^{k}(-1)^{i-1}<\mathbf{x},\mathbf{x}_i>\mathbf{x}_1\wedge\mathbf{x}_2\wedge\ldots\wedge\mathbf{x}_{i-1}\wedge\mathbf{x}_{i+1}\wedge\ldots\wedge\mathbf{x}_k.$$

4. $\mathbf{x} \rfloor (u\wedge v) = \mathbf{x} \rfloor ((\mathbf{u}_1\wedge\mathbf{u}_2\wedge\ldots\wedge\mathbf{u}_i)\wedge(\mathbf{v}_1\wedge\mathbf{v}_2\wedge\ldots\wedge\mathbf{v}_j))$

$$= \sum_{k=1}^{i}(\mathbf{x} \rfloor \mathbf{u}_k)(-1)^{k-1}\mathbf{u}_1\wedge\mathbf{u}_2\wedge\ldots\wedge\mathbf{u}_{k-1}\wedge\mathbf{u}_{k+1}\wedge\ldots\wedge\mathbf{u}_i\wedge v$$

$$+(-1)^i \sum_{k=1}^{j} (\mathbf{x} \rfloor \mathbf{v}_k)(-1)^{k-1} u \wedge \mathbf{v}_1 \wedge \mathbf{v}_2 \wedge \ldots \wedge \mathbf{v}_{k-1} \wedge \mathbf{v}_{k+1} \wedge \ldots \wedge \mathbf{v}_j$$

$$= (\mathbf{x} \rfloor u) \wedge v + (-1)^i u \wedge (\mathbf{x} \rfloor v).$$

5. $<(u \wedge v) \rfloor w, z> = <w, (u \wedge v)^{\sim} \wedge z> = <w, \tilde{v} \wedge \tilde{u} \wedge z> = <v \rfloor w, \tilde{u} \wedge z>$ $= <u \rfloor (v \rfloor w), z>$ and so $(u \wedge v) \rfloor w = u \rfloor (v \rfloor w)$. ∎

In the case of a non-degenerate Q we have verified the following properties of the contraction:

(a) $\quad \mathbf{x} \rfloor \mathbf{y} = <\mathbf{x}, \mathbf{y}> \quad$ for $\quad \mathbf{x}, \mathbf{y} \in V$

(b) $\quad \mathbf{x} \rfloor (u \wedge v) = (\mathbf{x} \rfloor u) \wedge v + \hat{u} \wedge (\mathbf{x} \rfloor v)$

(c) $\quad (u \wedge v) \rfloor w = u \rfloor (v \rfloor w) \quad$ for $\quad u, v, w \in \bigwedge V$

(see Helmstetter 1982). These properties also determine the contraction uniquely for an arbitrary, not necessarily non-degenerate Q. The identity (c) introduces a scalar multiplication on $\bigwedge V$ making it a left module over $\bigwedge V$. The identity (b) means that contraction by $\mathbf{x} \in V$ operates like a derivation. Evidently, $\mathbf{x} \rfloor \mathbf{a} \in \bigwedge^{k-1} V$ for $\mathbf{a} \in \bigwedge^k V$.

Introduce the *Clifford product* of $\mathbf{x} \in V$ and $u \in \bigwedge V$ by

$$\mathbf{x}u = \mathbf{x} \wedge u + \mathbf{x} \rfloor u$$

and extend this product by linearity and associativity to all of $\bigwedge V$, which then becomes, as an associative algebra, isomorphic to $\mathcal{C}\ell(Q)$. For instance, the product of a simple bivector $\mathbf{x} \wedge \mathbf{y} \in \bigwedge^2 V$ and an arbitrary element $u \in \bigwedge V$ is given by

$$(\mathbf{x} \wedge \mathbf{y})u = \mathbf{x} \wedge \mathbf{y} \wedge u + \mathbf{x} \wedge (\mathbf{y} \rfloor u) - \mathbf{y} \wedge (\mathbf{x} \rfloor u) + \mathbf{x} \rfloor (\mathbf{y} \rfloor u)$$

where we have first expanded $(\mathbf{x} \wedge \mathbf{y})u = (\mathbf{x}\mathbf{y} - \mathbf{x} \rfloor \mathbf{y})u$, then used

$$\mathbf{x}(\mathbf{y}u) = \mathbf{x} \wedge \mathbf{y} \wedge u + \mathbf{x} \wedge (\mathbf{y} \rfloor u) + \mathbf{x} \rfloor (\mathbf{y} \wedge u) + \mathbf{x} \rfloor (\mathbf{y} \rfloor u)$$

and the derivation rule $\mathbf{x} \rfloor (\mathbf{y} \wedge u) = (\mathbf{x} \rfloor \mathbf{y}) \wedge u - \mathbf{y} \wedge (\mathbf{x} \rfloor u)$.

Exercises 1,2,. . . ,10

Remark. Some authors use, instead of the left and right contractions, a more symmetric dot product in $\mathcal{C}\ell(Q)$, defined by the Clifford product for homogeneous elements as (char $\neq 2$)

$$\mathbf{a} \cdot \mathbf{b} = \langle \mathbf{ab} \rangle_{|i-j|}, \quad \mathbf{a} \in \overset{i}{\bigwedge} V,\ \mathbf{b} \in \overset{j}{\bigwedge} V,$$

and extended by linearity to all of $C\ell(Q)$. The relation between the dot product and the contractions,

$$u \cdot v = u \rfloor v + u \lfloor v - <u, \tilde{v}>,$$

shows that the dot product cannot be expected to have properties, which can be easily proved (using the more fundamental contractions). Some authors try to make the dot product look like derivation, and define exceptionally $\lambda \cdot u = 0$, $u \cdot \lambda = 0$ for $\lambda \in \mathbb{F}$. This only makes things worse, because for this dot product the relation to the contractions is still more complicated:

$$u \cdot v = u \rfloor v + u \lfloor v - <u, \tilde{v}> - \langle u \rangle_0 v - u \langle v \rangle_0 + \langle u \rangle_0 \langle v \rangle_0.$$

■

22.2 Chevalley's identification of $C\ell(Q) \subset \mathrm{End}(\bigwedge V)$

Chevalley 1954 pp. 38-42 tried to include the characteristic 2 by embedding the Clifford algebra $C\ell(Q)$ into the endomorphism algebra $\mathrm{End}(\bigwedge V)$ of the exterior algebra $\bigwedge V$. He introduced a linear operator $L'_{\mathbf{x}} = \varphi_{\mathbf{x}} \in \mathrm{End}(\bigwedge V)$ such that

$$\varphi_{\mathbf{x}}(u) = \mathbf{x} \wedge u + \mathbf{x} \rfloor u \quad \text{for} \quad \mathbf{x} \in V,\ u \in \bigwedge V.$$

From the derivation rule $\mathbf{x} \rfloor (\mathbf{x} \wedge u) = (\mathbf{x} \rfloor \mathbf{x}) \wedge u - \mathbf{x} \wedge (\mathbf{x} \rfloor u)$ and $\mathbf{x} \wedge \mathbf{x} \wedge u = 0$, $\mathbf{x} \rfloor (\mathbf{x} \rfloor u) = 0$ we can conclude the identity $(\varphi_{\mathbf{x}})^2 = Q(\mathbf{x})$. Chevalley's inclusion map $V \to \mathrm{End}(\bigwedge V)$, $\mathbf{x} \to \varphi_{\mathbf{x}}$ is then a Clifford map and can be extended to an algebra homomorphism $\psi : C\ell(Q) \to \mathrm{End}(\bigwedge V)$, whose image evaluated at $1 \in \bigwedge V$ yields the map $\phi : \mathrm{End}(\bigwedge V) \to \bigwedge V$. The composite linear map $\theta = \phi \circ \psi$ is the right inverse of the natural map $\bigwedge V \to C\ell(Q)$ and

$$\bigwedge V \to C\ell(Q) \xrightarrow{\psi} \mathrm{End}(\bigwedge V) \xrightarrow{\phi} \bigwedge V$$

is the identity mapping on $\bigwedge V$. The faithful representation ψ sends $C\ell(Q)$ onto an isomorphic subalgebra of $\mathrm{End}(\bigwedge V)$.

Chevalley's identification also works well with a contraction defined by an arbitrary – *not necessarily symmetric* – bilinear form B such that $B(\mathbf{x}, \mathbf{x}) = Q(\mathbf{x})$ and

(a) $\mathbf{x} \rfloor \mathbf{y} = B(\mathbf{x}, \mathbf{y})$ for $\mathbf{x}, \mathbf{y} \in V$

(b) $\mathbf{x} \rfloor (u \wedge v) = (\mathbf{x} \rfloor u) \wedge v + \hat{u} \wedge (\mathbf{x} \rfloor v)$

(c) $(u \wedge v) \rfloor w = u \rfloor (v \rfloor w)$ for $u, v, w \in \bigwedge V$

(see Helmstetter 1982). As before, $\mathbf{x} \lrcorner \mathbf{a} \in \bigwedge^{k-1} V$ for $\mathbf{a} \in \bigwedge^k V$ and

$$\mathbf{x} \lrcorner (\mathbf{x}_1 \wedge \mathbf{x}_2 \wedge \ldots \wedge \mathbf{x}_k) = \sum_{i=1}^{k} (-1)^{i-1} B(\mathbf{x}, \mathbf{x}_i)\, \mathbf{x}_1 \wedge \mathbf{x}_2 \wedge \ldots \wedge \mathbf{x}_{i-1} \wedge \mathbf{x}_{i+1} \wedge \ldots \wedge \mathbf{x}_k,$$

and the faithful representation ψ sends the Clifford algebra $C\ell(Q)$ onto an isomorphic subalgebra of $\mathrm{End}(\bigwedge V)$, which, however, as a subspace depends on B.

Remark. Chevalley introduced his identification $C\ell(Q) \subset \mathrm{End}(\bigwedge V)$ in order to be able to include the exceptional case of characteristic 2. In characteristic $\neq 2$ the theory of quadratic forms is the same as the theory of symmetric bilinear forms and Chevalley's identification gives the Clifford algebra of the *symmetric* bilinear form $<\mathbf{x}, \mathbf{y}> = \frac{1}{2}(B(\mathbf{x}, \mathbf{y}) + B(\mathbf{y}, \mathbf{x}))$ satisfying $\mathbf{xy} + \mathbf{yx} = 2<\mathbf{x}, \mathbf{y}>$. ∎

For arbitrary Q but $\mathrm{char}\,\mathbb{F} \neq 2$ there is the natural choice of the unique symmetric bilinear form B such that $B(\mathbf{x}, \mathbf{x}) = Q(\mathbf{x})$ giving rise to the canonical/privileged linear isomorphism $C\ell(Q) \to \bigwedge V$. The case $\mathrm{char}\,\mathbb{F} = 2$ is quite different. In general, there are no symmetric bilinear forms such that $B(\mathbf{x}, \mathbf{x}) = Q(\mathbf{x})$ and when there is such a symmetric bilinear form, it is not unique since any alternating [1] bilinear form is also symmetric and could be added to the symmetric bilinear form without changing Q. Hence the contraction is not unique if $\mathrm{char}\,\mathbb{F} \neq 2$, and there is an ambiguity in $\varphi_{\mathbf{x}}$.

In characteristic 2 the theory of quadratic forms is not the same as the theory of symmetric bilinear forms.

Example. Let $\dim_{\mathbb{F}} V = 2$, $B(\mathbf{x}, \mathbf{y}) = ax_1y_1 + bx_1y_2 + cx_2y_1 + dx_2y_2$ and $Q(\mathbf{x}) = B(\mathbf{x}, \mathbf{x})$. The contraction $\mathbf{x} \lrcorner \mathbf{y} = B(\mathbf{x}, \mathbf{y})$ gives the Clifford product $\mathbf{x}v = \mathbf{x} \wedge v + \mathbf{x} \lrcorner v$ of $\mathbf{x} \in V$, $v \in \bigwedge V$. We will determine the matrix of $v \to uv$, $u = u_0 + u_1\mathbf{e}_1 + u_2\mathbf{e}_2 + u_{12}\mathbf{e}_1 \wedge \mathbf{e}_2$ with respect to the basis $\{1, \mathbf{e}_1, \mathbf{e}_2, \mathbf{e}_1 \wedge \mathbf{e}_2\}$ for $\bigwedge V$. The matrix of $\mathbf{e}_1$ is obtained by the following computation:

$$\mathbf{e}_1 1 = \mathbf{e}_1 \wedge 1 = \mathbf{e}_1 \qquad \text{(first column} \quad = 0100)$$

$$\mathbf{e}_1\mathbf{e}_1 = \mathbf{e}_1 \lrcorner \mathbf{e}_1 = a \qquad \text{(second column} = a000)$$

$$\mathbf{e}_1\mathbf{e}_2 = \mathbf{e}_1 \wedge \mathbf{e}_2 + \mathbf{e}_1 \lrcorner \mathbf{e}_2 = \mathbf{e}_1 \wedge \mathbf{e}_2 + b \qquad \text{(third column} \quad = b001)$$

$$\mathbf{e}_1(\mathbf{e}_1 \wedge \mathbf{e}_2) = \mathbf{e}_1 \lrcorner (\mathbf{e}_1 \wedge \mathbf{e}_2) = (\mathbf{e}_1 \lrcorner \mathbf{e}_1) \wedge \mathbf{e}_2 - (\mathbf{e}_1 \lrcorner \mathbf{e}_2) \wedge \mathbf{e}_1 = a\mathbf{e}_2 - b\mathbf{e}_1.$$

1 Recall that antisymmetric means $B(\mathbf{x}, \mathbf{y}) = -B(\mathbf{y}, \mathbf{x})$ and alternating $B(\mathbf{x}, \mathbf{x}) = 0$; alternating is always antisymmetric, though in characteristic 2 antisymmetric is not necessarily alternating.

The matrix of $\mathbf{e}_1 \wedge \mathbf{e}_2$ is obtained by

$$(\mathbf{e}_1 \wedge \mathbf{e}_2)\,1 = \mathbf{e}_1 \wedge \mathbf{e}_2$$
$$(\mathbf{e}_1 \wedge \mathbf{e}_2)\mathbf{e}_1 = (\mathbf{e}_1 \wedge \mathbf{e}_2) \llcorner \mathbf{e}_1 = \mathbf{e}_1 \wedge (\mathbf{e}_2 \llcorner \mathbf{e}_1) - \mathbf{e}_2 \wedge (\mathbf{e}_1 \llcorner \mathbf{e}_1) = c\mathbf{e}_1 - a\mathbf{e}_2$$
$$(\mathbf{e}_1 \wedge \mathbf{e}_2)\mathbf{e}_2 = (\mathbf{e}_1\mathbf{e}_2 - \mathbf{e}_1 \lrcorner \mathbf{e}_2)\mathbf{e}_2 = \mathbf{e}_1\mathbf{e}_2^2 - (\mathbf{e}_1 \lrcorner \mathbf{e}_2)\mathbf{e}_2 = d\mathbf{e}_1 - b\mathbf{e}_2$$
$$(\mathbf{e}_1 \wedge \mathbf{e}_2)(\mathbf{e}_1 \wedge \mathbf{e}_2) = (\mathbf{e}_1\mathbf{e}_2 - \mathbf{e}_1 \lrcorner \mathbf{e}_2)(\mathbf{e}_1 \wedge \mathbf{e}_2)$$
$$= \mathbf{e}_1(\mathbf{e}_2 \wedge (\mathbf{e}_1 \wedge \mathbf{e}_2) + \mathbf{e}_2 \lrcorner (\mathbf{e}_1 \wedge \mathbf{e}_2)) - (\mathbf{e}_1 \lrcorner \mathbf{e}_2)(\mathbf{e}_1 \wedge \mathbf{e}_2)$$
$$= \mathbf{e}_1(c\mathbf{e}_2 - d\mathbf{e}_1) - b(\mathbf{e}_1 \wedge \mathbf{e}_2) = -ad + bc + (-b + c)(\mathbf{e}_1 \wedge \mathbf{e}_2).$$

So we have the following matrix representations:

$$\mathbf{e}_1 = \begin{pmatrix} 0 & a & b & 0 \\ 1 & 0 & 0 & -b \\ 0 & 0 & 0 & a \\ 0 & 0 & 1 & 0 \end{pmatrix}, \qquad \mathbf{e}_2 = \begin{pmatrix} 0 & c & d & 0 \\ 0 & 0 & 0 & -d \\ 1 & 0 & 0 & c \\ 0 & -1 & 0 & 0 \end{pmatrix},$$

$$\mathbf{e}_1 \wedge \mathbf{e}_2 = \begin{pmatrix} 0 & 0 & 0 & -ad + bc \\ 0 & c & d & 0 \\ 0 & -a & -b & 0 \\ 1 & 0 & 0 & -b + c \end{pmatrix},$$

or in general

$$u = \begin{pmatrix} u_0 & au_1 + cu_2 & bu_1 + du_2 & -(ad - bc)u_{12} \\ u_1 & u_0 + cu_{12} & du_{12} & -(bu_1 + du_2) \\ u_2 & -au_{12} & u_0 - bu_{12} & au_1 + cu_2 \\ u_{12} & -u_2 & u_1 & u_0 + (-b + c)u_{12} \end{pmatrix}.$$

Evidently, the commutation relations $\mathbf{e}_1\mathbf{e}_2 + \mathbf{e}_2\mathbf{e}_1 = b + c$ and $\mathbf{e}_1^2 = a$, $\mathbf{e}_2^2 = d$ are satisfied, and we have the following multiplication table:

	$\mathbf{e}_1$	$\mathbf{e}_2$	$\mathbf{e}_1 \wedge \mathbf{e}_2$
$\mathbf{e}_1$	a	$\mathbf{e}_1 \wedge \mathbf{e}_2 + b$	$-b\mathbf{e}_1 + a\mathbf{e}_2$
$\mathbf{e}_2$	$-\mathbf{e}_1 \wedge \mathbf{e}_2 + c$	d	$-d\mathbf{e}_1 + c\mathbf{e}_2$
$\mathbf{e}_1 \wedge \mathbf{e}_2$	$c\mathbf{e}_1 - a\mathbf{e}_2$	$d\mathbf{e}_1 - b\mathbf{e}_2$	$-ad + bc + (-b + c)\mathbf{e}_1 \wedge \mathbf{e}_2$

In characteristic $\neq 2$ we find

$$\frac{1}{2}(\mathbf{e}_1\mathbf{e}_2 - \mathbf{e}_2\mathbf{e}_1) = \mathbf{e}_1 \wedge \mathbf{e}_2 + \frac{1}{2}(b - c)$$

and more generally for $\mathbf{x} = x_1\mathbf{e}_1 + x_2\mathbf{e}_2$, $\mathbf{y} = y_1\mathbf{e}_1 + y_2\mathbf{e}_2$

$$\frac{1}{2}(\mathbf{xy} - \mathbf{yx}) = (x_1y_2 - x_2y_1)\,\mathbf{e}_1 \wedge \mathbf{e}_2 + \frac{1}{2}(b - c)(x_1y_2 - x_2y_1)$$

$$= \mathbf{x} \wedge \mathbf{y} + A(\mathbf{x}, \mathbf{y})$$

with an alternating scalar valued form $A(\mathbf{x}, \mathbf{y}) = \frac{1}{2}(B(\mathbf{x}, \mathbf{y}) - B(\mathbf{y}, \mathbf{x}))$. For nonzero $A(\mathbf{x}, \mathbf{y})$ the quotient $\mathbf{x} \wedge \mathbf{y}/A(\mathbf{x}, \mathbf{y})$ is independent of $\mathbf{x}, \mathbf{y} \in V$. Note that the matrix of $\frac{1}{2}(\mathbf{xy} - \mathbf{yx})$ is traceless. The symmetric bilinear form associated with $Q(\mathbf{x})$ is

$$\mathbf{x} \cdot \mathbf{y} = \frac{1}{2}(B(\mathbf{x}, \mathbf{y}) + B(\mathbf{y}, \mathbf{x})) = a x_1 y_1 + \frac{1}{2}(b + c)(x_1 y_2 + x_2 y_1) + d x_2 y_2$$

and we have $\mathbf{xy} + \mathbf{yx} = 2\mathbf{x} \cdot \mathbf{y}$ for $\mathbf{x}, \mathbf{y} \in V \subset C\ell(Q)$. Orthogonal vectors $\mathbf{x} \perp \mathbf{y}$ anticommute, $\mathbf{xy} = -\mathbf{yx}$ and $(\mathbf{xy})^2 = -\mathbf{x}^2\mathbf{y}^2$, even though $\mathbf{xy} = \mathbf{x} \wedge \mathbf{y} + A(\mathbf{x}, \mathbf{y})$. [In this special case $A(\mathbf{x}, \mathbf{y}) = B(\mathbf{x}, \mathbf{y}) \neq 0$ while $\mathbf{x} \cdot \mathbf{y} = 0$ implies $B(\mathbf{y}, \mathbf{x}) = -B(\mathbf{x}, \mathbf{y})$.] ∎

It is convenient to regard $\bigwedge V$ as the subalgebra of $\mathrm{End}(\bigwedge V)$ with the canonical choice of the symmetric $B = 0$. We may also regard $C\ell(Q)$ as a subalgebra of $\mathrm{End}(\bigwedge V)$ obtained with some B such that $B(\mathbf{x}, \mathbf{x}) = Q(\mathbf{x})$ and choose the symmetric B in char $\neq 2$.

The following example shows that for $Q = 0$ and $B = 0$, Chevalley's process results in the original multiplication of the exterior algebra $\bigwedge V$, but that for $Q = 0$ and alternating $B \neq 0$, the process gives an isomorphic but different exterior multiplication on $\bigwedge V$.

Example. Take a special case of the previous example, the Clifford algebra with $Q = 0$ and $B(\mathbf{x}, \mathbf{y}) = b(x_1 y_2 - x_2 y_1)$. Send the matrix of the exterior product (with the symmetric bilinear form $= 0$) to a matrix of the isomorphic Clifford product (determined by the alternating bilinear form $= B$):

$$\begin{pmatrix} u_0 & 0 & 0 & 0 \\ u_1 & u_0 & 0 & 0 \\ u_2 & 0 & u_0 & 0 \\ u_{12} & -u_2 & u_1 & u_0 \end{pmatrix} \overset{\beta}{\to} \begin{pmatrix} u_0 & -bu_2 & bu_1 & -b^2 u_{12} \\ u_1 & u_0 - bu_{12} & 0 & -bu_1 \\ u_2 & 0 & u_0 - bu_{12} & -bu_2 \\ u_{12} & -u_2 & u_1 & u_0 - 2bu_{12} \end{pmatrix}.$$

In this case

$$\beta(\mathbf{x})\beta(\mathbf{y}) = \beta(\mathbf{x} \wedge \mathbf{y} + B(\mathbf{x}, \mathbf{y})).$$

In particular, $\beta(\mathbf{e}_1)\beta(\mathbf{e}_2) = \beta(\mathbf{e}_1 \wedge \mathbf{e}_2 + b)$, $\beta(\mathbf{e}_2)\beta(\mathbf{e}_1) = -\beta(\mathbf{e}_1 \wedge \mathbf{e}_2 + b)$ and $\beta(\mathbf{e}_1 \wedge \mathbf{e}_2 + b)\beta(\mathbf{e}_i) = 0$, $\beta(\mathbf{e}_i)\beta(\mathbf{e}_1 \wedge \mathbf{e}_2 + b) = 0$. We have already met this situation in the chapter on the *Definitions of the Clifford Algebra* in the section on the *Uniqueness and the definition by generators and relations* except that here the exterior algebra and the Clifford algebra (determined by the alternating B) are regarded as different subspaces of $\mathrm{End}(\bigwedge V)$ [although they are isomorphic subalgebras of $\mathrm{End}(\bigwedge V)$]. ∎

The above example shows that those who do not accept the existence of k-vectors in a Clifford algebra $C\ell(Q)$ over $\mathbb{F}$, char $\mathbb{F} \neq 2$, should also exclude fixed subspaces $\bigwedge^k V \subset \bigwedge V$.

In general, consider two copies of $C\ell(Q)$ in $\mathrm{End}(\bigwedge V)$ so that $Q(\mathbf{x})$ equals $B_1(\mathbf{x},\mathbf{x}) = B_2(\mathbf{x},\mathbf{x})$, which determine $\beta_1(\mathbf{x})\beta_1(\mathbf{y}) = \beta_1(\mathbf{x}\wedge\mathbf{y} + B_1(\mathbf{x},\mathbf{y}))$ and $\beta_2(\mathbf{x})\beta_2(\mathbf{y}) = \beta_2(\mathbf{x}\wedge\mathbf{y} + B_2(\mathbf{x},\mathbf{y}))$. A transition between the two copies is given by an alternating bilinear form $B(\mathbf{x},\mathbf{y}) = B_1(\mathbf{x},\mathbf{y}) - B_2(\mathbf{x},\mathbf{y})$ and $\beta(\mathbf{x})\beta(\mathbf{y}) = \beta(\mathbf{xy} + B(\mathbf{x},\mathbf{y}))$.

In characteristic $\neq 2$ this means that the symmetric bilinear form such that $\langle\mathbf{x},\mathbf{x}\rangle = Q(\mathbf{x})$ gives rise to the natural choice $\mathbf{xy} = \mathbf{x}\wedge\mathbf{y} + \langle\mathbf{x},\mathbf{y}\rangle$ among the Clifford products $\mathbf{xy} + B(\mathbf{x},\mathbf{y})$ with an alternating B. In other words, the Clifford product $\mathbf{xy}$ has a distinguished decomposition into the sum $\mathbf{x}\wedge\mathbf{y} + \langle\mathbf{x},\mathbf{y}\rangle$ where $\langle\mathbf{x},\mathbf{y}\rangle = \frac{1}{2}(\mathbf{xy}+\mathbf{yx})$ is a scalar and $\mathbf{x}\wedge\mathbf{y} = \frac{1}{2}(\mathbf{xy}-\mathbf{yx})$ is a bivector [this decomposition is unique among all the possible decompositions with antisymmetric part $\mathbf{x}\wedge\mathbf{y} = \frac{1}{2}(\mathbf{xy}-\mathbf{yx})$ equaling a new kind of bivector $\mathbf{x}\dot\wedge\mathbf{y} = \mathbf{x}\wedge\mathbf{y} + B(\mathbf{x},\mathbf{y})$ where $\mathbf{x}\dot\wedge\mathbf{y} \in \dot{\bigwedge}^2 V$ and $\mathbf{x}\wedge\mathbf{y} \in \bigwedge^2 V$]. [Similarly, a completely antisymmetric product of three vectors equals a new kind of 3-vector $\mathbf{x}\dot\wedge\mathbf{y}\dot\wedge\mathbf{z} = \mathbf{x}\wedge\mathbf{y}\wedge\mathbf{z} + \mathbf{x}B(\mathbf{y},\mathbf{z}) + \mathbf{y}B(\mathbf{z},\mathbf{x}) + \mathbf{z}B(\mathbf{x},\mathbf{y})$.]

Example. Let $\mathbb{F} = \{0,1\}$, $\dim_{\mathbb{F}} V = 2$ and $Q(x_1\mathbf{e}_1 + x_2\mathbf{e}_2) = x_1x_2$. There are only two bilinear forms B_i such that $B_i(\mathbf{x},\mathbf{x}) = Q(\mathbf{x})$, namely $B_1(\mathbf{x},\mathbf{y}) = x_1y_2$ and $B_2(\mathbf{x},\mathbf{y}) = x_2y_1$, and neither is symmetric. The difference $A = B_1 - B_2$, $A(\mathbf{x},\mathbf{y}) = x_1y_2 - x_2y_1$ $(= x_1y_2 + x_2y_1)$ is alternating (and thereby symmetric). Therefore, there are two representations of $C\ell(Q)$ in $\mathrm{End}(\bigwedge V)$:

$$\text{for } B_1: \qquad u = \begin{pmatrix} u_0 & 0 & u_1 & 0 \\ u_1 & u_0 & 0 & -u_1 \\ u_2 & 0 & u_0 - u_{12} & 0 \\ u_{12} & -u_2 & u_1 & u_0 - u_{12} \end{pmatrix}$$

$$\text{for } B_2: \qquad u = \begin{pmatrix} u_0 & u_2 & 0 & 0 \\ u_1 & u_0 + u_{12} & 0 & 0 \\ u_2 & 0 & u_0 & u_2 \\ u_{12} & -u_2 & u_1 & u_0 + u_{12} \end{pmatrix}$$

These representations have the following multiplication tables with respect to the basis $\{1, \mathbf{e}_1, \mathbf{e}_2, \mathbf{e}_1\wedge\mathbf{e}_2\}$ for $\bigwedge V$:

B_1	$\mathbf{e}_1$	$\mathbf{e}_2$	$\mathbf{e}_1\wedge\mathbf{e}_2$
$\mathbf{e}_1$	0	$1+\mathbf{e}_1\wedge\mathbf{e}_2$	$-\mathbf{e}_1$
$\mathbf{e}_2$	$-\mathbf{e}_1\wedge\mathbf{e}_2$	0	0
$\mathbf{e}_1\wedge\mathbf{e}_2$	0	$-\mathbf{e}_2$	$-\mathbf{e}_1\wedge\mathbf{e}_2$

B_2	$\mathbf{e}_1$	$\mathbf{e}_2$	$\mathbf{e}_1 \wedge \mathbf{e}_2$
$\mathbf{e}_1$	0	$\mathbf{e}_1 \wedge \mathbf{e}_2$	0
$\mathbf{e}_2$	$1 - \mathbf{e}_1 \wedge \mathbf{e}_2$	0	$\mathbf{e}_2$
$\mathbf{e}_1 \wedge \mathbf{e}_2$	$\mathbf{e}_1$	0	$\mathbf{e}_1 \wedge \mathbf{e}_2$

In this case there are only two linear isomorphisms $\bigwedge V \to C\ell(Q)$ which are identity mappings when restricted to $\mathbb{F} \oplus V$ and which preserve parity (send even elements to even elements and odd to odd). It is easy to verify that the above multiplication tables actually describe the only representations of $C\ell(Q)$ in $\bigwedge V$. In this case **there are no canonical linear isomorphisms** $\bigwedge V \to C\ell(Q)$, in other words, neither of the above multiplication tables can be preferred over the other. In particular, $\bigwedge^2 V$ cannot be canonically embedded in $C\ell(Q)$, and **there are no bivectors in characteristic 2.** ∎

The need for a simplification of Chevalley's presentation is obvious. For instance, van der Waerden 1966 said that 'the ideas underlying Chevalley's proof (p. 40) are not easy to discern' and gave another proof, equivalent but easier to follow. [Also Crumeyrolle 1990 p. xi claims that 'Chevalley's book proved too abstract for most physicists' and in a *Bull. AMS* review Lam 1989 p. 122 admits that 'Chevalley's book on spinors is ... not the easiest book to read.'] It might be helpful to get acquainted with a simpler and more direct method of relating $\bigwedge V$ and $C\ell(Q)$ due to M. Riesz 1958/1993 pp. 61-67. Riesz introduced a second product in $C\ell(Q)$ making it isomorphic with $\bigwedge V$ without resorting to the usual completely antisymmetric Clifford product of vectors and constructed a privileged linear isomorphism $C\ell(Q) \to \bigwedge V$.

22.3 Riesz's introduction of an exterior product in $C\ell(Q)$

In the following we review a construction of M. Riesz 1958/1993 pp. 61-67. Start from the Clifford algebra $C\ell(Q)$ over $\mathbb{F}$, $\operatorname{char} \mathbb{F} \neq 2$. The isometry $\mathbf{x} \to -\mathbf{x}$ of V when extended to an automorphism of $C\ell(Q)$ is called the *grade involution* $u \to \hat{u}$. Define the *exterior product* of $\mathbf{x} \in V$ and $u \in C\ell(Q)$ by

$$\mathbf{x} \wedge u = \frac{1}{2}(\mathbf{x}u + \hat{u}\mathbf{x}), \qquad u \wedge \mathbf{x} = \frac{1}{2}(u\mathbf{x} + \mathbf{x}\hat{u})$$

and extend it by linearity to all of $C\ell(Q)$, which then becomes isomorphic to $\bigwedge V$. The exterior products of two vectors $\mathbf{x} \wedge \mathbf{y} = \frac{1}{2}(\mathbf{x}\mathbf{y} - \mathbf{y}\mathbf{x})$ are simple *bivectors* and they span $\bigwedge^2 V$. The exterior product of a vector and a bivector, $\mathbf{x} \wedge \mathbf{B} = \frac{1}{2}(\mathbf{x}\mathbf{B} + \mathbf{B}\mathbf{x})$, is a 3-vector in $\bigwedge^3 V$. The subspace of k-vectors is constructed recursively by

$$\mathbf{x} \wedge \mathbf{a} = \frac{1}{2}(\mathbf{xa} + (-1)^{k-1}\mathbf{ax}) \in \bigwedge^k V \quad \text{for} \quad \mathbf{a} \in \bigwedge^{k-1} V.$$

We may deduce associativity of the exterior product as follows. First, the definition implies for $\mathbf{x}$, $\mathbf{y}$, $\mathbf{z} \in V$

$$\mathbf{x} \wedge (\mathbf{y} \wedge \mathbf{z}) = \frac{1}{4}(\mathbf{xyz} - \mathbf{xzy} + \mathbf{yzx} - \mathbf{zyx})$$
$$(\mathbf{x} \wedge \mathbf{y}) \wedge \mathbf{z} = \frac{1}{4}(\mathbf{xyz} - \mathbf{yxz} + \mathbf{zxy} - \mathbf{zyx}).$$

Then $\mathbf{x} \wedge (\mathbf{y} \wedge \mathbf{z}) = (\mathbf{x} \wedge \mathbf{y}) \wedge \mathbf{z}$ since

$$\mathbf{xyz} - \mathbf{zyx} = \mathbf{xyz} - \mathbf{zyx} + (\mathbf{zy} + \mathbf{yz})\mathbf{x} - \mathbf{x}(\mathbf{yz} + \mathbf{zy}) = \mathbf{yzx} - \mathbf{xzy}.$$

This last result implies

$$\mathbf{x} \wedge \mathbf{y} \wedge \mathbf{z} = \frac{1}{6}(\mathbf{xyz} + \mathbf{yzx} + \mathbf{zxy} - \mathbf{zyx} - \mathbf{xzy} - \mathbf{yxz})$$

when $\operatorname{char}\mathbb{F} \neq 2, 3$ (note the resemblance with antisymmetric tensors). [Similarly, we may conclude that $\mathbf{xyz}+\mathbf{zyx} = \mathbf{x}(\mathbf{yz}+\mathbf{zy}) - (\mathbf{xz}+\mathbf{zx})\mathbf{y} + \mathbf{z}(\mathbf{xy}+\mathbf{yx})$ is a vector in V.] Riesz's construction shows that **bivectors exist in all characteristics** $\neq 2$.

Introduce the *contraction* of $u \in C\ell(Q)$ by $\mathbf{x} \in V$ so that

$$\mathbf{x} \lrcorner u = \frac{1}{2}(\mathbf{x}u - \hat{u}\mathbf{x})$$

and show that this contraction is a derivation of $C\ell(Q)$ since

$$\mathbf{x} \lrcorner (uv) = \frac{1}{2}(\mathbf{x}uv - \widehat{uv}\mathbf{x}) = \frac{1}{2}(\mathbf{x}uv - \hat{u}\hat{v}\mathbf{x})$$

$$= \frac{1}{2}(\mathbf{x}uv - \hat{u}\mathbf{x}v + \hat{u}\mathbf{x}v - \hat{u}\hat{v}\mathbf{x}) = (\mathbf{x} \lrcorner u)v + \hat{u}(\mathbf{x} \lrcorner v).$$

Thus one and the same contraction is indeed a derivation for both the exterior product and the Clifford product. [Kähler 1962 p. 435 (4.4) and p. 456 (10.3) was aware of the equations

$$\mathbf{x} \lrcorner (u \wedge v) = (\mathbf{x} \lrcorner u) \wedge v + \hat{u} \wedge (\mathbf{x} \lrcorner v) \quad \text{and}$$
$$\mathbf{x} \lrcorner (uv) = (\mathbf{x} \lrcorner u)v + \hat{u}(\mathbf{x} \lrcorner v).]$$

Provided with the scalar multiplication $(u \wedge v) \lrcorner w = u \lrcorner (v \lrcorner w)$, the exterior algebra $\bigwedge V$ and the Clifford algebra $C\ell(Q)$ are linearly isomorphic as left $\bigwedge V$-modules.

Exercises 11,12,...,20

Exercises

Show that

1. $\mathbf{x} \wedge (\mathbf{y} \,\lrcorner\, u) - \mathbf{y} \wedge (\mathbf{x} \,\lrcorner\, u) = \mathbf{x} \,\lrcorner\, (\mathbf{y} \wedge u) - \mathbf{y} \,\lrcorner\, (\mathbf{x} \wedge u)$ for $\mathbf{x}, \mathbf{y} \in V$.
2. $\mathbf{x} \wedge \mathbf{y} \wedge (\mathbf{z} \,\lrcorner\, u) - \mathbf{x} \wedge \mathbf{z} \wedge (\mathbf{y} \,\lrcorner\, u) + \mathbf{y} \wedge \mathbf{z} \wedge (\mathbf{x} \,\lrcorner\, u)$
$= \mathbf{x} \,\lrcorner\, (\mathbf{y} \wedge \mathbf{z} \wedge u) - \mathbf{y} \,\lrcorner\, (\mathbf{x} \wedge \mathbf{z} \wedge u) + \mathbf{z} \,\lrcorner\, (\mathbf{x} \wedge \mathbf{y} \wedge u)$.
3. $\mathbf{x} \wedge (\mathbf{y} \,\lrcorner\, (\mathbf{z} \,\lrcorner\, u)) - \mathbf{y} \wedge (\mathbf{x} \,\lrcorner\, (\mathbf{z} \,\lrcorner\, u)) + \mathbf{z} \wedge (\mathbf{x} \,\lrcorner\, (\mathbf{y} \,\lrcorner\, u))$
$= (\mathbf{x} \wedge \mathbf{y}) \,\lrcorner\, (\mathbf{z} \wedge u) - (\mathbf{x} \wedge \mathbf{z}) \,\lrcorner\, (\mathbf{y} \wedge u) + (\mathbf{y} \wedge \mathbf{z}) \,\lrcorner\, (\mathbf{x} \wedge u)$
$= \mathbf{x} \,\lrcorner\, (\mathbf{y} \,\lrcorner\, (\mathbf{z} \wedge u)) - \mathbf{x} \,\lrcorner\, (\mathbf{z} \,\lrcorner\, (\mathbf{y} \wedge u)) + \mathbf{y} \,\lrcorner\, (\mathbf{z} \,\lrcorner\, (\mathbf{x} \wedge u))$.
4. $(\mathbf{x} \wedge \mathbf{y} \wedge \mathbf{z})u = \mathbf{x} \wedge \mathbf{y} \wedge (\mathbf{z} \,\lrcorner\, u) - \mathbf{x} \wedge \mathbf{z} \wedge (\mathbf{y} \,\lrcorner\, u) + \mathbf{y} \wedge \mathbf{z} \wedge (\mathbf{x} \,\lrcorner\, u)$
$+\mathbf{x} \wedge (\mathbf{y} \,\lrcorner\, (\mathbf{z} \,\lrcorner\, u)) - \mathbf{y} \wedge (\mathbf{x} \,\lrcorner\, (\mathbf{z} \,\lrcorner\, u)) + \mathbf{z} \wedge (\mathbf{x} \,\lrcorner\, (\mathbf{y} \,\lrcorner\, u))$
$+\mathbf{x} \wedge \mathbf{y} \wedge \mathbf{z} \wedge u + \mathbf{x} \,\lrcorner\, (\mathbf{y} \,\lrcorner\, (\mathbf{z} \,\lrcorner\, u))$.
5. $\mathbf{a} \,\lrcorner\, \mathbf{b} \in \bigwedge^{j-i} V$ for $\mathbf{a} \in \bigwedge^{i} V$, $\mathbf{b} \in \bigwedge^{j} V$ $(\operatorname{char}\mathbb{F} \neq 2)$.

In the next five exercises we have a non-degenerate Q. Define the right contraction $u \,\llcorner\, v$ by $<u \,\llcorner\, v, w> = <u, w \wedge \tilde{v}>$ for all $w \in \bigwedge V$ (we say that u is contracted by the contractor v). Show that

6. $u \,\lrcorner\, v = v_0 \,\llcorner\, u_0 - v_0 \,\llcorner\, u_1 + v_1 \,\llcorner\, u_0 + v_1 \,\llcorner\, u_1 = v \,\llcorner\, u - 2v_0 \,\llcorner\, u_1$
$(v_0 = \operatorname{even}(v),\ u_1 = \operatorname{odd}(u))$.
7. $u \,\llcorner\, v = v_0 \,\lrcorner\, u_0 + v_0 \,\lrcorner\, u_1 - v_1 \,\lrcorner\, u_0 + v_1 \,\lrcorner\, u_1 = v \,\lrcorner\, u - 2v_1 \,\lrcorner\, u_0$
$(v_1 = \operatorname{odd}(v),\ u_0 = \operatorname{even}(u))$.

Show that (when $\operatorname{char}\mathbb{F} \neq 2$)

8. $\mathbf{a} \,\lrcorner\, \mathbf{b} = (-1)^{i(j-i)} \mathbf{b} \,\llcorner\, \mathbf{a}$ for $\mathbf{a} \in \bigwedge^{i} V$, $\mathbf{b} \in \bigwedge^{j} V$.
9. $\mathbf{a} \in \bigwedge^{k} V$, $\mathbf{a} \neq 0$, $\mathbf{x} \in V$, $\mathbf{x} \,\lrcorner\, \mathbf{a} = 0 \Leftrightarrow \mathbf{x} = \mathbf{a} \,\lrcorner\, \mathbf{b}$ for some $\mathbf{b} \in \bigwedge^{k+1} V$.
10. $\mathbf{b} \in \bigwedge^{k} V$ is simple $\Leftrightarrow (\mathbf{a} \,\lrcorner\, \mathbf{b}) \wedge \mathbf{b} = 0$ for all $\mathbf{a} \in \bigwedge^{k-1} V$.
11. $\mathbf{x}$ and $\mathbf{x} \wedge \mathbf{y}$ anticommute for vectors $\mathbf{x}, \mathbf{y} \in V$.
12. $\mathbf{x}$ and $\mathbf{x} \,\lrcorner\, \mathbf{B}$ anticommute for a bivector $\mathbf{B} \in \bigwedge^2 V$.
13. $(\mathbf{x} \wedge \mathbf{y})^2 = (\mathbf{x} \,\lrcorner\, \mathbf{y})^2 - \mathbf{x}^2\mathbf{y}^2$ (Lagrange's identity).
14. $(\mathbf{x} \wedge \mathbf{y} \wedge \mathbf{z}) \,\lrcorner\, u = (\mathbf{x} \wedge \mathbf{y}) \,\lrcorner\, (\mathbf{z} \,\lrcorner\, u) = \mathbf{x} \,\lrcorner\, (\mathbf{y} \,\lrcorner\, (\mathbf{z} \,\lrcorner\, u))$.
15. $(\mathbf{xyz} - \mathbf{zyx})^2 \in \mathbb{F}$, $\mathbf{x} \wedge \mathbf{y} \wedge \mathbf{z} = \frac{1}{2}(\mathbf{xyz} - \mathbf{zyx})$.
16. $\mathbf{a} \wedge \mathbf{b} = (-1)^{ij} \mathbf{b} \wedge \mathbf{a}$ for $\mathbf{a} \in \bigwedge^{i} V$ and $\mathbf{b} \in \bigwedge^{j} V$.
17. $u \wedge v = v_0 \wedge u_0 + v_0 \wedge u_1 + v_1 \wedge u_0 - v_1 \wedge u_1 = v \wedge u - 2v_1 \wedge u_1$ where $u, v \in \bigwedge V$, $u_0 = \operatorname{even}(u)$, $v_0 = \operatorname{even}(v)$, $u_1 = \operatorname{odd}(u)$, $v_1 = \operatorname{odd}(v)$.
18. $\mathbf{B}u = \mathbf{B} \wedge u + \frac{1}{2}(\mathbf{B}u - u\mathbf{B}) + \mathbf{B} \,\lrcorner\, u$ for $\mathbf{B} \in \bigwedge^2 V$.
[Hint: $(\mathbf{x} \wedge \mathbf{y}) \wedge u + (\mathbf{x} \wedge \mathbf{y}) \,\lrcorner\, u = \mathbf{x} \wedge (\mathbf{y} \wedge u) + \mathbf{x} \,\lrcorner\, (\mathbf{y} \,\lrcorner\, u)$.]
19. $Q(u) = \langle \tilde{u}u \rangle_0$, $<u, v> = \langle \tilde{u} \,\lrcorner\, v \rangle_0$ (= the scalar part of $\tilde{u} \,\lrcorner\, v$).

In the last exercise we have a non-degenerate Q:

20. Q on V extends to a neutral or anisotropic Q on $\mathcal{C}\ell(Q)$.

Bibliography

N. Bourbaki: *Algèbre, Chapitre 9, Formes sesquilinéaires et formes quadratiques.* Hermann, Paris, 1959.

C. Chevalley: *Theory of Lie Groups.* Princeton University Press, Princeton, NJ, 1946.

C. Chevalley: *The Algebraic Theory of Spinors.* Columbia University Press, New York, 1954.

A. Crumeyrolle: *Orthogonal and Symplectic Clifford Algebras, Spinor Structures.* Kluwer, Dordrecht, The Netherlands, 1990.

J. Helmstetter: Algèbres de Clifford et algèbres de Weyl. *Cahiers Math.* **25**, Montpellier, 1982.

E. Kähler: Der innere Differentialkalkül. *Rendiconti di Matematica e delle sue Applicazioni* (Roma) **21** (1962), 425-523.

M. Riesz: *Clifford Numbers and Spinors.* The Institute for Fluid Dynamics and Applied Mathematics, Lecture Series No. **38**, University of Maryland, 1958. Reprinted as facsimile (eds.: E.F. Bolinder, P. Lounesto) by Kluwer, Dordrecht, The Netherlands, 1993.

B.L. van der Waerden: On Clifford algebras. *Neder. Akad. Wetensch. Proc.* Ser. A**69** (1966), 78-83.

E. Witt: Theorie der quadratischen Formen in beliebigen Körpern. *J. Reine Angew. Math.* **176** (1937), 31-44.

23

Octonions and Triality

Complex numbers and quaternions form special cases of lower-dimensional Clifford algebras, their even subalgebras and their ideals

$$\mathbb{C} \simeq \mathcal{C}\ell_{0,1} \simeq \mathcal{C}\ell_2^+ \simeq \mathcal{C}\ell_{0,2}^+,$$
$$\mathbb{H} \simeq \mathcal{C}\ell_{0,2} \simeq \mathcal{C}\ell_3^+ \simeq \mathcal{C}\ell_{0,3}^+ \simeq \tfrac{1}{2}(1 \pm \mathbf{e}_{123})\mathcal{C}\ell_{0,3}.$$

In this chapter, we explore another generalization of $\mathbb{C}$ and $\mathbb{H}$, a non-associative real algebra, the Cayley algebra of octonions, $\mathbb{O}$. Like complex numbers and quaternions, octonions form a real division algebra, of the highest possible dimension, 8. As an extreme case, $\mathbb{O}$ makes its presence felt in classifications, for instance, in conjunction with exceptional cases of simple Lie algebras.

Like $\mathbb{C}$ and $\mathbb{H}$, $\mathbb{O}$ has a geometric interpretation. The automorphism group of $\mathbb{H}$ is $SO(3)$, the rotation group of $\mathbb{R}^3$ in $\mathbb{H} = \mathbb{R} \oplus \mathbb{R}^3$. The automorphism group of $\mathbb{O} = \mathbb{R} \oplus \mathbb{R}^7$ is not all of $SO(7)$, but only a subgroup, the exceptional Lie group G_2. The subgroup G_2 fixes a 3-vector, in $\bigwedge^3 \mathbb{R}^7$, whose choice determines the product rule of $\mathbb{O}$.

The Cayley algebra $\mathbb{O}$ is a tool to handle an esoteric phenomenon in dimension 8, namely triality, an automorphism of the universal covering group $\mathbf{Spin}(8)$ of the rotation group $SO(8)$ of the Euclidean space $\mathbb{R}^8$. In general, all automorphisms of $SO(n)$ are either inner or similarities by orthogonal matrices in $O(n)$, and all automorphisms of $\mathbf{Spin}(n)$ are restrictions of linear transformations $\mathcal{C}\ell_n \to \mathcal{C}\ell_n$, and project down to automorphisms of $SO(n)$. The only exception is the triality automorphism of $\mathbf{Spin}(8)$, which cannot be linear while it permutes cyclically the three non-identity elements $-1, \mathbf{e}_{12\ldots8}, -\mathbf{e}_{12\ldots8}$ in the center of $\mathbf{Spin}(8)$.

We shall see that triality is a restriction of a polynomial mapping $\mathcal{C}\ell_8 \to \mathcal{C}\ell_8$, of degree 2. We will learn how to compose trialities, when they correspond to different octonion products. We shall explore triality in terms of classical linear algebra by observing how eigenplanes of rotations transform under triality.

23.1 Division algebras

An algebra A over $\mathbb{R}$ is a linear (that is a vector) space A over $\mathbb{R}$ together with a bilinear map $A \times A \to A$, $(a, b) \to ab$, the algebra product. Bilinearity means distributivity $(a+b)c = ac+bc$, $a(b+c) = ab+ac$ and $(\lambda a)b = a(\lambda b) = \lambda(ab)$ for all $a, b, c \in A$ and $\lambda \in \mathbb{R}$. An algebra is without *zero-divisors* if $ab = 0$ implies $a = 0$ or $b = 0$. In a *division algebra* $\mathbb{D}$ the equations $ax = b$ and $ya = b$ have unique solutions x, y for all non-zero $a \in \mathbb{D}$. A division algebra is without zero-divisors, and conversely, every finite-dimensional algebra without zero-divisors is a division algebra. If a division algebra is associative, then it has unity 1 and each non-zero element has a unique inverse (on both sides).

An algebra with a unity is said to *admit inverses* if each non-zero element admits an inverse (not necessarily unique). An algebra is *alternative* if $a(ab) = a^2b$ and $(ab)b = ab^2$, and *flexible* if $a(ba) = (ab)a$. An alternative algebra is flexible. An alternative division algebra has unity and admits inverses, which are unique. The only alternative division algebras over $\mathbb{R}$ are $\mathbb{R}$, $\mathbb{C}$, $\mathbb{H}$ and $\mathbb{O}$.

An algebra A with a positive-definite quadratic form $N: A \to \mathbb{R}$, is said to *preserve norm*, or *admit composition*, if for all $a, b \in A$, $N(ab) = N(a)N(b)$. The dimension of a norm-preserving division algebra $\mathbb{D}$ over $\mathbb{R}$ is 1, 2, 4 or 8; if furthermore $\mathbb{D}$ has unity, then it is $\mathbb{R}$, $\mathbb{C}$, $\mathbb{H}$ or $\mathbb{O}$.

Examples. 1. Define in $\mathbb{C}$ a new product $a \circ b$ by $a \circ b = a\bar{b}$. Then $\mathbb{C}$ becomes a non-commutative and non-alternative division algebra over $\mathbb{R}$, without unity.
2. Consider a 3-dimensional algebra over $\mathbb{R}$ with basis $\{1, i, j\}$ such that 1 is the unity and $i^2 = j^2 = -1$ but $ij = ji = 0$. The algebra is commutative and flexible, but non-alternative. It admits inverses, but inverses of the elements of the form $xi + yj$ are not unique, $(xi + yj)^{-1} = \lambda(yi - xj) - \dfrac{ix + jy}{x^2 + y^2}$, where $\lambda \in \mathbb{R}$. It has zero-divisors, by definition, and cannot be a division algebra, although all non-zero elements are invertible.
3. Consider a 3-dimensional algebra over $\mathbb{Q}$ with basis $\{1, i, i^2\}$, unity 1 and multiplication table

	i	i^2
i	i^2	3
i^2	3	$-6i$

The algebra is commutative and flexible, but non-alternative. Each non-zero element has a unique inverse. Multiplication by $x + iy + i^2z$ has determinant $x^3 + 3y^3 - 18z^3$, which has no non-zero rational roots (Euler 1862). Thus, the algebra is a division algebra, $3D$ over $\mathbb{Q}$. ∎

23.2 The Cayley–Dickson doubling process

Complex numbers can be considered as pairs of real numbers with componentwise addition and with the product

$$(x_1, y_1)(x_2, y_2) = (x_1x_2 - y_1y_2, x_1y_2 + y_1x_2).$$

Quaternions can be defined as pairs of complex numbers, but this time the product involves complex conjugation

$$(z_1, w_1)(z_2, w_2) = (z_1z_2 - w_1\bar{w}_2, z_1w_2 + w_1\bar{z}_2).$$

Octonions can be defined as pairs of quaternions, but this time order of multiplication matters

$$(p_1, q_1) \circ (p_2, q_2) = (p_1p_2 - \bar{q}_2q_1, q_2p_1 + q_1\bar{p}_2).$$

This doubling process, of Cayley-Dickson, can be repeated, but the next algebras are not division algebras, although they still are simple and flexible (Schafer 1954). Every element in such a Cayley-Dickson algebra satisfies a quadratic equation with real coefficients.

Example. The quaternion $q = w+ix+jy+kz$ satisfies the quadratic equation

$$q^2 - 2wq + |q|^2 = 0.$$ ∎

The Cayley-Dickson doubling process

$$\mathbb{C} = \mathbb{R} \oplus \mathbb{R}i$$
$$\mathbb{H} = \mathbb{C} \oplus \mathbb{C}j$$
$$\mathbb{O} = \mathbb{H} \oplus \mathbb{H}\ell$$

provides a new imaginary unit ℓ, $\ell^2 = -1$, which anticommutes with i, j, k. The basis $\{1, i, j, k\}$ of $\mathbb{H}$ is complemented to a basis $\{1, i, j, k, \ell, i\ell, j\ell, k\ell\}$ of $\mathbb{O} = \mathbb{H} \oplus \mathbb{H}\ell$. Thus, $\mathbb{O}$ is spanned by $1 \in \mathbb{R}$ and the 7 imaginary units $i, j, k, \ell, i\ell, j\ell, k\ell$, each with square -1, so that $\mathbb{O} = \mathbb{R} \oplus \mathbb{R}^7$. Among subsets of 3 imaginary units, there are 7 triplets, which associate and span the imaginary part of a quaternionic subalgebra. The remaining 28 triplets anti-associate.

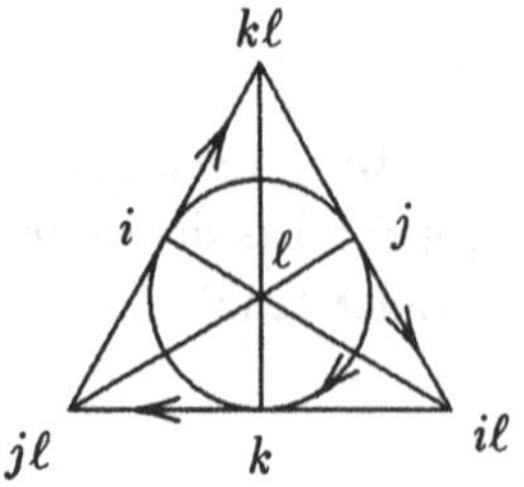

The multiplication table of the unit octonions can be summarized by the Fano plane, the smallest projective plane, consisting of 7 points and 7 lines, with orientations. The 7 oriented lines correspond to the 7 quaternionic/associative triplets.

23.3 Multiplication table of $\mathbb{O}$

Denote the product of $a, b \in \mathbb{O}$ by $a \circ b$. Let $1, \mathbf{e}_1, \mathbf{e}_2, \ldots, \mathbf{e}_7$ be a basis of $\mathbb{O}$. Define the product in terms of the basis by

$$\mathbf{e}_i \circ \mathbf{e}_i = -1, \quad \text{and} \quad \mathbf{e}_i \circ \mathbf{e}_j = -\mathbf{e}_j \circ \mathbf{e}_i \quad \text{for} \quad i \neq j,$$

and by the table

$$\begin{array}{lll} \mathbf{e}_1 \circ \mathbf{e}_2 = \mathbf{e}_4, & \mathbf{e}_2 \circ \mathbf{e}_4 = \mathbf{e}_1, & \mathbf{e}_4 \circ \mathbf{e}_1 = \mathbf{e}_2, \\ \mathbf{e}_2 \circ \mathbf{e}_3 = \mathbf{e}_5, & \mathbf{e}_3 \circ \mathbf{e}_5 = \mathbf{e}_2, & \mathbf{e}_5 \circ \mathbf{e}_2 = \mathbf{e}_3, \\ \quad\vdots & \quad\vdots & \quad\vdots \\ \mathbf{e}_7 \circ \mathbf{e}_1 = \mathbf{e}_3, & \mathbf{e}_1 \circ \mathbf{e}_3 = \mathbf{e}_7, & \mathbf{e}_3 \circ \mathbf{e}_7 = \mathbf{e}_1. \end{array}$$

The table can be condensed into the form

$$\mathbf{e}_i \circ \mathbf{e}_{i+1} = \mathbf{e}_{i+3}$$

where the indices are permuted cyclically and translated modulo 7.

If $\mathbf{e}_i \circ \mathbf{e}_j = \pm \mathbf{e}_k$, then $\mathbf{e}_i$, $\mathbf{e}_j$, $\mathbf{e}_k$ generate a subalgebra isomorphic to $\mathbb{H}$. The sign in $\mathbf{e}_i \circ \mathbf{e}_j = \pm \mathbf{e}_k$ can be memorized by rotating the triangle in the following picture by an integral multiple of $2\pi/7$:

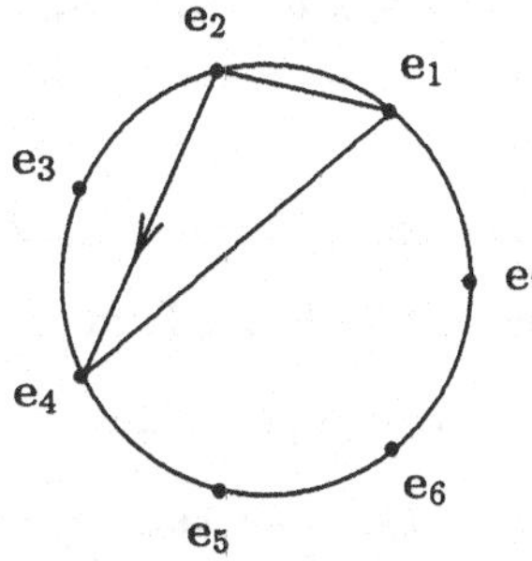

Example. The product $\mathbf{e}_2 \circ \mathbf{e}_5 = -\mathbf{e}_3$ corresponds to a triangle obtained by rotating the picture by $2\pi/7$. ∎

In the Clifford algebra $\mathcal{C}\ell_{0,7}$ of $\mathbb{R}^{0,7}$, octonions can be identified with paravectors, $\mathbb{O} = \mathbb{R} \oplus \mathbb{R}^{0,7}$, and the octonion product may be expressed in terms of the Clifford product as

$$a \circ b = \langle ab(1 - \mathbf{v}) \rangle_{0,1},$$

where $\mathbf{v} = \mathbf{e}_{124} + \mathbf{e}_{235} + \mathbf{e}_{346} + \mathbf{e}_{457} + \mathbf{e}_{561} + \mathbf{e}_{672} + \mathbf{e}_{713} \in \bigwedge^3 \mathbb{R}^{0,7}$. In $\mathcal{C}\ell_{0,7}$, the octonion product can be also written as

$$a \circ b = \langle ab(1 + \mathbf{w})(1 - \mathbf{e}_{12\ldots7}) \rangle_{0,1} \quad \text{for} \quad a, b \in \mathbb{R} \oplus \mathbb{R}^{0,7}$$

where $\frac{1}{8}(1+\mathbf{w})\frac{1}{2}(1-\mathbf{e}_{12\ldots7})$ is an idempotent, $\mathbf{w} = \mathbf{v}\mathbf{e}_{12\ldots7}^{-1} \in \bigwedge^4 \mathbb{R}^{0,7}$ and $\mathbf{e}_{12\ldots7}^{-1} = \mathbf{e}_{12\ldots7}$.

In the Clifford algebra $\mathcal{C}\ell_8$ of $\mathbb{R}^8$, we represent octonions by vectors, $\mathbb{O} = \mathbb{R}^8$. As the identity of octonions we choose the unit vector $\mathbf{e}_8$ in $\mathbb{R}^8$. The octonion product is then expressed in terms of the Clifford product as

$$\mathbf{a} \circ \mathbf{b} = \langle \mathbf{a}\mathbf{e}_8\mathbf{b}(1+\mathbf{w})(1-\mathbf{e}_{12\ldots8})\rangle_1 \quad \text{for} \quad \mathbf{a},\mathbf{b} \in \mathbb{R}^8$$

where $\frac{1}{8}(1+\mathbf{w})\frac{1}{2}(1-\mathbf{e}_{12\ldots8})$ is an idempotent, $\mathbf{w} = \mathbf{v}\mathbf{e}_{12\ldots7}^{-1} \in \bigwedge^4 \mathbb{R}^8$, $\mathbf{e}_{12\ldots7}^{-1} = -\mathbf{e}_{12\ldots7}$ and $\mathbf{v} = \mathbf{e}_{124} + \mathbf{e}_{235} + \mathbf{e}_{346} + \mathbf{e}_{457} + \mathbf{e}_{561} + \mathbf{e}_{672} + \mathbf{e}_{713} \in \bigwedge^3 \mathbb{R}^8$.

23.4 The octonion product and the cross product in $\mathbb{R}^7$

A product of two vectors is linear in both factors. A vector-valued product of two vectors is called a **cross product**, if the vector is orthogonal to the two factors and has length equal to the area of the parallelogram formed by the two vectors. A cross product of two vectors exists only in $3D$ and $7D$.

The cross product of two vectors in $\mathbb{R}^7$ can be constructed in terms of an orthonormal basis $\mathbf{e}_1, \mathbf{e}_2, \ldots, \mathbf{e}_7$ by antisymmetry, $\mathbf{e}_i \times \mathbf{e}_j = -\mathbf{e}_j \times \mathbf{e}_i$, and

$$\begin{array}{lll} \mathbf{e}_1 \times \mathbf{e}_2 = \mathbf{e}_4, & \mathbf{e}_2 \times \mathbf{e}_4 = \mathbf{e}_1, & \mathbf{e}_4 \times \mathbf{e}_1 = \mathbf{e}_2, \\ \mathbf{e}_2 \times \mathbf{e}_3 = \mathbf{e}_5, & \mathbf{e}_3 \times \mathbf{e}_5 = \mathbf{e}_2, & \mathbf{e}_5 \times \mathbf{e}_2 = \mathbf{e}_3, \\ \vdots & \vdots & \vdots \\ \mathbf{e}_7 \times \mathbf{e}_1 = \mathbf{e}_3, & \mathbf{e}_1 \times \mathbf{e}_3 = \mathbf{e}_7, & \mathbf{e}_3 \times \mathbf{e}_7 = \mathbf{e}_1. \end{array}$$

The above table can be condensed into the form

$$\mathbf{e}_i \times \mathbf{e}_{i+1} = \mathbf{e}_{i+3}$$

where the indices are permuted cyclically and translated modulo 7.

This cross product of vectors in $\mathbb{R}^7$ satisfies the usual properties, that is,

$$\begin{array}{ll} (\mathbf{a} \times \mathbf{b}) \cdot \mathbf{a} = 0, \ (\mathbf{a} \times \mathbf{b}) \cdot \mathbf{b} = 0 & \text{orthogonality} \\ |\mathbf{a} \times \mathbf{b}|^2 = |\mathbf{a}|^2|\mathbf{b}|^2 - (\mathbf{a} \cdot \mathbf{b})^2 & \text{Pythagorean theorem} \end{array}$$

where the second rule can also be written as $|\mathbf{a} \times \mathbf{b}| = |\mathbf{a}||\mathbf{b}| \sin \sphericalangle(\mathbf{a},\mathbf{b})$. Unlike the 3-dimensional cross product, the 7-dimensional cross product does not satisfy the Jacobi identity, $(\mathbf{a} \times \mathbf{b}) \times \mathbf{c} + (\mathbf{b} \times \mathbf{c}) \times \mathbf{a} + (\mathbf{c} \times \mathbf{a}) \times \mathbf{b} \neq 0$, and so it does not form a Lie algebra. However, the 7-dimensional cross product satisfies the Malcev identity, a generalization of Jacobi, see Ebbinghaus et al. 1991 p. 279.

In $\mathbb{R}^3$, the direction of $\mathbf{a} \times \mathbf{b}$ is unique, up to two alternatives for the orientation, but in $\mathbb{R}^7$ the direction of $\mathbf{a} \times \mathbf{b}$ depends on a 3-vector defining the cross product; to wit,

$$\mathbf{a} \times \mathbf{b} = -(\mathbf{a} \wedge \mathbf{b}) \lrcorner \mathbf{v} \quad [\neq -(\mathbf{a} \wedge \mathbf{b})\mathbf{v}]$$

depends on $\mathbf{v} = \mathbf{e}_{124} + \mathbf{e}_{235} + \mathbf{e}_{346} + \mathbf{e}_{457} + \mathbf{e}_{561} + \mathbf{e}_{672} + \mathbf{e}_{713} \in \bigwedge^3 \mathbb{R}^7$. In the 3-dimensional space $\mathbf{a} \times \mathbf{b} = \mathbf{c} \times \mathbf{d}$ implies that $\mathbf{a}, \mathbf{b}, \mathbf{c}, \mathbf{d}$ are in the same plane, but for the cross product $\mathbf{a} \times \mathbf{b}$ in $\mathbb{R}^7$ there are also other planes than the linear span of $\mathbf{a}$ and $\mathbf{b}$ giving the same direction as $\mathbf{a} \times \mathbf{b}$.

The 3-dimensional cross product is invariant under all rotations of $SO(3)$, while the 7-dimensional cross product is not invariant under all of $SO(7)$, but only under the exceptional Lie group G_2, a subgroup of $SO(7)$. When we let $\mathbf{a}$ and $\mathbf{b}$ run through all of $\mathbb{R}^7$, the image set of the simple bivectors $\mathbf{a} \wedge \mathbf{b}$ is a manifold of dimension $2 \cdot 7 - 3 = 11 > 7$ in $\bigwedge^2 \mathbb{R}^7$, $\dim(\bigwedge^2 \mathbb{R}^7) = \frac{1}{2}7(7-1) = 21$, while the image set of $\mathbf{a} \times \mathbf{b}$ is just $\mathbb{R}^7$. So the mapping

$$\mathbf{a} \wedge \mathbf{b} \to \mathbf{a} \times \mathbf{b} = -(\mathbf{a} \wedge \mathbf{b}) \lrcorner \mathbf{v}$$

is not a one-to-one correspondence, but only a method of associating a vector to a bivector.

The 3-dimensional cross product is the vector part of the quaternion product of two pure quaternions, that is,

$$\mathbf{a} \times \mathbf{b} = \operatorname{Im}(\mathbf{ab}) \quad \text{for} \quad \mathbf{a}, \mathbf{b} \in \mathbb{R}^3 \subset \mathbb{H}.$$

In terms of the Clifford algebra $\mathcal{C}\ell_3 \simeq \operatorname{Mat}(2, \mathbb{C})$ of the Euclidean space $\mathbb{R}^3$ the cross product could also be expressed as

$$\mathbf{a} \times \mathbf{b} = -\langle \mathbf{abe}_{123} \rangle_1 \quad \text{for} \quad \mathbf{a}, \mathbf{b} \in \mathbb{R}^3 \subset \mathcal{C}\ell_3.$$

In terms of the Clifford algebra $\mathcal{C}\ell_{0,3} \simeq \mathbb{H} \times \mathbb{H}$ of the negative definite quadratic space $\mathbb{R}^{0,3}$ the cross product can be expressed not only as

$$\mathbf{a} \times \mathbf{b} = -\langle \mathbf{abe}_{123} \rangle_1 \quad \text{for} \quad \mathbf{a}, \mathbf{b} \in \mathbb{R}^{0,3} \subset \mathcal{C}\ell_{0,3}$$

but also as [1]

$$\mathbf{a} \times \mathbf{b} = \langle \mathbf{ab}(1 - \mathbf{e}_{123}) \rangle_1 \quad \text{for} \quad \mathbf{a}, \mathbf{b} \in \mathbb{R}^{0,3} \subset \mathcal{C}\ell_{0,3}.$$

Similarly, the 7-dimensional cross product is the vector part of the octonion product of two pure octonions, that is, $\mathbf{a} \times \mathbf{b} = \langle \mathbf{a} \circ \mathbf{b} \rangle_1$. The octonion algebra $\mathbb{O}$ is a norm-preserving algebra with unity 1, whence the vector part $\mathbb{R}^7$ in $\mathbb{O} = \mathbb{R} \oplus \mathbb{R}^7$ is an algebra with cross product, that is, $\mathbf{a} \times \mathbf{b} = \frac{1}{2}(\mathbf{a} \circ \mathbf{b} - \mathbf{b} \circ \mathbf{a})$ for $\mathbf{a}, \mathbf{b} \in \mathbb{R}^7 \subset \mathbb{O} = \mathbb{R} \oplus \mathbb{R}^7$. The octonion product in turn is given by

$$a \circ b = \alpha\beta + \alpha\mathbf{b} + \mathbf{a}\beta - \mathbf{a} \cdot \mathbf{b} + \mathbf{a} \times \mathbf{b}$$

1 This expression is also valid for $\mathbf{a}, \mathbf{b} \in \mathbb{R}^3 \subset \mathcal{C}\ell_3$, but the element $1 - \mathbf{e}_{123}$ does not pick up an ideal of $\mathcal{C}\ell_3$. Recall that $\mathcal{C}\ell_3$ is simple, that is, it has no proper two-sided ideals.

for $a = \alpha + \mathbf{a}$ and $b = \beta + \mathbf{b}$ in $\mathbb{R} \oplus \mathbb{R}^7$. If we replace the Euclidean space $\mathbb{R}^7$ by the negative definite quadratic space $\mathbb{R}^{0,7}$, then not only

$$a \circ b = \alpha\beta + \alpha\mathbf{b} + \mathbf{a}\beta + \mathbf{a} \cdot \mathbf{b} + \mathbf{a} \times \mathbf{b}$$

for $a, b \in \mathbb{R} \oplus \mathbb{R}^{0,7}$, but also

$$a \circ b = \langle ab(1 - \mathbf{v}) \rangle_{0,1}$$

where $\mathbf{v} = \mathbf{e}_{124} + \mathbf{e}_{235} + \mathbf{e}_{346} + \mathbf{e}_{457} + \mathbf{e}_{561} + \mathbf{e}_{672} + \mathbf{e}_{713} \in \bigwedge^3 \mathbb{R}^{0,7}$.

23.5 Definition of triality

Let $n \geq 3$. All automorphisms of $SO(n)$ are of the form $U \to SUS^{-1}$ where $S \in O(n)$. All automorphisms of $\mathbf{Spin}(n)$, $n \neq 8$, are of the form $u \to sus^{-1}$ where $s \in \mathbf{Pin}(n)$. The group $\mathbf{Spin}(8)$ has exceptional automorphisms, which permute the non-identity elements $-1, \mathbf{e}_{12\ldots8}, -\mathbf{e}_{12\ldots8}$ in the center of $\mathbf{Spin}(8)$:

$$\begin{array}{ccc} -1 & \longrightarrow & \mathbf{e}_{12\ldots8} \\ & \nwarrow \quad \swarrow & \\ & -\mathbf{e}_{12\ldots8} & \end{array} \qquad \rho(\pm\mathbf{e}_{12\ldots8}) = -I.$$

Such an automorphism of $\mathbf{Spin}(8)$, of order 3, is said to be a *triality automorphism,* denoted by $\operatorname{trial}(u)$ for $u \in \mathbf{Spin}(8)$.

Regard $\mathbf{Spin}(8)$ as a subset of $\mathcal{C}\ell_8$. In $\mathcal{C}\ell_8$, triality sends the lines through $1, -\mathbf{e}_{12\ldots8}$ and $-1, \mathbf{e}_{12\ldots8}$, which are parallel, to the lines through $1, -1$ and $\mathbf{e}_{12\ldots8}, -\mathbf{e}_{12\ldots8}$, which intersect each other. Thus, a triality automorphism of $\mathbf{Spin}(8)$ cannot be a restriction of a linear mapping $\mathcal{C}\ell_8 \to \mathcal{C}\ell_8$.

A non-linear automorphism of $\mathbf{Spin}(8)$ might also interchange -1 with either of $\pm\mathbf{e}_{12\ldots8}$. Such an automorphism of $\mathbf{Spin}(8)$, of order 2, is said to be a *swap automorphism,* denoted by $\operatorname{swap}(u)$ for $u \in \mathbf{Spin}(8)$.

On the Lie algebra level, triality acts on the space of bivectors $\bigwedge^2 \mathbb{R}^8$, of dimension 28. Triality stabilizes point-wise the Lie algebra $\mathcal{G}_2$ of G_2, which is the automorphism group of $\mathbb{O}$. The dimension of $\mathcal{G}_2$ is 14. In the orthogonal complement $\mathcal{G}_2^\perp$ of $\mathcal{G}_2$, triality is an isoclinic rotation, turning each bivector by the angle 120°. A swap stabilizes point-wise not only $\mathcal{G}_2$ but also a 7-dimensional subspace of $\mathcal{G}_2^\perp$, and reflects the rest of the Lie algebra $so(8) \simeq D_4$, that is, another 7-dimensional subspace of $\mathcal{G}_2^\perp$. For a bivector $\mathbf{F} \in \bigwedge^2 \mathbb{R}^8$, we denote triality by $\operatorname{Trial}(\mathbf{F})$ and swap by $\operatorname{Swap}(\mathbf{F})$.

On the level of representation spaces, triality could be viewed as permuting the vector space $\mathbb{R}^8$ and the two even spinor spaces, that is, the minimal left ideals $\mathcal{C}\ell_8^+ \frac{1}{8}(1 + \mathbf{w})\frac{1}{2}(1 \pm \mathbf{e}_{12\ldots8})$, which are sitting in the two-sided ideals

$C\ell_8^+\frac{1}{2}(1\pm \mathbf{e}_{12\ldots 8}) \simeq \text{Mat}(8,\mathbb{R})$ of $C\ell_8^+ \simeq {}^2\text{Mat}(8,\mathbb{R})$. This means a 120° rotation of the Coxeter–Dynkin diagram of the Lie algebra D_4 :

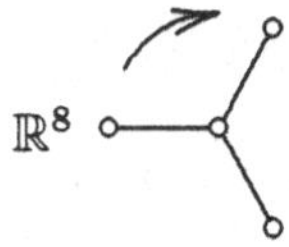

$$C\ell_8^+ f, \quad f = \tfrac{1}{8}(1+\mathbf{w})\tfrac{1}{2}(1+\mathbf{e}_{12\ldots 8})$$

$$C\ell_8^+ \hat{f}, \quad \hat{f} = \tfrac{1}{8}(1+\mathbf{w})\tfrac{1}{2}(1-\mathbf{e}_{12\ldots 8})$$

Rather than permuting the representation spaces, triality permutes elements of **Spin**(8), or their actions on the vector space and the two spinor spaces.

Because of its relation to octonions, it is convenient to view triality in terms of the Clifford algebra $C\ell_{0,7} \simeq {}^2\text{Mat}(8,\mathbb{R})$, the paravector space $\$R^8 = \mathbb{R}\oplus\mathbb{R}^{0,7}$, having an octonion product, the spin group

$$\$pin(8) = \{u \in C\ell_{0,7} \mid u\bar{u} = 1;\ \text{for all } x \in \$R^8 \text{ also } ux\hat{u}^{-1} \in \$R^8\},$$

the minimal left ideals $C\ell_{0,7}\frac{1}{8}(1+\mathbf{w})\frac{1}{2}(1\mp \mathbf{e}_{12\ldots 7})$ of $C\ell_{0,7} \simeq {}^2\text{Mat}(8,\mathbb{R})$, and the primitive idempotents

$$f = \tfrac{1}{8}(1+\mathbf{w})\tfrac{1}{2}(1-\mathbf{e}_{12\ldots 7}), \quad \hat{f} = \tfrac{1}{8}(1+\mathbf{w})\tfrac{1}{2}(1+\mathbf{e}_{12\ldots 7}).$$

For $u \in \$pin(8)$, define two linear transformations U_1, U_2 of $\$R^8$ by

$$U_1(x) = 16\langle uxf\rangle_{0,1}, \quad U_2(x) = 16\langle ux\hat{f}\rangle_{0,1}.$$

The action of u on the left ideal $C\ell_{0,7}\frac{1}{8}(1+\mathbf{w})$ of $C\ell_{0,7}$ results in the matrix representation [2] [3]

$$\$pin(8) \ni u \simeq \begin{pmatrix} U_1 & 0 \\ 0 & U_2 \end{pmatrix} \quad \text{where} \quad U_1, U_2 \in \$O(8).$$

For $U \in SO(8)$, [4] define the *companion* $\check{U}$ by

$$\check{U}(x) = \widehat{U(\hat{x})} \quad \text{for all} \quad x \in \$R^8.$$

2 Choose the bases $(\mathbf{e}_1 f, \mathbf{e}_2 f, \ldots, \mathbf{e}_7 f, f)$ for $C\ell_{0,7}f$ and $(\mathbf{e}_1\hat{f}, \mathbf{e}_2\hat{f}, \ldots, \mathbf{e}_7\hat{f}, \hat{f})$ for $C\ell_{0,7}\hat{f}$, where $f = \frac{1}{8}(1+\mathbf{w})\frac{1}{2}(1-\mathbf{e}_{12\ldots 7})$ and $\hat{f} = \frac{1}{8}(1+\mathbf{w})\frac{1}{2}(1+\mathbf{e}_{12\ldots 7})$. Then the matrices of U_1 and U_2 are the same as in the basis $(\mathbf{e}_1, \mathbf{e}_2, \ldots, \mathbf{e}_7, 1)$ of $\$R^8$. Denoting $f_i = \mathbf{e}_i f$, $i = 1, 2, \ldots, 7$, and $f_8 = f$, $(U_1)_{ij} = 16\langle \bar{f}_i u f_j\rangle_0$, and denoting $g_i = \mathbf{e}_i\hat{f}$, $i = 1, 2, \ldots, 7$, and $g_8 = \hat{f}$, $(U_2)_{ij} = 16\langle \bar{g}_i u g_j\rangle_0$.

3 If we had chosen the bases $(f_1, f_2, \ldots, f_7, f)$ for $C\ell_{0,7}f$ and $(\hat{f}_1, \hat{f}_2, \ldots, \hat{f}_7, \hat{f})$ for $C\ell_{0,7}\hat{f}$, where $f_i = \mathbf{e}_i f$ and $\hat{f}_i = -\mathbf{e}_i\hat{f}$ for $i = 1, 2, \ldots, 7$, then we would have obtained the following matrix representation

$$u \simeq \begin{pmatrix} U_1 & 0 \\ 0 & \check{U}_2 \end{pmatrix},$$

where $U_1(x) = 16\langle uxf\rangle_{0,1}$ as before but $\check{U}_2(x) = 16\langle \hat{u}xf\rangle_{0,1}$. This representation is used by Porteous 1995.

4 Or, for $U \in \text{Mat}(8,\mathbb{R})$.

The *companion* $\breve{u}$ of $u \in \$pin(8)$ is just its main involution, $\breve{u} = \hat{u}$, [5] and corresponds to the matrix

$$\breve{u} \simeq \begin{pmatrix} \breve{U}_2 & 0 \\ 0 & \breve{U}_1 \end{pmatrix}.$$

For a paravector $a \in \$R^8$, define the linear transformation A of $\$R^8$ by [6]

$$A(x) = 16\langle axf\rangle_{0,1}, \quad \text{that is,} \quad A(x) = a \circ x,$$

making $\$R^8$ the Cayley algebra $\mathbb{O}$. Since $\breve{A}^{\mathsf{T}}(x) = 16\langle ax\hat{f}\rangle_{0,1}$, we have the correspondence

$$a \simeq \begin{pmatrix} A & 0 \\ 0 & \breve{A}^{\mathsf{T}} \end{pmatrix}, \quad \text{abbreviated as} \quad a \sim A.$$

Computing the matrix product

$$U(a) = ua\hat{u}^{-1} \simeq \begin{pmatrix} U_1 & 0 \\ 0 & U_2 \end{pmatrix} \begin{pmatrix} A & 0 \\ 0 & \breve{A}^{\mathsf{T}} \end{pmatrix} \begin{pmatrix} \breve{U}_2^{-1} & 0 \\ 0 & \breve{U}_1^{-1} \end{pmatrix},$$

we find the correspondence $U(a) \sim U_1 A \breve{U}_2^{-1}$. Denote $U_0 = \breve{U}$, and let $\breve{U}_0(a)$ operate on $x \in \$R^8$, to get

$$\breve{U}_0(a) \circ x = (U_1 A \breve{U}_2^{-1})(x) = U_1(a \circ \breve{U}_2^{-1}(x)).$$

The ordered triple (U_0, U_1, U_2) in $SO(8)$ is called a *triality triplet* with respect to the octonion product of $\mathbb{O}$.

If (U_0, U_1, U_2) is a triality triplet, then also (U_1, U_2, U_0), (U_2, U_0, U_1) and $(\breve{U}_2, \breve{U}_1, \breve{U}_0)$ are a triality triplets. This results in

$$\breve{U}_0(x \circ y) = U_1(x) \circ U_2(y) \quad \text{for all} \quad x, y \in \mathbb{O} = \$R^8,$$

referred to as *Cartan's principle of triality*. Conversely, for a fixed $U_0 \in \$O(8)$, the identity $\breve{U}_0(x \circ y) = U_1(x) \circ U_2(y)$ has two solutions U_1, U_2 in $\$O(8)$, resulting in the triality triplet (U_0, U_1, U_2) and its *opposite* $(U_0, -U_1, -U_2)$.

5 Recall that for $x \in \$R^8 = \mathbb{R} \oplus \mathbb{R}^{0,7}$, $U(x) = ux\hat{u}^{-1}$, and so $\breve{U}(x) = \hat{u}\hat{x}u^{-1}$.

6 The matrix of A can be computed as $A_{ij} = 16\langle \bar{f}_i a f_j\rangle_0$. The paravector $a = a_0 + a_1 e_1 + \cdots + a_7 e_7$ has the matrix

$$A = \begin{pmatrix} a_0 & -a_4 & -a_7 & a_2 & -a_6 & a_5 & a_3 & a_1 \\ a_4 & a_0 & -a_5 & -a_1 & a_3 & -a_7 & a_6 & a_2 \\ a_7 & a_5 & a_0 & -a_6 & -a_2 & a_4 & -a_1 & a_3 \\ -a_2 & a_1 & a_6 & a_0 & -a_7 & -a_3 & a_5 & a_4 \\ a_6 & -a_3 & a_2 & a_7 & a_0 & -a_1 & -a_4 & a_5 \\ -a_5 & a_7 & -a_4 & a_3 & a_1 & a_0 & -a_2 & a_6 \\ -a_3 & -a_6 & a_1 & -a_5 & a_4 & a_2 & a_0 & a_7 \\ -a_1 & -a_2 & -a_3 & -a_4 & -a_5 & -a_6 & -a_7 & a_0 \end{pmatrix}.$$

Thus, U_0 corresponds to two triality triplets (U_0, U_1, U_2) and $(U_0, -U_1, -U_2)$, while $-U_0$, corresponds to $(-U_0, -U_1, U_2)$ and $(-U_0, U_1, -U_2)$.

The rotations $U_1, U_2 \in \mathcal{S}O(8)$ are represented by $\pm u_1, \pm u_2 \in \mathcal{S}pin(8)$. We choose the signs so that

$$\hat{u}_0 \simeq \begin{pmatrix} U_1 & 0 \\ 0 & U_2 \end{pmatrix}, \quad \hat{u}_1 \simeq \begin{pmatrix} U_2 & 0 \\ 0 & U_0 \end{pmatrix}, \quad \hat{u}_2 \simeq \begin{pmatrix} U_0 & 0 \\ 0 & U_1 \end{pmatrix},$$

where $u_0 = \hat{u}$ and $U_0 = \check{U}$. Using the notion of triality triplets,

$$u_0 \simeq (U_0, U_1, U_2), \quad u_1 \simeq (U_1, U_2, U_0), \quad u_2 \simeq (U_2, U_0, U_1).$$

The rotation U_0 in $\mathcal{S}O(8)$ corresponds to $u_0 \simeq (U_0, U_1, U_2)$ and its opposite $-u_0 \simeq (U_0, -U_1, -U_2)$ in $\mathcal{S}pin(8)$, and the opposite rotation $-U_0$ corresponds to $\mathbf{e}_{12\ldots7}u_0 \simeq (-U_0, -U_1, U_2)$ and $-\mathbf{e}_{12\ldots7}u_0 \simeq (-U_0, U_1, -U_2)$. **Triality** is defined as the mapping

$$\text{trial} : \mathcal{S}pin(8) \to \mathcal{S}pin(8), \; u_1 \simeq \begin{pmatrix} \check{U}_0 & 0 \\ 0 & \check{U}_2 \end{pmatrix} \to u_2 \simeq \begin{pmatrix} \check{U}_1 & 0 \\ 0 & \check{U}_0 \end{pmatrix}.$$

Triality is an automorphism of $\mathcal{S}pin(8)$; it is of order 3 and permutes the non-identity elements $-1, \mathbf{e}_{12\ldots7}, -\mathbf{e}_{12\ldots7}$ in the center of $\mathcal{S}pin(8)$.

Example. Take a unit paravector $a \in \mathcal{S}R^8 = \mathbb{R} \oplus \mathbb{R}^{0,7}$, $|a| = 1$. The action $x \to ax\hat{a}^{-1}$ is a simple rotation of $\mathcal{S}R^8$. [7] Thus, $a \in \mathcal{S}pin(8)$. Denote $a_0 = \hat{a}$, $a_1 = \text{trial}(a_0)$ and $a_2 = \text{trial}(a_1)$ so that $16\langle\hat{a}_1 x f\rangle_{0,1} = a_2 x \hat{a}_2^{-1}$, $16\langle\hat{a}_2 x \hat{f}\rangle_{0,1} = a_1 x \hat{a}_1^{-1}$. Then

$$a \circ x = a_1 x \hat{a}_1^{-1} \quad \text{and} \quad x \circ a = a_2 x \hat{a}_2^{-1}$$

represent isoclinic rotations of $\mathcal{S}R^8$. Left and right multiplications by $a \in S^7 \subset \mathbb{O}$ are positive and negative isoclinic rotations of $\mathcal{S}R^8 = \mathbb{O}$. [8] The Moufang identity

$$a \circ (x \circ y) \circ a = (a \circ x) \circ (y \circ a)$$

results in a special case of Cartan's principle of triality [9]

$$\hat{a}_0 (x \circ y) a_0^{-1} = (a_1 x \hat{a}_1^{-1}) \circ (a_2 y \hat{a}_2^{-1}).$$

In this special case, a_0, a_1, a_2 commute, $a_2 = \tilde{a}_1$ $(= \hat{a}_1^{-1})$ and $a_0 a_1 a_2 = 1$ implying $a = a_1 a_2 = a_1 \hat{a}_1^{-1} = a_2 \hat{a}_2^{-1}$. ∎

---

7 Note that $a \circ x \circ a = ax\tilde{a}$, $\tilde{a} = \hat{a}^{-1}$ and $a \circ x \circ a^{-1} = sxs^{-1}$ where $s = a_1 a_2^{-1} \in \mathcal{S}pin(7)$.

8 Any four mutually orthogonal invariant planes of an isoclinic rotation of $\mathcal{S}R^8$ induce the same orientation on $\mathcal{S}R^8$.

9 To prove Cartan's principle of triality, in the general case, iterate the Moufang identity, like $b \circ a \circ (x \circ y) \circ a \circ b = (b \circ (a \circ x)) \circ ((y \circ a) \circ b)$. Observe the nesting $b \circ (a \circ x) = sx\hat{s}^{-1}$, where $s = \text{trial}(\hat{b})\text{trial}(\hat{a}) = \text{trial}(\widehat{ba})$.

Triality sends a simple rotation to a positive isoclinic rotation and a positive isoclinic rotation to a negative isoclinic rotation. The isoclinic rotations can be represented by octonion multiplication having neutral axis in the rotation plane of the simple rotation:

$$\begin{array}{ccc} & \nearrow & \text{positive isoclinic rotation} \\ \text{simple rotation} & & \downarrow \\ & \nwarrow & \text{negative isoclinic rotation} \end{array}$$

23.6 Spin(7)

Let $u_0 \in \mathbf{Spin}(7) \subset C\ell_8$, and $u_1 = \text{trial}(u_0)$, $u_2 = \text{trial}(u_1)$. Then $u_2 = \check{u}_1$, that is, $\text{trial}(\text{trial}(u_0)) = \mathbf{e}_8 \text{trial}(u_0) \mathbf{e}_8^{-1}$. [10] Thus, $u_1 \check{u}_2^{-1} = 1$ and $u_1 u_2^{-1} = u_1 \mathbf{e}_8 u_1^{-1} \mathbf{e}_8^{-1} \in \mathbb{R} \oplus \mathbb{R}^7 \mathbf{e}_8$, being a product of two vectors, represents a simple rotation. [11] [12] Since $\check{U}_0 = U_0$, $U_2 = \check{U}_1$,

$$u_0 \simeq \begin{pmatrix} U_1 & 0 \\ 0 & \check{U}_1 \end{pmatrix}, \quad u_1 \simeq \begin{pmatrix} U_0 & 0 \\ 0 & U_1 \end{pmatrix}, \quad u_2 \simeq \begin{pmatrix} U_2 & 0 \\ 0 & U_0 \end{pmatrix}.$$

Comparing matrix entries of $u_1 u_0^{-1} u_2$, we find $u_1 u_0^{-1} u_2 \in \mathbf{Spin}(7)$ and so $U_1 U_0^{-1} U_2 \in SO(7)$.

Let the rotation angles of $U_0 \in SO(7)$ be $\alpha_0, \beta_0, \gamma_0$ so that $\alpha_0 \geq \beta_0 \geq \gamma_0 \geq 0$. Then the rotation angles of $U_1 \in SO(8)$ are

$$\begin{cases} \alpha_1 = \frac{1}{2}(\alpha_0 + \beta_0 + \gamma_0) \\ \beta_1 = \frac{1}{2}(\alpha_0 + \beta_0 - \gamma_0) \\ \gamma_1 = \frac{1}{2}(\alpha_0 - \beta_0 + \gamma_0) \\ \delta_1 = \frac{1}{2}(\alpha_0 - \beta_0 - \gamma_0). \end{cases}$$

Since eigenvalues change in $U_0 \to U_1$, triality cannot be a similarity, $U_1 \neq SU_0S^{-1}$. Represent the rotation planes of $U_0 \in SO(7) \subset SO(8)$ by unit bivectors $\mathbf{A}_0, \mathbf{B}_0, \mathbf{C}_0$, and choose the orientation of $\mathbf{D}_0 = \mathbf{u}\mathbf{e}_8$, $\mathbf{u} \in \mathbb{R}^7$, $|\mathbf{u}| = 1$ so that $\mathbf{A}_0 \wedge \mathbf{B}_0 \wedge \mathbf{C}_0 \wedge \mathbf{D}_0 = \mathbf{e}_{12\ldots 8}$. The rotation planes of U_1 can be expressed as unit bivectors [13]

$$\begin{cases} \mathbf{A}_1 = \frac{1}{2}\text{Trial}(\mathbf{A}_0 + \mathbf{B}_0 + \mathbf{C}_0 - \mathbf{D}_0) \\ \mathbf{B}_1 = \frac{1}{2}\text{Trial}(\mathbf{A}_0 + \mathbf{B}_0 - \mathbf{C}_0 + \mathbf{D}_0) \\ \mathbf{C}_1 = \frac{1}{2}\text{Trial}(\mathbf{A}_0 - \mathbf{B}_0 + \mathbf{C}_0 + \mathbf{D}_0) \\ \mathbf{D}_1 = \frac{1}{2}\text{Trial}(\mathbf{A}_0 - \mathbf{B}_0 - \mathbf{C}_0 - \mathbf{D}_0). \end{cases}$$

10 Note that $\text{trial}(\text{trial}(u_1)) \neq \mathbf{e}_8 \text{trial}(u_1) \mathbf{e}_8^{-1}$, $\text{trial}(\text{trial}(u_2)) \neq \mathbf{e}_8 \text{trial}(u_2) \mathbf{e}_8^{-1}$.
11 In $C\ell_{0,7}$, $u_2 = \hat{u}_1$, and so $u_1 \hat{u}_2^{-1} = 1$, but $u_1 u_2^{-1} \in \mathbb{R} \oplus \mathbb{R}^{0,7}$.
12 Recall that for $u \in \mathbf{Spin}(8)$, $u^{-1} = \tilde{u}$, and for $u \in \$pin(8)$, $u^{-1} = \bar{u}$.
13 Trial : $\bigwedge^2 \mathbb{R}^8 \to \bigwedge^2 \mathbb{R}^8$ sends negative isoclinic bivectors to simple bivectors.

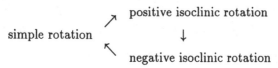

The rotation angles and planes of U_2 are

$$\alpha_2 = \alpha_1,\ \beta_2 = \beta_1,\ \gamma_2 = \gamma_1,\ \delta_2 = -\delta_1$$
$$\mathbf{A}_2 = \check{\mathbf{A}}_1,\ \mathbf{B}_2 = \check{\mathbf{B}}_1,\ \mathbf{C}_2 = \check{\mathbf{C}}_1,\ \mathbf{D}_2 = -\check{\mathbf{D}}_1.$$

The rotation planes of U_0, U_1, U_2 induce the same orientation on $\mathbb{R}^8$, that is,

$$\mathbf{A}_0 \wedge \mathbf{B}_0 \wedge \mathbf{C}_0 \wedge \mathbf{D}_0 = \mathbf{A}_1 \wedge \mathbf{B}_1 \wedge \mathbf{C}_1 \wedge \mathbf{D}_1 = \mathbf{A}_2 \wedge \mathbf{B}_2 \wedge \mathbf{C}_2 \wedge \mathbf{D}_2.$$

For $u_0 \in \mathbf{Spin}(7)$, $u_1, u_2 \in \mathbf{Spin}(8)$, so that

$$u_0 = \exp(\tfrac{1}{2}(\alpha_0\mathbf{A}_0 + \beta_0\mathbf{B}_0 + \gamma_0\mathbf{C}_0))$$
$$u_1 = \exp(\tfrac{1}{2}(\alpha_1\mathbf{A}_1 + \beta_1\mathbf{B}_1 + \gamma_1\mathbf{C}_1 + \delta_1\mathbf{D}_1))$$
$$u_2 = \exp(\tfrac{1}{2}(\alpha_1\mathbf{A}_2 + \beta_1\mathbf{B}_2 + \gamma_1\mathbf{C}_2 - \delta_1\mathbf{D}_2)).$$

23.7 The exceptional Lie algebra G_2

A rotation $U \in SO(7)$ such that

$$U(\mathbf{x} \circ \mathbf{y}) = U(\mathbf{x}) \circ U(\mathbf{y}) \quad \text{for all} \quad \mathbf{x}, \mathbf{y} \in \mathbb{O}$$

is an automorphism of the Cayley algebra $\mathbb{O}$. The automorphisms form a group G_2 with Lie algebra $\mathcal{G}_2 \subset \bigwedge^2 \mathbb{R}^8$, $\dim \mathcal{G}_2 = 14$. A bivector $\mathbf{U} \in \mathcal{G}_2$ acts on the octonion product as a derivation

$$\mathbf{U} \llcorner (\mathbf{x} \circ \mathbf{y}) = (\mathbf{U} \llcorner \mathbf{x}) \circ \mathbf{y} + \mathbf{x} \circ (\mathbf{U} \llcorner \mathbf{y}) \quad \text{for all} \quad \mathbf{x}, \mathbf{y} \in \mathbb{O} = \mathbb{R}^8.$$

The double cover of $G_2 \subset SO(7)$ in $\mathbf{Spin}(7)$ consists of two components, $\mathbf{G}_2$ and $-\mathbf{G}_2$. The groups $\mathbf{G}_2$ and $G_2 = \rho(\mathbf{G}_2)$ are isomorphic, $G_2 \simeq \mathbf{G}_2$. [14]

A rotation $U_0 \in G_2 \subset SO(7)$ has only one preimage in $\mathbf{G}_2 \subset \mathbf{Spin}(7)$, say u_0, $\rho(u_0) = U_0$. Since $\text{trial}(u_0) = u_0$, $u_1 = \text{trial}(u_0)$ equals u_0, and $U_1 = \rho(u_1)$ equals U_0. The rotation angles $\alpha_0, \beta_0, \gamma_0$ of U_0, such that $\alpha_0 \geq \beta_0 \geq \gamma_0 \geq 0$, satisfy the identities

$$\begin{cases} \alpha_1 = \frac{1}{2}(\alpha_0 + \beta_0 + \gamma_0) = \alpha_0 \\ \beta_1 = \frac{1}{2}(\alpha_0 + \beta_0 - \gamma_0) = \beta_0 \\ \gamma_1 = \frac{1}{2}(\alpha_0 - \beta_0 + \gamma_0) = \gamma_0 \\ \delta_1 = \frac{1}{2}(\alpha_0 - \beta_0 - \gamma_0) = 0 \end{cases}$$

each of which implies

$$\alpha_0 = \beta_0 + \gamma_0.$$

14 Note that $-I \notin G_2$ because $-I \notin SO(7)$ and $-1 \notin \mathbf{G}_2$ because triality stabilizes pointwise $\mathbf{G}_2$ but sends -1 to $\pm\mathbf{e}_{12\ldots8}$.

This can also be expressed by saying that the signed rotation angles α, β, γ of $U \in G_2$ satisfy

$$\alpha + \beta + \gamma = 0.$$

Represent the rotation planes of U by unit bivectors $\mathbf{A}, \mathbf{B}, \mathbf{C}$ and choose orientations so that $u = \exp(\frac{1}{2}(\alpha\mathbf{A} + \beta\mathbf{B} + \gamma\mathbf{C}))$, when $U = \rho(u)$. Then $\mathbf{A} \rfloor \mathbf{w} = \mathbf{B} + \mathbf{C}$. Conversely, for an arbitrary rotation $U \in SO(7)$ to be in G_2 it is sufficient that

$$\mathbf{A} \rfloor \mathbf{w} = \mathbf{B} + \mathbf{C} \quad \text{and} \quad \alpha + \beta + \gamma = 0.$$

In order to construct a bivector $\mathbf{U} \in \mathcal{G}_2$, pick up a unit bivector $\mathbf{A} \in \bigwedge^2 \mathbb{R}^7$, $\mathbf{A}^2 = -1$, decompose the bivector $\mathbf{A} \rfloor \mathbf{w}$ into a sum of two simple unit bivectors $\mathbf{B} + \mathbf{C}$ (this decomposition is not unique), choose $\alpha, \beta, \gamma \in \mathbb{R}$ so that $\alpha + \beta + \gamma = 0$, and write $\mathbf{U} = \alpha\mathbf{A} + \beta\mathbf{B} + \gamma\mathbf{C}$.

For $u \in \mathbf{G}_2$, $\text{trial}(u) = u$, and for $\mathbf{U} \in \mathcal{G}_2$, $\text{Trial}(\mathbf{U}) = \mathbf{U}$, in other words, triality stabilizes point-wise $\mathbf{G}_2$ and $\mathcal{G}_2$. Multiplication by $u \in \mathbf{G}_2$ stabilizes the idempotent $\frac{1}{8}(1 + \mathbf{w})$, $u\frac{1}{8}(1 + \mathbf{w}) = \frac{1}{8}(1 + \mathbf{w})u = \frac{1}{8}(1 + \mathbf{w})$, while a bivector $\mathbf{U} \in \mathcal{G}_2$ annihilates $\frac{1}{8}(1+\mathbf{w})$, $\mathbf{U}\frac{1}{8}(1+\mathbf{w}) = \frac{1}{8}(1+\mathbf{w})\mathbf{U} = 0$, and thus $\mathbf{U}\frac{1}{8}(7 - \mathbf{w}) = \frac{1}{8}(7 - \mathbf{w})\mathbf{U} = \mathbf{U}$. Conversely, a rotation $U \in SO(7)$ is in G_2 if it fixes the 3-vector

$$\mathbf{v} = \mathbf{e}_{124} + \mathbf{e}_{235} + \mathbf{e}_{346} + \mathbf{e}_{457} + \mathbf{e}_{561} + \mathbf{e}_{672} + \mathbf{e}_{713}$$

for which $\mathbf{w} = \mathbf{v}\mathbf{e}_{12\ldots7}^{-1} = \mathbf{e}_{1236} - \mathbf{e}_{1257} - \mathbf{e}_{1345} + \mathbf{e}_{1467} + \mathbf{e}_{2347} - \mathbf{e}_{2456} - \mathbf{e}_{3567}$.

A bivector $\mathbf{F} \in \bigwedge^2 \mathbb{R}^8$, $\dim(\bigwedge^2 \mathbb{R}^8) = 28$, can be decomposed as

$$\mathbf{F} = \mathbf{G} + \mathbf{H} \quad \text{where} \quad \mathbf{G} \in \mathcal{G}_2 \quad \text{and} \quad \mathbf{H} = \frac{1}{3}\mathbf{w} \rfloor (\mathbf{w} \wedge \mathbf{F}) \in \mathcal{G}_2^\perp.$$

Under triality, $\mathbf{F}$ goes to $\text{Trial}(\mathbf{F}) = \mathbf{G} + \text{Trial}(\mathbf{H})$, $\text{Trial}(\mathbf{H}) \in \mathcal{G}_2^\perp$, where the angle between $\mathbf{H}$ and $\text{Trial}(\mathbf{H})$ is $120°$. In particular, triality is an isoclinic rotation when restricted to $\mathcal{G}_2^\perp$, $\dim(\mathcal{G}_2^\perp) = 14$.

A bivector $\mathbf{F} \in \bigwedge^2 \mathbb{R}^7$ can be decomposed as $\mathbf{F} = \mathbf{G} + \mathbf{H}$, where $\mathbf{G} \in \mathcal{G}_2$,

$$\mathbf{H} \in \mathcal{G}_2^\perp \cap \overset{2}{\bigwedge} \mathbb{R}^7, \quad \dim(\mathcal{G}_2^\perp \cap \overset{2}{\bigwedge} \mathbb{R}^7) = 7.$$

For a vector $\mathbf{a} \in \mathbb{R}^7$, $\mathbf{v} \llcorner \mathbf{a} \in \mathcal{G}_2^\perp \cap \bigwedge^2 \mathbb{R}^7$. The mapping $\mathbf{a} \to \mathbf{v} \llcorner \mathbf{a}$ is one-to-one, since $\mathbf{a} = \frac{1}{3}\mathbf{v} \llcorner (\mathbf{v} \llcorner \mathbf{a})$. The element $u = \exp(\mathbf{v} \llcorner \mathbf{a}) \in \mathbf{Spin}(7)$ induces a rotation of $\mathbb{R}^7$, which has $\mathbf{a}$ as its axis and is isoclinic in $\mathbf{a}^\perp = \{\mathbf{x} \in \mathbb{R}^7 \mid \mathbf{x} \cdot \mathbf{a} = 0\}$. A miracle happens when $|\mathbf{a}| = 2\pi/3$. Then the rotation angles of $U = \rho(u)$ are $4\pi/3$, which is the same as $4\pi/3 - 2\pi = -2\pi/3$ in the opposite sense of rotation. For the signed rotation angles we can choose $\alpha = 4\pi/3$, $\beta = \gamma = -2\pi/3$ which satisfy $\alpha+\beta+\gamma=0$. Since also $\mathbf{A} \lrcorner \mathbf{w} = \mathbf{B}+\mathbf{C}$, it follows that $u \in \mathbf{G}_2$. Therefore, $u = \exp(\mathbf{v} \llcorner \mathbf{a})$, where $\mathbf{a} \in \mathbb{R}^7$ and $|\mathbf{a}| = 2\pi/3$, belongs to

$$\exp(\mathcal{G}_2) \cap \exp(\mathcal{G}_2^\perp \cap \textstyle\bigwedge^2 \mathbb{R}^7) \simeq S^6, \quad \text{where} \quad \exp(\mathcal{G}_2) = \mathbf{G}_2.$$

Note that $\alpha+\beta+\gamma=0$ in $\mathbf{G}_2$ while $\alpha=\beta=\gamma$ in $\exp(\mathcal{G}_2^\perp \cap \bigwedge^2 \mathbb{R}^7)$. An element $u \in \mathbf{G}_2 \cap \exp(\mathcal{G}_2^\perp \cap \bigwedge^2 \mathbb{R}^7)$ can be also constructed by choosing a unit bivector $\mathbf{A} \in \bigwedge^2 \mathbb{R}^7$, $\mathbf{A}^2 = -1$, decomposing $\mathbf{A} \lrcorner \mathbf{w} = \mathbf{B}+\mathbf{C}$, constructing bivectors

$$\frac{2\pi}{3}(2\mathbf{A}-\mathbf{B}-\mathbf{C}) = \mathbf{G} \in \mathcal{G}_2 \quad \text{and} \quad \frac{2\pi}{3}(-\mathbf{A}-\mathbf{B}-\mathbf{C}) = \mathbf{H} \in \mathcal{G}_2^\perp \cap \textstyle\bigwedge^2 \mathbb{R}^7$$

and exponentiating

$$u = e^{\mathbf{G}} = e^{\mathbf{H}} = -\tfrac{1}{8} + \cdots$$

The elements u are extreme elements in $\mathbf{G}_2$ in the sense that $\langle u\rangle_0 = -\frac{1}{8}$, while for all other $g \in \mathbf{G}_2$, $\langle g\rangle_0 > -\frac{1}{8}$.

The elements $u = \exp(\mathbf{v} \llcorner \mathbf{a})$, where $\mathbf{a} \in \mathbb{R}^{0,7}$ and $|\mathbf{a}| = 2\pi/3$, satisfy $u^3 = 1$, and they are the only non-identity solutions of $u^3 = 1$ in $\mathbf{G}_2$. The octonion $a = e^{\circ \mathbf{a}}$ $(= e^{\mathbf{a}})$ satisfies $a^{\circ 3} = 1$ and $a^{\circ -1} \circ x \circ a = uxu^{-1}$ for all $x \in \$R^8 = \mathbb{R} \oplus \mathbb{R}^{0,7}$. Conversely, the only unit octonions $a \in S^7 \subset \mathbb{O} = \$R^8$ satisfying

$$a^{\circ -1} \circ (x \circ y) \circ a = (a^{\circ -1} \circ x \circ a) \circ (a^{\circ -1} \circ y \circ a) \quad \text{for all} \quad x, y \in \$R^8$$

are solutions of $a^{\circ 3} = \pm 1$.

23.8 Components of the automorphism group of Spin(8)

In general, the only exterior automorphisms of $\mathbf{Spin}(n)$, $n \neq 8$, are of type $u \to sus^{-1}$, where $s \in \mathbf{Pin}(n) \backslash \mathbf{Spin}(n)$. Thus, $\mathrm{Aut}(\mathbf{Spin}(n))/\mathrm{Int}(\mathbf{Spin}(n)) \simeq \mathbb{Z}_2$, when $n \neq 8$. However, in the case $n = 8$, the following sequence is exact

$$1 \longrightarrow \mathrm{Int}(\mathbf{Spin}(8)) \longrightarrow \mathrm{Aut}(\mathbf{Spin}(8)) \longrightarrow S_3,$$

that is, $\mathrm{Aut}(\mathbf{Spin}(8))/\mathrm{Int}(\mathbf{Spin}(8)) \simeq S_3$, a non-commutative group of order 6.

For $u \in \mathbf{Spin}(8)$, denote $\mathrm{swap}_1(u) = \mathbf{e}_8 \mathrm{trial}(u) \mathbf{e}_8^{-1} = \mathrm{trial}(\mathrm{trial}(\mathbf{e}_8 u \mathbf{e}_8^{-1}))$ and $\mathrm{swap}_2(u) = \mathbf{e}_8 \mathrm{trial}(\mathrm{trial}(u)) \mathbf{e}_8^{-1} = \mathrm{trial}(\mathbf{e}_8 u \mathbf{e}_8^{-1})$. Then trial, swap_1, swap_2 generate S_3 :

$$\mathrm{trial} \circ \mathrm{trial} \circ \mathrm{trial} = \mathrm{swap}_1 \circ \mathrm{swap}_1 = \mathrm{swap}_2 \circ \mathrm{swap}_2 = \mathrm{identity}$$
$$\mathrm{swap}_1 \circ \mathrm{swap}_2 = \mathrm{trial} \quad \text{and} \quad \mathrm{swap}_2 \circ \mathrm{swap}_1 = \mathrm{trial} \circ \mathrm{trial}.$$

The automorphism group of **Spin**(8) contains 6 components, represented by the identity, trial, trial ∘ trial, swap_1, swap_2 and the companion. [15] In the component of trial, all automorphisms of order 3 are trialities for some octonion product.

23.9 Triality is quadratic

Triality of $u \in \mathbf{Spin}(8) \subset \mathcal{C}\ell_8$ is a restriction a polynomial mapping $\mathcal{C}\ell_8 \to \mathcal{C}\ell_8$, of degree 2,

$$\begin{aligned}
\text{trial}(u) &= \text{trial}_1(u)\text{trial}_2(u)\\
\text{trial}_1(u) &= \tfrac{1}{2}(1+\mathbf{e}_{12\ldots8})[\langle u(1+\mathbf{w})(1-\mathbf{e}_{12\ldots8})\rangle_{0,6} \wedge \mathbf{e}_8]\mathbf{e}_8^{-1}\\
&\quad +\tfrac{1}{2}(1-\mathbf{e}_{12\ldots8})\\
\text{trial}_2(u) &= (\mathbf{w}-3)[(u(1+\mathbf{e}_{12\ldots8}))\wedge \mathbf{e}_8]\mathbf{e}_8^{-1}(\mathbf{w}-3)^{-1}.
\end{aligned}$$

The first factor is affine linear and the second factor is linear. To verify that trial is a triality, it is sufficient to show that it is an automorphism of order 3 sending -1 to $\mathbf{e}_{12\ldots8}$.

In the Lie algebra level, the triality automorphism of a bivector $\mathbf{F} \in \bigwedge^2 \mathbb{R}^8$ is

$$\begin{aligned}
\text{Trial}(\mathbf{F}) &= \mathbf{e}_8\langle \mathbf{F} - \tfrac{1}{2}\mathbf{F}(1+\mathbf{w})(1+\mathbf{e}_{12\ldots8})\rangle_2 \mathbf{e}_8^{-1}\\
&= \tfrac{1}{2}\mathbf{e}_8(\mathbf{F} - \mathbf{F}\lrcorner\mathbf{w} - (\mathbf{F}\wedge\mathbf{w})\lrcorner\mathbf{e}_{12\ldots8})\mathbf{e}_8^{-1}.
\end{aligned}$$

The triality automorphism of a para-bivector $F \in \mathbb{R}^{0,7} \oplus \bigwedge^2 \mathbb{R}^{0,7}$ is

$$\begin{aligned}
\text{Trial}(F) &= \langle F - \tfrac{1}{2}F(1+\mathbf{w})(1-\mathbf{e}_{12\ldots7})\rangle_{1,2}^{\wedge}\\
&= \tfrac{1}{2}(F - \langle F\rangle_2\lrcorner\mathbf{w} + (F\wedge\mathbf{w})\lrcorner\mathbf{e}_{12\ldots7})^{\wedge}.
\end{aligned}$$

For $u \in \$pin(8)$, triality is

$$\begin{aligned}
\text{trial}(u) &= \text{trial}_1(u)\text{trial}_2(u)\\
\text{trial}_1(u) &= \tfrac{1}{2}(1-\mathbf{e}_{12\ldots7})\langle u(1+\mathbf{w})(1+\mathbf{e}_{12\ldots7})\rangle_{0,6}\\
&\quad +\tfrac{1}{2}(1+\mathbf{e}_{12\ldots7})\\
\text{trial}_2(u) &= (\mathbf{w}-3)\,\text{even}(u(1-\mathbf{e}_{12\ldots7}))(\mathbf{w}-3)^{-1}.
\end{aligned}$$

23.10 Triality in terms of eigenvalues and invariant planes

Triality can also be viewed classically, without Clifford algebras, by inspection of changes in eigenvalues and invariant planes of rotations. Consider $U_0 \in SO(8)$ and a triality triplet (U_0, U_1, U_2). Let the rotation angles $\alpha_0, \beta_0, \gamma_0, \delta_0$

15 The subgroup of linear automorphisms contains 2 components, represented by the identity and the companion, $u \to \mathbf{e}_8 u \mathbf{e}_8^{-1}$.

of U_0 be such that $\alpha_0 \geq \beta_0 \geq \gamma_0 \geq \delta_0 \geq 0$. Represent the rotation planes of U_0 by the unit bivectors $\mathbf{A}_0, \mathbf{B}_0, \mathbf{C}_0, \mathbf{D}_0$. Then the rotation angles of U_1 and U_2 are

$$\begin{cases} \alpha_1 = \frac{1}{2}(\alpha_0 + \beta_0 + \gamma_0 - \delta) \\ \beta_1 = \frac{1}{2}(\alpha_0 + \beta_0 - \gamma_0 + \delta) \\ \gamma_1 = \frac{1}{2}(\alpha_0 - \beta_0 + \gamma_0 + \delta) \\ \delta_1 = \frac{1}{2}(\alpha_0 - \beta_0 - \gamma_0 - \delta) \end{cases} \quad \text{and} \quad \begin{cases} \alpha_2 = \frac{1}{2}(\alpha_0 + \beta_0 + \gamma_0 + \delta) \\ \beta_2 = \frac{1}{2}(\alpha_0 + \beta_0 - \gamma_0 - \delta) \\ \gamma_2 = \frac{1}{2}(\alpha_0 - \beta_0 + \gamma_0 - \delta) \\ \delta_2 = \frac{1}{2}(-\alpha_0 + \beta_0 + \gamma_0 - \delta) \end{cases}$$

and the rotation planes are

$$\begin{cases} \mathbf{A}_1 = \frac{1}{2}\text{Trial}(\mathbf{A}_0 + \mathbf{B}_0 + \mathbf{C}_0 - \mathbf{D}_0) \\ \mathbf{B}_1 = \frac{1}{2}\text{Trial}(\mathbf{A}_0 + \mathbf{B}_0 - \mathbf{C}_0 + \mathbf{D}_0) \\ \mathbf{C}_1 = \frac{1}{2}\text{Trial}(\mathbf{A}_0 - \mathbf{B}_0 + \mathbf{C}_0 + \mathbf{D}_0) \\ \mathbf{D}_1 = \frac{1}{2}\text{Trial}(\mathbf{A}_0 - \mathbf{B}_0 - \mathbf{C}_0 - \mathbf{D}_0) \end{cases}$$

and

$$\begin{cases} \mathbf{A}_2 = \frac{1}{2}\text{Trial}(\text{Trial}(\mathbf{A}_0 + \mathbf{B}_0 + \mathbf{C}_0 + \mathbf{D}_0)) \\ \mathbf{B}_2 = \frac{1}{2}\text{Trial}(\text{Trial}(\mathbf{A}_0 + \mathbf{B}_0 - \mathbf{C}_0 - \mathbf{D}_0)) \\ \mathbf{C}_2 = \frac{1}{2}\text{Trial}(\text{Trial}(\mathbf{A}_0 - \mathbf{B}_0 + \mathbf{C}_0 - \mathbf{D}_0)) \\ \mathbf{D}_2 = \frac{1}{2}\text{Trial}(\text{Trial}(-\mathbf{A}_0 + \mathbf{B}_0 + \mathbf{C}_0 - \mathbf{D}_0)). \end{cases}$$

23.11 Trialities with respect to different octonion products

An arbitrary 4-vector $\mathbf{w} \in \bigwedge^4 \mathbb{R}^8$, for which $\frac{1}{8}(1+\mathbf{w})$ is an idempotent in $\mathcal{C}\ell_8^+$, is called a calibration. [16] A direction $\mathbf{n} \in \mathbb{R}^8$, $|\mathbf{n}| = 1$, fixed by the calibration, $\mathbf{wn} = \mathbf{nw}$, is called the neutral axis of the calibration. A calibration together with its neutral axis can be used to introduce an octonion product on $\mathbb{R}^8$ and a triality of $\mathbf{Spin}(8)$.

Let $\mathbf{w}_1, \mathbf{w}_2$ be calibrations, with neutral axes $\mathbf{n}_1, \mathbf{n}_2 \in \mathbb{R}^8$. Denote the octonion products by

$$\mathbf{x} \circ_{\mathbf{w}_1,\mathbf{n}_1} \mathbf{y} = \langle \mathbf{x}\mathbf{n}_1\mathbf{y}(1 + \mathbf{w}_1)(1 - \mathbf{e}_{12\ldots8})\rangle_1,$$
$$\mathbf{x} \circ_{\mathbf{w}_2,\mathbf{n}_2} \mathbf{y} = \langle \mathbf{x}\mathbf{n}_2\mathbf{y}(1 + \mathbf{w}_2)(1 - \mathbf{e}_{12\ldots8})\rangle_1$$

and the trialities by $\text{trial}_{\mathbf{w}_1,\mathbf{n}_1}(u)$, $\text{trial}_{\mathbf{w}_2,\mathbf{n}_2}(u)$. Denote the opposite of the composition of the trialities by $\text{trial}_{\mathbf{w}_{12},\mathbf{n}_{12}}(u)$ so that

$$\text{trial}_{\mathbf{w}_{12},\mathbf{n}_{12}}(\text{trial}_{\mathbf{w}_{12},\mathbf{n}_{12}}(u)) = \text{trial}_{\mathbf{w}_1,\mathbf{n}_1}(\text{trial}_{\mathbf{w}_2,\mathbf{n}_2}(u)).$$

Then

$$\begin{aligned} \mathbf{w}_{12} &= \tfrac{1}{2}(\mathbf{w}_1 + \mathbf{w}_2) + \tfrac{1}{2}(-\mathbf{w}_1 + \mathbf{w}_2)\mathbf{e}_{12\ldots8} \\ &= \tfrac{1}{2}(1 - \mathbf{e}_{12\ldots8})\mathbf{w}_1 + \tfrac{1}{2}(1 + \mathbf{e}_{12\ldots8})\mathbf{w}_2 \end{aligned}$$

16 Note that $\mathbf{w}$ satisfies $\mathbf{w}^2 = 7 + 6\mathbf{w}$.

and

$$\mathbf{n}_{12} = \frac{\mathbf{y}}{|\mathbf{y}|} \quad \text{for a non-zero} \quad \mathbf{y} = \mathbf{x} - \tfrac{1}{4}\mathbf{w}_{12} \,\llcorner\, (\mathbf{w}_{12} \,\llcorner\, \mathbf{x}), \quad \text{where} \quad \mathbf{x} \in \mathbb{R}^8.$$

23.12 Factorization of $u \in \mathbf{Spin}(8)$

Take $u \in \mathbf{Spin}(8) \setminus \mathbf{Spin}(7)$. Denote $s = (u \wedge \mathbf{e}_8)\mathbf{e}_8^{-1}$, and $u_7 = \frac{s}{|s|}$. Then $u_7 \in \mathbf{Spin}(7)$, and $u = u_8 u_7 = u_7 u_8'$, where $u_8, u_8' \in \mathbb{R} \oplus \mathbb{R}^7\mathbf{e}_8$. [17] These factorizations are unique, up to a sign (-1 is a square root of 1 in $\mathbf{Spin}(7)$) :

$$u = u_8 u_7 = (-u_8)(-u_7) = u_7 u_8' = (-u_7)(-u_8').$$

The following factorizations are unique, up to a cube root of 1 in $\mathbf{G}_2$:

$$u_7 = h_0 g_0 = h_1 g_1 = h_2 g_2 = g_0 h_0' = g_1 h_1' = g_2 h_2',$$

where

$$h_0^3 = u_7 \mathrm{trial}(u_7)^{-1} u_7 \mathrm{trial}(\mathrm{trial}(u_7))^{-1},$$
$$h_0'^3 = \mathrm{trial}(u_7)^{-1} u_7 \mathrm{trial}(\mathrm{trial}(u_7))^{-1} u_7,$$

and

$$h_1 = h_0 g, \; h_2 = h_0 g^2, \quad h_1' = g' h_0', \; h_2' = g'^2 h_0',$$
$$g = \exp\left(\tfrac{2\pi}{3}\mathbf{v} \,\llcorner\, \tfrac{\mathbf{h}_0}{|\mathbf{h}_0|}\right), \quad g' = \exp\left(\tfrac{2\pi}{3}\mathbf{v} \,\llcorner\, \tfrac{\mathbf{h}_0'}{|\mathbf{h}_0'|}\right),$$
$$\mathbf{h}_0 = (\mathbf{H}_0 \wedge \mathbf{H}_0 \wedge \mathbf{H}_0)\mathbf{e}_{12\ldots7}, \quad \mathbf{h}_0' = (\mathbf{H}_0' \wedge \mathbf{H}_0' \wedge \mathbf{H}_0')\mathbf{e}_{12\ldots7},$$
$$h_0 = e^{\mathbf{H}_0}, \quad h_0' = e^{\mathbf{H}_0'}.$$

In this factorization, $g_0, g_1, g_2 \in \mathbf{G}_2$ and $h_0, h_1, h_2, h_0', h_1', h_2' \in \exp(\mathcal{G}_2^\perp \cap \bigwedge^2 \mathbb{R}^7)$ and $g, g' \in \mathbf{G}_2 \cap \exp(\mathcal{G}_2^\perp \cap \bigwedge^2 \mathbb{R}^7) \simeq S^6$. These factorizations are unique, up to a factor $g, g' \in \mathbf{G}_2$, $g^3 = g'^3 = 1$. [18]

Appendix: Comparison of formalisms in $\mathbb{R}^8$ and $\mathbb{R} \oplus \mathbb{R}^{0,7}$

We use the 3-vector

$$\mathbf{v} = \mathbf{e}_{124} + \mathbf{e}_{235} + \mathbf{e}_{346} + \mathbf{e}_{457} + \mathbf{e}_{561} + \mathbf{e}_{672} + \mathbf{e}_{713}$$

in $\bigwedge^3 \mathbb{R}^8$ or $\bigwedge^3 \mathbb{R}^{0,7}$, and the 4-vector $\mathbf{w} = \mathbf{v}\mathbf{e}_{12\ldots7}^{-1}$ in $\bigwedge^4 \mathbb{R}^8$ or $\bigwedge^4 \mathbb{R}^{0,7}$,

$$\mathbf{w} = \mathbf{e}_{1236} - \mathbf{e}_{1257} - \mathbf{e}_{1345} + \mathbf{e}_{1467} + \mathbf{e}_{2347} + \mathbf{e}_{2456} - \mathbf{e}_{3467}.$$

Note that $\mathbf{e}_{12\ldots7}^{-1} = -\mathbf{e}_{12\ldots7}$ in $\mathcal{C}\ell_8$ while $\mathbf{e}_{12\ldots7}^{-1} = \mathbf{e}_{12\ldots7}$ in $\mathcal{C}\ell_{0,7}$.

---

17 Note that $u_8 = \sqrt{u(\mathbf{e}_8 u \mathbf{e}_8^{-1})^{-1}}$ and $u_8' = \sqrt{(\mathbf{e}_8 u \mathbf{e}_8^{-1})^{-1} u}$.
18 Recall that $-1 \notin \mathbf{G}_2$.

We use the octonion product

$$\mathbf{x} \circ \mathbf{y} = \langle \mathbf{x}\mathbf{e}_8\mathbf{y}(1+\mathbf{w})(1-\mathbf{e}_{12\ldots8})\rangle_1 \quad \text{for vectors} \quad \mathbf{x},\mathbf{y} \in \mathbb{R}^8,$$
$$x \circ y = \langle xy(1+\mathbf{w})(1-\mathbf{e}_{12\ldots7})\rangle_{0,1} \quad \text{for paravectors} \quad x,y \in \mathbb{R}\oplus\mathbb{R}^{0,7}.$$

Note that in $\mathcal{C}\ell_{0,7}$, also $x \circ y = \langle xy(1-\mathbf{v})\rangle_{0,1}$.

The bivector $\mathbf{F} = \mathbf{A}+\mathbf{B} \in \bigwedge^2\mathbb{R}^8$, with $\mathbf{B} \in \bigwedge^2\mathbb{R}^7$ and $\mathbf{A} = \mathbf{a}\mathbf{e}_8$, $\mathbf{a} \in \mathbb{R}^7$, corresponds to the para-bivector $F = \mathbf{a}-\mathbf{B} \in \mathbb{R}^{0,7}\oplus\bigwedge^2\mathbb{R}^{0,7}$, with $\mathbf{a} = \mathbf{A}\mathbf{e}_8^{-1} \in \mathbb{R}^{0,7}$. Let $u = u_+ + u_-\mathbf{e}_8 \in \mathbf{Spin}(8)$, where $u_\pm \in \mathcal{C}\ell_7^\pm$. Then $u \in \mathbf{Spin}(8)$ corresponds to

$$(\widetilde{u_+ + u_-}) = (\widehat{u_+ + u_-})^{-1} \in \$pin(8).$$

The companion $\breve{u}$ of u is

$$\breve{u} = \mathbf{e}_8 u \mathbf{e}_8^{-1} \qquad \text{for } u \text{ in } \mathbf{Spin}(8) \text{ or } \mathcal{C}\ell_8^+ \text{ (or } \mathcal{C}\ell_8\text{)},$$
$$\breve{u} = \hat{u} \qquad \text{for } u \text{ in } \$pin(8) \text{ or } \mathcal{C}\ell_{0,7}.$$

For u_0, $u_1 = \operatorname{trial}(u_0)$, $u_2 = \operatorname{trial}(u_1)$, Cartan's principle of triality says

$$\breve{u}_0(\mathbf{x}\circ\mathbf{y})\breve{u}_0^{-1} = (u_1\mathbf{x}u_1^{-1})\circ(u_2\mathbf{y}u_2^{-1}) \qquad \text{in} \quad \mathbf{Spin}(8),$$
$$\hat{u}_0(x\circ y)u_0^{-1} = (u_1 x\hat{u}_1^{-1})\circ(u_2 y\hat{u}_2^{-1}) \qquad \text{in} \quad \$pin(8).$$

In the Lie algebra level, Freudenthal's principle of triality says [19]

$$(\mathbf{x}\circ\mathbf{y})\lrcorner\breve{\mathbf{F}}_0 = (\mathbf{x}\lrcorner\mathbf{F}_1)\circ\mathbf{y} + \mathbf{x}\circ(\mathbf{y}\lrcorner\mathbf{F}_2) \quad \text{in} \quad \textstyle\bigwedge^2\mathbb{R}^8,$$
$$\langle (x\circ y)\lrcorner F_0\rangle_{0,1} = \langle x\lrcorner\hat{F}_1\rangle_{0,1}\circ y + x\circ\langle y\lrcorner\hat{F}_2\rangle_{0,1} \quad \text{in} \quad \mathbb{R}^{0,7}\oplus\textstyle\bigwedge^2\mathbb{R}^{0,7}.$$

The non-identity central elements of the Lie groups are permuted as follows

$$\operatorname{trial}(-1) = \mathbf{e}_{12\ldots8},\ \operatorname{trial}(\mathbf{e}_{12\ldots8}) = -\mathbf{e}_{12\ldots8} \quad \text{in} \quad \mathbf{Spin}(8),$$
$$\operatorname{trial}(-1) = -\mathbf{e}_{12\ldots7},\ \operatorname{trial}(-\mathbf{e}_{12\ldots7}) = \mathbf{e}_{12\ldots7} \quad \text{in} \quad \$pin(8).$$

Exercises

Show that

1. For $\mathbf{U} \in \mathcal{G}_2$, $\mathbf{U}\lrcorner\mathbf{w} = -\mathbf{U}$, $\mathbf{U}\mathbf{w} = -\mathbf{U}$, $\mathbf{U}\lrcorner\mathbf{U} = -|\mathbf{U}|^2$, $|\mathbf{U}\wedge\mathbf{U}| = |\mathbf{U}|^2$.
2. For $\mathbf{U}_0 \in \bigwedge^2\mathbb{R}^8$, $\mathbf{U}_1 = \operatorname{Trial}(\mathbf{U}_0)$, $\mathbf{U}_2 = \operatorname{Trial}(\mathbf{U}_1)$, $\mathbf{U}_0+\mathbf{U}_1+\mathbf{U}_2 \in \mathcal{G}_2$, $2\mathbf{U}_0-\mathbf{U}_1-\mathbf{U}_2 \in \mathcal{G}_2^\perp$.
3. In $\mathcal{C}\ell_{0,7}$, $(1+\mathbf{w})(1-\mathbf{e}_{12\ldots7}) = (1-\mathbf{e}_{124})(1-\mathbf{e}_{235})(1-\mathbf{e}_{346})(1-\mathbf{e}_{457})$, $(1+\mathbf{w})(1-\mathbf{e}_{12\ldots7}) = (1-\mathbf{v})(1-\mathbf{e}_{12\ldots7})$.
4. $\mathbf{w}^2 = 7+6\mathbf{w}$.
5. $\mathbf{w}^n = \begin{cases} \frac{1}{8}(7^n-1)+1+\frac{1}{8}(7^n-1)\mathbf{w}, & n \text{ even}, \\ \frac{1}{8}(7^n+1)-1+\frac{1}{8}(7^n+1)\mathbf{w}, & n \text{ odd}. \end{cases}$

19 Note that $\mathbf{x}\lrcorner\mathbf{F}$ corresponds to $\langle x\lrcorner\hat{F}\rangle_{0,1}$.

6. For $x \in \mathbb{R}$, $f(x\mathbf{w}) = f(-x)\frac{1}{8}(7-\mathbf{w}) + f(7x)\frac{1}{8}(1+\mathbf{w})$. Hint: the minimal polynomial of $\mathbf{w}$, $x^2 = 7 + 6x$, has roots $x = -1$, $x = 7$.
7. For $\mathbf{G} \in \mathcal{G}_2$, $\frac{1}{8}(1+\mathbf{w})\mathbf{G}\frac{1}{8}(1+\mathbf{w}) = 0$, $\frac{1}{8}(7-\mathbf{w})\mathbf{G}\frac{1}{8}(7-\mathbf{w}) = \mathbf{G}$, $e^{\mathbf{G}} = \frac{1}{8}(7-\mathbf{w})e^{\mathbf{G}}\frac{1}{8}(7-\mathbf{w}) + \frac{1}{8}(1+\mathbf{w})$.
8. $\mathbf{v}^2 = -7 - 6\mathbf{w}$ in $\mathcal{C}\ell_8$, $\mathbf{v}^2 = 7 + 6\mathbf{w}$ in $\mathcal{C}\ell_{0,7}$.
9. For $\mathbf{v} \in \bigwedge^3 \mathbb{R}^7$, $e^{\pi\mathbf{v}} = -1$. Hint: $\mathbf{v}^4 + 50\mathbf{v}^2 + 49 = 0$, and $x^4 + 50x^2 + 49 = 0$ has roots $\pm i$, $\pm 7i$.
10. For $\mathbf{v} \in \bigwedge^3 \mathbb{R}^{0,7}$, $\cos(\pi\mathbf{v}) = -1$, $\sin(\pi\mathbf{v}) = 0$. Hint: $\mathbf{v}^4 - 50\mathbf{v}^2 + 49 = 0$, and $x^4 - 50x^2 + 49 = 0$ has roots ± 7, ± 1.
11. For $\mathbf{F} \in \bigwedge^2 \mathbb{R}^8$, $\mathbf{F} = \mathbf{G} + \mathbf{H}$, $\mathbf{G} \in \mathcal{G}_2$, $\mathbf{H} \in \mathcal{G}_2^\perp$: $\mathbf{H} = \frac{1}{3}\mathbf{w} \lrcorner (\mathbf{w} \wedge \mathbf{F})$.
12. For $\mathbf{H} \in \mathcal{G}_2^\perp$, $\mathbf{H} = \frac{1}{4}\mathbf{w} \llcorner (\mathbf{w} \llcorner \mathbf{H})$.
13. For $u_0 \in \$pin(8)$, $u_1 = \text{trial}(u_0)$, $u_2 = \text{trial}(u_1)$: $u_0\hat{u}_1^{-1}u_2 \in \$pin(7)$. Hint: $a = u_0^{-1}\hat{u}_0 \in \mathbb{R} \oplus \mathbb{R}^{0,7}$ and so $a = \widehat{\text{trial}(a)}\text{trial}(a)^{-1}$.
14. For the opposite product $\mathbf{x} \bullet \mathbf{y} = \mathbf{y} \circ \mathbf{x}$ of $\mathbf{x}, \mathbf{y} \in \mathbb{O} = \mathbb{R}^8$, $\check{u}_0(\mathbf{x} \bullet \mathbf{y})\check{u}_0^{-1} = (u_2\mathbf{x}u_2^{-1}) \bullet (u_1\mathbf{y}u_1^{-1})$.
15. $u_0(\mathbf{x} \circ \mathbf{y})u_0^{-1} = (\check{u}_2\mathbf{x}\check{u}_2^{-1}) \circ (\check{u}_1\mathbf{y}\check{u}_1^{-1})$.
16. $\langle \hat{u}_0 x(1+\mathbf{w})(1-\mathbf{e}_{12\ldots7})\rangle_{0,1} = u_1 x \hat{u}_1^{-1}$, $\langle \hat{u}_0 x(1+\mathbf{w})(1+\mathbf{e}_{12\ldots7})\rangle_{0,1} = u_2 x \hat{u}_2^{-1}$.
17. $\langle \hat{u}_0 xy(1+\mathbf{w})(1-\mathbf{e}_{12\ldots7})\rangle_{0,1} = u_1(x \circ y)\hat{u}_1^{-1}$, $\langle \hat{u}_0 xy(1+\mathbf{w})(1+\mathbf{e}_{12\ldots7})\rangle_{0,1} = u_2(y \circ x)\hat{u}_2^{-1}$.
18. For $\mathbf{B} \in \bigwedge^2 \mathbb{R}^8 \cap \mathbf{Spin}(8)$, $\langle \text{trial}(\mathbf{B})\rangle_0 = \frac{1}{4}$. For $\mathbf{C} \in \bigwedge^4 \mathbb{R}^8 \cap \mathbf{Spin}(8)$, $\langle \text{trial}(\mathbf{C})\rangle_0 = 0$. For $\mathbf{D} \in \bigwedge^6 \mathbb{R}^8 \cap \mathbf{Spin}(8)$, $\langle \text{trial}(\mathbf{D})\rangle_0 = -\frac{1}{4}$.
19. For $\mathbf{C} \in \bigwedge^4 \mathbb{R}^8 \cap \mathbf{Spin}(8)$, $\text{trial}(\mathbf{C}) \in \bigwedge^4 \mathbb{R}^8$.
20. For $u \in \mathbf{Spin}(8)$, inducing a simple rotation $U = \rho(u)$: $\langle \text{trial}(u)\rangle_8 \mathbf{e}_{12\ldots8} > 0$ and $\langle \text{trial}(\text{trial}(u))\rangle_8 \mathbf{e}_{12\ldots8} < 0$.
21. $\langle \mathbf{G}_2 \rangle_0 \geq -\frac{1}{8}$, $\langle \text{trial}(\mathbf{Spin}(7))\rangle_0 \geq -\frac{1}{4}$.
22. $\mathbf{G}_2 \cap \exp(\mathcal{G}_2^\perp \cap \bigwedge^2 \mathbb{R}^7)$ is homeomorphic but not isometric to S^6.
23. $\text{diam}(\mathbf{G}_2) = \frac{3}{2}$, $\text{diam}(\text{trial}(\mathbf{Spin}(7))) = \sqrt{\frac{5}{2}}$.
24. Triality does not extend to an automorphism of $\mathbf{Pin}(8)$.
25. $[\frac{1}{4}(\mathbf{w} - 3)]^2 = 1$.

Determine

26. The matrix of $\frac{1}{4}(\mathbf{w} - 3)$ in the basis $(f_1, f_2, \ldots, f_8)$ of $\mathcal{C}\ell_8^+ f$, where $f_i = \mathbf{e}_i\mathbf{e}_8 f$, $i = 1, 2, \ldots, 8$, and $f = \frac{1}{8}(1+\mathbf{w})\frac{1}{2}(1+\mathbf{e}_{12\ldots8})$.

Solutions

13. $u_0^{-1}\hat{u}_0 = \hat{u}_1^{-1}u_2\hat{u}_2^{-1}u_1$ so $1 = u_0\hat{u}_1^{-1}u_2(\widehat{\hat{u}_2^{-1}u_1\hat{u}_0^{-1}}) = u_0\hat{u}_1^{-1}u_2(\hat{u}_0 u_1^{-1}\hat{u}_2)^{-1}$ which implies $u_0\hat{u}_1^{-1}u_2 = \hat{u}_0 u_1^{-1}\hat{u}_2 = (u_0\hat{u}_1^{-1}u_2)\widehat{}$.
24. Triality sends $-1 \in \text{Cen}(\mathbf{Pin}(8))$ to $\mathbf{e}_{12\ldots8} \notin \text{Cen}(\mathbf{Pin}(8))$.

26.

$$16\langle \tilde{f}_i \tfrac{1}{4}(\mathbf{w}-3) f_j \rangle_0 = \begin{pmatrix} -1 & 0 & 0 & 0 & 0 & 0 & 0 & 0 \\ 0 & -1 & 0 & 0 & 0 & 0 & 0 & 0 \\ 0 & 0 & -1 & 0 & 0 & 0 & 0 & 0 \\ 0 & 0 & 0 & -1 & 0 & 0 & 0 & 0 \\ 0 & 0 & 0 & 0 & -1 & 0 & 0 & 0 \\ 0 & 0 & 0 & 0 & 0 & -1 & 0 & 0 \\ 0 & 0 & 0 & 0 & 0 & 0 & -1 & 0 \\ 0 & 0 & 0 & 0 & 0 & 0 & 0 & 1 \end{pmatrix}.$$

Bibliography

J.F. Adams: On the non-existence of elements of Hopf invariant one. *Ann. of Math.* **72** (1960), 20-104.

J.F. Adams: Vector fields of spheres. *Ann. of Math.* **75** (1962), 603-632.

F. van der Blij: History of the octaves. *Simon Stevin* **34** (1961), 106-125.

C. Chevalley: *The Algebraic Theory of Spinors.* Columbia University Press, New York, 1954. *The Algebraic Theory of Spinors and Clifford Algebras.* Springer, Berlin, 1997.

G.M. Dixon: *Division Algebras: Octonions, Quaternions, Complex Numbers and the Algebraic Design of Physics.* Kluwer, Dordrecht, 1994.

H.-D. Ebbinghaus et al. (eds.): *Numbers.* Springer, New York, 1991.

A.J. Hahn: Cayley algebras and the automorphisms of $PO_8'(V)$ and $P\Omega_8'(V)$. *Amer. J. Math.* **98** (1976), 953-987.

A.J. Hahn: Cayley algebras and the isomorphisms of the orthogonal groups over arithmetic and local domains. *J. Algebra* **45** (1977), 210-246.

F.R. Harvey: *Spinors and Calibrations.* Academic Press, San Diego, 1990.

M.-A. Knus, A. Merkurjev, M. Rost, J.-P. Tignol: *The Book of Involutions.* American Mathematical Society, Colloquium Publications **44**, 1998.

W.S. Massey: Cross products of vectors in higher dimensional Euclidean spaces. *Amer. Math. Monthly* **90** (1983), #10, 697-701.

A. Micali, Ph. Revoy: *Modules quadratiques.* Cahiers Mathématiques **10**, Montpellier, 1977. *Bull. Soc. Math. France* **63**, suppl. (1979), 5-144.

S. Okubo: *Octonions and Non-associative Algebras in Physics.* Cambridge University Press, Cambridge, 1994.

I.R. Porteous: *Clifford Algebras and the Classical Groups.* Cambridge University Press, Cambridge, 1995.

R.D. Schafer: On the algebras formed by the Cayley-Dickson process. *Amer. J. Math.* **76** (1954), 435-446.

T.A. Springer, F.D. Veldkamp: *Octonions, Jordan Algebras and Exceptional Groups.* Springer, Berlin, 2000.

G.P. Wene: A construction relating Clifford algebras and Cayley-Dickson algebras. *J. Math. Phys.* **25** (1984), 2351-2353.

A History of Clifford Algebras

Clifford's geometric algebras were created by William K. Clifford in 1878/1882, when he introduced a new multiplication rule into Grassmann's exterior algebra $\bigwedge \mathbb{R}^n$, by means of an orthonormal basis $(\mathbf{e}_1, \mathbf{e}_2, \ldots, \mathbf{e}_n)$ of $\mathbb{R}^n$. Clifford also classified his algebras into four classes according to the signs in $(\mathbf{e}_1\mathbf{e}_2 \ldots \mathbf{e}_n)^2 = \pm 1$ and $(\mathbf{e}_1\mathbf{e}_2 \ldots \mathbf{e}_n)\mathbf{e}_i = \pm \mathbf{e}_i(\mathbf{e}_1\mathbf{e}_2 \ldots \mathbf{e}_n)$. In the special case of $n = 3$, Clifford's construction embodied Hamilton's quaternions, as bivectors $i = \mathbf{e}_2\mathbf{e}_3$, $j = \mathbf{e}_3\mathbf{e}_1$, $k = \mathbf{e}_1\mathbf{e}_2$. Clifford algebras were independently rediscovered by Lipschitz 1880/1886, who also presented their first application to geometry, while exploring rotations of $\mathbb{R}^n$, in terms of $\mathbf{Spin}(n)$, a normalized subgroup of the Lipschitz group $\mathbf{\Gamma}_n^+$.

Spinor representations of the orthogonal Lie algebras, $B_\ell = so(2n+1)$ and $D_\ell = so(2n)$, were observed by E. Cartan in 1913, but without using the term "spinor". Two-valued spinor representations of the rotation groups $SO(n)$ were re-constructed recursively by Brauer & Weyl in 1935, but without using the term "Clifford algebra".

In the Schrödinger equation, Pauli 1927 replaced $\pi^2 = \vec{\pi} \cdot \vec{\pi}$, where $\vec{\pi} = -i\hbar\nabla - e\vec{A}$, by

$$\vec{\pi}^2 = \vec{\pi} \cdot \vec{\pi} + \vec{\pi} \wedge \vec{\pi},$$

where the exterior part does not vanish: $(\vec{\pi} \wedge \vec{\pi})\psi = -\hbar e\vec{B}\psi$. Pauli explained interaction of a spinning electron with a magnetic field $\vec{B}$ by means of a spinor-valued wave function $\psi : \mathbb{R}^3 \times \mathbb{R} \to \mathbb{C}^2$. Thus, besides tensors, nature required new kinds of objects: spinors, whose construction calls for Clifford algebras.

Clifford algebras are not only necessary but also offer advantages: their multivector structure enables controlling of subspaces without losing information about their orientations. Physicists are familiar with this advantage in the special case of 1-dimensional oriented subspaces, which they manipulate by vectors, not by projection operators, which forget orientations.

23.13 Algebras of Hamilton, Grassmann and Clifford

The first step towards a Clifford algebra was taken by Hamilton in 1843 (first published 1844), when he studied products of sums of squares and invented his quaternions while searching for multiplicative compositions of triplets in $\mathbb{R}^3$. The present formulation of vector algebra was extracted from the quaternion product of triplets/vectors $\mathbf{xy} = -\mathbf{x}\cdot\mathbf{y}+\mathbf{x}\times\mathbf{y}$ by Gibbs 1881-84 (first published 1901). Hamilton regarded quaternions as quotients of vectors and wrote a rotation in the form $\mathbf{y} = a\mathbf{x}$ using a unit quaternion $a \in \mathbb{H}$, $|a| = 1$. In such a rotation, the axis $\mathbf{a}$ had to be perpendicular to the vector $\mathbf{x}$ (which turned in the plane orthogonal to $\mathbf{a}$). Cayley 1845 published the formula for rotations

$$\mathbf{y} = a\mathbf{x}a^{-1}, \quad a = \cos(\frac{\alpha}{2}) + \frac{\mathbf{a}}{\alpha}\sin(\frac{\alpha}{2}),$$

about an arbitrary axis $\mathbf{a} \in \mathbb{R}^3$ by angle $\alpha = |\mathbf{a}|$ (Cayley ascribed the discovery to Hamilton). Cayley thus came into contact with half-angles and spin representation of rotations in $\mathbb{R}^3$. However, in 1840 Olinde Rodrigues had already recognized the relevancy of half-angles in his study on the composition of rotations in $\mathbb{R}^3$. Cayley 1855 also discovered the quaternionic representation $q \to aqb$ of rotations in $\mathbb{R}^4$, equivalent to the decomposition $\mathbf{Spin}(4) \simeq \mathbf{Spin}(3) \times \mathbf{Spin}(3)$.

The quaternion algebra $\mathbb{H}$ is isomorphic to the even Clifford algebras $\mathcal{C}\ell_3^+ \simeq \mathcal{C}\ell_{0,3}^+$ as well as to the proper ideals $\mathcal{C}\ell_{0,3}\frac{1}{2}(1 \pm \mathbf{e}_{123})$ of $\mathcal{C}\ell_{0,3}$. Hamilton also considered complex quaternions $\mathbb{C}\otimes\mathbb{H}$, isomorphic to the Clifford algebra $\mathcal{C}\ell_3$ on $\mathbb{R}^3$. Note the algebra isomorphisms $\mathcal{C}\ell_3 \simeq \mathrm{Mat}(2,\mathbb{C})$ and $\mathcal{C}\ell_{0,3} \simeq \mathbb{H}\oplus\mathbb{H}$.

Bivectors were introduced by H. Grassmann, when he created his exterior algebras in 1844. The exterior product of two vectors, the bivector $\mathbf{a}\wedge\mathbf{b}$, was interpreted geometrically as the parallelogram with $\mathbf{a}$ and $\mathbf{b}$ as edges, and two such exterior products were equal if their parallelograms lay in parallel planes and had the same area with the same sense of rotation (from $\mathbf{a}$ to $\mathbf{b}$). Thus the exterior product of two vectors was anticommutative, $\mathbf{a}\wedge\mathbf{b} = -\mathbf{b}\wedge\mathbf{a}$. Using a basis $(\mathbf{e}_1, \mathbf{e}_2, \ldots, \mathbf{e}_n)$ for $\mathbb{R}^n$, the exterior algebra $\bigwedge\mathbb{R}^n$ had a basis

$$\begin{array}{l} 1 \\ \mathbf{e}_1, \mathbf{e}_2, \ldots, \mathbf{e}_n \\ \mathbf{e}_1\wedge\mathbf{e}_2, \mathbf{e}_1\wedge\mathbf{e}_3, \ldots, \mathbf{e}_1\wedge\mathbf{e}_n, \mathbf{e}_2\wedge\mathbf{e}_3, \ldots, \mathbf{e}_{n-1}\wedge\mathbf{e}_n \\ \quad\vdots \\ \mathbf{e}_1\wedge\mathbf{e}_2\wedge\ldots\wedge\mathbf{e}_n \end{array}$$

and was thereby of dimension 2^n.

W. K. Clifford studied compounds (tensor products) of two quaternion algebras, where quaternions of one algebra commuted with the quaternions of the other algebra, and applied exterior algebra to grade these tensor products of quaternion algebras. Clifford coined his **geometric algebra** in 1876 (first published in 1878). Clifford's geometric algebra was generated by n orthogonal unit vectors $\mathbf{e}_1, \mathbf{e}_2, \ldots, \mathbf{e}_n$ which anticommuted $\mathbf{e}_i\mathbf{e}_j = -\mathbf{e}_j\mathbf{e}_i$ (like Grassmann) and satisfied all $\mathbf{e}_i^2 = -1$ (like Hamilton) [or then all $\mathbf{e}_i^2 = 1$ as in Clifford's paper 1882]. The number of independent products $\mathbf{e}_i\mathbf{e}_j = \mathbf{e}_i \wedge \mathbf{e}_j$, $i < j$, of degree 2 was $\frac{1}{2}n(n-1) = \binom{n}{2}$. Clifford summed up the numbers of independent products of various degrees 0, 1, 2, ..., n and thereby determined the dimension of his geometric algebra to be

$$1 + n + \binom{n}{2} + \ldots + 1 = 2^n.$$

Clifford distinguished four classes of these geometric algebras characterized by the signs of $(\mathbf{e}_1\mathbf{e}_2\ldots\mathbf{e}_n)\mathbf{e}_i = \pm\mathbf{e}_i(\mathbf{e}_1\mathbf{e}_2\ldots\mathbf{e}_n)$ and $(\mathbf{e}_1\mathbf{e}_2\ldots\mathbf{e}_n)^2 = \pm 1$. He also introduced two algebras of lower dimension 2^{n-1}, namely, the subalgebra of even elements and, for odd n, a reduced (non-universal) algebra obtained by putting $\mathbf{e}_1\mathbf{e}_2\ldots\mathbf{e}_n = \pm 1$ [instead of letting $\mathbf{e}_1\mathbf{e}_2\ldots\mathbf{e}_n = \mathbf{e}_1 \wedge \mathbf{e}_2 \wedge \ldots \wedge \mathbf{e}_n$ be of degree n with $(\mathbf{e}_1\mathbf{e}_2\ldots\mathbf{e}_n)^2 = 1$].

23.14 Rudolf Lipschitz

Clifford's geometric algebra was reinvented in 1880, just two years after its first publication, by Rudolf Lipschitz, who later acknowledged Clifford's prior discovery in his book, see R. Lipschitz: *Untersuchungen über die Summen von Quadraten*, 1886. In his study on sums of squares, Lipschitz considered a representation of rotations by complex numbers and quaternions and generalized this to higher dimensional rotations in $\mathbb{R}^n$. Lipschitz thus gave the first geometric application of Clifford algebras in 1880. He expressed a rotation

$$\mathbf{y} = (I + A)(I - A)^{-1}\mathbf{x}, \quad \mathbf{x} \in \mathbb{R}^n,$$

(written here in modern notation with an antisymmetric matrix A) in the form $\mathbf{y} - A\mathbf{y} = \mathbf{x} + A\mathbf{x}$ or as $\mathbf{y} + \mathbf{y} \rfloor \mathbf{B} = \mathbf{x} + \mathbf{B} \llcorner \mathbf{x}$ where $\mathbf{B} \in \bigwedge^2 \mathbb{R}^n$ is the bivector determined by $A\mathbf{x} = \mathbf{B} \llcorner \mathbf{x}$ $(= -\mathbf{x} \rfloor \mathbf{B})$. Lipschitz rewrote $(I - A)\mathbf{y} = (I + A)\mathbf{x}$ using an even Clifford number $a \in \mathcal{C}\ell_{0,n}^+$ in the form $\mathbf{y}a = a\mathbf{x}$, thus representing the rotation as

$$\mathbf{y} = a\mathbf{x}a^{-1}, \quad a \in \mathbf{\Gamma}_{0,n}^+.$$

[Lipschitz wrote $ya_1 = ax$ where $x = \mathbf{x}\mathbf{e}_1^{-1}$, $y = \mathbf{y}\mathbf{e}_1^{-1}$, $a_1 = \mathbf{e}_1 a \mathbf{e}_1^{-1}$.] In modern terms (introduced by Chevalley), Lipschitz used the exterior exponential

$$a = e^{\wedge \mathbf{B}}, \quad e^{\wedge \mathbf{B}} = 1 + \mathbf{B} + \frac{1}{2}\mathbf{B} \wedge \mathbf{B} + \frac{1}{6}\mathbf{B}^{\wedge 3} + \cdots,$$

of the bivector **B** so that the normalized element $a/|a|$ was in the spin group **Spin**(n).

23.15 Theodor Vahlen

Vahlen 1897 found an explicit expression for the multiplication rule of two basis elements in $\mathcal{C}\ell_{0,n}$

$$(\mathbf{e}_1^{\alpha_1}\mathbf{e}_2^{\alpha_2}\ldots\mathbf{e}_n^{\alpha_n})(\mathbf{e}_1^{\beta_1}\mathbf{e}_2^{\beta_2}\ldots\mathbf{e}_n^{\beta_n}) = (-1)^{\sum_{i\geq j}\alpha_i\beta_j}\mathbf{e}_1^{\alpha_1+\beta_1}\mathbf{e}_2^{\alpha_2+\beta_2}\ldots\mathbf{e}_n^{\alpha_n+\beta_n}$$

where the exponents are 0 or 1 (added here modulo 2, although for Vahlen 1+1 = 2, so that summation was over $i > j$). Vahlen's formula has frequently been reinvented afterwards: for positive metrics by Brauer & Weyl 1935, for arbitrary metrics by Deheuvels 1981 p. 294, disguised with index sets as in Chevalley 1946 p. 62, Artin 1957 p. 186 and Brackx & Delanghe & Sommen 1982 p. 2, or hidden among permutations as in Kähler 1960/62 and Delanghe & Sommen & Souček 1992, pp. 58-59.

Although Brauer & Weyl 1935 reinvented (in the case of the Clifford algebra $\mathcal{C}\ell_n$) the above explicit multiplication formula of Vahlen, they did not observe the connection to the Walsh functions (discovered in the meantime by Walsh 1923). The connection to the Walsh functions was observed by Hagmark & Lounesto 1986.

In 1902, Vahlen initiated the study of Möbius transformations of vectors in $\mathbb{R}^n$ (or paravectors in $\mathbb{R}\oplus\mathbb{R}^n$) by 2×2-matrices with entries in $\mathcal{C}\ell_{0,n}$. This study was re-initiated by Ahlfors in the 1980's.

23.16 Elie Cartan

Besides detecting spinors in 1913 (and pure spinors in 1938), Cartan made two other contributions to Clifford algebras: their periodicity of 8 and the triality of **Spin**(8).

Cartan 1908 p. 464, identified the Clifford algebras $\mathcal{C}\ell_{p,q}$ as matrix algebras with entries in $\mathbb{R}$, $\mathbb{C}$, $\mathbb{H}$, $\mathbb{R}\oplus\mathbb{R}$, $\mathbb{H}\oplus\mathbb{H}$ and found a periodicity of 8. To decipher Cartan's notation:

h		
±1	S_m	$\mathrm{Mat}(m,\mathbb{R})$
±2	IS_m	$\mathrm{Mat}(m,\mathbb{C})$
±3	QS_m	$\mathrm{Mat}(m,\mathbb{H})$
0	$2S_m$	${}^2\mathrm{Mat}(m,\mathbb{R})$
4	$2Q_m$	${}^2\mathrm{Mat}(m,\mathbb{H})$

where $h = 1 - p + q \pmod 8$. Clifford's original notion of 4 classes was thus refined to 8 classes (and generalized from $C\ell_n$ and $C\ell_{0,n}$ to $C\ell_{p,q}$). [20]

Cartan's periodicity of 8 for real Clifford algebras, with an involution, was extended by C.T.C Wall 1968 and Porteous 1969 (rediscovered by Harvey 1990). Wall considered real graded Clifford algebras, with an anti-involution, and found a 2-way periodicity of type $(8 \times 8)/2$, like the movements of a bishop on a chessboard. Porteous used the anti-involution to induce a scalar product for spinors, and classified the scalar products of spinors into 32 classes, according to the signature types (p, q) of real quadratic spaces $\mathbb{R}^{p,q}$.

In 1925, Cartan came into contact with the triality automorphism of **Spin**(8). Lounesto 1997 (in the first edition of this book) showed that triality is a restriction of a polynomial mapping $C\ell_8 \to C\ell_8$, of degree 2.

23.17 Ernst Witt

Witt 1937 started the modern algebraic theory of quadratic forms. He identified Clifford algebras of non-degenerate quadratic forms over arbitrary fields of characteristic $\neq 2$. The Witt ring $W(\mathbb{F})$, of a field $\mathbb{F}$, consists of similarity classes of non-singular quadratic forms over $\mathbb{F}$ (similar quadratic forms have isometric anisotropic parts). In characteristic $\neq 2$, the structure of Clifford algebras of certain quadratic forms was studied by Lee 1945/48 ($\mathbf{e}_i^2 = 1$), Chevalley 1946 ($\mathbf{e}_i^2 = -1$), and Kawada & Iwahori 1950 ($\mathbf{e}_i^2 = \pm 1$). These authors did not benefit the Witt ring (although they already had it at their disposal), and so they did not consider all the isometry classes of anisotropic quadratic forms.

Example. The Witt ring $W(\mathbb{F}_5)$ of the finite field $\mathbb{F}_5 = \{0, 1, 2, 3, 4\}$ of characteristic 5 contains four isometry classes 0, $\langle 1 \rangle$, $\langle s \rangle$, $\langle 1, s \rangle$ where $s = 2$ or $s = 3$. Chevalley 1946, Lee and Kawada & Iwahori did not notice that none of the quadratic forms $x_1^2 + x_2^2$, $x_1^2 - x_2^2$, $-x_1^2 - x_2^2$ on the plane $\mathbb{F}_5^2$ is isometric with $x_1^2 + s x_2^2 \simeq \langle 1, s \rangle$ (but in fact they are all neutral, and thereby in the same isometry class as 0). A simpler example is the line $\mathbb{F}_5$ where the Clifford algebra of $2x^2 \simeq \langle 2 \rangle$ is the quadratic extension $\mathbb{F}_5(\sqrt{2})$ whereas the Clifford algebras of both $\pm x^2 \simeq \langle \pm 1 \rangle$ split $\mathbb{F}_5 \times \mathbb{F}_5$. ∎

20 Cartan's periodicity of 8 for Clifford algebras, from 1908, is often attributed to Bott, who was born in 1923 and proved his periodicity of homotopy groups of rotation groups in 1959.

23.18 Claude Chevalley

Chevalley 1954 constructed Clifford algebras as subalgebras of the endomorphism algebra of the exterior algebra, $\mathcal{C}\ell(Q) \subset \mathrm{End}(\bigwedge V)$, by means of a not necessarily symmetric bilinear form B on V such that $Q(\mathbf{x}) = B(\mathbf{x},\mathbf{x})$. By this construction, Chevalley managed to include the exceptional case of characteristic 2, and thus amended the work of Witt.

Chevalley 1954 went on further and gave the most general definition, as a factor algebra of the tensor algebra, $\mathcal{C}\ell(Q) = \otimes V/\mathcal{I}_Q$, valid also when ground fields are replaced by commutative rings. From the pedagogical point of view, this approach is forbidding, while it refers to the infinite-dimensional tensor algebra $\otimes V$.

Chevalley 1954 introduced exterior exponentials of bivectors and used them to scrutinize the Lipschitz group, unfairly naming it a 'Clifford group'. Thus, there were two exponentials such that

$$e^{\mathbf{B}} \in \mathbf{Spin}(n) \quad \text{and} \quad \frac{e^{\wedge\mathbf{B}}}{|e^{\wedge\mathbf{B}}|} \in \mathbf{Spin}(n) \quad \text{for} \quad \mathbf{B} \in \bigwedge^2 \mathbb{R}^n.$$

23.19 Marcel Riesz

Marcel Riesz 1958, pp. 61-67, reconstructed Grassmann's exterior algebra from the Clifford algebra, in any characteristic $\neq 2$, by

$$\mathbf{x} \wedge u = \frac{1}{2}(\mathbf{x}u + (-1)^k u\mathbf{x}),$$

where $\mathbf{x} \in V$ and $u \in \bigwedge^k V$. Riesz' contribution enhanced Chevalley's result of 1946, which related exterior products of vectors to antisymmetrized Clifford products of vectors (Chevalley's result was valid only in characteristic 0).

Clifford algebras admit a **parity grading** or **even-odd grading** with even and odd parts, $\mathcal{C}\ell(Q) = \mathcal{C}\ell^+(Q) \oplus \mathcal{C}\ell^-(Q)$. Riesz's construction of 1958 showed that there also exists a **dimension grading**

$$\mathcal{C}\ell(Q) = \mathbb{K} \oplus V \oplus \bigwedge^2 V \oplus \ldots \oplus \bigwedge^n V$$

in all characteristics $\neq 2$ (because there is a privileged linear isomorphism between a Clifford algebra and the exterior algebra). Thus, bivectors exist in all characteristics $\neq 2$.

M. Riesz 1947 expressed the Maxwell stress tensor as $T_{\mu\nu} = -\frac{1}{2}\langle \mathbf{e}_\mu \mathbf{F} \mathbf{e}_\nu \mathbf{F} \rangle_0$.

The first one to consider spinors as elements in a **minimal left ideal** of a Clifford algebra was M. Riesz 1947 (although the special case of pure spinors had been considered earlier by Cartan 1938).

23.20 Atiyah & Bott & Shapiro

In 1964, Atiyah & Bott & Shapiro reconsidered spinor spaces as modules over a Clifford algebra, instead of regarding them as minimal left ideals in the Clifford algebra. This permitted differentiation of spinor valued functions on manifolds, not just on flat spaces.

They re-identified the definite real Clifford algebras $\mathcal{C}\ell_n$ and $\mathcal{C}\ell_{0,n}$ as matrix algebras with entries in $\mathbb{R}$, $\mathbb{C}$, $\mathbb{H}$, $\mathbb{R} \oplus \mathbb{R}$, $\mathbb{H} \oplus \mathbb{H}$ (identified by Cartan in 1908). They rediscovered the periodicity of 8 (found by Cartan in 1908) with respect to the graded tensor product (used earlier by Chevalley in 1955). They emphasized the importance of the $\mathbb{Z}_2$-graded structure (the even-odd parity in Clifford's works) and used it to simplify Chevalley's approach to Lipschitz groups (the role of parity grading and/or grade involution in reflections was not observed by Chevalley).

Atiyah & Bott & Shapiro's paper of 1964 is known for its dirty joke: they attributed the introduction of $\mathbf{Pin}(n)$ to Serre from France, where pronunciation of 'pin group' in English sounds the same as 'pine groupe' in French: pine [pin] is a slang expression for male genitals.

E. Kähler 1960/62 introduced a second product for Cartan's exterior differential forms making Grassmann's exterior algebra isomorphic with a Clifford algebra. This re-interpretation of Chevalley's definition of the Clifford algebra (extending $\mathbf{x}u = \mathbf{x} \rfloor u + \mathbf{x} \wedge u$) was renamed as *Kähler–Atiyah algebra* by W. Graf 1978 (Graf's Kähler–Atiyah algebra was again reinvented and applied to the Kähler–Dirac equation by Salingaros & Wene 1985).

23.21 The Maxwell Equations

In the special case of a homogeneous and isotropic medium, Maxwell equations can be condensed into a single equation. This has been done by means of complex vectors (Silberstein 1907), complex quaternions (Silberstein 1912/1914), spinors (Laporte & Uhlenbeck 1931) and in terms of Clifford algebras. Juvet & Schidlof 1932, Mercier 1935 and Riesz 1958 condensed the Maxwell equations into a single equation by bivectors in the Clifford algebra $\mathcal{C}\ell_{1,3}$ of $\mathbb{R}^{1,3}$. In the Clifford algebra $\mathcal{C}\ell_{3,1} \simeq \mathrm{Mat}(4, \mathbb{R})$, the single Maxwell equation

$$\partial \mathbf{F} = \mathbf{J},$$

where $\partial = \nabla - \mathbf{e}_0 \partial_0$ and $\mathbf{F} = \mathbf{E}\mathbf{e}_0 - \mathbf{B}\mathbf{e}_{123} \in \bigwedge^2 \mathbb{R}^{3,1}$, could be decomposed into two parts

$$\partial \rfloor \mathbf{F} = \mathbf{J}, \quad \partial \wedge \mathbf{F} = 0.$$

Similarly, $\partial\mathbf{A} = -\mathbf{F}$ could be decomposed into two parts, $\partial\wedge\mathbf{A} = -\mathbf{F}$ and the Lorenz gauge/condition $\partial\cdot\mathbf{A} = 0$ discovered by the Danish physicist Ludwig Lorenz and not by the Dutch physicist H. A. Lorentz. [21] Marcel Riesz 1947 expressed the Maxwell stress tensor as

$$T_{\mu\nu} = -\frac{1}{2}\langle\mathbf{e}_\mu\mathbf{F}\mathbf{e}_\nu\mathbf{F}\rangle_0$$

and Hestenes 1966, p. 31, introduced the vectors $\mathbf{T}_\mu = -\frac{1}{2}\mathbf{F}\mathbf{e}_\mu\mathbf{F}$ for which $T_{\mu\nu} = \mathbf{T}_\mu\cdot\mathbf{e}_\nu = \mathbf{e}_\mu\cdot\mathbf{T}_\nu$ and the mapping $T\mathbf{x} = -\frac{1}{2}\mathbf{F}\mathbf{x}\mathbf{F}$ where $(T\mathbf{x})^\mu = T^\mu{}_\nu x^\nu$. [Juvet & Schidlof 1932, p. 141, gave

$$T_{\mu\nu} = F_\mu{}^\lambda F_{\lambda\nu} + \frac{1}{4}g_{\mu\nu}F_{\alpha\beta}F^{\alpha\beta}$$

but did not observe that $T\mathbf{x} = -\frac{1}{2}\mathbf{F}\mathbf{x}\mathbf{F}$.] Note that $\mathbf{T}_0 = \frac{1}{2}(\mathbf{E}^2+\mathbf{B}^2)\mathbf{e}_0+\mathbf{E}\times\mathbf{B}$.

23.22 Spinors in ideals, as operators and recovered

Juvet 1930 and Sauter 1930 replaced column spinors by square matrices in which only the first column was non-zero – thus spinor spaces became minimal left ideals in a **matrix algebra**. Marcel Riesz 1947 was the first one to consider spinors as elements in a minimal left ideal of a **Clifford algebra** (although the special case of pure spinors had been considered earlier by Cartan 1938).

Gürsey 1956/58 rewrote the Dirac equation with 2×2 quaternion matrices in $\mathrm{Mat}(2,\mathbb{H})$ (see also Gsponer & Hurni 1993). Kustaanheimo 1964 presented the spinor regularization of the Kepler motion, the KS-transformation, which emphasized the operator aspect of spinors. This led David Hestenes 1966-74 to a reformulation of the Dirac theory, where the role of spinors [in columns $\mathbb{C}^4$ or in minimal left ideals of the complex Clifford algebra $\mathbb{C}\otimes\mathcal{C}\ell_{1,3}\simeq\mathrm{Mat}(4,\mathbb{C})$] was taken over by operators in the even subalgebra $\mathcal{C}\ell^+_{1,3}$ of the real Clifford algebra $\mathcal{C}\ell_{1,3}\simeq\mathrm{Mat}(2,\mathbb{H})$.

Spinors were reconstructed from their bilinear covariants by Y. Takahashi 1983 and J. Crawford 1985, in the case of the electron. Lounesto 1993 generalized the reconstruction of spinors to the null case of the neutron, and predicted existence of a new particle residing in between electrons and neutrons.

Bibliography

E. Artin: *Geometric Algebra*. Interscience, New York, 1957, 1988.

21 J. van Bladel: *Lorenz or Lorentz?* IEEE Antennas and Propagation Magazine **33** (1991) p. 69 and The Radioscientist **2** (1991) p. 55.

M.F. Atiyah, R. Bott, A. Shapiro: Clifford modules. *Topology* **3**, suppl. 1 (1964), 3-38. Reprinted in R. Bott: *Lectures on* $K(X)$. Benjamin, New York, 1969, pp. 143-178. Reprinted in *Michael Atiyah: Collected Works*, Vol. 2. Clarendon Press, Oxford, 1988, pp. 301-336.

R. Brauer, H. Weyl: Spinors in n dimensions. *Amer. J. Math.* **57** (1935), 425-449. Reprinted in *Selecta Hermann Weyl*, Birkhäuser, Basel, 1956, pp. 431-454.

E. Cartan (exposé d'après l'article allemand de E. Study): Nombres complexes, pp. 329-468 in J. Molk (red.): *Encyclopédie des sciences mathématiques*, Tome **I**, vol. **1**, Fasc. **4**, art. **I5**, 1908.

E. Cartan: Sur les groupes projectifs qui ne laissent invariante aucune multiplicité plane. *Bull. Soc. Math. France* **41** (1913), 53-96.

E. Cartan: Le principe de dualité et la théorie des groupes simples et semi-simples. *Bull. Sci. Math.* **49** (1925), 161-174.

A. Cayley: On certain results relating to quaternions. *Phil. Mag.* (3) **26** (1845), 141-145.

A. Cayley: Recherches ultérieures sur les déterminants gauches. *J. Reine Angew. Math.* **50** (1855), 299-313.

C. Chevalley: *Theory of Lie Groups*. Princeton University Press, Princeton, 1946.

C. Chevalley: *The Algebraic Theory of Spinors*. Columbia University Press, New York, 1954. *The Algebraic Theory of Spinors and Clifford Algebras*. Springer, Berlin, 1997.

C. Chevalley: *The Construction and Study of Certain Important Algebras*. Mathematical Society of Japan, Tokyo, 1955.

W.K. Clifford: Applications of Grassmann's extensive algebra. *Amer. J. Math.* **1** (1878), 350-358.

W.K. Clifford: On the classification of geometric algebras; pp. 397-401 in R. Tucker (ed.): *Mathematical Papers by William Kingdon Clifford*. Macmillan, London, 1882. (Reprinted by Chelsea, New York, 1968.) Title of talk announced already on p. 135 in *Proc. London Math. Soc.* **7** (1876).

J.P. Crawford: On the algebra of Dirac bispinor densities: Factorization and inversion theorems. *J. Math. Phys.* **26** (1985), 1439-1441.

P.A.M. Dirac: The quantum theory of the electron. *Proc. Roy. Soc. A* **117** (1928), 610-624.

A. Gsponer, J.-P. Hurni: Lanczos' equation to replace Dirac's equation? in pp. 509-512 of J.D. Brown (ed.): *Proceedings of the Cornelius Lanczos Centenary Conference*, (Raleigh, NC, 1993).

F. Gürsey: Correspondence between quaternions and four-spinors. *Rev. Fac. Sci. Univ. Istanbul* **A21** (1956), 33-54.

F. Gürsey: Relation of charge independence and baryon conservation to Pauli's transformation. *Nuovo Cimento* **7** (1958), 411-415.

P.-E. Hagmark, P. Lounesto: Walsh functions, Clifford algebras and Cayley-Dickson process, pp. 531-540 in J.S.R. Chisholm, A.K. Common (eds.): *Clifford Algebras and their Applications in Mathematical Physics*. Reidel, Dordrecht, The Netherlands, 1986.

J. Helmstetter: *Algèbres de Clifford et algèbres de Weyl*. Cahiers Math. **25**, Montpellier, 1982.

D. Hestenes: *Space-Time Algebra*. Gordon and Breach, New York, 1966, 1987, 1992.

D. Hestenes: Real spinor fields. *J. Math. Phys.* **8** (1967), 798-808.

D. Hestenes, G. Sobczyk: *Clifford Algebra to Geometric Calculus*. Reidel, Dordrecht, 1984, 1987.

B. Jancewicz: *Multivectors and Clifford Algebra in Electrodynamics*. World Scientific Publ., Singapore, 1988.

G. Juvet: Opérateurs de Dirac et équations de Maxwell. *Comment. Math. Helv.* **2** (1930), 225-235.

G. Juvet, A. Schidlof: Sur les nombres hypercomplexes de Clifford et leurs applications à l'analyse vectorielle ordinaire, à l'électromagnetisme de Minkowski et à la théorie de Dirac. *Bull. Soc. Neuchat. Sci. Nat.*, **57** (1932), 127-147.

E. Kähler: Der innere Differentialkalkül. *Rendiconti di Matematica e delle sue Applicazioni (Roma)* **21** (1962), 425-523.

Y. Kawada, N. Iwahori: On the structure and representations of Clifford algebras. *J. Math. Soc. Japan* **2** (1950), 34-43.

P. Kustaanheimo, E. Stiefel: Perturbation theory of Kepler motion based on spinor regularization. *J. Reine Angew. Math.* **218** (1965), 204-219.

O. Laporte, G. E. Uhlenbeck: Application of spinor analysis to the Maxwell and Dirac equations. *Phys. Rev.* **37** (1931), 1380-1397.

H. C. Lee: On Clifford algebras and their representations. *Ann. of Math.* **49** (1948), 760-773.

R. Lipschitz: Principes d'un calcul algébrique qui contient comme espèces particulières le calcul des quantités imaginaires et des quaternions. *C.R. Acad. Sci. Paris* **91** (1880), 619-621, 660-664. Reprinted in *Bull. Soc. Math.* (2) **11** (1887), 115-120.

R. Lipschitz: *Untersuchungen über die Summen von Quadraten*. Max Cohen und Sohn, Bonn, 1886, pp. 1-147. (The first chapter of pp. 5-57 translated into French by J. Molk: Recherches sur la transformation, par des substitutions réelles, d'une somme de deux ou troix carrés en elle-même. *J. Math. Pures Appl.* (4) **2** (1886), 373-439. French résumé of all three chapters in *Bull. Sci. Math.* (2) **10** (1886), 163-183.)

R. Lipschitz (signed): Correspondence. *Ann. of Math.* **69** (1959), 247-251. Reprinted on pages 557-561 of A. Weil: *Œuvres Scientifiques, Collected Papers*, Volume II. Springer, 1979.

P. Lounesto: Clifford algebras and Hestenes spinors. *Found. Phys.* **23** (1993), 1203-1237.

A. Mercier: *Expression des Équation de l'Électromagnetisme au Moyen des Nombres de Clifford*. Thèse, Université de Genéve, 1935.

W. Pauli: Zur Quantenmechanik des magnetischen Elektrons. *Z. Phys.* **42** (1927), 601-623.

I.R. Porteous: *Topological Geometry*. Van Nostrand Reinhold, London, 1969. Cambridge University Press, Cambridge, 1981.

I.R. Porteous: *Clifford Algebras and the Classical Groups*. Cambridge University Press, Cambridge, 1995.

M. Riesz: Sur certain notions fondamentales en théorie quantique relativiste. *C. R.* 10e *Congrès Math. Scandinaves, (Copenhagen, 1946)*. Jul. Gjellerups Forlag, Copenhagen, 1947, pp. 123-148. Reprinted in L. Gårding, L. Hörmander (eds.): *Marcel Riesz, Collected Papers*. Springer, Berlin, 1988, pp. 545-570.

M. Riesz: *Clifford Numbers and Spinors*. University of Maryland, 1958. Reprinted as facsimile by Kluwer, Dordrecht, 1993.

F. Sauter: Lösung der Diracschen Gleichungen ohne Spezialisierung der Diracschen Operatoren. *Z. Phys.* **63** (1930), 803-814.

Y. Takahashi: A passage between spinors and tensors. *J. Math. Phys.* **24** (1983), 1783-1790.

K. Th. Vahlen: Über höhere komplexe Zahlen. *Schriften der phys.-ökon. Gesellschaft zu Königsberg* **38** (1897), 72-78.

K. Th. Vahlen: Über Bewegungen und complexe Zahlen. *Math. Ann.* **55** (1902), 585-593.

C.T.C. Wall: Graded algebras, antiinvolutions, simple groups and symmetric spaces. *Bull. Amer. Math. Soc.* **74** (1968), 198-202.

B.L. van der Waerden: On Clifford algebras. *Nederl. Akad. Wetensch. Proc. Ser. A* **69** (1966), 78-83.

E. Witt: Theorie der quadratischen Formen in beliebigen Körpern. *J. Reine Angew. Math.* **176** (1937), 31-44.

Selected Reading

R. Abłamowicz, P. Lounesto, J. Maks: Conference Report, Second Workshop on 'Clifford Algebras and their Applications in Mathematical Physics' (Montpellier, 1989). *Found. Phys.* **21** (1991), 735-748.

L. Ahlfors, P. Lounesto: Some remarks on Clifford algebras. *Complex Variables, Theory and Application* **12** (1989), 201-209.

S.L. Altmann: *Rotations, Quaternions and Double Groups*. Clarendon Press, Oxford, 1986.

E. Artin: *Geometric Algebra*. Interscience, New York, 1957, 1988.

M.F. Atiyah, R. Bott, A. Shapiro: Clifford modules. *Topology* **3**, suppl. 1 (1964), 3-38. Reprinted in R. Bott: *Lectures on* $K(X)$. Benjamin, New York, 1969, pp. 143-178. Reprinted in *Michael Atiyah: Collected Works*, Vol. 2. Clarendon Press, Oxford, 1988, pp. 301-336.

W.E. Baylis: *Electrodynamics: A Modern Geometric Approach*. Birkhäuser, Boston, 1999.

I.M. Benn, R.W. Tucker: *An Introduction to Spinors and Geometry with Applications in Physics*. Adam Hilger, Bristol, 1987.

E.F. Bolinder: Unified microwave network theory based on Clifford algebra in Lorentz space; pp. 25-35 in *Conference Proceedings, 12th European Microwave Conference (Helsinki 1982)*. Microwave Exhibitions and Publishers, Tunbridge Wells, Kent, 1982.

E.F. Bolinder: Clifford algebra: what is it? *IEEE Antennas and Propagation Society Newsletter* **29** (1987), 18-23.

N. Bourbaki: *Algèbre, Chapitre 9, Formes sesquilinéaires et formes quadratiques*. Hermann, Paris, 1959.

F. Brackx, R. Delanghe, F. Sommen: *Clifford Analysis*. Research Notes in Mathematics **76**, Pitman, London, 1982.

R. Brauer, H. Weyl: Spinors in n dimensions. *Amer. J. Math.* **57** (1935), 425-449. Reprinted in *Selecta Hermann Weyl,* Birkhäuser, Basel, 1956, pp. 431-454.

P. Budinich, A. Trautman: *The Spinorial Chessboard*. Springer, Berlin, 1988.

E. Cartan (exposé d'après l'article allemand de E. Study): Nombres complexes; pp. 329-468 in J. Molk (red.): *Encyclopédie des sciences mathématiques*, Tome **I**, vol. **1**, Fasc. **4**, art. **I5** (1908). Reprinted in E. Cartan: *Œuvres Complètes*, Partie II. Gauthier-Villars, Paris, 1953, pp. 107-246.

C. Chevalley: *Theory of Lie Groups*. Princeton University Press, Princeton, NJ, 1946.

C. Chevalley: *The Algebraic Theory of Spinors.* Columbia University Press, New York, 1954.

C. Chevalley: *The Construction and Study of Certain Important Algebras.* Mathematical Society of Japan, Tokyo, 1955.

J.S.R. Chisholm, A.K. Common (eds.): *Proceedings of the NATO and SERC Workshop on 'Clifford Algebras and their Applications in Mathematical Physics' (Canterbury, 1985).* Reidel, Dordrecht, The Netherlands, 1986.

W.K. Clifford: Applications of Grassmann's extensive algebra. *Amer. J. Math.* **1** (1878), 350-358.

W.K. Clifford: On the classification of geometric algebras, pp. 397-401 in R. Tucker (ed.): *Mathematical Papers by William Kingdon Clifford*, Macmillan, London, 1882. Reprinted by Chelsea, New York, 1968. Title of talk announced already in *Proc. London Math. Soc.* **7** (1876), p. 135.

J. Crawford: On the algebra of Dirac bispinor densities: factorization and inversion theorems. *J. Math. Phys.* **26** (1985), 1439-1441.

A. Crumeyrolle: *Orthogonal and Symplectic Clifford Algebras, Spinor Structures.* Kluwer, Dordrecht, The Netherlands, 1990.

R. Deheuvels: *Formes quadratiques et groupes classiques.* Presses Universitaires de France, Paris, 1981.

R. Delanghe, F. Sommen, V. Souček: *Clifford Algebra and Spinor Valued Functions: A Function Theory for the Dirac Operator.* Kluwer, Dordrecht, The Netherlands, 1992.

V.L. Figueiredo, E. Capelas de Oliveira, W.A. Rodrigues, Jr.: Covariant, algebraic and operator spinors. *Internat. J. Theoret. Phys.* **29** (1990), 371-395.

L. Gårding, L. Hörmander (eds.): *Marcel Riesz, Collected Papers.* Springer, Berlin, 1988.

J. Gilbert, M. Murray: *Clifford Algebras and Dirac Operators in Harmonic Analysis.* Cambridge Studies in Advanced Mathematics, Cambridge University Press, Cambridge, 1991.

W. Greub: *Multilinear Algebra*, 2nd edn. Springer, Berlin, 1978.

P.-E. Hagmark, P. Lounesto: Walsh functions, Clifford algebras and Cayley-Dickson process, pp. 531-540 in J.S.R. Chisholm, A.K. Common (eds.): *Clifford Algebras and their Applications in Mathematical Physics.* Reidel, Dordrecht, The Netherlands, 1986.

J.D. Hamilton: The Dirac equation and Hestenes' geometric algebra. *J. Math. Phys.* **25** (1984), 1823-1832.

F.R. Harvey: *Spinors and Calibrations.* Academic Press, San Diego, 1990.

J. Helmstetter: Algèbres de Clifford et algèbres de Weyl. *Cahiers Math.* **25**, Montpellier, 1982.

D. Hestenes: *Space-Time Algebra.* Gordon and Breach, New York, 1966, 1987, 1992.

D. Hestenes: Real spinor fields. *J. Math. Phys.* **8** (1967), 798-808.

D. Hestenes: Observables, operators, and complex numbers in the Dirac theory. *J. Math. Phys.* **16** (1975), 556-571.

D. Hestenes, G. Sobczyk: *Clifford Algebra to Geometric Calculus.* Reidel, Dordrecht, The Netherlands, 1984, 1987.

D. Hestenes: *New Foundations for Classical Mechanics.* Reidel, Dordrecht, 1986, (2nd ed.) 1999.

P.R. Holland: Relativistic algebraic spinors and quantum motions in phase space. *Found. Phys.* **16** (1986), pp. 708-709.

D.J. Hurley, M.A. Vandyck: *Geometry, Spinors, and Applications.* Springer, Berlin, 1999.

B. Jancewicz: *Multivectors and Clifford Algebra in Electrodynamics*. World Scientific, Singapore, 1988.

E. Kähler: Der innere Differentialkalkül, *Rendiconti di Matematica e delle sue Applicazioni* (Roma) **21** (1962), 425-523.

J. Keller, S. Rodríguez-Romo: A multivectorial Dirac equation. *J. Math. Phys.* **31** (1990), 2502.

M.-A. Knus: *Quadratic Forms, Clifford Algebras and Spinors*. Univ. Estadual de Campinas, Brazil, 1988.

T.Y. Lam: *The Algebraic Theory of Quadratic Forms*. Benjamin, Reading, MA, 1973, 1980.

H.B. Lawson, M.-L. Michelsohn: *Spin Geometry*. Universidade Federal do Ceará, Brazil, 1983. Princeton University Press, Princeton, NJ, 1989.

R. Lipschitz: Principes d'un calcul algébrique qui contient comme espèces particulières le calcul des quantités imaginaires et des quaternions. *C.R. Acad. Sci. Paris* **91** (1880), 619-621, 660-664. Reprinted in *Bull. Soc. Math.* (2) **11** (1887), 115-120.

R. Lipschitz: *Untersuchungen über die Summen von Quadraten*. Max Cohen und Sohn, Bonn, 1886, pp. 1-147. The first chapter of pp. 5-57 translated into French by J. Molk: Recherches sur la transformation, par des substitutions réelles, d'une somme de deux ou troix carrés en elle-même. *J. Math. Pures Appl.* (4) **2** (1886), 373-439. French résumé of all three chapters in *Bull. Sci. Math.* (2) **10** (1886), 163-183.

R. Lipschitz (signed): Correspondence. *Ann. of Math.* **69** (1959), 247-251. Reprinted on pages 557-561 of A. Weil: *Œuvres Scientifiques, Collected Papers*, Volume II. Springer, Berlin, 1979.

P. Lounesto: Scalar products of spinors and an extension of Brauer-Wall groups. *Found. Phys.* **11** (1981), 721-740.

P. Lounesto: Report of Conference, NATO and SERC Workshop on 'Clifford Algebras and their Applications in Mathematical Physics' (Canterbury, 1985). *Found. Phys.* **16** (1986), 967-971.

P. Lounesto, P. Bergh: Axially symmetric vector fields and their complex potentials. *Complex Variables: Theory and Application* **2** (1983), 139-150.

P. Lounesto, A. Springer: Möbius transformations and Clifford algebras of Euclidean and anti-Euclidean spaces, pp. 79-90 in J. Lawrynowicz (ed.): *Deformations of Mathematical Structures*. Kluwer, Dordrecht, The Netherlands,1989.

P. Lounesto, G.P. Wene: Idempotent structure of Clifford algebras. *Acta Applic. Math.* **9** (1987), 165-173.

A. Micali, R. Boudet, J. Helmstetter (eds.): *Proceedings of the 2nd Workshop on 'Clifford Algebras and their Applications in Mathematical Physics' (Montpellier, 1989)*. Kluwer, Dordrecht, The Netherlands, 1992.

A. Micali, Ph. Revoy: Modules quadratiques. *Cahiers Math.* **10**, Montpellier, 1977. *Bull. Soc. Math. France* **63**, suppl. (1979), 5-144.

R. Penrose, W. Rindler: *Spinors and Space-Time*. Vol. 1. Cambridge University Press, Cambridge, 1984.

I.R. Porteous: *Topological Geometry*. Van Nostrand Reinhold, London, 1969. Cambridge University Press, Cambridge, 1981.

I.R. Porteous: *Clifford Algebras and the Classical Groups*. Cambridge University Press, Cambridge, 1995.

M. Riesz: Sur certaines notions fondamentales en théorie quantique relativiste; pp. 123-148 in *C.R. 10ᵉ Congrès Math. Scandinaves, Copenhagen, 1946*. Jul. Gjellerups Forlag, Copenhagen, 1947. Reprinted in L. Gårding, L. Hörmander (eds.): *Marcel Riesz, Collected Papers*, Springer, Berlin, 1988, pp. 545-570.

M. Riesz: L'équation de Dirac en relativité générale; pp. 241-259 in *Tolfte[=12.] Skandinaviska Matematikerkongressen i Lund, 1953*. Håkan Ohlssons boktryckeri, Lund, 1954. Reprinted in L. Gårding, L. Hörmander (eds.): *Marcel Riesz, Collected Papers*, Springer, Berlin, 1988, pp. 814-832.

M. Riesz: *Clifford Numbers and Spinors*. The Institute for Fluid Dynamics and Applied Mathematics, Lecture Series No. **38**, University of Maryland, 1958. Reprinted as facsimile (eds.: E.F. Bolinder, P. Lounesto) by Kluwer, Dordrecht, The Netherlands, 1993.

W.A. Rodrigues, Jr., E. Capelas de Oliveira: Dirac and Maxwell equations in the Clifford and spin-Clifford bundles. *Internat. J. Theoret. Phys.* **29** (1990), 397-412.

A. Ronveaux, D. Lambert: *Le probleme de factorisation de Hurwitz. Approche historique, solutions, applications en physique*. Faculte Universitaires N.D. de la Paix, Namur, 1991.

N.A. Salingaros, G.P. Wene: The Clifford algebra of differential forms. *Acta Applic. Math.* **4** (1985), 271-292.

L. Silberstein: *The Theory of Relativity.* Macmillan, London, 1914.

J. Snygg: *Clifford Algebra, a Computational Tool for Physicists*. Oxford University Press, Oxford, 1997.

K. Th. Vahlen: Über höhere komplexe Zahlen. *Schriften der phys.-ökon. Gesellschaft zu Königsberg* **38** (1897), 72-78.

K. Th. Vahlen: Über Bewegungen und complexe Zahlen. *Math. Ann.* **55** (1902), 585-593.

B.L. van der Waerden: *A History of Algebra*. Springer, Berlin, 1985.

E. Witt: Theorie der quadratischen Formen in beliebigen Körpern. *J. Reine Angew. Math.* **176** (1937), 31-44.

Index